Synthetic Lubricants 1972

Dr. M.W. Ranney

Thirty-Six Dollars

NOYES DATA CORPORATION
Noyes Building
Park Ridge, New Jersey 07656, U.S.A.

Library of Congress Catalog Card Number: 72-180891
ISBN: 0-8155-0423-3
Printed in the United States

FOREWORD

The detailed, descriptive information in this book is based on U.S. patents since 1967 relating to the production and application of synthetic lubricants.

This book serves a double purpose in that it supplies detailed technical information and can be used as a guide to the U.S. patent literature in this field. By giving all the information that is significant, and eliminating legalistic phraseology, this book presents an advanced, technically oriented review of synthetic lubricants.

The U.S. patent literature is the largest and most comprehensive collection of technical information in the world. There is more practical, commercial, timely process information assembled here than is available from any other source. The technical information obtained from a patent is extremely reliable and comprehensive; sufficient information must be included to avoid rejection for "insufficient disclosure".

The patent literature covers a substantial amount of information not available in the journal literature. The patent literature is a prime source of basic commercially utilizable information. This information is overlooked by those who rely primarily on the periodical journal literature. It is realized that there is a lag between a patent application on a new process development and the granting of a patent, but it is felt that this may roughly parallel or even anticipate the lag in putting that development into commercial practice.

Many of these patents are being exploited commercially. Whether used or not, they offer opportunities for technological transfer. Also, a major purpose of this book is to describe the number of technical possibilities available, which may open up profitable areas of research and development.

These publications are bound in paper in order to close the time gap between "manuscript" and "completed book". Industrial technology is progressing so rapidly that hard cover books do not always reflect the latest developments in a particular field, due to the longer time required to produce a hard cover book.

The Table of Contents is organized in such a way as to serve as a subject index. Other indexes by company, inventor, and patent number help in providing easy access to the information contained in this volume.

CONTENTS AND SUBJECT INDEX

INTRODUCTION

The ever increasing severity of operating conditions in both industrial and military equipment has necessitated the development of high performance petroleum and synthetic lubricating fluids. Polymerized olefins were produced commercially in 1929 and the military demands of World Wall II resulted in the manufacture of ester lubricants in Germany, followed by polyglycols and silicones in the United States. In recent years, fire resistant phosphates, fluorocarbons, polyphenyl ethers and silicate esters have all been developed for specialty lubrication uses.

The organic esters represent the largest single segment of the synthetic lube market and their development has closely paralleled the advances being made in jet engine design, both in Britain and the United States. Di(2-ethylhexyl) sebacate is typical and is widely used to meet the MIL-L-7808 specification for jet engine oils. More severe conditions require the use of triesters, based on trimethylolpropane and pentaerythritol, which must function over the -65° to +400°F. temperature range. Additive research has been of prime concern in recent years leading to improved load bearing, rust and oxidation inhibitors.

The viscosity-temperature characteristics of the silicones have led to their use in aircraft and other special applications where very low temperatures are encountered, combined with high temperature stability up to 500°F. The silicones are also used in many high temperature greases. The phosphates, such as tricresyl phosphate, provide fire resistant fluids for die casting, air compressors and aircraft hydraulic systems. Systems employing a large proportion of chlorinated biphenyl mixed with a phosphate fluid are available commercially, providing somewhat improved hydrolytic stability. Silicate esters which are operable up to temperatures of 400°F. and higher have found considerable use as electronic coolants and general military hydraulic fluids.

The polyglycols have low pour-points, high viscosity indexes and show very little deleterious effect on most rubber compounds and gasket materials. These properties make them eminently suitable for brake fluids and various industrial hydraulic applications. They are also widely used as rubber lubricants and as processing aids in the manufacture of textiles. Fire resistant fluids, containing 30 to 60% water are used in die casting machines, furnace controls, Navy hydraulic catapults and missile ground handling equipment.

The polyphenyl ethers are extremely effective in high temperature hydraulic systems, nuclear power plants and in high performance greases. While somewhat limited in low temperature performance, these fluids are operable up to 500°F. The fluorochemical oils find use in applications requiring maximum resistance to oxidation; particularly in equipment for manufacturing and handling liquid oxygen and hydrogen peroxide. Improvements in petroleum based oils have come about as a result of research efforts directed to improved additive components. Viscosity index improvers, dispersants, oxidation and rust inhibitors are routinely used in automotive oils. Turbine and gear oils may contain oxidation and rust inhibitors, along with a variety of sulfur, phosphorus and chlorine-containing additives to improve scuffing and wear. This book presents information on 200 processes related to the patent literature on synthetic lubricating fluids over the period from 1966 to 1971.

ORGANIC ESTERS

GENERAL SYNTHESIS

Pentaerythritol and Mixed Alkanoic Acids

M. Warman; U.S. Patent 3,360,465; December 26, 1967; assigned to Drew Chemical Corporation describes lubricant compositions which are liquid and possess lubricity at temperatures between -40° and 210°F., containing as an essential ingredient an esterification product of pentaerythritol and a mixture of 2 to 6 monocarboxylic alkanoic acids having from 5 to 9 carbon atoms, in which mixture the average number of carbon atoms is in the range of 6.0 to 7.25 and the maximum number contributed by the straight chain 8- and 9-carbon acids is 60% and the maximum number contributed by branched chain acids is 45% of the average number of carbon atoms.

The pentaerythritol mixed esters have satisfactory initial kinematic viscosities at 210°F. and below 13,000 centistokes at -40°F. if the mixed acid reactants, of from 5 to 9 carbon atoms, are present in such molar quantities that the total number of carbon atoms contributed by acids in the mixture having straight chain 8 and 9 carbon atoms is below 60% of the average number of carbon atoms and the total contributed by all branched acids is below 45%. In other words, upon complete esterification, the maximum number of carbon atoms contributed by the straight chain 8- and 9-carbon acids is about 4.25 and that contributed by the branched acids is about 3.25. The following sample calculation demonstrates the method of computing the average number of carbon atoms and the percent carbon atom contributions.

An acid mixture to be reacted with pentaerythritol contains 25 mol percent of caprylic acid (denoted C_8), 15 mol percent of pelargonic acid (C_9), 45 mol percent of a commercial isopentanoic acid (C_5) containing 29.25 mol percent of normal valeric and 15.75 mol percent of 2-methyl and 3-methylbutyric acids, and 15 mol percent of 2-methyl pentanoic acid (C_6).

Caprylic acid	8 x 0.25	= 2.0000
Pelargonic acid	9 x 0.15	= 1.3500
Valeric acid	5 x 0.2925	= 1.4625
Methyl butyric	5 x 0.1575	= 0.7875
2-Methyl pentanoic acid	6 x 0.15	= 0.9000
Average number of carbon atoms		6.5000

The sum of carbon atoms contributed by the methyl butyric and the 2-methyl pentanoic acids (0.7875 + 0.9000 = 1.6875) and the sum contributed by caprylic and pelargonic acids (2.00 + 1.35 = 3.35) are each divided by the average carbon atom content, 6.50, and multiplied by 100%.

Carbon Atom Contribution	Percent
Branched acids	(1.6875/6.50) x 100 = 25.93
C_8 + C_9 acids	(3.35/6.50) x 100 = 51.5

The initial kinematic viscosity at -40°F. of an ester made from a technical grade of pentaerythritol and this acid mixture is 8,425 centistokes; at 210°F., the initial kinematic viscosity of 5.10 centistokes. The following examples illustrate the process.

Example 1: Into a reaction vessel equipped with a thermometer, an agitator and a condenser were added 273.3 parts (2.0 mol) of pentaerythritol and a mixture consisting of 629.1 parts (6.16 mols) of a commercial isopentanoic acid (containing 65% by weight of n-valeric acid and 35% by weight of 2-methyl and 3-methyl butyric acids) and 417.7 parts of pelargonic acid (2.64 mols). This acid mixture contains 24.5 mol percent of 2-methyl and 3-methyl butyric acid, 45.5 mol percent of normal valeric acid and 30 mol percent of pelargonic acid. The average carbon atom content is 6.2, the percent C_9 contribution is 43.5, and the percent branched acid contribution is 19.8.

The reaction vessel was closed and nitrogen gas was passed through at atmospheric pressure. The vessel was then heated from about 300° to 450°F. for 7 to 9 hours under agitation until about 144 parts of water were condensed, indicating substantially complete esterification. Excess acid was removed in vacuo, using a nitrogen gas blanket, at about 3 mm. Hg and a temperature of 392°F. The remaining liquid reaction mass was cooled to 158°F., treated with dilute caustic solution followed by a water wash, slurried with 1% by weight of activated charcoal, dried at 230°F. and 3 mm. Hg. for about 1 hour, and filtered to yield a clear, straw-colored liquid product. Upon analysis, the product was determined to have the following characteristics.

Specific gravity (ASTM D-1298) at 20/20°C.	0.995 to 1.000
Color (ASTM-1500)	1.0
Neutralization number (ASTM D-664), max.	0.15
Hydroxyl value, max.	10.0
Saponification number	410±10
Water content, percent (with Karl Fischer reagent)	0.05

Samples of the product were tested for specific properties with the following results:

Kinematic viscosity, centistokes (ASTM D-445):	
at 210°F.	5.0 to 5.2
at -40°F.	8,200
Flash Point, °F. (ASTM D-92), min.	480
Fire Point, °F. (ASTM D-92), min.	535
Pour Point, °F. (ASTM D-97)	<-65
SOD Lead Corrosion Test, mg./in.^{2}:	
Weight loss (with 0.5% phenothiazine by weight for 1 hour at 325°F.), max.	1.0

The above properties show that this ester meets the requirements of present jet-engine military specifications, including Government specification Navy XWS-2994 (PP).

A lubricant composition containing essentially the ester of Example 1 was subjected to a 150 hour test run in a Pratt and Whitney JT-3D test engine following the procedure of Pratt and Whitney Type II Lubricating Oil Specification for Aircraft Turbine Engines. This specification calls for adding the sample lubricant to the engine and running the engine through a series of 5 hour and 10 hour cycles, operating at 20,000 pounds of thrust at a temperature in excess of 600°F. The lubricant is periodically tested for thermal and oxidative stability. When the test was complete, the lubricant was found to comply with acceptable standards. It showed satisfactory thermal and oxidative stability, and the bearings and other metal parts in contact with the lubricant were clean and free of corrosion, wear or cracking.

Example 2: Using the same procedure and apparatus as in Example 1, 272.3 parts (2 mols) of pentaerythritol (using technical grade pentaerythritol) were mixed with an acid mixture consisting of 584.2 parts (5.72 mols) of commercial isopentanoic acid (containing 65% of n-valeric acid and 35% by weight of 2-methyl and 3-methyl butyric acid) and 487.4 parts (3.08 mols) of pelargonic acid. This mixture thus consists of 22.75 mol percent of 2-methyl and 3-methyl butyric acids, 42.25 mol percent of normal valeric, and 35 mol percent of pelargonic. The average carbon atom content is 6.4, the percent C9 contribution is 49.3 and the percent branched acid contribution is 17.8. The reactants were added to the reaction vessel and heated. When a total of 144 parts of water was collected in the condensate receiver, the the heat was removed. The clear liquid reaction product was refined as described above. The viscosities of this product were measured at the extreme temperatures required in the military specifications with the following results: at 210°F, 5.23 centistokes and at -40°F., 8,425 centistokes.

Pentaerythritol Neoalkyl Acid Esters

A process described by T.S. Chao and M. Kjonaas; U.S. Patent 3,562,300; February 9, 1971; assigned to Sinclair Research, Inc. involves certain neoalkylpolyol esters which are prepared by stepwise esterification of mixtures of straight or branched chain fatty acids other than neo acids and neoalkyl fatty acids and which possess superior low temperature properties and oxidation resistance.

In the process, the esters are prepared by esterifying a neoalkyl polyol of up to 10 or 12 or more carbon atoms and 2 to 6 hydroxy groups and preferably not more than 7 carbon atoms and 2 to 4 hydroxy groups, such as, for example, pentaerythritol, dipentaerythritol, anhydroenneaheptitol, or 1,1,1-trimethylolalkanes of 5 to 7 carbon atoms with straight chain alkanoic carboxylic acids having from 4 to 12 carbon atoms or branched chain alkanoic carboxylic acids of 4 to 12 carbon atoms other than neo acids and a neoalkanoic acid having 5 to 20 carbon atoms and preferably 5 to 10 carbons, and the general formula:

$$R_1-\overset{\overset{\displaystyle R_2}{|}}{\underset{\underset{\displaystyle R_3}{|}}{C}}-COOH$$

in which R_1, R_2 and R_3 are alkyl groups of 1 to 6 carbon atoms and R_2 and R_3 are preferably 1 to 4 carbons, particularly methyls. Examples of suitable neoalkyl acids are 2,2-dimethylpropanoic, 2,2-dimethylbutanoic, 2,2-dimethylpentanoic, 2,2-dimethylhexanoic and 2,2-dimethyloctanoic acids. The esterification of the neoalkyl polyol with the neo or straight or branched chain acids is carried out using a mol ratio of neoalkyl polyol to straight or branched chain acid to neo acid of 1:1.5-2:3-8 and is allowed to proceed until 80 to 90% completion as judged from the quantity of water formed. At this point the remaining unreacted acids are removed by distillation, preferably under reduced pressure. Following removal of the unreacted acids, sufficient straight chain alkanoic carboxylic acid of from 4 to 12 carbon atoms is added to complete the esterification.

The esterification is preferably carried out in the presence of an inert gas such as N_2 and at temperatures of 150° to 210°C. and in the presence of a solvent to facilitate mixing during the earlier stages of esterification and to aid water removal via azeotropic distillation. Suitable solvents include xylene, toluene, ethylbenzene and aliphatic hydrocarbons having comparable boiling points. Purification, if desired of the reaction mixture can then be accomplished by distillation under vacuum to remove any excess acid, and washing successively with aqueous Na_2CO_3, or other suitable alkali reagents, and water, drying under vacuum, and filtering to give a clear liquid ester.

Example 1: The preparation pf pentaerythritol neoheptanoate n-valerate (Ester A) is as follows. A mixture of 272 g. (2.0 mols) of pentaerythritol, 408 g. (4.0 mols) n-valeric acid, 1,040 g. (8.0 mols) of neoheptanoic acid and 200 g. of xylene was placed in a 2 l., four necked flask equipped with a mechanical stirrer, a reflux condenser and a Dean-Stark azeotrope trap. The mixture was refluxed at 156° to 178°C., with constant stirring and under the protection of N_2, for a period of 22 hours. During this period a total of 115.5 ml. (80% of theory) of water was collected from the trap. The mixture was distilled under vacuum to remove the unreacted acids, the final temperature and pressure being 175°C. and 1 mm. Hg. To the residue was added 204 g. (2 mols) of n-valeric acid and 200 g. of xylene.

Esterification was continued at 173° to 180°C. for another 33 hours during which an additional 28.5 ml. (20% of theory) of water was collected. The reaction mixture was vacuum distilled to remove remaining acid and then washed twice with 5% Na_2CO_3 solution. The ester was then stirred with 6 g. of $NaBH_4$ and 100 ml. of 0.1 N NaOH at 50° to 60°C. for about 20 hours under a N_2 atmosphere. Upon cooling and settling, the aqueous layer was removed and the ester layer was diluted with xylene and washed twice with water. The washed solution was vacuum distilled and the residue was treated with 50 g. of Attapulgus Fines (a decolorizing clay) for 2 hours at 100° to 110°C. and filtered through Hyflo Super-Cel (a diatomite filter aid). There was obtained about 750 g. of a light yellow clear viscous liquid which will be called Ester A in subsequent discussions.

Example 2: The preparation of pentaerythritol neoheptanoate n-caproate (Ester B) is as follows. A mixture of 2 mols of pentaerythritol, 4 mols of n-caproic acid, 8 mols of neoheptanoic acid and 200 g. of xylene was similarly refluxed at 160° to 176°C. until 122 ml. (84.6%) of water was collected in the Dean-Stark trap (43 hours). After removal of unreacted acids by vacuum distillation 116 g. (1 mol) of n-caproic acid and 100 g. of xylene were added. Esterification was continued at 175° to 177°C. until no more water was collected (18 hours). The reaction mixture was worked up in the same manner as described in Example 1, yielding 940 g. of light yellow clear liquid which will be called Ester B in subsequent discussion.

Example 3: The preparation of a liquid pentaerythritol ester (Ester C) from a mixture of neoheptanoic, n-valeric, isovaleric, n-caproic and n-heptanoic acids is as follows. A mixture of 1 mol (136 g.) of pentaerythritol, 3 mols of neoheptanoic acid and 0.5 mol each of n-valeric, isovaleric, n-caproic and n-heptanoic acids was refluxed with 125 g. of xylene in the same manner as described above. After 39.5 hours at 173° to 182°C., a total of 58 ml. (80% of theory) of water was collected in the trap.

After removal of the unreacted acids and xylene, 163 g. (1.5 mols) of n-valeric acid and 125 g. of xylene were added and esterification was completed after another 31.5 hours at 171° to 181°C. The reaction mixture was worked up in the same manner as described in Example 1, yielding a light yellow clear liquid which will be referred to as Ester C. The yield before the $NaBH_4$ and clay treatment was 507 g.

In addition to these three esters, two other esters, D and E, were also prepared employing conventional, one step, preparative procedures. Ester D was prepared by the same procedure as shown in Example 1, except that esterification was completed in one step and a larger excess of neo-C7 acid was used.

The ester was a liquid when first prepared, but crystals started to appear after a few days' storage at room temperature. When a 4 ounce sample was placed in a refrigerator (temperature 20° F.), solidification occurred in a few hours. The mass consisted of a mixture of needle-like crystals and viscous liquid and could not be poured without stirring. The same crystalline mass was observed after storing in low temperature baths for 60 hours at 0°, -30° and -60°F. Under the same low temperature storage conditions Ester A remained as clear pourable liquid.

Ester E was prepared as shown in Example 4 in order to insure that the low temperature storage stability attributed to Ester A was not due to the reduced amount of excess neoheptanoic acid used in its preparation.

Example 4: The preparation of pentaerythritol neoheptanoate n-valerate (Ester E) is as follows. A mixture of 272 g. (2 mols) of pentaerythritol, 408 g. (4.0 mols) of n-valeric acid, 1,040 g. (8.0 mols) of neoheptanoic acid and 200 g. xylene was placed in a 3 l., four necked flask and stirred at reflux temperature until 14 ml. (100% of theory) of water was collected from the trap. The mixture was distilled under vacuum to remove the solvent and the excess acid. The residue was diluted with xylene and washed twice with 5% Na_2CO_3 solution and twice with water. The washed solution was topped to 165°F./1 mm., leaving 913 g. of residue which was then filtered through diatomaceous earth.

The product showed a slightly better low temperature property than Ester D. However, the same low temperature instability was observed. Crystallization took place when the ester was placed in refrigerator or in low temperature baths at 0°, -30° and -60°F. The pour points shown by Esters D and E are due to super cooling resulting from the short period of time given to the esters in carrying out the pour point determination. When sufficient time was given, crystallization took place and the esters can no longer be poured.

Table 1 below, shows physical properties of Esters A, B and C prepared by the method and Esters D and E prepared by conventional procedures. It may be seen that the Esters A, B and C have superior low temperature properties to Esters D and E or the neo acid esters shown in Table 1. The pour points are lower, the viscosity at 0° and -40°F. are lower and the esters of this process remained clear liquids during low temperature storage tests at 20°, 0°, -30° and -60°F.

TABLE I: PHYSICAL PROPERTIES OF PENTAERYTHRITOL NEO ACID ESTERS PREPARED BY DIFFERENT METHODS

Identity	Ester A	Ester B	Ester C	Ester D	Ester E
Acids other than neo-C_7	n-C_5	n-C_6	n-C_5, i-C_5, n-C_6, n-C_7.	n-C_5	n-C_5.
Method of preparation	New	New	New	Conventional	Conventional.
Kinematic vis., c.s.:					
210° F	4.744	4.988	4.854	5.245	5.218.
100°F	25.29	30.14	29.12	35.89	35.37.
0° F	980.4	1,429.0	1,303.1	(1)	2,516.
−40° F	18,325	42,414	27,498	(1)	80,245.
Pour point, ° F	−70	−75	−70	[2]−45	[2]−60.
Flash point, ° F	470	[3] 475	[3] 450	450	470.
Fire point, ° F	520	[3] 540	[3] 540−	520	520.
Appearance after 60 hrs	Clear liquid	Clear liquid	Clear ilquid	Crystalized	Crystalized.
20° F	do	do	do	do	Do.
0° F	do	do	do	do	Do.
−30° F	do	do	do	do	Do.
−60° F	do	do	do	do	Do.

[1] Too viscous to measure.
[2] Supercooled.
[3] Micro-scale.

Table 2 below, shows the oxygen absorption test results of the three esters (A, B and C) prepared by the process. A typical ester of a straight chain fatty acid used as a base fluid for Type 2 synthetic lubricants, Ester F, is shown for comparison.

TABLE 2: OXYGEN ABSORPTION TEST RESULTS OF PENTAERYTHRITOL NEO ACID ESTERS

[Temperature=450° F.; Oxygen rate=1 ft.³/hr.; Inhibitor=1% N-phenyl-1-naphthylamine.]

Base Fluid	Ester A	Ester B	Ester C	Ester D	Ester F [1]
Results:					
T_i, min	413	390	353	442	200
T_t, min	473	437	393	482	242
V_i, ml	575	537	537	600	466
V_t, ml	2,500	2,500	2,500	2,500	2,500
Δ (KV/100), percent	37.5	41.8	37.3	36.1	49.7
Δ (a.n.)	11.2	14.1	11.8	13.1	10.1

[1] Hercolube A.

It can be seen that all three esters of this process have T_i and T_t considerably longer than those of Ester F and the viscosity increases, Δ(KV/100), were considerably lower. This indicates that the three base fluids prepared by the process have a lower rate of oxidation in presence of the common inhibitor and, when oxidized to the same extent (2,500 ml. O_2 absorbed), gave lower increase in viscosity than the Type 2 base fluid.

A comparison between results of Esters A and D also indicates that the process, while greatly improving the low temperature properties of the mixed esters, did not detract from the good oxidation resistance inherent in the mixed neo-C_7-n-C_5 esters.

Tetramethylcyclobutanediol Diesters of Linolenic Acid

J.P. Friedrich; U.S. Patent 3,373,176; March 12, 1968; assigned to the U.S. Secretary of Agriculture has found that the diesters formed by reacting a dearomatized cyclized linolenic acid isomer mixture with two equivalents of tetramethylcyclobutanediol exhibit low pour points coupled with high resistance to thermal oxidation. The preparation of the diester low temperature lubricant base, which has the following formula, is set forth in Example 1

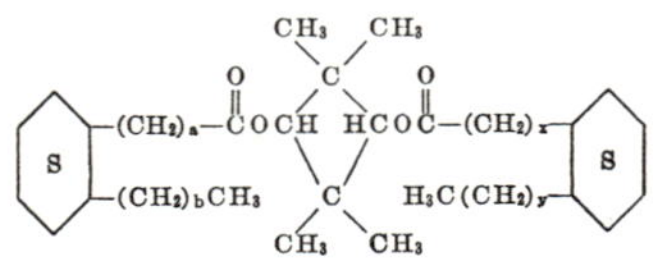

where the sum of (a+b)=10 and the sum of (x+y)=10.

Example 1: Into a 1 l. round-bottom flask equipped with a water trap, a reflux condenser, and boiling chips, were introduced 36 g. (0.25 mol) of 2,2,4,4-tetramethyl-1,3-cyclobutanediol, 282 g. (1 mol) of the highly purified hydrogenated C_{18} cyclic acid mixed isomers, and 300 ml. of toluene. The flask was heated to a boil (120°C.) and the contents refluxed until 8.5 g. of water (9.0 g. theoretical) had collected in the trap. After removing the toluene solvent under vacuum, the mixture of diester product and unreacted cyclic acid starting material was subjected to molecular distillation at 10 microns pressure; approximately 140 g. of unreacted cyclic acid, BP 90° to 110°C. and 145 g. of the cyclic acid diester product, BP 155° to 165°C., were recovered. The diester having an acid number of 2 was passed through an alumina column thereby lowering the acid number to practically zero. The water-white diester also had an iodine number of zero.

The viscosities at 100° and 210°F. shown in the table below, were determined in Ostwald-Fenske pipettes and were plotted on ASTM standard viscosity charts for liquid petroleum products (D341), thereby permitting extrapolation of the viscosity at -40°F. The viscosity index for each ester was obtained from the viscosities at 100° and 210°F. in accordance with ASTM Method D567-41.

Ester	Viscosity cs. 100° F. Pre Δ	Viscosity cs. 100° F. Percent Increase after Δ	Viscosity cs. 210° F. Pre Δ	Acid number Pre Δ	Acid number After Δ	Pour point, °F.	Sludge formed formed after Δ	Viscosity index
(1) 2.2-4,4-tetramethyl-1,3-cyclobutanediol	169	377	16.4	0	<1	−36	None	108
(2) 1,4-cyclohexane-bis methanol	220	[1] 70	19 6	0	[1] 6.7	−35	Large amount	108
(3) 2,2-dimethyl-1,3-propanediol (discontinued at 24 hrs. of 400° F.).	170	216	11.7	0	6.7	−44	None so far	43
(4) 1,4-benzene dimethanol	122	([2])	14.1	0	([2])	−38	All sludge	118
(5) 2,2-dimethylpentanol	19.9	([2])	3.71	0	([2])	−79	do	97
(6) "Herclube C" [3]	17.7	36	3.8	0	17.3	<−100	None	120
(7) Blend 24% of (1) and 76% of (6)	28.1	71	5.23	0	14	−71	do	130

[1] Filtrate.
[2] Impossible because of massive sludge.
[3] Proprietary diester low temp. lube base stock of Hercules Powder Co., Wilmington, Delaware.

Δ Blown as req'd in MIL-L-23699.
NOTE.—MIL-L-23699—limits viscosity at 210° F. to 5.0-5.5 cs.; percent viscosity incr. on heating −5% to +25%; acid No., max. 3.0; sludge, not in excess of 1 g. per 100 ml.

Since the equipment required for determining pour points by ASTM Method D97-47 was not available, the following procedure was employed. The esterified mixed isomers were packed in Dry Ice overnight. The intensely cold samples (-79°C.) were then placed in an air bath and the air bath in turn immersed in a Dewar flask containing isopropanol and Dry Ice. The temperature was allowed to rise at the rate of 2°C. per hour, and the temperature at which the surface of the sample would deform in 5 seconds when tipped was arbitrarily called the pour point.

The oxidative stabilities of the esters were determined by the following procedure. A 10 g. sample of ester containing 100 mg. of an equal mixture of phenyl-α-naphthylamine and p-p'-dioctyldiphenylamine was placed in a 30 ml. vial. A capillary tube was inserted to the bottom of the vial, and filtered compressed air was metered at 500 cc per hour for 72 hours, the vial being immersed in a stirred oil bath held at 400° ± 1.0°F. After, filtration to remove traces of solids, the acid value and viscosity of the sample were redetermined.

Example 2: n-Pentyl Ester — Purified saturated C_{18} cyclic acid isomer mixture, 25 g., n-pentyl alcohol, 50 g., and BF_3-etherate complex, 6 g. were introduced into a loosely stoppered 125 ml. Erlenmeyer flask and heated for 2 hours at

60° to 80°C. The cooled contents were added to a large excess of water in a separatory funnel, and petroleum ether, 50 ml., was then added to facilitate phase separation. After washing the organic layer twice with distilled water and dry drying, the ether was removed under vacuum. Unreacted pentyl alcohol was distilled off and the crude ester containing about 2% of unreacted cyclic acids was placed on a column of basic alumina. The ester was eluted with petroleum ether which was then removed under vacuum. Distillation of the ester at 0.1 mm. and 135° to 155°C. gave 33 g. of the ester product.

The viscosities at 100° to 210°F. shown in the above table, were determined in Ostwald-Fenske pipettes and were plotted on ASTM standard viscosity charts for liquid petroleum products (D341), thereby permitting extrapolation of the viscosity at -65°F. The viscosity index for each ester was obtained from the viscosities at 100° and 210°F. in accordance with ASTM Method D567-41.

Liquid Esters of Neoalkyl Polyols and Neoalkyl Fatty Acids

T.S. Chao, M. Kjonaas and W.D. Hoffman; U.S. Patent 3,441,600; April 29, 1969 and U.S. Patent 3,523,084; August 4, 1970; both assigned to Sinclair Research, Inc. have found that the neoalkyl polyol esters of certain neoalkyl fatty acids having from 7 to as high as 9 carbon atoms have excellent susceptibility to amine antioxidants and the attendant advantages of ester lubricants and yet exist as liquids at normal temperatures. In this process, the normally liquid esters useful as lubricants, or as synthetic lubricant components, are those resulting from the esterification of a neoalkyl polyol selected from the group consisting of pentaerythritol and 1,1,1-trimethylolalkanes of 5 to 7, preferably 5 to 6, carbon atoms with one or more neoalkyl fatty acids selected from the group consisting of 3,3-dimethylpentanoic acid; 3,3,4-trimethylpentanoic acid; 4,4-dimethylhexanoic acid; 4,4,5-trimethylhexanoic acid and 5,5-dimethylhexanoic acid. The esters are those in which at least three, preferably essentially all, of the hydroxyl groups of the polyol are esterified with the neoalkanoic acid.

Preferred polyols are pentaerythritol (PE) and trimethylolpropane (TMP). With respect to the acid component, normally solid esterification products result when the alkanoic acid contains a neo carbon atom adjacent the carboxyl group, that is, when the alkanoic acid is 2,2,2-trialkyl-substituted, thus forming what is commonly referred to as a "neo acid". For example, the 2,2-dimethylpentanoic acid (a neo acid) tetraester of pentaerythritol is a high melting solid (MP 189° to 192°F.), whereas the 3,3-dimethylpentanoic acid tetraester of pentaerythritol exists either as a supercooled liquid or as a low melting solid at room temperature and, in its liquid state, is a preferred lubricant of the process.

Some of the acids used in the preparation of the esters are commercially available, while others can be prepared according to known methods. For example, 3,3-dimethylpentanoic acid can be prepared from 3,3-dimethylpentanol by oxidation with potassium permanganate and 5,5-dimethylhexanoic acid can be prepared by free radical addition of acetic acid to 3,3-dimethylbutene-1. The following example shows a procedure for the preparation of 5,5-dimethylhexanoic acid.

Example 1: A 5 l., four-necked flask was fitted with two reflux condensers, a mechanical stirrer and a thermometer. The flask was charged with 2,860 ml. (50 mols) of glacial acetic acid, 84 g. (1 mol) of neohexene and 36.3 (0.15 mol) of benzoyl peroxide. After heating for 24 hours at 90°C., an aliquot showed that all of the peroxide had been consumed. The mixture was then distilled to remove excess acetic acid. No unreacted neohexane was detected. The residue in the flask was washed with water to remove any remaining acetic acid and then extracted with aqueous K_2CO_3. The extract was acidified, whereupon an organic layer separated, and was washed and dried. The crude 5,5-dimethylhexanoic acid was distilled to remove a small amount of forerun. The bulk of the product distilled at 125°C. It first appeared as a colorless liquid, but solidified on standing. MP, 37° to 39°C.; yield, 91.7 g. or 64% of theory. Analysis — Calculated for $C_8H_{10}O_2$ (percent): C, 66.62; H, 11.8. Found: C, 66.57; H, 11.0.

Example 2: This example shows the preparation of trimethylolpropane tris(5,5-dimethylhexanoate). A mixture of 54 g. (0.40 mol) of trimethylolpropane, 180 g. (1.25 mols) of 5,5-dimethylhexanoic acid and 100 g. of xylene was charged into a 1 l., four necked flask equipped with mechanical stirrer, thermometer, reflux condenser and a Dean-Stark azeotrope trap. The mixture was stirred and refluxed for a period of 28 hours during which 20 cc of water was collected in the trap, the pot temperature being 161° to 169°C. The reaction mixture was stripped under vacuum (to 180°C. at 1 mm. Hg) to remove xylene and unreacted acid. The residue was diluted with hexane and washed three times with a 5% aqueous solution of Na_2CO_3 and once with water. It was then topped to 130°C. at 1 mm. Hg and filtered through Hyflo Super-Cel to give 190 g. of a light yellow, viscous liquid having the following properties.

KV at 210°F., cs.	6.828
100°F., cs.	65.64
Pour point, °F.	-25
Flash point, °F.	475
Sap. No.	319.7
Acid No.	0.08
Hydroxyl No.	22.4

Example 3: This example illustrates the preparation of pentaerythritol tetrakis(3, 3-dimethylpentanoate). A mixture of 38 g. (0.265 mol) of pentaerythritol, 160 g. (1.21 mols) of 3, 3-dimethylpentanoic acid and 60 ml. of xylene was stirred at reflux temperature (167° to 210°C.) for 32 hours in a four necked flask equipped with reflux condenser and Dean-Stark trap. A total of 21.0 ml. of water was collected in the trap. The reaction mixture was topped to 175°C. at 1 mm. Hg pressure. The residue was diluted with xylene and washed twice with a 5% aqueous solution of Na_2CO_3 and twice with water. The washed ester was topped to 165°C. at 1 mm. Hg pressure, leaving 144 g. of a viscous liquid having the following properties.

KV at 210°F., cs.	14.91
100°F., cs.	277
Pour Point, °F.	-10
Sap. No.	338.7

That the ester thus prepared existed as a supercooled liquid at ambient temperatures was evidenced by treating a later liquid sample, prepared from a heart cut of 3, 3-dimethylpentanoic acid, with a few broken crystals of pentaerythritol tetrakis(neoheptanoate). The seeding of the sample resulted in its solidification when cooled to about 20°F. The solid melted at 80° to 82°F.

A number of other esters were made using this same general procedure and are reported below. In Table 1, the viscosity characteristics and other observations made on a number of samples are reported; included in Table 1, for comparison purposes are esters of so-called neo acids having the same number of carbon atoms as those of the process, and which were found to be solids at room temperatures.

TABLE 1: PHYSICAL PROPERTIES OF TRIMETHYLOLPROPANE (TMP) TRI- AND PENTAERYTHRITOL (PE) TETRAESTERS OF BRANCHED CHAIN ACIDS

Sample	Polyol	Acid	Kinematic Viscosity, cs. 210°F.	100°F.	Pour Point, °F.	Flash Point, °F.	Physical State of Ester	Melting Point, °F.
374	TMP	4, 4-Dimethylhexanoic	9.042	103.7	-35	465	Liquid	-
426	TMP	4, 4, 5-Trimethylhexanoic	13.97	226.7	-	-	Liquid	-
375	PE	4, 4-Dimethylhexanoic	17.01	340.3	-10	530	Liquid	-
220	PE	5, 5-Dimethylhexanoic	11.16	169.3	5	480	Liquid	-
425	PE	4, 4, 5-Trimethylhexanoic	28.75	959.3	20	-	Liquid	-
025	PE	3, 3-Dimethylpentanoic	14.91	277.0	-10	-	Liquid*	-
700	PE	"	15.09	273.2	-20	490	Liquid*	80-82
006	PE	2, 2-Dimethylbutyric	-	-	-	-	Solid	300-310
007	PE	2, 2-Dimethylpentanoic	-	-	-	440	Solid	189-192
005	PE	2, 2-Dimethylpropanoic	-	-	-	-	Solid	258-260
003	TMP	5, 5-Dimethylhexanoic	6.828	65.64	-25	475	Liquid	-
238	TMP	4, 4-Dimethylpentanoic	8.162	-	-	-	Solid	-
239	PE	"	15.84	-	-	-	Solid	167-172
	PE	2, 2-Dimethylhexanoic	13.19	-	-	470	Solid	149-153
669	DPE**	3, 3-Dimethylpentanoic	47.40	-	-	-	Solid	189-192

*Supercooled.
**Dipentaerythritol.

A perusal of Table 1 shows that the neoalkyl fatty acids of this process give esters with neoalkyl polyols having favorable viscosity properties, whereas nearly homologous or isomeric acids yield esters which exist only as solids at room temperature.

A number of normally liquid esters were prepared and tested for oxidation resistance and antioxidant susceptibility by a standard test called the Oxygen Absorption Test. In this test, a 75 g. sample of the ester containing a standard antioxidant is oxidized at 450°F. with 1 $ft.^3$/hr. of oxygen in a tube about 9" long and 1 1/4 i.d. The exit gas passes through a reflux condenser, an entrainment trap and two absorption tubes. The first tube is filled with active charcoal, Drierite and Ascarite to remove organic vapor, water and CO_2. The gases come out from the first tube then pass through a catalytic oxidation tube filled with CuO, in which CO and hydrocarbons are converted to CO_2. The exit gas from the oxidation tube then passes through a second absorption tube filled with Drierite and Ascarite and is circulated with a tubing pump back into the sample tube. Thus oxygen is continuously circulated through the system until it is used up by the sample.

The decrease in the volume of oxygen in the system, which represents the volume of O_2 absorbed, is continuously measured by a Statham gauge and recorded on a Brown recorder. The recorder is so calibrated that the slope of the curve represents the rate of oxygen absorption. During the earlier part of the test, oxidation of the oil is effectively inhibited by the antioxidant and the rate of O_2 absorption is usually small. The rate usually stays constant until a time is reached at which the inhibitor is either exhausted or is no longer effective.

At this point, there is generally a fast increase in the rate of oxidation. The time in minutes required for this change in rate to happen is called the induction period (T_i). The volume of O_2 absorbed during this period of time is represented by V_i. The test is continued until a total volume (V_t) of 2,500 ml. of O_2 has been absorbed. The time in minutes for this to take place is called the total time (T_t). A good base fluid for lubricants should exhibit a high T_i value (i.e., a long induction period), a high T_t value (i.e., a long period for total absorption) and little increase in acid number and viscosity as a result of the oxygen absorption.

Table 2 shows the O_2 Absorption Test results on the neoalkyl polyol esters of neoalkyl fatty acids of this process. Results on esters of straight chain and monoalkyl-substituted fatty acids are included for purposes of comparison. The same concentration (1% by weight) of the same PAN antioxidant (N-phenyl-α-naphthylamine) was used in all cases.

TABLE 2

	Sample No.						
	635	448	660	410	409	553	277
	Acid						
	Valeric	Isovaleric	Caproic	4-methyl-pentanoic	2-methyl-pentanoic	n-Heptanoic	
	Number of C atoms						
	5	5	6	6	6	7	7
Polyol	PE	PE	PE	PE	PE	TMP	PE
O_2 Absorption:							
Induction time (min.), T_i	358	2,371	260	322	175	239	156
Total time (min.), T_t	413	2,392	298	350	201	278	202
Induction volume (ml.), V_i	346	1,712	474	985	705	614	307
Total volume (ml.), V_t	2,500	2,500	2,500	2,500	2,500	2,500	2,50
Used Oil Properties:							
KV at 100° F. (cs.)	24.67	50.74	28.77	40.48	33.36	20.49	35.94
Percent increase over original	39.8	16.4	43.95	18.85	38.6	32.7	52.3
Acid No.[1]	12.08	9.99	8.85	6.64	10.27	9.81	7.64
(Increase over original)		9.90	8.84	6.09	10.22	9.75	7.63
n-Pentane insolubles	0.314	0.019	0.048	0.095	0.140	0.088	0.055

TABLE 3

	Sample No.								
	085	701	370	278	583	412	411	233	467
	Acid								
	3,3-dimethylpentanoic		Caprylic		2-ethylhexanoic	4,4-dimethylhexanoic		5,5-dimethylhexanoic	4,4,5-trimethylhexanoic
	Number of C atoms								
	7	7	8	8	8	8	8	8	9
Polyol	PE	PE	TMP	PE	TMP	TMP	PE	PE	TMP
O_2 Absorption:									
Induction time (min.), T_i	528	733	208	71	131	309	271	332	159
Total time (min.), T_t	581	772	250	112	165	351	330	368	212
Induction volume (ml.), V_i	717	1,360	486	345	448	486	524	1,200	449
Total volume (ml.), V_t	2,500	2,500	2,500	2,500	2,500	2,500	2,500	2,500	2,500
Used Oil Properties:									
KV at 100° F. (cs.)	311.0		25.47	45.39	37.58	126.2	432.2	248.8	
Percent increase over original	16.9	8.33	38.1	60.6	22.9	21.8	32.8	45.2	22.9
Acid No.[1]	20.38		12.45	7.00	13.64	12.64	10.82	14.1	
(Increase over original)	20.33	9.39	12.45	6.99	13.52	12.49	10.61	14.0	9.26
n-Pentane insolubles	0.693		0.052	0.030	0.120	0.210	0.077	0.057	

[1] Acid number prior to O_2 absorption was about zero in all samples.

It can readily be seen from the data that the esters of the process exhibit a combination of advantageous properties not found in any one of those esters prepared from straight chain or mono-substituted acids. Thus, the esters of the process (Sample Nos. 085, 412, 411 and 233) compare favorably with those esters prepared from 5 and 6 carbon atoms (Sample Nos. 635, 448, 660, 410 and 409) as regards oxidation resistance and acid number increase, yet the esters of the process, by virtue of their much lower volatility, find utility in extreme temperature-low pressure applications for which the lower molecular weight esters are unsuited. On the other hand, those unbranched and mono-substituted acid esters having the same carbon atom content as the neoalkyl esters of the process (Sample Nos. 553, 277, 370, 278 and 583), while possessing the requisite property of low volatility, are markedly inferior to the neoalkyl esters of this process in oxidation resistance properties.

Low Viscosity (5.5 cs.) Fluid

A process described by S.W. Critchley, H. Barns and P.M. Blanchard; U.S. Patent 3,328,302; June 27, 1967; assigned to The Geigy Company Limited and The British Petroleum Company Limited, England involves a thermally and oxidatively stable lubricant which has a flash point not less than 500°F., a 210°F. viscosity not greater than 5.5 cs. and a -40°F. viscosity not greater than 13,000 cs. The complex ester is prepared by reacting together under esterification conditions and in one or more stages: (A) 1 molecular proportion of 1,1,1-trimethylolpropane, (B) 2.75 to 2.85 molecular proportions of caprylic acid and (C) 0.125 to 0.075 molecular proportions of at least one straight chain, dicarboxylic acid having 9 to 10 carbon atoms per molecule.

In the preparation of the complex ester, the reaction is preferably carried out at a temperature not exceeding 230°C. and until the product has an acidity of about 5 mg. KOH/g. or less, and the product should be stripped under reduced pressure (e.g., 1 mm. Hg. or less) to reduce the proportion of unreacted or partially reacted materials in the product. The stripped product is then preferably given a finishing treatment to reduce its acidity, e.g., to 0.05 mg KOH/g. or less.

Suitable finishing treatments include water or alkali washing or percolation through alumina or an ion-exchange resin or stirring with the latter materials followed by filtration. The esterification reaction may be carried out in the presence of a catalyst, e.g., toluene-p-sulfonic acid, sodium bisulfate, a tetraalkyl titanate or titanium tetrachloride.

Example: Four complex esters (El, E2, E3 and E4) were prepared according to the process. In all cases the esters were prepared from caprylic acid, 1,1,1-trimethylolpropane and sebacic acid in the molecular ratio 2.80:1:0.10 and the esterification reactions were carried out in conventional manner using toluene as a water entrainer. The variables in the reactions are shown in Table 1.

TABLE 1

	E1	E2	E3	E4
Catalyst	None	None	$Ti(OBu)_4$	$Ti(OBu)_4$
Reaction temperature: range °C	130-200	130-230	130-200	130-200
Period of reaction, hr	120	120	72	72
Acidity at end of reaction, mg. KOH/g	5.4	1.4	0.45	1.0
Stripping temperature, °F	217	220	216	216
Stripping pressure, mm. Hg	0.2	0.2	0.4	0.3
Final treatment	Alumina	Alumina	Alumina	Resin
Final Acidity, mg. KOH/g	0.05	0.05	0.05	0.05
Viscosity of ester at—				
210° F., cs	5.1	5.0	4.9	4.9
−40° F., cs	6,200	5,790	-	5,390
Open flash point of ester, °F	515	510	-	515

The resin was an anionic ion-exchange resin of the polystyrene type containing quaternary ammonium groups. $Ti(OBu)_4$= titanium tetrabutoxide. In the case of esters prepared using a titanium catalyst the esterification product was stirred for 4 hours at a temperature not exceeding 50°C. with 10% weight of a clay used in refining mineral lubricating oils. The clay was then filtered off and the clay treatment was repeated. This reduced the titanium content of the ester to less than 10 ppm.

The properties of complex esters E1, E2, E3 and E4 blended with antioxidants are given in Table 2. Each ester was blended with 2% weight (based on the ester) of iminodibenzyl (IDB), an antioxidant which was found to be particularly suitable for lubricants intended for use at very high temperatures in contact with metals including magnesium alloy, aluminum alloy, copper, silver and steel. Ester E1 was also blended with 2% weight of various other antioxidants, including N,N'-diphenyl-m-phenylenediamine (DPPDA).

TABLE 2

	Ester							
	E1	E2	E3	E4	E1	E1	E1	E1
	Antioxidant, 2% wt. in each case							
	IDB	IDB	IDB	IDB	DPPDA	UOP 278	PANA	PTZ
Viscosity at—								
400° F., cs	1.4	1.4	-	1.4	-	-	-	-
210° F., cs	5.1	5.0	4.9	4.9	-	-	-	-
100° F., cs	26.0	25.4	24.5	24.8	26.7	25.4	26.0	25.8
−40° F., cs	7,310	6,830	-	6,525	-	-	-	-
Open Flash point, °F	505	530	-	505	-	-	-	-
Pour point, °F	<−75	-	-	<−75	-	-	-	-
Total acidity, mg. KOH/g	<0.05	0.05	0.05	<0.05	-	-	-	-
Oxidation/Corrosion Test at 450° F.:								
Metal weight changes, mg./cm.^{2}:								
Magnesium alloy	+0.1	+0.3	+0.1	+0.1	−0.3	0.0	0.0	0.0
Aluminum alloy	+0.2	+0.1	+0.2	+0.1	+0.1	0.0	0.0	0.0
Copper	+0.1	−0.1	−0.3	+0.3	0.0	0.0	−0.4	−0.4
Silver	+0.2	+0.1	+0.1	+0.2	+0.2	+0.1	0.0	0.0
Steel	+0.1	+0.1	+0.1	0.0	+0.1	+0.1	0.0	+0.1
Increase in 100° F. viscosity, percent	36	41	33	39	31	51	76	134
Thermal Stability test at 325° C.:								
210° F. visc. change after 6 hr., percent	−2	-	-	-	-	-	-	-
210° F. visc. change after 12 hr., percent	Nil	-	-	-	-	-	-	-
210° F. visc. change after 18 hr., percent	+1	-	-	-	-	-	-	-
210° F. visc. change after 24 hr., percent	+2	-	-	-	-	-	-	-
Acidity incr. after 24 hr., mg. KOH/g	5.45	-	-	-	-	-	-	-

UOP 278 is a commercially available experimental antioxidant. PANA=Phenyl-alpha-naphthylamine. PTZ=Phenothiazine.

The oxidation/corrosion test referred to in Table 2 is similar to that detailed in Pratt and Whitney's "Type II Lubricant" specification. One inch square strips of the five metals indicated are immersed in 100 ml. of the oil under test which is heated to 450°F. for 48 hours; during this period, air is bubbled through the oil at a rate of 5 l. per hour. To meet the specification requirements, a lubricant must give a viscosity increase not in excess of 50% and the metal weight changes must not be in excess of ±0.3 mg./cm.^{2}. It will be seen that suitably inhibited lubricants according to the process give good results in this test. The thermal stability test referred to is that laid down by the British Ministry of Aviation specification DERD 2497 for high temperature lubricants. According to specification DERD 2497, the change in the 210°F. viscosity of the oil after 6, 12, 18 and 24 hours at 325°C. should in no case be outside the limits -10 to +20%. It will be seen that ester E1 was extremely thermally stable as assessed by this test.

Blends to Meet Temperature Viscosity Requirements

R.D. Aylesworth and R.H. Boehringer; U.S. Patent 3,309,318; March 14, 1967; assigned to Emery Industries, Inc. have found that blends of certain hindered polyol esters provide gas turbine lubricants which have suitably functional viscosities at -65°F., e.g., 13,000 cs. or less, without sacrifice of the desired oxidation stability and corrosion resistance.

The particular blends of the esters found to be suitable are those having from 30 to 60% of a neopentyl glycol ester of one or more monocarboxylic alkanoic acids having from 7 to 10 carbon atoms, and from about 40 to 70% of at least one other ester of the fully hindered polyol type selected from the group consisting of trimethylol ethane, trimethylol propane, trimethylol butane, pentaerythritol and mixtures with one or more mono-carboxylic alkanoic acids having from 5 to 10 carbon atoms. The percentages are based on the total weight of the blend of hindered polyol esters. When a pentaerythritol ester is used with the neopentyl glycol ester, carbon chain length of the esterifying acid is preferably from 5 to 10, while with the trimethylol alkane ester, the acid preferably has a chain length of 7 to 10 carbon atoms.

The process primarily overcomes the viscosity deficiencies of the otherwise desirable hindered esters. Specifically, diesters of neopentyl glycol prepared from mono-carboxylic alkanoic acids containing from 7 to 10 carbon atoms when used alone cannot meet the viscosity limits set by the U.S. Specification MIL-L-7808F because they have pour points above -65°F. Esters of trimethylol propane, and adjacent homologs prepared from alkanoic monobasic acids containing from 7 to 10 carbon atoms when used as the sole ester base fluid fail to meet Specification MIL-L-7808F in that the -65°F. viscosity of these esters is well in excess of the 13,000 cs. maximum required. The esters of pentaerythritol or dipentaerythritol or mixtures thereof, prepared from aliphatic monobasic acids containing from 5 to 10 carbon atoms again are unsuitable for application against Specification MIL-L-7808F by reason of excessive -65°F. viscosity.

It is significant therefore, that by blending the abovementioned esters in various essential proportions synthetic lubricant base fluids can be prepared which meet the low temperature viscosity requirements of MIL-L-7808F. Table 1 provides an illustration of the drastic reduction in low temperature viscosity obtainable with the blended esters of this process.

TABLE 1:

Ester	Viscosity −65° F. cs.	Viscosity 210° F. cs.
(1) Neopentyl glycol dipelargonate $(C_9)_2$	(1)	2.6
(2) Trimethylol propane dipleargonate $(C_9)_2$, monoheptanoate (C_7)	28,000	4.2
(3) Pentaerythritol tetravalerate $(C_5)_4$	25,000	3.35
(4) 55% TMP $(C_7)(C_9)_2$,[2] 45% NPG $(C_9)_2$[3]	9,685	3.36
(5) 55% TMP $(C_7)(C_9)_2$, 45% NPG $(C_8)(C_{10})$	9,760	3.19
(6) 60% PE $(C_5)_4$,[4] 40% NPG $(C_7)(C_9)$	8,350	3.0
(7) 55% PE $(i\text{-}C_5)_2(C_9)_2$,[5] 45% NPG $(C_7)(C_9)$	11,500	3.15
(8) 48% PE $(C_5)_4$, 32% NPG $(C_7)(C_9)$, 20% 2,2,4-trimethyl pentanediol $(C_9)_2$	7,625	2.98

[1] Freezes at −50° F.
[2] TMP—trimethylol propane.
[3] NPG—neopentyl glycol (2,2 dimethyl propanediol).
[4] PE—pentaerythritol.
[5] $(i\text{-}C_5)$—diisovalerate.

Compounded lubricants prepared from the blended hindered polyol esters of the process have greater oxidation-corrosion stability than lubricants from blends of polyol esters with diesters of alkanedioic acids, e.g., azelaic acid with an alkanol. The latter blends are suggested in British Patent 927,794. This is illustrated by the data of Table 2, shown on the following page, using blended esters of the following compositions and properties compounded with the additives shown.

Blend 1

Pentaerythritol tetravalerate*, %	60
Neopentyl glycol dipelargonate, %	40
Viscosity at 210°F., cs.	3.0
Viscosity at -65°F., cs.	8,350
Acid value	0.06
Additives:	
Phenyl-α-naphthylamine**, %	1.0
Dioctyldiphenylamine, %	0.66
Dihydroxyanthraquinone, %	0.02

Blend 2

Di-2-ethylhexyl azelate*, %	78.4
Trimethylolpropane dipelargonate monoheptanoate, %	21.6
Viscosity at 210°F., cs.	3.22
Viscosity at -65°F., cs.	8,206
Acid value	0.06
Additives:	
Phenyl-α-naphthylamine**, %	1.0
Dioctyldiphenylamine, %	0.66
Dihydroxyanthraquinone, %	0.02

*Percentage by weight of ester blend.
**Percentage by weight of base fluid.

Test Procedure — 100 ml. of the test lubricant were placed in a 50 mm. o.d. test tube. The five test metals, wired together, were placed in the test liquid. A water jacketed condenser and an air delivery tube were placed on the test tube. The apparatus was placed in a constant temperature heating block for 72 hours at 400° ± 2°F. with air being delivered to the bottom of the test tube at a rate of 5 l. per hour during the test period. At the conclusion of the test the viscosity, acid value and metal surface corrosion were measured and are shown in Table 2.

TABLE 2: OXIDATION-CORROSION, MIL-L-23699 (WEP) TEST RUN 72 HOURS AT 400°F.

Test Results	Blend 1, Percent	Blend 2, Percent
Viscosity change, max. cs. at 100° F.	−5.6	+16.8
Acid value change, mgs. KOH/gm.	1.68	11.8
Copper corrosion, mg./cm.^{2}	−0.05	−0.23
Steel corrosion, mg./cm.^{2}	+0.01	+0.05
Silver corrosion, mg./cm.^{2}	−0.20	+0.08
Aluminum corrosion, mg./cm.^{2}	−0.03	+0.15
Magnesium corrosion, mg./cm.^{2}	0.0	−16.20

This data clearly shows the hindered ester blends to be superior to the linear aliphatic diester-hindered ester blend in regard to viscosity degradation, acid formation and metal corrosion.

Example 1: 134 g. (1.0 mol) of pentaerythritol and 510 g. (5.0 mols) of valeric acid were charged into a reaction vessel and heated to 210°C., removing the water of reaction as it formed by means of an azeotrope with xylene. Reacted 6 hours until the hydroxyl value was below 3.0. Stripped the excess valeric acid and xylene under high vacuum to 234°C. cooled to room temperature and broke vacuum. Alkali refined using 50% excess 10% aqueous NaOH solution followed by distilled water washing until the wash water was neutral to phenolphthalein. Vacuum dried to 90°C./1 mm. Hg. Final acid value was 0.04; hydroxyl value 0.0.

Example 2: 102 g. (1.0 mol) of neopentyl glycol was reacted with 347 g. (2.2 mols) of pelargonic acid to 230°C. removing the water of reaction by means of a xylene azeotrope. Reacted 4 hours to a hydroxyl value below 3.0. Stripped the excess pelargonic acid and xylene under high vacuum to 230°C. Alkali refined the product using 50% excess of 10% NaOH solution followed by distilled water washes until the wash water was neutral to phenolphthalein. Vacuum dried the product to 90°C./1 mm. Hg. Final acid value was 0.02; hydroxyl value 0.0.

Example 3: 600 g. of pentaerythritol tetravalerate and 400 g. of neopentyl glycol dipelargonate were charged into a 2 l. beaker. The components were thoroughly mixed by agitating for 1/2 hour with a mechanical agitator. This resultant blend is the base ester for the following compounded lubricant.

989.5 g. of the above ester were charged into a beaker. 5.0 g. of phenyl-α-naphthylamine, 5.0 g. dioctyl diphenylamine and 0.5 g. dihydroxy anthraquinone were added to the ester and blended by heating to 125° to 150°C. with constant agitation for 1/2 hour. The final properties of the finished lubricant are listed below.

Kinematic viscosity at 210°F., cs.	3.0
Kinematic viscosity at 100°F., cs.	11.8
Kinematic viscosity at -65°F., cs.	8,350
Acid value	0.10
Flash point, °F.	415
Pour point, °F.	-100

Diesters Containing Adamantane Nuclei

A process described by I.N. Duling and A. Schneider; U.S. Patent 3,398,165; August 20, 1968; assigned to Sun Oil Company relates to diesters containing one or two adamantane nuclei per molecule. Each adamantane nucleus contains not more than one unsubstituted bridgehead carbon atom or in other words contains not more than one tertiary hydrogen atom. The diesters are either oily liquids or waxy solids. They are characterized by good hydrolytic stability, oxidative stability and thermal stability and have particular utility as lubricants or as components in lubricating oil and grease compositions. The diesters thus comprise two types as follows. (1) Diesters of alkyl-substituted adamantane 1,3-diols with alkanoic or cycloalkanoic acids. (2) Diesters of alkyl-substituted adamantane 1-monools with alkanedioic or cycloalkanedioic acids. The following examples illustrate the process.

Example 1: This example shows the preparation of the Type (1) diester described above, 1,3-(5,7-dimethyl)adamantyl dipelargonate from pelargonic acid and 1,3-dihydroxy-5,7-dimethyladamantane. Oxalyl chloride was used to convert the pelargonic acid to its acyl chloride. Specifically, 110 g. of oxalyl chloride were slowly added to 125 g. of pelargonic acid and the mixture was refluxed for 2 hours. Excess oxalyl chloride was removed under aspirator vacuum at room temperature. The resulting acid chloride was heated to 105° to 110°C. and 70 g. of 1,3-dihydroxy-5,7-dimethyladamantane were cautiously added in small portions to the well stirred mixture. The oil was then stirred and heated at 150°C. for 1/2 hour to complete the reaction.

After this a solution of 50 g. sodium carbonate in 300 ml. water was added and the mixture was stirred for 2 to 3 days, then diluted to 1,200 ml. with water and extracted with 500 ml. ether. The ether layer was washed twice with sodium chloride solution, stirred with anhydrous magnesium sulfate and charcoal, and filtered to give a light yellow solution. Evaporation of the ether gave 175 g. of the 1,3-(5,7-dimethyl)adamantyl dipelargonate. Identification was accomplished by infrared, NMR, and mass spectra of the broad VPC peak. The diester product was a pale yellow oil and its properties are listed in Table 1 below.

Examples 2 through 5: For Examples 2, 3, 4 and 5 other Type (1) diesters were made from 1,3-dihydroxy-5,7-dimethyladamantane in substantially the same manner as described in Example 1 except that the following aliphatic monoacids, respectively, were used in place of pelargonic acid: acetic, caproic, caprylic and lauric. Properties of the resulting diesters are also listed in Table 1.

TABLE 1: PROPERTIES OF TYPE (1) DIESTERS

Acid Used (RCOOH)	No. of C in R	Density, g./l. at 20° C.	R.I., 20° C. / D	Kv. at 100° F., cs.	Kv. at 210° F., cs.	ASTM D-567, V.I.	Glass Transition Temp. (T_g),° C.[1]	Melting Point, ° C.[1]
Acetic	1	[2] 1.065	----------	124.7	4.46	−1,076	----------	----------
Caproic	5	1.009	1.4744	34.6	4.73	30	−86	(3)
Caprylic	7	0.972	1.4723	42.0	5.79	82	−91	(3)
Pelargonic	8	0.973	1.4735	54.1	7.05	95	−87	(3)
Lauric	11	0.942	1.4702	51.1	7.47	116	[4] −47	10

[1] T_g and M.P. obtained by differential thermal analysis.
[2] At 30° C.
[3] None.
[4] Higher glass transition temperature for the dilaurate due to crystallinity.

The data in Table 1 show that a variety of Type (1) diester oils can be made of different viscosities and varying widely in viscosity index. Some of the oils will have relatively high VI values, as shown by the dipelargonate and dilaurate which have ASTM VI's of 95 and 116, respectively. It can also be seen that products with low glass transition temperatures (T_g) can be obtained. The pour points of these oils will be somewhat higher than T_g but not greatly.

Example 6: The preparation of bis-(3,5-dimethyladamantyl)dodecanedicarboxylate is as follows. In this example the Type (2) diester was prepared from the diterminal diacid, dodecanedicarboxylic acid and 1,3-dihydroxy-5,7-dimethyladamantane. The diacid was first converted to the alkane dioyl chloride by means of oxalyl chloride. Specifically 10 ml. of oxalyl chloride were added dropwise to 5 g. of the diacid in a 100 ml. flask fitted with a reflux condenser and a dropping funnel. Vigorous reaction and evolution of gas occurred. The mixture was gently refluxed for 2 hours and allowed to cool to room temperature. Excess oxalyl chloride was removed at room temperature by maintaining the flask under aspirator vacuum for 2 hours. Then 1-hydroxy-3,5-dimethyladamantane (7.6 g.) was added in three portions, an aspirator vacuum being maintained between additions to facilitate removal of by-product HCl.

The mixture was stirred for 1/2 hour with gentle heating, then cautiously added to 10% sodium carbonate solution (500 ml.). The mixture was stirred overnight and extracted twice with a total of 500 ml. ether. The combined ether layers were washed three times with water (sodium chloride sometimes was added to break up an emulsion), then stirred with anhydrous magnesium sulfate and decolorizing carbon. Filtration and evaporation of the ether gave 11.5 g. of a clear yellow oil. The oil was chromatographed on alumina. Combination of center fractions gave 6.5 g. of a colorless viscous oil after removal of the elution solvent (5% ether in hexane). Characterization of the single broad VPC peak by infrared, NMR and mass spectra identified the product as bis-(3,5-dimethyladamantyl)dodecanedicarboxylate. This oil had properties as shown in Table 2 below.

Examples 7 and 8: In these examples two other Type (2) diesters were prepared in substantially the same manner as described in Example 6, except that the diacids used were, respectively, adipic acid and azelaic acid. These products likewise were oils having the properties shown in Table 2.

TABLE 2: PROPERTIES OF TYPE (2) DIESTERS

Diacid Used [R(COOH)2]	No. of C in R	Density, g./l. at 20° C.	R.I., 20° C. / D	Kv. at 100° F., cs.	Kv. at 210° F., cs.	ASTM D-567, V.I.	Glass Transition Temp. (T_g),° C.[1]	Melting Point, ° C.[1]
Adipic	4	----------	----------	132.6	9.76	35	−62	(2)
Azelaic	7	1.036	1.4977	245.1	15.3	55	−64	(2)
Dodecanedioic	10	1.008	1.4944	370.6	23.6	88	−67	(2)

[1] T_g and M.P. obtained by differential thermal analysis.
[2] None.

As shown in Table 2, the three representative oils of Type (2) were more viscous than those of Type (1). The data for both types of diesters show that VI increases as the length of the linear chains derived from the aliphatic acid increases. Products of the process have better low temperature properties than would be predicted from the ASTM viscosity-temperature slope between 100° and 210°F. This is shown, for example, by data listed in Table 3 on the following page, for the dipelargonate product of Example 1.

Table 3 shows measured values of kinematic viscosities at various temperatures compared with values predicted from the ASTM slope. It also shows flash point, T_g and pour point values for this product.

TABLE 3: PROPERTIES OF DIPELARGONATE

	Measured	Predicted
Viscosity properties:		
Kv. at 500° F	0.85	1.07
Kv. at 210° F	7.05	----------
Kv. at 100° F	54.1	----------
Kv. at 0° F	2,924	4,450
Kv. at −30° F	22,076	51,000
Pour point, ° F	−90	----------
T_g, ° F	−125	----------
Flash point, ° F	480	----------

As shown here, viscosities at temperature below 0°F. are considerably less than the values that would be predicted from the ASTM slope.

Example 9: The dipelargonate oil of Example 1 was subjected to intensive purification in the following manner. A hexane solution of the oil was hydrogenated at room temperature for 2 hours at 2,000 psig H_2 pressure in the presence of Raney nickel catalyst. Some uptake of hydrogen and a temperature rise occurred but only to small extent. The hydrogenation product was poured into 10% HCl solution (1 l.). The mixture was filtered through diatomaceous earth, pentane being used to wash the oil therefrom. The pentane layer was separated from the green aqueous layer and washed three times with water. Most of the pentane was stripped off and anhydrous sodium carbonate added to the oil to make a thick paste. This was maintained for 2 hours at 70°C. under nitrogen.

The sodium carbonate was filtered from the oil and washed with pentane. Alumina (5 g.) was added to the pentane solution and the mixture was stirred 2 hours at 70°C. under nitrogen. The pentane was stripped off and the oil was distilled. The first 5% and the last 7% of distillate were discarded. A small amount of dark residue was obtained and also discarded. The product boiled at 212.5°C. at 0.12 mm. Hg. The oil was filtered through a 0.8μ fiberglass filter covered with 1/2 inch of alumina. The resulting clear, colorless product had a faint pleasant ester odor, acid number <0.2 and iodine number <0.1.

The highly purified dipelargonate product was subjected to two kinds of stability tests. One was a thermal stability test involving subjecting the oil in a nitrogen atmosphere to 600°F. for 6 hours. For comparison, the conventional ester lubricant, di-(2-ethylhexyl)sebacate, was also tested by this procedure. The other was a bench scale test for corrosion and oxidation stability substantially like that described in WAD Tech. Rep. 60-794 (Wright Air Development Center). This involved contacting a 20 ml. sample in the presence of a copper washer for 24 hours at 400°F. with air at a rate of 20 l./hour. Again for comparison, di-(2-ethylhexyl)sebacate was tested in the same manner. For these corrosion and oxidation stability tests both oils contained added inhibitors in conventional amounts.

TABLE 4: THERMAL STABILITY (600°F.) TESTS

	1,3-(5,7-dimethyl)-adamantyl dipelargonate	di-(2-ethylhexyl)-sebacate
Kv. at 100° F.:		
Before test	54.1	12.7
After test	42.2	(1)
Acid No. mg. KOH/g.:		
Before test	0.01	0.03
After test	24.0	75.3

1 Mainly solid.

The tabulated results are indicative of better thermal stability for the dipelargonate product than for the conventional ester lubricant. Results of the comparative corrosion and oxidation stability tests are set forth in Table 5.

TABLE 5: CORROSION AND OXIDATION STABILITY TESTS

	1,3-(5,7-dimethyl)-adamantyl dipelargonate (inhibited)	di-(2-ethylhexyl) sebacate (inhibited)
Kv. at 100° F.:		
Before test	53.1	13.1
After test	56.4	13.7
Percent increase	6.0	4.4
Acid No., mg. KOH/g.:		
Before test	0.06	0.07
After test	0.16	1.64
Percent increase	0.10	1.57
Evaporation Loss, wt. percent	5.8	16.2
Appearance after test	(1)	(2)
Percent Insolubles	0.41	0.4
Effect on Copper:		
Weight loss, mg./sq. cm	0.02	0.16
A.S.T.M. D-130 rating	(3)	(4)

1 Brown; no precipitation.
2 Brown; moderate precipitation.
3 Sl. tarnish.
4 Corrosion.

These results show that the dipelargonate product had a distinctly better oxidation stability than the conventional ester lubricant. The considerably lower values for acidity and evaporation loss for the dipelargonate are particularly significant.

Dicarboxylic Acid, Boric Acid and Polyhydric Alcohol

N.C. Schwartz; U.S. Patent 3,544,614; December 1, 1970; assigned to R.H. Miller Company, Inc. describes complex esters which are prepared by combining, for example:

(a) Boric acid, a polyhydroxylic alcohol, and a saturated or unsaturated aliphatic dicarboxylic acid or anhydride.
(b) Boric acid, a polyhydroxylic alcohol, and an aromatic dicarboxylic acid or anhydride.
(c) Boric acid, an alkyolamine, and a saturated or unsaturated aliphatic dicarboxylic acid or anhydride.
(d) Orthophosphoric acid, a polyhydroxylic alcohol, and a saturated or unsaturated aliphatic dicarboxylic acid or anhydride.
(e) Orthophosphoric acid, a polyhydroxylic alcohol, and an aromatic diccrboxylic acid or anhydride.

The polyhydroxylic alcohols include the glycols, triols, tetrols and hexols, and encompass the entire general family of polyhydroxylic alcohols. Among the particularly suitable specific polyhydroxylic alcohols are glycerin ethylene glycol, hexylene glycol, polyethylene glycols, pentcerythritol, sorbitol, and 2,4,7,9-tetramethyl-5-decyne-4,7-diol. Typical examples of formulations for the preparation of useful complex esters of this process are as follows.

Examples 1 through 14:

Example	Starting ingredients	Weight, grams	Mole ratio	Hours heated at 110° C.	Acid value calc.	Acid value observed
1	Boric acid	22.8	1.00			
	Glycerin, U.S.P	69.0	1.96	300	200	125
	Sebacic acid	146.0	1.89			
2	Boric acid	22.8	1.00			
	Glycerin, U.S.P	69.0	1.96	300	210	125
	Azelaic acid	136.0	1.89			
3	Boric acid	22.8	1.00			
	Glycerin, U.S.P	69.0	1.96	300	226	146
	Azelaic acid	68.0	0.98			
	Phthalic acid	60.0	1.10			
4	Boric acid	30.2	1.00			
	Glycerin, U.S.P	92.0	2.05	300	179	195
	Phthalic anhydride	89.0	1.10			
5	Boric acid	20.0	1.00			
	Ethylene glycol	62.0	3.09	300	188	30
	Azelaic acid	90.0	1.48			
6	Boric acid	20.0	1.00			
	Hexylene glycol	88.6	3.09	150	217	208
	Azelaic aicd	94.0	1.54			
7	Boric acid	30.9	1.00			
	Pentaerythritol	136.00	2.00	150	136	154
	Sabacic acid	100.00	1.00			
8	Boric acid	12.5	1.00			
	60% sorbitol	85.0	1.62	300	203	195
	Azelaic acid	50.0	1.31			
	Phthalic acid	40.0	1.33			
9	85% phosphoric acid	19.0	1.00			
	200 MW polyethlene glycol	110.0	3.00	300	75	114
	Phthalic anhydride	46.0	1.50			
10	85% phosphoric acid	43.3	1.00			
	Glycerin, U.S.P	69.0	2.00	150	196	268
	Azelaic acid	68.0	0.96			
	Phthalic anhydride	60.0	0.96			
11	Boric acid	30.9	1.00			
	Diethanolamine	106.0	2.01	300	170	120
	Azelaic acid	141.0	1.50			
12	Boric acid	30.9	1.00			
	Triethanolamine	101.0	0.677	300	210	160
	Azelaic acid	141.0	1.50			
13	Boric acid	18.0	1.00			
	Glycerin, U.S.P	63.0	2.35	300	195	180
	Maleic anhydride	54.0	1.88			
14	Boric acid	5.0	1.00			
	Surfynol 475	100.0	1.37	300	96	26
	Sebacic acid	35.0	2.14			

The ingredients named above may be introduced to a rotary-bladed mixer and thoroughly mixed together, or otherwise thoroughly mixed together in any conventional way. The 60% sorbitol of Example 8 is an aqueous solution of 60% by weight sorbitol and 40% by weight water; the phosphoric acid of Example 10 is an aqueous solution of 85% by weight phosphoric acid and 15% by weight water; and the Surfynol 475, is an ethoxylated 2,4,7,9-tetramethyl-5-decyne-4,7-diol containing 75% by weight ethylene oxide and 25% by weight 2,4,7,9-tetramethyl-5-decyne-4,7-diol.

A number of tests were run in connection with the rust-inhibiting characteristics of the lubricants of this process. These included outdoor exposure of 1 1/2 x 2 inch cold rolled 18 gauge mild steel panels coated with a solution of 0.05% ester of this process in isopropyl alcohol. Exposure to air and rain disclosed only slight rusting over a period of a month's exposure, while a like panel, similarly exposed, but uncoated, typically started rusting in one day and was badly rusted in about four days. As a further test, a lubricant was used as a wire-drawing lubricant to draw bright basic wire in two holes from 11/32 inch rod down to 0.2437 inch. Exposed to the air, snow and rain, and compared with a like sample containing a lubricating film of approximately 20% potassium oleate, 5% lime and the balance water, the first-named wire had scattered areas of slight rusting at the end of a period from mid-November to mid-December, while the second-named wire had rust areas after three days and a uniform coating of light rust at the end of the same test period and under similar exposure conditions.

In another test, cold-rolled carbon steel panels were highly polished, cleaned with distilled water, alcohol and Freon TF. They were then quickly dipped into a fluidized sample of selected lubricant and allowed to drain. The samples were then placed in a humidity cabinet maintained at 95°F. and 95% relative humidity. The samples were periodically observed and, when significant corrosion appeared, they were removed. The lubricants showed no rusting of significance until an elapse of 72 hours while the panels coated with a conventional lubricant showed rusting after 48 hours. With the humidity cabinet maintained at 140°F. and 95% relative humidity like panels were tested. The panels containing lubricants of this process showed first rusting at 84 hours while those containing conventional lubricants, showed first corrosion at 34 hours. The tests employing the humidity cabinet were carried out essentially the same as that of the ASTM D-1748 "Rust Preventative Evaluation in the Humidity Cabinet".

Cyclic Keto Esters

A process described by K. Coupland, J. Pennington and S.J. Branch; U.S. Patent 3,313,732; April 11, 1967; assigned to The Distillers Company Limited, Scotland provides a synthetic lubricant base containing at least one keto-ester of formula:

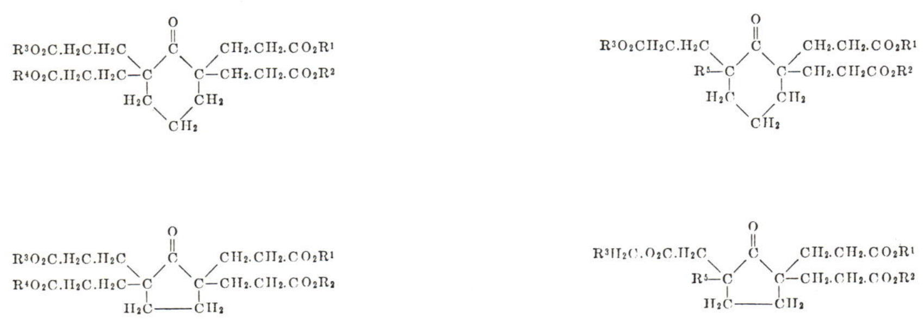

where R^1, R^2, R^3 and R^4 are the same or different alkyl groups, each of preferably from 4 to 16 carbon atoms, and R^5 is an alkyl group such as the methyl group. These compounds may, with advantage, be used as synthetic lubricant bases in a mixture with a compound having at least two ester groupings in the molecule, such as the alkyl esters of organic dicarboxylic acids and monocarboxylic acid esters of polyalcohols. It is contemplated that halogen-substituted esters are also useful as compounds having at least two ester groupings in the molecule.

Suitable compounds having at least two ester groupings in the molecule include dioctyl sebacate, dinonyl sebacate, esters of trimethylol propane such as trimethylol propane tricaprylate and esters of pentaerythritol. The proportions of keto-ester and compound having at least two ester groupings in the molecule in the synthetic lubricant bases will, naturally, vary within wide limits according to the requirements for the particular base to be formulated. Suitable proportions are within the range 85 to 10%, especially 70 to 25%, particularly 30 to 35% of keto-ester by weight of the total base.

The keto-esters employed in the synthetic lubricant bases can be prepared for example, by cyanoethylation of the appropriate cycloalkanone or alkyl-substituted cycloalkanone, followed by hydrolysis of the resulting tetra- or tri-nitrile, to give the corresponding acid. Such a reaction is described for specific compounds in JACS, 64, 2850-2858. The acid is then esterified with a small excess of the appropriate alcohol together with entrainer and an esterification catalyst. The reaction conditions are selected to give esterification temperatures within the range of 90° to 180°C., preferably between 120° and 160°C.

The process is illustrated by the following examples in which oxidative stability was assessed by the method of ASTM D972-56, except that for Examples 4 through 6 the test temperature used was 500°F. Viscosities and pour points were determined by conventional methods, and the measurement of thermal volatility was made by passing nitrogen through the sample at 2 l. per hour for 6 1/2 hours (6 hours in the cases of Examples 4 through 6) at a temperature of 300°C.; the loss in weight, the appearance and the increase in viscosity of the samples were noted at the end of the test.

Examples 1 through 6: A comparison of the properties of keto-esters having a structural formula containing a cyclohexanone ring with those of dinonyl sebacate and those of another keto-ester of the process are shown in Table 1 below. It may be seen from the table that keto-esters have very good oxidative stability and thermal volatility and possess adequate viscosity at 210°F., and adequate low temperature fluidity.

TABLE 1

Example/ Comparison Number	Ester	Oxidative Stability Weight loss (percent)	Thermal volatility		Viscosity at 210° F. Centistokes	Pour point, °F.
			Weight loss (percent)	Viscosity increase		
A	Dinonyl sebacate	79	12	7	4.6	−80
	Esters of this invention:					
1	$R^1=R^2=R^3=R^4$=n-Butyl	15	15	Nil	10.6	−35
2	R^1,R^2,R^3 and R^4 mixed C_7, C_9 alkyl	15	6	Nil	17.0	−15
3	$R^1=R^2=R^3=R^4$=2-ethylhexyl	16	6	Nil	14.4	−20
4	R^1,R^2,R^3 and R^4 mixed C_8, C_{10} alkyl	14	6	Nil	12.0	−30
5	R^1,R^2,R^3=n-butyl, R^5=methyl	28				−40
6	1,1,3,3 tetra(n-butyl-β-propionate) cyclopentan-2-one	6				

Example 7: A mixture of 70% of the keto-ester of Example 3 with 30% of dioctyl sebacate was evaluated according to the foregoing procedure. It had the following properties.

Oxidative stability, percent w./w. loss	26
Thermal volatility, percent w./w. loss	3
Viscosity increase	0
Viscosity at 210°F., cs.	8.6
Pour point, °F.	-65

Example 8: A mixture of 55% of the keto-ester of Example 3 with 45% dinonyl sebacate was evaluated according to the procedure described in Example 1 in respect of oxidative stability, viscosity and pour point. The results were as follows.

Oxidative stability, percent weight loss	35
Viscosity at 210°F., cs.	7.8
Viscosity at 100°F., cs.	52.1
Pour point, °F., below	-60

The normal volatility of this material was assessed in similar manner to that used in Example 1 but under more severe conditions, since the temperature of the ester was maintained at 340°C. The thermal properties of this material at this temperature are shown in Table 2, together with the properties of dinonyl sebacate.

TABLE 2

Ester	Thermal volatility (340° C.)		After test	
	Weight loss, percent	Viscosity increase	Acid No., Mg. KoH/g.	Appearance
Dinonyl sebacate	36	Material solidified	238	Solid.
55% Ester 4, 45% dinonyl sebacate	27	Small	36	Slight darkening.

Mixed Esters of an Alkoxylated Polyol

J.A. Wickings, E.J. Percy and D.J. Rosevear; U.S. Patent 3,530,070; September 22, 1970; assigned to BP Chemicals (U.K.) Limited, England describe lubricant compositions which have base compounds which may be made by esterifying with a mixture of acids the product of reaction of an alkylene oxide with a polyol. The compounds are found to have high viscosity indices and good resistance to shear stresses, and are therefore suitable for use in transmissions and gear boxes.

Suitable polyols are trimethylol propane, glycerol, pentaerythritol, dipentaerythritol, sorbitol and neopentylglycol. A preferred polyglycol is trimethylolpropane. The preferred alkylene oxide is propylene oxide, although it is possible to use mixtures of two or more different alkylene oxides for reaction with the polyol. The alkoxylated polyol may contain from 1 to 24, preferably 1 to 6 substituent alkoxy groups per hydroxyl group of the polyol.

The preferred carboxylic acids are aliphatic monobasic acids. Suitable acids may include high molecular weight acids such as heptoic, caprylic, pelargonic, capric and lauric acid and lower molecular weight acids such as propionic, butyric and valeric acid. It is preferred to use mixtures of lower and higher molecular weight acid for esterifying the alkoxylated polyol. By using mixtures of acids it is possible to exercise finer control over the viscosity of the final product. The following examples illustrate the process.

Examples 1 through 3: The preparation of the polyol/alkylene oxide adduct is as follows. The polyol, dried to a water content of less than about 0.3% (solvent, where used) and flake sodium hydroxide catalyst were charged to the autoclave. The vessel was sealed, evacuated and the vacuum broken with nitrogen. This cycle was repeated once more. The vessel was then evacuated and the contents heated to 130° to 140°C. The desired quantity of alkylene oxide was forced in, with stirring, the rate of addition being controlled so that the pressure in the vessel did not exceed ~40 psig. Heating and stirring were continued until no further reduction in pressure took place. The reaction mixture was cooled to 50°C. and vacuum applied to the vessel to remove any unreacted alkylene oxide. The details of the individual examples and the results are given in Table 1 below.

TABLE 1

	Charge quantities						
	Polyol (g.)	Propylene oxide, (g.) (PrO)	NaOH (g.)	Reaction temp. (°C.)	Reaction time (hr.)	Product	Intermediate for preparation of example
Example							
1	Glycerol 920	~8,120	~20	140	22	Glycerol/14 propylene oxide	5
2	Trimethylol propane 1,206.	~7,350	~8.5	135–140	10	Trimethylolpropane/14 propylene oxide	6
3	Trimethylol propane/4 propylene oxide 358.	705	0.7	145	8	Trimethylolpropane/16 propylene oxide	7

Examples 4 through 7: The esterification of the alkoxylated polyol is as follows. The alkoxylated polyol, higher fatty acid(s), catalyst and entrainer were charged to a suitable vessel. Refluxing was continued with the removal of water until the acid number of the reaction mixture was below about 5. The lower molecular weight acid was then added and reaction continued until the hydroxyl number of the product had fallen to a suitable level. Excess acid and entrainer were then removed under vacuum and the product purified by normal techniques used to purify esters. The details and results are given in Table 2 below.

TABLE 2

	Product			
	Example 4	Example 5	Example 6	Example 7
	Size of preparation (moles)			
	1.0	1.0	0.5	0.75
Polyol	Sorbitol/10 PrO	Glycerol/14 PrO adduct	Trimethyol propane/14 PrO adduct	Trimethylol propane/16 PrO adduct
Moles used	1.0	1.0	0.5	0.75
Grams used	680	871	465	772
Acid(s) in initial charge	(1)	(2)	(2)	(2)
Moles used	5.5	2.5	1.25	1.875
Grams used	787	500	250	375
Benzene sulphonic acid (g. of 33% solution)	11.1	31.8	12.0	28.5
Entrainer used	(3)	(3)	(3)	(3)
Grams used	220	137	230	230
Butyric acid (added later):				
Moles	1.0	0.5	1.25	0.375
Grams	88	2×44	110	2×33
Total reaction time (hrs.)	47	58	18	22
Reaction temp. range (° C.)	150–170	180–190	150–160	136–168
Additional entrainer added (ml.)	350	Nil	50	40
Analysis of product:				
Acid number	0.34	0.36	0.91	0.31
Ester number	262	184.8	136	119
Hydroxyl number	1.5	4.6	5.8	2.4
Water content (percent)	0.05	0.092	0.07	0.042
SG at 20° C	0.987		0.969	0.970
RI at 25° C			1.448	
Sodium content (p.p.m.)	20	8	2	Nil
Pour point (° F.)	−58	−27	−45	−51
Viscosity and miscibility characteristics:				
Viscosity at 210° F. (cs.)	6.85	7.95	11.40	12.01
Viscosity at 100° F. (cs.)	37.61	39.65	67.37	67.16
Viscosity index	142	153	142	146
Compatibility 1:1:				
With Pool 12	Yes	Yes	Yes	Yes
With Brightstock	Yes	Yes	Yes	Yes
Carbon to oxygen ratio	3.73	3.85	4.0	3.91

1 Mixed (C^{5}–C^{11}).
2 Lauric.
3 Toluene.

Wear Load Characteristics — A sample of the product of Example 6 was tested using the Shell four ball tester and steel on steel specimens. For comparison, the test was run against an SAE 20, high VI aliphatic mineral oil. The results are given in Table 3 below.

TABLE 3

	Product of Example 6 Wear Scar diameter (mm.)	SAE 20 mineral oil Wear Scar diameter (mm.)
Load (kg.):		
45	0.339	0.345
89	2.17	2.36
112	2.29	2.39
141	(1)	(1)
158	(2)	(2)

1 Catastrophic damage.
2 Weld point.

From these tests it is considered that the two fluids are practically identical in wear load characteristics.

Viscosity Characteristics After Shearing — Shear resistance was assessed by measuring the viscosity index before and after shearing. Shearing was carried out by circulating the fluid through a diesel fuel injector for 45 minutes. It was calculated that the time of circulation was equivalent to 200 passes through the injector. The fluids tested were the product of Example 6 and a solvent neutral 150 oil containing 5% of a commercial VI imprcver, Hitec 960. The results are given in Table 4 below. It will be noted that the effect of shearing on the viscosity and viscosity index of the lubricants was very much greater with solvent neutral 150 + Hitec 960.

TABLE 4

	Product of Example 6		Solvent neutral 150 +50% Hitec 960	
	Before shearing	After shearing	Before shearing	After shearing
Viscosity at 210° F. (cs.)	11.40	10.78	10.74	7.12
Viscosity at 100° F. (cs.)	67.37	64.72	56.55	41.58
Viscosity index	142	141	150	134

Polyol Esters of Fatty Acids

B.H. Kress and D.E. Alexander; U.S. Patent 3,526,596; September 1, 1970; assigned to Quaker Chemical Corporation have found that the polyol esters of fatty acids having 12 to 22 carbon atoms, or mixtures, are excellent lubricants for hot metal working operations. The preferred polyols have 2 to 12 hydroxy groups, including for example, pentaerythritol or polymers, trimethylol ethane, trimethylol propane and glycols containing from 2 to 40 carbon atoms. Among the fatty acids useful for making the polyol esters are the following.

(A) Natural fatty acids both saturated and unsaturated containing 12 to 22 carbon atoms including the following acids: lauric, lauroleic, myristic, myristoleic, arachidic, eicosapolyenoic, pentadecanoic, palmitic, palmitoleic, margaric, stearic, oleic and linoleic.

(B) Acidic phosphate esters of ethylene oxide condensates. These condensates may include those with fatty alcohol, phenolic and substituted phenolic, oxo process alcohol and acetylenic alcohol hydrophobes. The hydrophilic portion may include chains of ethylene glycol moieties, propylene glycol moieties or those containing a mixture thereof. The acid portion is contributed by an unesterified phosphoric acid group and may be neutralized with ammonia, amine, an alkali metal hydroxide, an alkaline earth hydroxide or unneutralized.

(C) Fatty acid modified alkyd resins. These include condensation products of fatty acids, usually mixtures, polyol or polyols, and a dicarboxylic acid. These include phthalic, isophthalic, terephthalic, maleic, fumaric, tetrahydrophthalic, succinic, adipic, azelaic, sebacic acids, and condensation products of maleic anhydride with various compositions such as rosin.

(D) Oxyalkylene copolymers. These copolymers for example, are formed by interaction between ethylene oxide and propylene oxide, either as random copolymer, or as grafts. Both of these classes may be formed by reacting the oxide with a compound containing a reactive hydrogen such as butanol or similar composition.

Example 1:

	Parts
Tallow fatty acids	840
Trimethylol ethane	120

These components were mixed and heated with nitrogen passing through the bottom of a reaction flask to remove water of reaction. Temperatures of 500° to 525°F. were used with agitation of the reaction mixture. When about 54 parts of water had been removed the reaction was stopped. The product was brown liquid consisting essentially of tritallowate of trimethylol ethane.

Example 2:

	Parts
Oleic acid	1,120
Pentaerythritol	136
Toluene sulfonic acid	0.1
Xylene	50

This mixture was heated at temperatures of up to 400°F. while water of reaction was distilled off azeotropically, the xylene layer being returned continuously. When 70 parts of water of reaction had been removed, the xylene was evaporated in vacuo leaving a yellow liquid product which was essentially tetraoleate of pentaerythritol.

Example 3:

	Parts
Tall oil-distilled	855
Trimethylol propane	135
Boron trifluoride etherate	0.1

This mixture was heated to 325°F. and maintained at this temperature while water of reaction was distilled off. When 52 parts of water had been removed, vacuum being used at the end of the reaction, the product formed was a yellow oil which was essentially tritallate of trimethylol propane.

The lubricants are useful in the continuous production of hot-rolled flat steel plate, strip, sheet and bars and other metals and alloys such as copper, brass, aluminum, magnesium, titanium, etc. The lubricants are also useful in other hot metal working operations such as extrusion, shaping, forging, continuous casting, etc. The lubricants have the property of wetting and spreading on the surface of the working elements as well as on the metal being worked to provide a boundary medium between the surface of the working elements and the metal being worked.

Esters of Tetramethyltrihydroxybiphenyl

D.W. Fox; U.S. Patent 3,336,350; August 15, 1967; assigned to General Electric Company describes functional fluid compositions consisting of triesters of substituted trihydroxybiphenyls and mixtures of triesters of substituted trihydroxybiphenyls with other functional fluids.

The triesters of trihydroxybiphenyls are relatively nonvolatile, pale yellow liquids having excellent hydrolytic stability and lubricity. They are compatible with a large number of other functional fluids and contribute to reduced volatility, increased lubricity and improved hydrolytic stability when used in combination with other functional fluids. The quantity of triester that may be mixed with other functional fluids depends upon the compatibility of the triester and the other functional fluid. For example, with diester functional fluids, the triesters are compatible over the entire concentration range. The same is true with the methylphenyl silicones. In addition, the triesters are excellent plasticizers for many polymeric compositions.

Example 1: A triester was formed by reacting 38.4 g. (0.1 mol) of 2,2,4'-triacetoxy-3,3',5,5'-tetramethylbiphenyl with 129.6 g. (0.9 mol) of caprylic acid. The mixture of triacetate and caprylic acid was heated under a nitrogen blanket at a temperature of approximately 244°C. for a period of 1 hour. The temperature was then elevated slightly and heating was continued for an additional 9 1/2 hours. The temperature fluctuated between 245° and 270°C. during this period.

The mixture was allowed to cool and then subjected to distillation to remove acetic acid. Distillation was first conducted at atmospheric pressure up to a temperature of 245°C. and then at 20 mm. to a temperature of 125°C. The residue was 79.4 g. This residue was then further distilled up to a pot temperature of 250°C. at 0.25 mm. leaving 38.5 g. of an amber oil. This oil was treated with fuller's earth and passed through a column containing alumina and charcoal to improve its color. The oil was identified as the tricaprylate of 3,3',5,5'-tetramethyl-2,4,4'-trihydroxybiphenyl.

Example 2: In this example, 20 g. of 3,3',5,5'-tetramethyl-2,4,4'-triacetoxybiphenyl were mixed with 88 g. of 2-ethylhexanoic acid. The reaction mixture was refluxed under a nitrogen blanket at a temperature of approximately 233°C. for a total of 46 hours. During the course of the reaction, samples of distillate were removed from the reflux condenser and identified as acetic acid. The liberation of acetic acid indicates triester formation. Following the reaction period, a vacuum strip was performed on the residue remaining in the reaction vessel. Acetic acid and excess ethylhexanoic acid were removed. Following the vacuum strip, 30.52 g. of a pale yellow liquid residue were left in the reaction vessel. This was identified as the triethylhexanoate of 3,3',5,5'-tetramethyl-2,4,4'-trihydroxybiphenyl.

Example 3: The procedure of Example 2 was repeated, but 5 drops of boron fluoride ethyl ether complex were added as a catalyst. The reaction was allowed to proceed for a total of 22 hours. The residue after vacuum stripping amounted to 20.82 g. and was identified as the triethylhexanoate of 3,3',5,5'-tetramethyl-2,4,4'-trihydroxybiphenyl.

Example 4: The procedure of Example 2 was repeated but 1 1/2 g. of toluene sulfonic acid were used as the catalyst. The reaction mixture of 20 g. of 3,3',5,5'-tetramethyl-2,4,4'-triacetoxybiphenyl and 88 g. of 2-ethylhexanoic acid was refluxed at a temperature of approximately 200°C. for approximately 1 hour and 55 minutes. The residue was vacuum stripped and 22 g. of 2-triethylhexanoate of 3,3',5,5'-tetramethyl-2,4,4'-trihydroxybiphenyl remained in the reaction vessel. This example indicates the effectiveness of the toluene sulfonic acid as a catalyst in this type of reaction.

Example 5: A 50-50 mixture of tri-2-ethylhexanoate of a 3,3',5,5'-tetramethyl-2,4,4'-trihydroxybiphenyl and a methyl phenyl silicone oil was prepared and tested for volatility. A sample of a commercial diester fluid identified as Emery 3383-D was tested in the same manner for purposes of comparison. After heating for 120 hours at 250°C., the triester-silicone sample had lost approximately 44% of its total weight while the Emery sample lost over 73% of its total weight.

Example 6: This sample illustrates the effectiveness of the triesters as additives for improving the lubricity properties of other functional fluids. Samples were prepared consisting of a methyl phenyl silicone oil and a tri-2-ethylhexanoate of a 3,3',5,5'-tetramethyl-2,4,4'-trihydroxybiphenyl. The mixtures were subjected to a Shell four-ball wear scar test. Spindle speed was 600 rpm and the load was 50 kg. The test was run at room temperature and a ball having a size of 0.52/0.52 was used. The following results were obtained.

TABLE: LUBRICITY TEST FOR TRIESTER-SILICONE MIXTURES

Composition (Wt. Percent)		Wear Scar Diameter (mm.)
Silicone	Triester	
100	0	>3.50
80	20	3.20
50	50	0.61
20	80	0.56
0	100	0.70

A value of 0.50 to 0.70 constitutes an acceptable value. The smaller the wear scar, the better the lubricity of the compound being tested. From the above, it is apparent that addition of triester substantially reduces the wear scar diameter of the silicone oil. Mixtures of 50% silicone oil and 50% triester allow for a wear scar diameter of only 0.61 mm. as compared with greater than 3.5 mm. for the wear scar diameter of the pure silicone oil.

Resorcinol Esters

W.H. Voris; U.S. Patent 3,347,884; October 17, 1967; assigned to Koppers Company, Inc. have found that the ester of resorcinol still residue is a good synthetic lubricant. Resorcinol still residue is the black, tarry, resinous material which remains as a residue after the technical grade resorcinol has been removed, as a distillate, from the reaction mass that is produced in the manufacture of resorcinol by the fusion of sodium meta-benzene disulfonate with caustic soda. The resorcinol still residue comprises from 5 to 30% by weight of the reaction mass.

The resorcinol still residue is a mixture of high molecular weight compounds. The compound that is found in a predominate amount is isomeric trihydroxydiphenyl with small amounts of dihydroxydiphenyls. The compound occurring in the greatest known amount is 2,3',4-trihydroxydiphenyl

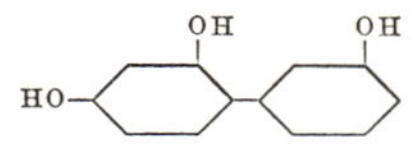

The dihydroxydiphenyl is usually the 3,4' isomer.

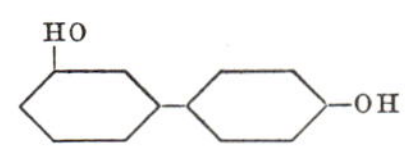

The useful compounds of resorcinol still residue are made by esterifying the resorcinol still residue with an alkanoic acid having from 2 to 20 carbon atoms. The process is illustrated by the following examples.

Example 1: Pivalic acid (250 g., 2.45 mols) was charged to a 1 l., three-necked flask equipped with a stirrer, thermometer, reflux condenser and dropping funnel. Thionyl chloride (574 g., 482 mols) was then added through the dropping funnel over a period of 1 hour while the reaction mass was stirred and heated to maintain it at a temperature of 20° to 32°C. The temperature of the mass was then maintained at 20° to 40°C. and the stirring was continued for a period of 6 1/2 hours. Then, the temperature of the reaction mass was raised over a period of 2 hours to reflux at 89°C. The product was fractionally distilled to yield 235 g. (80% of theory) pivaloyl chloride, BP 103° to 105°C. at atmospheric pressure.

Resorcinol still residue was subjected to vacuum distillation in a Rodney-Hunt evaporator until 40% by weight of the material had been removed as overhead. This overhead of resorcinol still residue (60.2 g., 0.8 eq.), toluene (324 ml.), and dimethyl formamide (6 ml.) were charged to a 1 l., three-necked flask equipped with a stirrer, thermometer, reflux condenser, and dropping funnel. The liquid was heated to 84°C. and pivaloyl chloride (106 g., 0.88 eq.) added over a period of 15 minutes while the reaction mass was stirred and maintained at a temperature of 84° to 70°C.

Thereafter, heating of the reaction mass was then continued for 5 1/4 hours while the temperature was maintained at 70° to 80°C. The mass was cooled to 20°C., and 60 ml. pyridine was added. The heating was again continued for 4 1/2 hours with the temperature at about 106°C. The reaction mass was then cooled. The reaction solution was washed six times with 60 ml. of 12.5% HCl each time, then seven times with 120 ml. of 20% KOH each time and finally five times with 140 ml. of water each time. The final water wash was neutral to pH paper.

Toluene was removed from the reaction mass by vacuum distillation. Then the distillation was continued until the temperature of the residue reached 203°C. at 3.5 mm. Hg. Only a few drops of distillate with a camphor-like odor appeared, identified by IR and nuclear magnetic resonance (NMR) to contain N, N-dimethylneopentanamide. The tripivalate of resorcinol still residue which remained as the residue (66.01 g., 51.7% yield) was a dark brown liquid whose viscosity was such that it barely flowed at room temperature.

Example 2: Neoheptanoic (2,2'-dimethylpentanoic) acid (520 g., 4.0 mols) was charged to a 3 l., four-necked flask. Thionyl chloride (769 g., 6.4 mols) was added over a period of 70 minutes while the solution was being stirred and maintained at a temperature of 30° to 40°C. The temperature of the reaction mass was then raised over a period of 2 1/4 hours to reflux at about 75°C. and maintained at 65° to 70°C. for 8 hours. Fractional distillation of the crude product yielded 488 g. (82% of theory) neoheptanoyl chloride, BP 147° to 152°C. at atmospheric pressure.

The overhead from resorcinol still residue as obtained in Example 1 (22.6% OH, 75.2 eq. wt.) (89.2 g., 1.19 eqs.), xylene (410 ml.) and dimethyl formamide (8 ml.) were charged to a 1 l., three-necked flask equipped with a dropping funnel, thermometer, a reflux condenser and stirrer. The mixture was heated to 70°C. and neoheptanoyl chloride (196 g., 1.32 eqs.) added through the dropping funnel over a period of 35 minutes as the mass was being stirred and its temperature was being maintained at 70° to 90°C. Then the reaction mass was maintained at a temperature of 90° to 121°C. for 5 hours. Thereafter, the mass was cooled to room temperature and 80 ml. of pyridine was added. The reaction mass was heated and the temperature kept at 100° to 120°C. for 2 hours. The mass was then cooled to room temperature.

The reaction mass was washed in a separatory funnel with four successive portions, 190 ml. each, of 12.5% HCl, and then eight times with 125 ml. each time of 20% KOH, and finally four times with 175 ml. portions of water. The final wash was neutral to pH paper. The xylene was removed by vacuum distillation. The distillation was continued and 5.7 g. (3%) of a colorless, very fluid distillate with a camphor-like odor was obtained at a final pot temperature of 210°C. at the subatmospheric pressure of 5 mm. Hg. This distillate was identified by IR and NMR as consisting mostly of N, N-dimethylneoheptanamide. The trineoheptanoate residue (201 g., 91.8% yield) was a dark, transparent, viscous oil. (Resorcinol still residue trineoheptanoate.)

A quantity (85.2 g.) of the trineoheptanoate was dissolved in 500 ml. acetone and was stirred with 10 g. Darco activated carbon for 10 minutes. The mixture was then filtered through Whatman No. 42 filter paper. This treatment with Darco carbon was repeated, the carbon filter cakes were extracted with 50 ml. acetone, and the combined acetone solution evaporated on the steam bath to a pot temperature of 100°C. at 2 mm. Hg. Trineoheptanoate of a light yellow color was recovered in an amount of 83.5 g., or 98% of the charged crude product.

Another series of experiments was conducted in which the crude trineoheptanoate was dissolved respectively in benzene and in ether and activated carbon was added afterwards. Such treatments changed the color of the trineoheptanoate from the original dark brown to a pale yellow. Acetone and ether appeared to be equally effective and both were slightly better than benzene as the solvent to provide a product with the best color.

Example 3: Resorcinol still residue (20.1% OH, eq. wt. 84.5) (84.5 g., 1.0 eq.), xylene (410 ml.), and dimethyl formamide (6 ml.) were charged to a 1 l., three-necked flask. The mixture was heated to 94°C. and neoheptanoyl chloride (163.5 g., 1.1 eqs.) then added through a dropping funnel over a 30 minute period while the liquid was stirred and its temperature was maintained at 94° to 100°C. The temperature was then raised to 131°C. and maintained there for 1 1/2 hours. The mass was allowed to cool overnight and pyridine (90 ml.) was added. The material was again heated and the temperature maintained at 90° to 124°C. for 3 hours. The material was cooled to room temperature. The xylene solution was washed in portions, with 900 ml. water in five portions, then with 1,200 ml. of 20% KOH in twelve equal portions, and thereafter with 900 ml. water in five portions. The final wash was neutral to pH paper.

The xylene was removed from the mass by vacuum distillation. The temperature was increased and 2.4 g. (2%) of the colorless distillate with a camphor-like odor was collected at a maximum pot temperature of 120° to 206°C. (max.) and a subatmospheric pressure of 1.5 mm. Hg. The trineoheptanoate residue (141 g. of 71.5% yield) was a clear brown, viscous liquid.

A 50.3 g. quantity of trineoheptanoate residue was dissolved with heat in 500 ml. acetone. The 10 g. activated carbon was stirred with the solution for 10 minutes, and the mixture filtered through Whatman No. 42 filter paper. The filtrate was a deep red solution. Another treatment of the filtrate was made with activated carbon, and the color of the solution changed to a dark red.

After four more repeat treatments with activated carbon, the color changed to a light red; the last treatment, however, resulted in little change in color. The filter cakes were washed with 50 ml. portions of actone and the combined acetone solution evaporated to a pot temperature of 100°C. at 3 mm. Hg. The resorcinol still residue trineoheptanoate had an amber color and amounted to 41.5 g., or 83% of the crude product.

Example 4: Neotridecanoic acid (689 g., 3.1 eq. wts.) was charged to a 2 l., three-necked flask equipped with a stirrer, dropping funnel, thermometer, and reflux condenser. Thionyl chloride (786 g., 6.6 mols) was added through the dropping funnel over a period of 2 3/4 hours while the mass was being stirred and maintained at a temperature of 32° to 40°C. The temperature of the reaction mass was then raised over a 2 hour period to reflux at about 80°C. and was maintained at 79° to 87°C. for 12 hours longer. Excess thionyl chloride was removed by distillation under a 10 inch Vigreaux column at atmospheric pressure. The crude product was then distilled at reduced pressure to yield 483 g. (64.6% of theory) neotridecanoyl chloride, BP 138° to 144°C. at 24 to 26 mm. Hg (distillation being discontinued because of bumping). A residue of 258 g. (34.6% of theory for neotridecanoyl chloride) remained.

The overhead of resorcinol still residue obtained as in Example 1 (75.2 g., 1.0 eq.), toluene (400 ml.), and dimethyl formamide (15 ml.) were charged to a 1 l., three-necked flask. The mixture was heated to 86°C. and neotridecanoyl chloride (229 g., 0.96 eq.) added through a dropping funnel over a 30 minute period while the mass was being stirred and was being maintained at this temperature. Heating was continued for 2 1/4 hours more keeping the temperature at 87° to 110°C. An immiscible dark red liquid formed which rimmed the flask wall at the liquid surface. The mixture was cooled to 25°C. overnight and pyridine (85 ml.) then added. The mass was heated to a temperature of 110°C. over a period of an hour and the temperature was kept at 110° to 113°C. for 4.3 hours more. The mass was cooled to room temperature.

The dark solid on the bottom of the flask was insoluble in the 600 ml. of 12.5% HCl used in three portions to wash the reaction mixture. Washing with 750 ml. of 20% KOH in seven portions, dissolved the solid phase; very dark emulsions formed at first. Water washing with 800 ml. in seven portions provided a final wash neutral to pH paper. Toluene was removed by vacuum distillation. The distillation was continued and 6.0 g. (3%) of yellow, oily distillate was collected by heating the still pot to a final temperature of 200°C. at about 1 mm. Hg. The trineotridecanoate residue (209 g., 88% yield) was a dark brown, transparent, viscous liquid (resorcinol still residue trineotridecanoate). The properties of the products from Examples 2 through 4 are tabulated below.

TABLE 1

	Viscosity, cs.		ASTM Slope	Viscosity Index	13,000 cs. T., °F.	Pour Point T., °F.	Thermal Stability TGA, T., °F.			Refractive Index
	100° F.	210° F.					10%	50%	100%	
Ex. 2	455	18.4	0.80	8	44	20	518	630	--------	1.5050
Ex. 3	2,180	47	0.75	28	70	35	550	667	752	1.5075
Ex. 4	7,276	77	0.79	--------	92	48	------	------	--------	1.5012

An important characteristic of fluids, of course, is their lubricity. Accordingly, the products of Examples 2 and 3, as set out below were tested in a conventional Shell Four-Ball Wear Tester under test conditions which included operating for 1 hour at a temperature of 167°F. and a speed of 620 rpm with 52 to 100 steel bearings under loads of 1, 10 and 40 kilograms. The results of the product with and without an additive (0.5 weight percent diisopropyl phosphite) are compared below with a standard ester lubricant di-2-ethylhexyl sebacate.

TABLE 2

Fluid	Average Wear Scar Diameter, mm.		
	1 kg.	10 kg.	40 kg.
Ex. 2	0.19	0.27	1.05
Plus Additive	----------	----------	0.40
Ex. 3	0.20	0.22	0.82
Plus Additive	----------	----------	0.45
Standard	0.43	0.56	0.80

Benzhydryl Esters of Dimer Acid

In a process described by W.E. Parker, H.B. Knight, R.E. Koos and W.C. Ault; U.S. Patent 3,393,214; July 16, 1968; assigned to the U.S. Secretary of Agriculture benzyl and benzhydryl esters of the dimerized linoleic acid and hydrogenated dimerized linoleic acid are prepared by reacting the dimer acid with the appropriate alcohol in the presence of an acid catalyst. The esters were evaluated and found to have viscosity characteristics, lubricating properties and thermal stabilities desired in lubricants, lubricant additives and hydraulic fluids.

In a typical procedure for preparation of the esters, a ratio of 2 mols of alcohol to 1 mol dimer acid and a small amount of an acid catalyst were dissolved in an inert organic solvent such as benzene and refluxed until the reaction was complete, as indicated by the amount of water collected. The benzene solution was washed untill free of acid, dried over calcium sulfate, the benzene removed, and the crude product purified by molecular distillation. It has been found that the dilinoleate esters of the process are liquids, even though total molecular weight of a compound is as high as about 900.

The dimer acid employed in the preparation of the esters contained 95% dilinoleic acid, 4% trimer acid and 1% monobasic acid and had characteristics of acid number, 191.2; neutralization equivalent, 293.4; and molecular weight, 565. The hydrogenated dimer acid had an acid number of 192.8; neutralization equivalent, 288.0; and iodine number 29.0. The alcohols were commercial products and only the anisyl alcohol was redistilled before use. Best results were obtained with the use of naphthalene-2-sulfonic acid as the catalyst. Naphthalene-1-sulfonic acid, methane sulfonic acid and sulfuric acid gave poorer results than naphthalene-2-sulfonic acid. Preparation of the esters of the process is illustrated by the following examples.

Example 1: Dimer acid, 149 g., benzyl alcohol, 58 g., and 3 g. naphthalene-2-sulfonic acid were combined with 200 ml. benzene in a 1 l., three-necked flask. The flask was equipped with a thermometer, a magnetic stirrer, and a Dean and Stark tube with a water cooled condenser. The mixture was heated at 90°C. reflux temperature until the calculated amount of water (9.5 ml.) was separated, reaction time 3 hours. After cooling the reaction mixture was washed until acid-free. The benzene solution was dried over calcium sulfate, filtered, and the benzene evaporated under vacuum. The crude product, an amber fluid, weighed 189 g., a yield of 95%. The product has an acid number of 0, and infrared spectra indicated absence of acid and presence of ester carbonyl. The crude product was purified by molecular distillation, the main fraction being collected between 280° and 300°C. at 9 to 15 microns pressure. This fraction weighed 156.2 g. and had a molecular weight of 750.

Example 2: The process of Example 1 was repeated except that 151 g. of hydrogenated dimer acid was used in place of the dimer acid and the mixture was refluxed 4 hours. Yield of crude ester was 90.4%.

Example 3: Following the typical procedure, 282.5 g. dimer acid, 150.2 g. p-isopropylbenzyl alcohol and 5.65 g. naphthalene-2-sulfonic acid were refluxed for 5 hours and the crude bis(p-isopropylbenzyl)dilinoleate was separated from the reaction mixture in 99% yield. The product was purified by molecular distillation.

Example 4: In a procedure similar to previous examples, dimer acid, 28.3 g.; benzhydryl alcohol (α-phenylbenzyl alcohol) 18.3 g., and 0.65 g. catalyst were refluxed 8 hours, the crude ester separated from the reaction mixture in 90% yield and the bis(benzhydryl)dilinoleate purified by molecular distillation.

Example 5: Dimer acid, 45.6 g.; p-nitrobenzyl alcohol, 25.0 g.; and 1.82 g. of naphthalene-2-sulfonic acid were combined in benzene solution and refluxed 8 hours, the crude ester separated from the reaction mixture (90% yield) and the bis(p-nitrobenzyl)dilinoleate purified by molecular distillation.

Example 6: Dimer acid, 10.2 g.; anisyl alcohol, 5.0 g.; and 0.4 g. catalyst were combined in benzene and refluxed 6 hours, the crude ester separated as described in Example 1 and bis(anisyl)dilinoleate purified by molecular distillation. Yield was 80%.

Example 7: As a control material, dibenzyl sebacate was prepared from sebacic acid and benzyl alcohol. The typical procedure for preparation of esters was followed. Physical and chemical data pertaining to the compounds of the examples were determined. Some of the data are presented in Table 1 below.

TABLE 1: CHARACTERISTICS OF DILINOLEATE ESTERS AND SEBACATE CONTROL

Ex. No.	Compounds	B.P.[1]		N_{D30}	Carbon, percent		Hydrogen, percent		Saponification Number		Saponification Equivalent	
		°C.	Microns		Calcd.	Found[2]	Calcd.	Found	Calcd.	Found	Calcd.	Found
1	Bis(benzyl)-dilinoleate	280	9	1.5028	81.03	81.11	10.35	10.54	151.4	146.4	370.6	383.2
2	Hydrogenated bis(benzyl)-dilinoleate	275	10	1.5013	80.59	80.68	10.52	10.56	150.6	147.0	371.8	381.5
3	Bis(p-isopropylbenzyl)-dilinoleate	300	10	1.5020	81.50	80.95	10.74	10.84	136.0	138.3	412.7	406.0
4	Bis(benzhydryl)-dilinoleate	(3)	-----	1.5295	83.36	82.72	9.48	9.53	125.6	126.2	446.8	444.6
5	Bis(p-nitrobenzyl)-dilinoleate[4]	>200	10	1.5177	72.25	72.33	8.98	9.32	135.0	139.5	415.6	403.0
6	Bis(anisyl)-dilinoleate	>225	6	1.5230	77.95	77.80	10.06	9.60	140.1	143.5	400.6	391.0
7	Dibenzyl sebacate[5]	180	7	[6] 1.5162	75.36	75.49	7.91	8.03	-----	-----	-----	-----

[1] Molecularly distilled. [2] Based on calcd. molecular weight of dilinoleic acid (M.W. 561), not on molecular weight derived from commercial dimer acid (M.W. 565). [3] Alcohol insol. fraction. [4] Nitrogen, percent calcd. 3.37; found, 3.12. [5] Control sample. [6] Solid at room temperature.

The compounds of the examples were then evaluated. Among the evaluations for usefulness of the compounds of the process were properties of thermal stability, viscosity characteristics and lubrication as indicated by wear scar test. In these evaluations bis(2-ethylhexyl)sebacate, dibenzyl-sebacate and paraffin oil having a viscosity of 100 Saybolt seconds at 100°F. were included as reference materials. The data are summarized in Table 2 below. Thermal stability was determined by thermogravimetry, heating the compound in a specially designed sample holder suspended from a weighing mechanism.

The sample holder was made from 4 mm. Pyrex glass tubing, 24 cm. long. A small bulb (capacity approximately 0.5 ml., 1.0 cm. diameter) was formed at one end of the tube and a flared lip was formed at the other. The size of the bulb may be varied for the weight of the sample being used. The bulb, with the above dimensions, will accommodate a sample weighing 100 mg. The sample was placed in the holder with a long, thin dropper, being careful not to deposit liquid on the wall of the tube. The holder was then attached to the eyelet of the weighing mechanism (support rod) with a hook made from 22 gauge nichrome wire. Using this arrangement, the sample holder tube extends approximately 7.5 cm. above the top of the furnace while the bulb is located approximately 6 mm. above the sample thermocouple well. This distance can be adjusted by the wire hook.

The samples were heated to 525° to 575°C. at a heating rate of 3°C. per minute in a dry air atmosphere which flowed at a rate of 10 ml. per minute at atmospheric pressure. Decomposition took place at a maximum of 1/3 of the distance up the sample holder tube as evidenced by observation of carbonaceous matter. Onset of decomposition is defined as the point where (a) the material shows the first detectable weight loss as determined by a deviation from a blank run and/or (b) the slope, dw/dT shows a radical change from some minimum value.

The viscosity index was determined according to the procedure given in ASTM D567-53, using semimicro viscometers. Lubricating property was evaluated using the four ball wear test as described by Peale et al., Lubrication Eng. Trans. 3, 48(1960). In the wear test, the smaller value denotes the best lubricating property. Several of the compounds were noticeably superior to the controls and the remainder compared favorably.

TABLE 2: THERMAL STABILITY, VISCOSITY, AND LUBRICATING PROPERTIES OF DILINOLEATE ESTERS AND CONTROLS

Ex. No.	Sample	Onset of Decomposition, Temp., °C.[1]	Viscosity Index	Wear Scar, Dia., mm.
1	Bis(benzyl)	384	124.3	0.653
2	Hydrogenated bis(benzyl)	392	126.0	0.748
3	Bis(p-isopropylbenzyl)	368	110.2	0.843
4	Bis(benzhydryl)	306	78.4	0.829
5	Bis(p-nitrobenzyl)	273	95.4	0.863
6	Bis(anisyl)	345	----	0.473
	Bis(2-ethylhexyl)sebacate	289	148.8	0.830
7	Dibenzyl sebacate	300	----	0.950
	100 Paraffin Oil		112.7	0.803

[1] Temperatures determined on average of 3 runs: Accuracy, ±4°.

Trimethylolpropane Trineoheptanoate

W.E. Taylor, H.H. Thigpen and E.R. Witt; U.S. Patent 3,341,574; September 12, 1957; assigned to Celanese Corporation have found that the ester trimethylolpropane trineoheptanoate is an excellent compound for use in synthetic high temperature lubricant compositions. For example, when compared to trimethylolpropane triheptanoate, it has been found that trimethylolpropane trineoheptanoate displays a greatly increased resistance to oxidation and undergoes thermal degradation at a temperature about 100°F. higher than does trimethylolpropane triheptanoate. In order to compare the oxidation resistance of trimethylolpropane trineoheptanoate to that of trimethylolpropane triheptanoate, each ester was maintained at 347°F. for 3 hours and then tested for the rate of oxygen absorption. The data from this test appears in the table below, along with the temperatures at which the respective esters display thermal degradation.

Ester	Temp. at which ester Undergoes Thermal Deg., °F.	O_2 Absorption Rate After 3 Hours at 347 in cc. of O_2 per minute
Trimethylolpropane trineoheptanoate	700	0.2
Trimethylolpropane triheptanoate	600	3.8

Example 1: Neoheptanoyl chloride is prepared by reacting equimolar proportions of neoheptanoic acid with phosphorus pentachloride. Then to 88 parts by weight (0.66 mol) of trimethylolpropane there is slowly added 321 parts by weight (2.2 mols) of the neoheptanoyl chloride over a period of 60 minutes while the mixture is maintained at 60°C. The mixture is then refluxed at 120°C. for 4 hours, after which the product is vacuum stripped free of all volatiles coming off up

to 230°C. pot temperature and 223°C. vapor temperature at 2 mm. Hg absolute. The ester product is then removed by distilling the residue from 230° to 244°C. pot temperature and 223° to 240°C. vapor temperature at a pressure of 2 mm. Hg absolute to produce 261 parts by weight of crude ester product. The product is diluted with an equal volume of toluene and washed five times with dilute caustic. Then the product was water-washed to neutrality. The washed ester was then dehydrated, treated with charcoal and filtered to give 208 parts by weight of finished material.

Example 2: Trimethylolpropane trineodecanoate was prepared by dissolving 67 parts by weight (0.5 mol) of redistilled trimethylolpropane and 4.7 parts by weight of dimethyl formamide in 86.5 parts by weight of toluene; heating the solution to 70°C.; adding 310 parts by weight (1.66 mols) of neodecanoyl chloride; heating the solution to 95°C. for 2 hours; cooling to room temperature; and then adding 49 parts by weight of pyridine. After permitting the reaction mixture to stand overnight, it was washed with hydrochloric acid, caustic and water; the solvent was evaporated, under vacuum; the product was decolorized with charcoal, and then filtered to give 240 parts by weight of product. The yield was 82% based upon the weight of trimethylolpropane charged.

INHIBITED FLUID FORMULATIONS

Arylamine, s-Triazine and Amine Salt of Phosphonic Acid

According to a process described by E. Wittner and K.T. Wendler; U.S. Patent 3,432,433; March 11, 1969; assigned to Shell Oil Company a synthetic ester lubricant having excellent oxidation stability which comprises a major amount of synthetic ester lubricating oil and minor amounts sufficient to increase the oxidation stability of the oil of an aryl amine, an amine salt of a monohaloalkylphosphonic acid and an s-triazine having the formula

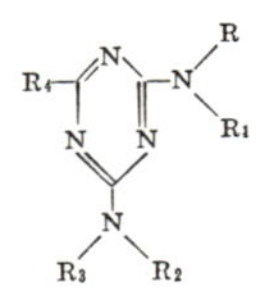

where R, R_1, R_2 and R_3 are hydrogen, C_1 to C_{20} hydrocarbyl, or pyridyl and R_4 is C_1 to C_{14} hydrocarbyl, C_1 to C_{20} hydrocarbylamino, pyridyl, or pyridylamino. R, R_1, R_2, R_3 and R_4 may all be different, or some or all may be alike.

To illustrate the efficacy of the additive combination of the process, air oxidation tests were conducted on samples of various ester-base lubricants. The tests were carried out at 450°F. by passing a stream of air at the rate of 4.2 l. per hour through a 20 g. sample of the appropriate ester fluid. To more closely approximate actual operating conditions, 20 ppm each of iron or iron and copper were added as octoates. The base stock for all tests was a mixture of 67% C_5 to C_9 acid esters of pentaerythritol and 33% C_5 to C_{10} acid esters of dipentaerythritol. Comparative runs were conducted using the base stock alone, others were conducted using the base stock plus one additive, i.e., the salt of monochloromethylphosphonic acid and C_{18-22} primary amine (MCMPA), phenyl-α-naphthylamine (PAN) or diphenylamine (DPA). Other runs combined the MCMPA with either PAN or DPA. The majority of the runs were made using as an oxidation inhibitor 1.0% weight of a triazine alone or in combination with MCMPA, PAN and DPA. In all cases, the tabulated results are mean values of duplicate tests.

TABLE 1: 450°F. MICRO AIR OXIDATION TESTS

s-Triazine	Oxidation Life, Hours					
	No Other Additive [1] (5)	0.2% Wt. MCMPA [1] (5)	1% PAN [1] (7)	1% DPA [1] (7)	0.2% MCMPA plus 1% Wt. DPA [1] (7)	0.2% MCMPA plus 1% Wt. PAN [1] (9)
None	2	-	-	2	2	5
2-amino-4-n-butylamino-6-phenyl-s-triazine	2	3	-	6	19	11
2,4-bis(n-butylamino)-6-phenyl-s-triazine	3	3	-	7	20	22
Mixture of 2-amino-4-sec-butylamino-6-phenyl-s-triazine and 2,4-bis-(sec-butylamino)-6-phenyl-s-triazine	3	2	-	-	16	-
2,4,6-tris(n-butylamino)-s-triazine	3	9	-	16	[1] (35) 51	[1] (30) 43
2-anilino-4,6-bis(n-butylamino)-s-triazine	2	3	-	8	[1] (35) 35	[1] (27) 33
2-n-butylamino-4,6-dianilino-s-triazine	3	5	-	6	[1] (32) 15	[1] (26) 13
2-(2-pyridylamino)-4,6-bis(n-butylamino)-s-triazine	-	-	-	-	[1] 36	-
2-methylamino-4,6-dianilino-s-triazine	-	-	-	-	[1] 31	-
2-n-octylamino-4,6-dianilino-s-triazine	-	-	-	-	[1] 27	-
2-sec-butylamino-4,6-dianilino-s-triazine	-	-	-	-	[1] 23	-
2-isobutylamino-4,6-dianilino-s-triazine	-	-	-	-	[1] 26	-
2-tertbutylamino-4,6-dianilino-s-triazine	-	-	-	-	[1] 19	-
2-anilino-4,6-dibenzylamino-s-triazine	-	-	-	-	[1] 29	-
2-amino-4-anilino-6-benzylamino-s-triazine	-	-	-	-	[1] 42	-

[1] 100 gm. sample, 9 liters/in air flow, other conditions same.

Results of the air oxidation tests are given in Table 1 above. The only runs not containing 1.0% weight triazine are those listed in the first line of Table 1.

Oxidation stability was determined by the length of time (oxidation life) for absorption of 3 millimols of oxygen per gram of oil. This criterion was used instead of the more conventional "induction period" measurement since oils of the process exhibited unconventional oxidation characteristics; in most cases there was a short period of rapid oxidation followed by a period in which little oxidation occurred, followed by another period of rapid oxidation.

These results show that the s-triazines themselves have essentially little or no antioxidant activity, but that they act to complement the aryl amine additives to impart improved stability to ester oils. In addition, it is apparent that the antioxidant activity of the triazine-arylamine combination is greatly enhanced by the presence of the MCMPA salt which is known as an extreme pressure agent, which by itself shows essentially no antioxidant activity. It is evident that the s-triazines, aryl amines and MCMPA salt coact in such a manner as to significantly enhance the antioxidant properties of the ester oils over the properties obtained using any of the additives alone or any of the additives in combination.

Additional oxidation tests were conducted to more accurately determine oxidation stability of oils of the process by measuring the acidity and viscosity increase of the oils under oxidative conditions. The base oil used for these tests was the same mono- and dipentaerythritol ester mixture described above. All samples contained 1.0% weight of a diarylamine (diphenylamine) and 0.2% weight MCMPA amine salt. In these tests, 9 l. per hour of air were flowed through a 200 g. oil sample at 450°F. for 40 hours. The oils contained either 10 ppm iron or copper catalysts, added as naphthenate, as indicated. The results shown in Table 2 below, are averages of duplicate determinations.

TABLE 2: SYNTHETIC OIL OXIDATION TESTS

s-Triazine	Iron Catalyst		Copper Catalyst	
	Δ Acid No.	Δ Vis at 100° F. percent	Δ Acid No.	Δ Vis. at 100° F., percent
None	27.3	294	23.7	198
2-amino-4-n-butylamino-6-phenyl-s-triazine	8.8	77	3.0	33
2,4-bis(n-butylamion)-6-phenyl-s-triazine	11.0	142	6.0	88
Mixture of 2-amino-4-sec-butylamino-6-phenyl-s-briazine and 2,4-bis(sec-butylamino)-6-phenyl-s-triazine	10.3	73	6.9	52
2,4,6-tris(n-butylamino)-s-triazine	9.9	54	5.4	50
2-anilino-4,6-bis(n-butyl-amino)-s-triazine	7.2	55	9.2	74
2-n-butylamino-4,6-dianilino-s-triazine	4.7	40	8.7	83

These results show that the triazines are insensitive to the presence of iron and copper oxidation catalysts. Tests using iron only as catalyst are considered to be very severe, more so even than combined iron-copper tests; nevertheless, triazine-containing lubricants performed very satisfactorily in these tests.

N-(C_1 to C_{14} Alkyl)-p,p'-Dioctyldiphenylamines

T.G. Davis and J.W. Thompson; U.S. Patent 3,368,975; February 13, 1968; assigned to Eastman Kodak Company describe the use of C_1-C_{14} N-alkyl-substituted-p,p'-dioctyldiphenylamines as antioxidants for lubricating oils, especially synthetic lubricants such as jet lubricants.

Example 1: N-ethyl-p,p'-dioctyldiphenylamine is prepared by the following procedure. 980 g. (2.5 mol) p,p'-dioctyldiphenylamine, 1,820 g. (15 mol) triethylphosphate and 50 g. p-toluene-sulfonic acid were combined in a 5 l. flask, heated between 190° to 200°C., with stirring, for 5 hours. The mixture was then cooled to 100°C. and poured into 4 l. of 88% isopropanol and stirred until the temperature reached 30°C. The solution was then chilled in a cold bath, and the solids filtered out. After standing approximately 72 hours the solid material was recrystallized twice from ethanol to yield 738 g. of tan colored crystals with a melting point of 66° to 67°C., 71% yield.

Example 2: N-methyl-p,p'-dioctyldiphenylamine is prepared by the procedure of Example 1 using trimethylphosphate as the alkylating agent.

Example 3: Table 2 on the following page, illustrates use of the N-alkyl-p,p'-dioctyldiphenylamine as antioxidant in a pentaerythritol mixed ester. A modified version of the Wright Air Development Division (WADD) oxidation test was used. The results demonstrate the improved oxidative stability of the high temperature antioxidants of this process over related additives. The antioxidants are also shown to be cleaner than those of the prior art. The original test is described in a paper entitled "Materials for High Temperature Jet Engine Lubricants" (H.E. Adams) in Wright Air Development Command (WADC) Technical Report 59-244 (U.S. Air Force, 1959). In the instant test a 250 ml. sample of lube oil was heated to 500°F. in a Pyrex test tube while 5 l. of air per hour were blown through the oil. Freshly polished 1" x 1" x 1/32" coupons of steel, aluminum, titanium and silver were assembled in accordance with method 5308.4 of Federal Test Method Standard 791 and immersed in the oil. The tube was capped with a Pyrex fitting which allows volatile products to escape. The oil was heated in 20 hour cycles. At the end of each cycle the tube was cooled, a 10 ml. oil sample withdrawn for analysis and the oil lost by evaporation replaced. The test was continued to approximately a 100% increase in viscosity. The 10 ml. sample was analyzed for viscosity in centistokes at 100°F., acid number and sludge in oil. The sludge was determined by filtering 2 ml. of the 10 ml. sample through Whatman No. 2 filter paper and washing with naphtha to remove the oil. The sludge on the pad was then rated according to a rating schedule where 0 is clean, 2 equals trace, etc., for increasing amounts of sludge. The results of this test are recorded in Table 1 below. Increase in antioxidant potency is shown by longer time to 100% viscosity increase.

TABLE 1: ANTIOXIDANT COMPARISON 500°F. IN WADC TYPE TEST[1]

Antioxidant at 2.0% w./w. in pentaerythritol tetra(octanoate/2-ethyl hexanoate 50/50	Hours to 100% viscosity increase	Varnish, deposits sludge rating
N-pheny-1-naphthylamine	45	Heavy.
p,p′-dioctyldiphenylamine	28	Do.
N-ethyl-p,p′-dioctyldiphenylamine	50	Light.
N-methyl-p,p′-dioctyldiphenylamine	40	Do.
Base oil, no antioxidant	25	Heavy.

[1] 250 ml. oil, 5 liters air/hr., Al, Fe, Ag, Ti, stainless steel.

Example 4: This example illustrates the use of the antioxidants where the oils have a tendency to form carbonaceous deposits or coke when in contact with a hot metal surface in the presence of air. This is an important factor in selection of antioxidants for oils to lubricate engines of a supersonic transport where heat rejection rates of the oil will be high due to increased engine temperatures at high speeds.

For this example a Model C panel coker was used with certain variations. The standard Model C panel coker test is outlined in the "Detailed Handbook on Test Procedures in Support of Turbojet and Turboprop Lubricants", WADC, March 1957. This procedure was varied to regulate the sump oil to a controlled 400°F.

Table 2 below, illustrates the benefits of use of the antioxidants in this type test. This example also shows the antioxidant utility of the compound in different esters and at different concentrations. It is to be noted that in each instance the N-ethyl-p, p'-dioctyldiphenylamine gave less coke than N-phenyl-1-naphthylamine or p, p'-dioctyldiphenylamine. The compound N-methyl-p, p'-dioctyldiphenylamine also gave a cleaner lubricant in this test. Results are essentially the same as for the N-ethyl and N-methyl product in the other base stocks.

TABLE 2: PANEL COKER TESTS AT 700°F. PANEL

Antioxidant at wt. percent	Coke on panel in grams	Sump and cover dirtiness rating
(A) Base-Pentaerythritol tetra(pelargonate/2-Ethyl hexanoate 75/25)+ 0.065% K(EDTA-Primene 81R Triamide)[1]		
2.5% N-phenyl-1-naphthylamine	0.398	Sump=Tr. dep., Cover=Light dep.
2.5% N-ethyl-p,p′-dioctyldiphenyl-amine.	0.052	Sump=Clean, Cover=Light.
B. Mixed ester base stock—Pentaerythritol tetra(pelargonate/2-Ethyl hexanoate 75/25)+25% Bis(2,2,4-trimethylpentyl)azelate+0.13% K (EDTA-Primene 81R Triamide)		
1.0% N-phenyl-1-naphthylamine	0.254	Sump=Light, Cover=Moderate.
1.0% p,p′-dioctyldiphenylamine (Vanlube 81).	0.255	Sump=Moderate, Cover=Moderate.
1.0% N-ethyl-p,p′-dioctyldiphenyl-amine.	0.155	Sump=Moderate, Cover=Light.
C. Base stock—Bis(2-ethylhexyl)sebacate "Plexol 201"		
2.5% N-phenyl-1-naphthylamine	0.482	Sump=Heavy, Cover=Heavy.
2.5% p,p′-dioctyldiphenylamine (Vanlube 81).	0.243	Sump=Clean, Cover=Moderate.
2.5% N-ethyl-p,p′-dioctyldiphenyl-amine.	0.062	Do.
2.5% N-methyl-p,p′-dioctyldiphenyl-amine.	0.066	Do.

[1] Primene 81R is a tert-alkyl primary amine (C 12–15).

Alkali Metal and Aromatic Amines

T.S. Chao, H.J. Matson and M. Kjonaas; U.S. Patent 3,436,348; April 1, 1969; assigned to Sinclair Research, Inc. describe an ester base synthetic lubricant having superior resistance toward oxidative and thermal degradation. It is a further object to protect and improve this resistance against the detrimental effects of metals with the use of suitable additives.

The process provides ester-based lubricant compositions containing small amounts of an aromatic amine and an alkali metal which are effective in retarding oxidation. In order to be dissolved in the base oil, the alkali metal is added as an alkali metal organic compound. Thus, the active component in combination with the aromatic amine is a soluble form of an alkali metal such as Na, K, Li, Cs and Rb. Typical organic compounds which form suitable alkali metal derivatives are carboxylic acids, phenols, alcohols, ketones, aldehydes, various chelating agents, etc., often of about 2 to 50 or more carbon atoms, preferably about 4 to 30 carbon atoms. Although the alkali metal is an essential active component, it has been found that certain classes of alkali metal derivatives or salts of acidic organic compounds are especially effective in enhancing oxidation inhibition. Exemplary of these acidic organic compounds are unsubstituted and S, O, N and halogen substituted carboxylic acids, phenols, including thiophenols, beta diketones and Schiff's bases.

The carboxylic acids which may be employed to make the alkali metal salts include mono- and polycarboxylic acids, for instance, those having straight, branched or cyclic hydrocarbon structures, including alkylated benzoic acids. The carboxylic acids may be saturated or unsaturated and may often contain from about 4 to 22 or even up to 50 or more carbon atoms. Examples of suitable fatty acids are caprylic, isodecanoic, isostearic, cerotic, myristic, cyclohexanoic, cyclooctanoic and ethyl cyclooctanoic acids. A group of monocarboxylic acids which can be used is represented by the formula RCOOH where R is an alkyl, aryl, aralkyl or alicyclic hydrocarbon radical of 3 to 21 carbon atoms.

The amine and alkali metal additives are incorporated in the base oil in small amounts sufficient to retard oxidation of the oil at temperatures in excess of 400°F. and the concentrations employed for optimum results may be dependent on the particular base oil and second additive component selected. The amount of the additives is insufficient to destroy the fluidity of the base ester oil, i.e., the oil remains essentially liquid at normal conditions and the alkali metal component is not used in an amount which produces a grease or gel-like structure. Ordinarily about 0.005 to 5%, preferably about 0.01 to 2%, by weight of the alkali metal derivative provides satisfactory results. The amine additive component provides satisfactory results when present in amounts of about 0.05 to 5% and preferably about 0.1 to 2%. The effectiveness of the organic alkali metal derivatives is shown by the following example.

Example: The following lubricant formulation was prepared by mixing the components at 150° to 160°F. The sodium perfluorobutyrate was first dissolved at 250°F. in about 100 times its weight of Hercolube A and then blended with the other components.

	Weight Percent
Hercolube A	71.99
Hercolube F	24.25
Phenyl-α-naphthylamine (PAN)	1.00
Di(p-octylphenyl)amine	1.00
Tetra-n-butyl ester of ethylene diamine tetraacetic acid	1.50
Half amide of hydrogenated dimer of linoleic acid***	0.20
C_3F_7COONa	0.06
Dow Corning fluid (200-60,000)*	+0.001

The sample was filtered through paper and submitted for tests indicated. The results were as follows.

(1)	O_2 Absorption Test, 450°F., 1 ft.^{3} O_2/hr., 75 g. sample.	
	T_i**, minutes	1,145
	T_t**, minutes	1,145
	V_t**, mls.	2,500
(2)	Type II Bearing Rig Test:	
	Acid No., final	0.93
	Viscosity rise, final, (KV/100), %	50.0
	WADD Demerit rating	58
	100 mesh filter deposit, g.	1.24
(3)	Ryder Gear Scuff Test	
	Scuff load:	
	A side, lb./in.	2,590
	B side, lb./in.	2,860
	Relative rating, %	100
(4)	Gear Fatigue Test — more than 110 hours.	
(5)	Corrosion-Oxidation Test (425°F., 48 hours)	

Weight change, mg./cm.^{2}:

Cu	-0.30	Al	+0.007	Viscosity rise, (KV/100), %	22.2
Ag	+0.024	Mg	-0.77	Acid No. increase	1.27
Steel	+0.054	Ti	+0.038		

(6) SOD Pb corrosion, 325°F., 1 hr., $mg./in.^2$, +6.48, +4.46.
(7) Carbon seal wear, in./min., 0.017×10^{-8}
(8) Inspection tests:

KV/210°F., cs.	5.104
KV/100°F., cs.	27.63
KV/-40°F., cs.	9.824
Flash point, °F.	495
Pour point, °F.,	<-80
Acid No. (pH 11)	0.3
Foam Test	0-0-0 0-0-0 0-0-0
Percent swell, H rubber	26.1
Percent swell, Viton	18.4
Durometer, Viton	65
Evaporation loss, 6.5 hours:	
400°F., sea level, %	3.41
450°F., 5.5" Hg, %	18.62
Percent sonic shear	0
-40°F., 72 hour storage	Clear and fluid

*An antifoam silicone antifoaming agent having a viscosity of 200 cs. at 250°C. and a molecular weight of 60,000.
**Described in U.S. Patent 3,493,510 below.
***An extreme pressure agent.

J.W. Thompson and T.G. Davis; U.S. Patent 3,347,791; October 17, 1967; assigned to Eastman Kodak Company describe an oil composition particularly useful as a lubricant at temperatures as high as 400° to 550°F. The oil composition consists essentially of a synthetic ester type lube oil containing a protective concentration of a mixture consisting essentially of an arylamine portion that consists essentially of at least one arylamine and, as a high temperature activator for the arylamine portion, at least one alkali metal salt selected from the group consisting of alkali metal salts of carboxylic acids and alkali metal salts of hydroxy aromatic compounds. An especially preferred combination of arylamines is N-phenyl-1-naphthylamine and 5-ethyl-10,10-diphenylphenazasiline at about a 2:1 weight ratio.

Preferred types of alkali metal salts of this process and the order of their preference are: alkali metal salts derived from amino carboxylic acids, alkali metal salts of naphthenic acids, alkali metal salts of branched chain aliphatic monocarboxylic acids, alkali metal salts of straight chain aliphatic monocarboxylic acids, alkali metal salts of substituted phenols, alkali metal salts of carboxy terminated polyesters and alkali metal salts of dibasic acid half esters.

N-Phenylnaphthylamine and Tetrahydroxy Benzophenone

In a process described by T.S. Chao; U.S. Patent 3,493,510; February 3, 1970; assigned to Sinclair Research, Inc. normally liquid ester base oil lubricating compositions of high temperature oxidative resistance are provided containing minor amounts, effective to retard oxidation, of an ester base oil-soluble aromatic amine, such as N-phenyl-1-naphthylamine or p,p'-dioctyl-diphenylamine, and particular aromatic ketones having at least 3 hydroxy groups on the aromatic ring, such as 2,2',4,4'-tetrahydroxy benzophenone or 2,4,4'-trihydroxy benzophenone. These synthetic ester base oil lubricating compositions exhibit increased resistance to oxidation under high temperatures, e.g., in excess of 400°F. They have the further advantages of being noncorrosive to metals, possess excellent thermal stability and have compatibility with seal materials used in jet engines.

Samples of ester-based fluids containing only the aromatic amine or the hydroxy benzophenone and samples made according to the process containing both the aromatic amine and the benzophenone were subjected to oxygen absorption tests. The tests were conducted at 450°F. using 12 ppm Fe (as iron octoate) as a catalyst, by passing a stream of oxygen at the rate of 1 cubic foot per hour through 75 g. of the ester fluid containing the inhibitors and comparing the amount of oxygen absorbed versus time. The induction period is signalled by a marked increase in the rate of oxygen adsorption. The results are shown in Table 1 on the following page.

In the tables, the notation T_i indicates time of induction, T_t indicates total time, V_i indicates volume absorbed during induction, and V_t indicates total volume absorbed. As it can be seen from Table 1, time required for the absorption of 2,500 ml. of oxygen for a synthetic ester lubricant formulation (Example 1) is 87 minutes. When as in Example 2, 0.5% of 2,2',4,4'-tetrahydroxy benzophenone is added, the time required for the absorption of the same amount of oxygen is increased to 352 minutes. Example 3 differs from Example 2 in the amount of benzophenone added to the lubricant composition and Example 4 shows the effectiveness of using 2,4,4'-trihydroxy benzophenone.

TABLE 1

Ex. No.	Basic Components	Aromatic Ketone	Wt., Percent	O_2 Absorption Test Data			
				Ti (min.)	Tt (min.)	Vi (ml.)	Vt (ml.)
1	A [1]			23	87	204	2,500
2	A [1]	2,2',4,4'-tetrahydroxybenzophenone	+0.50	327	352	1,860	2,500
3	A [1]	do	+0.30	230	264	1,150	2,500
4	A [1]	2,4,4'-trihydroxybenzophenone	+0.50	238	272	1,530	2,500
5	B [2]	2,2',4,4'-tetrahydroxybenzophenone	+0.50	0	69		2,500
6	C [3]			46	93	410	2,500
7	C [3]	2,2',4,4'-tetrahydroxybenzophenone	+0.50	266	304	1,060	2,500

[1] Base formulation A contains 72.22% Hercolube A and 24.08% Hercolube F to which was added 1% N-phenyl-1-naphthylamine 1% dioctyldiphenyl-amine, 1.5% tetrabutyl ester of ethylene diamine tetraacetic acid and 0.20% of a half amide of hydrogenated dimer of linoleic acid.
[2] Base formulation B contains 75% Hercolube A and 25% Hercolube F.
[3] Base formulation C contains 75% Hercolube A and 25% Hercolube F, 1% N-phenyl-1-naphthylamine and 1% dioctyldiphenylamine.

Hercolube A is an essentially complete pentaerythritol ester of fatty acids having an average of 5–6 carbon atoms and with the following approximate inspection data: Acid No. 0.01; saponification No. 420–430; viscosity at 210° F., 4.25–4.35 cs.; viscosity at 100° F., 20.5–21.5 cs.
Hercolube F is an essentially complete ester of dipentaerythritol and a mixture of alkanoic acids containing an average of 5 to 6 carbon atoms per molecule and characterized by an acid number of about 0.01, a saponification number of about 390–398, 0.3% hydroxyl, and a kinematic viscosity at 210° F. of 8.7–8.9 cs. and at 100° F. of 55.5–57.5 cs.

The same base oil containing 0.5% 2,2',4,4'-tetrahydroxy benzophenone as the sole antioxidant (Example 5) lasted only 69 minutes in the test. Examples 6 and 7 show similar results to Examples 1 through 4 in the absence of tetrabutyl ester of ethylene diamine tetraacetic acid and half amide of hydrogenated dimer acid, which serve respectively as corrosion inhibitor and antiscuff and antifatigue additives in the formulation. Table 2 below, shows a comparison between the benzophenones of the process with conventional phenolic antioxidants.

TABLE 2

Ex. No.	Basic Components	Aromatic Ketone	Wt., Percent	O_2 Absorption Test Data			
				Ti (min.)	Tt (min.)	Vi (ml.)	Vt (ml.)
1	A [1]			23	87	204	2,500
2	A [1]	2,2',4,4'-tetrahydroxybenzophenone	+0.50	327	352	1,860	2,500
8	A [1]	3,3',5,5'-tetra-tert-butyl-4,4'-dihydroxybiphenyl.	1.0	43	86	765	2,500
9	A [1]	Bis(3,5-di-tert-butyl-4-hydroxyphenyl)-methane.	1.0	61	101	950	2,500

[1] Hercolube A is an essentially complete pentaerythritol ester of fatty acids having an average of 5–6 carbon atoms and with the following approximate inspection data: Acid No. 0.01; saponification No. 420–430; viscosity at 210° F., 4.25–4.35 cs., viscosity at 100° F., 20.5–21.5 cs.

As can be seen from Table 2, the use of the conventional dihydroxy phenolic antioxidants in lubricants (Examples 8 and 9) compares very unfavorably with the benzophenones of the process (Example 2).

Table 3 below, shows the results of a 400°F., 72 hour Oxidation-Corrosion Test for a lubricant containing 0.3% 2,2',4,4'-tetrahydroxy benzophenone in addition to the components given in Example 1. No weight losses were shown for metal pieces made of copper, silver, steel, aluminum and magnesium. The slight weight increases are well within the limits of the specification. The excellent oxidation resistance of the lubricant is also reflected in the low viscosity increase and low acid number of the used oil.

TABLE 3

Material	Experimental Weight Change, mg./cm.²	Specification Requirements Weight Change, mg./cm.²
Copper	+0.10	±0.8
Silver	+0.031	±0.2
Steel	+0.038	±0.2
Aluminum	+0.015	±0.2
Magnesium	+0.007	±0.2
Used oil properties:		
Viscosity at 100° F., percent increase	19.65	−5 to 25
Acid No. change	0.96	3.0

Table 4 below, shows the results of a number of other bench tests on the lubricant containing the additives of the process. As can be seen, the lubricant causes no excessive swelling of H-rubber and Viton A, causes no hardening of Viton A, and has a very low SOD lead corrosion. It passed the Navy 7 day, 221°F., oven storage test, the Allison -40°F., 72 hour stability test and the Seq. I, II, III foam tests.

TABLE 4

Formulation:	Weight Percent
Hercolube A	72.69
Hercolube F	23.81
N-phenyl-1-naphthylamine	1.00
Dioctyldiphenylamine	1.00
Tetrabutyl ester of EDTA	1.00
2,2',4,4'-Tetrahydroxy benzophenone	0.30
Half amide of hydrogenated dimer acid*	0.20
Dow Corning fluid (200-60,000)**	+0.001

*An extreme pressure agent.
**A silicone antifoaming agent, viscosity 200 cs., 250°C., MW of 60,000.

The sample (formulation) was filtered and submitted to the following tests.

		Required Specification
(1) Inspection tests:		
KV/210° F. cs	5.385	5.0–5.5.
KV/−40° cs	28.86	100 max.
KV/−40° F. cs	10,675	13,000 max.
Flash point, ° F	500	400 min.
Fire point, ° F	555	
Pour point, ° F	Below −80	−75 max.
Cloud point, ° F	Below −80	
Acid No. pH 11	0.45	0.50.
Sonic shear: Δ(KV/210), percent	<1	4 max.
Rubber swell, wt. percent:		
H-rubber	21.2	12–35.
Viton A	−1.1	
Durometer reading	65	
Foam test:		
Sequence I	0–0–0	25–0.
Sequence II	0–0–0	25–0.
Sequence III	0–0–0	25–0.
(2) SOD lead corrosion, 325° F., 1 hour mg./in.^{2}	−0.62	−6.0 max.
(3) Navy oven storage, 7 days 221° F.:		
Appearance	Clear	Clear.
Centrifuge deposit, ml./200 ml	Trace	0.005 max.

N-Phenylnaphthylamine and Ethyldiphenylazasiline

A process described by H.W. Adams; U.S. Patent 3,287,269; November 22, 1966; assigned to the U.S. Secretary of the Air Force involves the addition to an ester lubricant, of about 1.25 to 3.0% by weight of an antioxidant mixture which mixture comprises about 0.5 to 2.0% by weight of N-phenyl-1-naphthylamine and about 0.75 to 1.0% by weight of 5-ethyl-10,10-diphenylphenazasiline. A sacrifice in terms of acid buildup occurs when the N-phenyl-1-naphthylamine is employed at the 0.5% level, especially when the concentration of 5-ethyl-10,10-diphenylphenazasiline is at a 1.0% level. Excessive sludge formation during engine operation often results from the use of 2.0% N-phenyl-1-naphthylamine. Thus, the optimum concentration of N-phenyl-1-naphthylamine is at the 1.0% level. Reduction of 5-ethyl-10,10-diphenylphenazasiline from the preferred 1.0% level to 0.75% may be effected without reducing the oxidative resistance of the lubricant. No advantage is achieved by employing the 5-ethyl-10,10-diphenylphenazasiline component at a level greater than 1.0%.

The additive mixture has been found to be especially effective with trimethylolpropane esters of aliphatic saturated monocarboxylic acids having 4 or more carbon atoms in the chain, such as valeric, enanthic and pelargonic acid; and aliphatic saturated dicarboxylic acids having 2 or more carbon atoms in the chain, such as adipic, pimelic, azelaic and sebacic acid.

Example: The following compositions were prepared.

Composition A	Parts
Trimethylolpropane enanthate	100.0
5-Ethyl-10,10-diphenylphenazasiline	1.0
N-phenyl-1-naphthylamine	1.0
Composition B	
Trimethylolpropane monopelargonic, divalerate	100.0
5-Ethyl-10,10-diphenylphenazasiline	0.75
N-phenyl-1-naphthylamine	1.00
Composition C	
Di(2-ethylhexyl)sebacate	100.0
5-Ethyl-10,10-diphenylphenazasiline	1.0
N-phenyl-1-naphthylamine	1.5
Composition D	
Trimethylolpropane valerate	100.0
5-Ethyl-10,10-diphenylphenazasiline	1.0
N-phenyl-1-naphthylamine	2.0

The above compositions, when tested for lubricating quality, gave outstanding results when tested following use in a gas turbine engine over prolonged periods of operation at elevated temperatures, especially with respect to oxidation resistance.

Referring to Figure 1.1, on the following page, there is described a graphical representation of the oxidative resistance achieved by the ester of Composition A, with and without the antioxidant mixture of this process.

FIGURE 1.1: OXIDATION INHIBITION

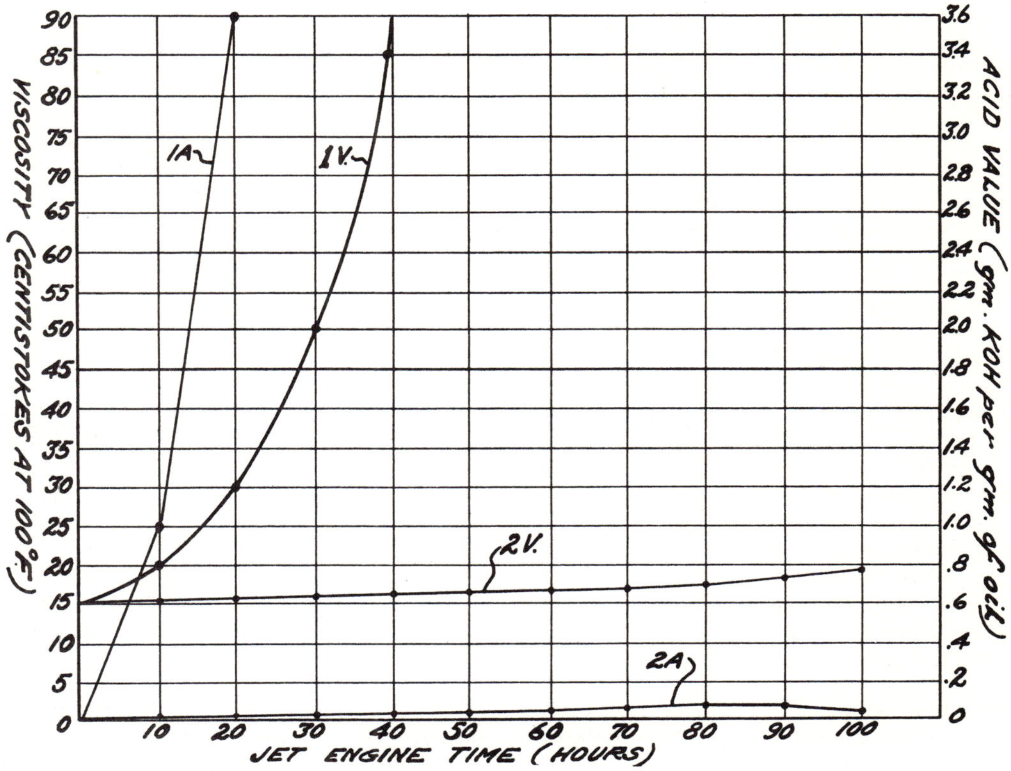

Source: H.W. Adams; U.S. Patent 3,287,269; November 22, 1966

Curves are plotted, according to ASTM requirements, for the viscosity and the acid value as a function of engine time. Curve 1A represents the acid value of a trimethylol propane enanthate ester without the oxidation inhibitor of this process, while curve 2A represents the acid value with the oxidation inhibotor incorporated in the ester. Curve 1V represents the viscosity index of a trimethylolpropane enanthate ester without the oxidation inhibitor while curve 2V represents the viscosity index of the ester containing the oxidation inhibitor.

It will be noted that curves 2A and 2V, that is the ester lubricant containing the antioxidant mixture of this process, amply demonstrate the increase in oxidative resistance achieved over prolonged periods of engine operation. The base ester when used alone has a useful life in the engine for less than 15 hours, while the incorporation of the additive of the process increases that usefulness for a period of time in excess of 100 hours.

N-Acyl Derivatives of 2-Hydroxyphenylamines

In a process described by K. Coupland; U.S. Patent 3,574,743; April 13, 1971; assigned to Distillers Chemicals and Plastics, Limited, England N-acyl derivatives of 2-hydroxyphenylamines are produced from the corresponding disubstituted catechols and are useful as lubricant additives.

As specific compounds which, according to the process are used as additives in lubricant compositions, the N-acetyl and N-benzoyl derivatives of the following are mentioned by way of example:

3,5-di-tert-butyl-2-hydroxy-4'-methyldiphenylamine
3,5-di-tert-butyl-2-hydroxy-4'-chlorodiphenylamine
3,5-di-tert-butyl-2-hydroxy-4'-ethoxydiphenylamine
N,N'-di(3,5-di-tert-butyl-2-hydroxyphenyl)orthophenylenediamine,

and the compounds of the formulas shown on the following page.

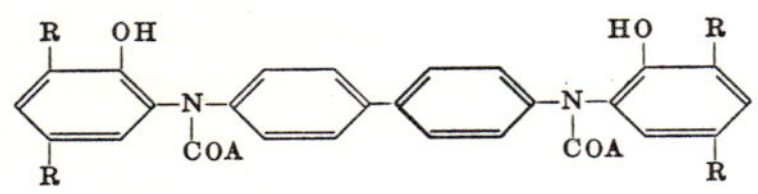

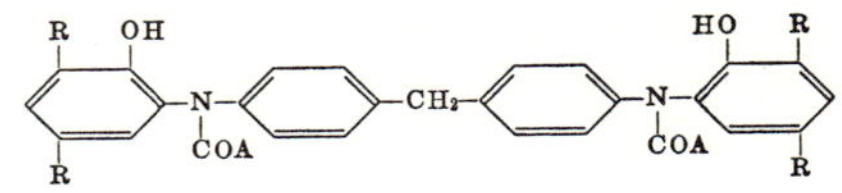

where R is a tert-butyl radical and A is a methyl or phenyl radical.

Example 1: 22.2 g. di-tert-butyl catechol was dissolved in n-pentane and mixed with a solution of 10.5 g. p-toluidine in the same solvent. A few drops of triethylamine were added, and the mixture set aside to crystallize. When the deposition of products was complete, the mixture was filtered to give 2-hydroxy-3,5,di-tert-butyl-4'-methyl diphenylamine as the product. The product was washed with a little heptane and dried. The yield was almost quantitative, and the structure of the product was confirmed by infrared spectroscopy, mass spectrometry and nuclear magnetic resonance as well as by elemental analysis.

Example 2: 32 g. of material of Example 1 was dissolved in acetic acid and 12 g. of acetic anhydride added. The mixture was brought to reflux with stirring. After 2 hours reflux the mixture was poured into water and the solid filtered off. The solid was recrystallized from ethanol to give a white powder in high yield. This material was identified by molecular weight, infrared spectroscopy and chemical analysis as:

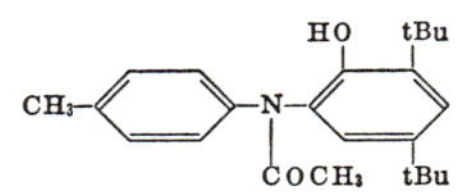

Example 3: 32 g. of material of Example 1 was dissolved in pyridine and 14.5 g. of benzoyl chloride added. Rapid reaction ensued, and pyridine hydrochloride was precipitated. The pyridine salt was removed by filtration, and the product isolated by precipitation from water. An almost quantitative yield of off-white crystals was obtained. This material was identified as:

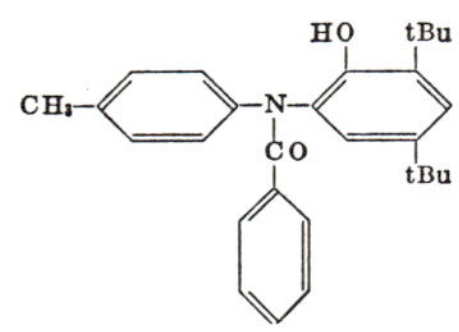

An oxidation/corrosion test in the presence of copper was used to assess the properties of the materials of Examples 1, 2, and 3. In this test air is passed at a rate of 2 l. per hour into a 40 g. sample of the test fluid maintained at 425°F. for 48 hours. A copper test piece of surface area 10 $cm.^2$ is immersed in the test fluid throughout the test. The metal weight loss, and the viscosity increase and weight loss of the test fluid, are reported in the following table.

			Properties of the oxidised material		
Base fluid	Antioxidant (percent)	Corrosion inhibitor (percent)	Copper corrosion ($mg./cm.^2$)	Fluid viscosity increase (percent)	Fluid weight loss (percent)
Trimethylolpropane tricaprylate (TMPTC).	None	None	0.02	260	16.1
	Material Ex. 1 (0.5)	do	1.60	130	8.0
	Material Ex. 2 (0.5)	do	0.68	85	12.1
	Material Ex. 2 (2.0)	do	0.55	85	10.5
	Material Ex. 2 (0.5)	Benzotriazole (0.1)	0.16	250	14.6
	Material Ex. 2 (2.0)	do	0.29	100	10.8
66% TMPTC+34% Ketoester of Example 1 of B.P. 1,036,692.	None	None	0.01	102	8.0
	Dioctylphenothiazine.	do	5.3	------	------
	do	Benzotriazole (0.1)	0.35	52	5.7
	Material Ex. 2 (0.5)	None	2.8	16	6.8
	do	Benzotriazole (0.2)	0.19	63	6.6
	Material Ex. 3 (0.5)	None	3.8	38	5.2
	do	Benzotriazole (0.1)	2.4	54	3.6

Phenothiazine and Dibutyl Tin Sulfide

In a process described by R.E. Malec; U.S. Patent 3,489,683; January 13, 1970; assigned to Ethyl Corporation ester lubricants are stabilized against the effect of oxygen by the synergistic combination of a phenothiazine and a dihydrocarbyl tin sulfide. For example, the stabilizing effect of phenothiazine on di(2-ethylhexyl)sebacate is synergized by dibutyl tin sulfide.

Tests were conducted to demonstrate the synergistic effect of dihydrocarbyl tin sulfides on phenothiazine stabilizers. The test used was the Panel Coker Test. This test is described in "Aeronautical Standards of the Departments of Navy and Air Force," Spec. MIL-L 7808C, November 2, 1955. In the test, di(2-ethylhexyl)sebacate, a commonly used ester lubricant, was placed in a sump under an aluminum plate which was heated to 600°F. The lubricant was periodically splashed against the heated plate and allowed to drain back into the sump. The oil was splashed for 5 seconds and drained for 55 seconds. This cycle was repeated over a 10 hour period. Following this, the aluminum plate was washed with hexanes and the weight gain determined. Any gain in weight is due to thermal and oxidative breakdown of the ester lubricant leaving a carbonaceous deposit. The test was carried out with the unstabilized ester lubricant and with the same ester lubricant containing various stabilizers and stabilizer combinations. The results obtained are shown in the following table.

	Additive	Conc. (wt. percent)	Deposit wt. (mg.)
1	None		190
2	Phenothiazine	0.12	36
3	do	0.25	12
4	Dibutyl tin sulfide	0.01	176
5	do	0.25	195
6	Phenothiazine Dibutyl tin sulfide	0.12 0.01	8

As the above results show, dibutyl tin sulfide had very little effect on the stability of the lubricant. However, when used in an amount of only 0.01 weight percent in combination with 0.12 weight percent of phenothiazine only 8 mg. of deposit formed. This represents a deposit weight much less than that obtained using twice the amount of phenothiazine. Therefore, the dihydrocarbyl tin sulfide has greatly enhanced the stabilizing effect of phenothiazine even when used in extremely low concentration.

Phenothiazine, Phenol Antioxidant and Benzoguanamine

R.A. Cupper and C.A. Wolf, Jr.; U.S. Patent 3,296,135; January 3, 1967; assigned to Union Carbide Corporation found that improved nonoxidative and noncorrosive properties may be achieved in synthetic lubricating compositions, by the addition of an additive system which comprises at least one amine antioxidant, at least one phenolic antioxidant, and a metal corrosion inhibitor of the formula:

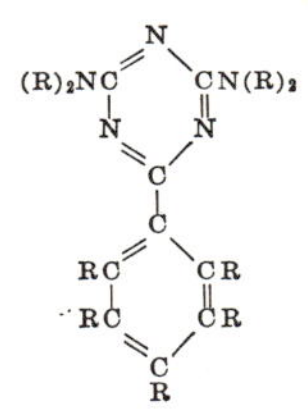

where R represents hydrogen or an alkyl group having 1 to 4 carbon atoms.

The amine antioxidants which are suitable for use in the additive package include phenothiazine, N-methylphenothiazine, N-phenyl-α-naphthylamine, N-phenyl-β-naphthylamine, p-phenylenediamine, 2,4'-diaminodiphenylamine and α,β-dinaphthylamine.

The phenolic antioxidants which are useful in the additive package are 2,6-di-t-butyl-4-methylphenol, 2,6-di-t-butyl-4-benzylphenol, 3-t-butyltetrahydro-2-naphthol, 6-butyl-5-indanol, diphenylol propane and 2,2'-methylene-bis(4-ethyl-6-t-butylphenol).

Metal corrosion inhibitors which may be used in the additive systems have been described above by formula. These compounds include benzoguanamine and the alkyl substituted benzoguanamines. Benzoguanamine, itself, is preferably employed. The following examples are illustrative.

Example 1: A blend was prepared having the following composition (all parts are percent by weight):

Di-2-ethylhexyl azelate	58.8	Ucon lubricant LB-1715**	7.84
Ucon lubricant LB-625*	31.36	2,6-Di-t-butyl-4-methylphenol	1.0
		Benzoguanamine	0.50

*Monobutyl ether of polypropylene glycol having a SUS viscosity of 625 at 100°F.
**Monobutyl ether of polypropylene glycol having a SUS viscosity of 1,715 at 100°F.

Example 2: A blend was made having the following composition (all parts are percent by weight):

Di-2-ethylhexyl azelate	54.2	Phenothiazine	0.5
Uncon lubricant LB-625	44.0	2,6-Di-t-butyl-4-methylphenol	1.0
		Benzoguanamine	0.3

Properties of this product and the properties of the product of Example 1 are shown in Table 1 below.

TABLE 1

	Allison EMS 35F Limits	Examples 1	Examples 2
Kinematic Viscosity, cs. at:			
210° F	7.5 (min.)	7.93	7.8
100° F		36.4	26.47
−40° F	13,000 (max.)	12,950	
Corrosion and Oxidation, 352° F./72 hrs., Weight Change, mg./sq. cm.:			
Fe	±0.2	±0.03	Nil
Cu	±0.4	−0.06	−0.06
Pb		−.06	−0.32
Viscosity Change at 100° F	% −5 to +12	−0.2	+0.19
Total Acid No. Increase	1.5 (max.)	1.3	1.07

Table 2 below, summarizes various lubricating oil compositions and the effect of omitting one or more of the critical components of the additive system of this process.

TABLE 2

Example	Base Fluid	Weight Percent Composition, Additives: Amine	Phenolic	Corrosion Inhibitor	Corrosion-Oxidation Test 72 hrs. at 352° F.: Percent Viscosity Change, 100° F.	Metal Weight Change, mg./cm.[3] Fe	Cu	Pb
3	32% Polyglycol AU [1], 66.3% Di-2-ethylhexyl azelate.	0.5 PHTZ [2]	1.0 DBMP [3]	0.2 BG [4]	−4.1	0.05	Nil	Nil
4	32% Polyglycol AU, 66% Di-2-ethylhexyl azelate.	0.5 PHTZ	1.0 DBMP	0.5 BG	−0.99	−0.03	−0.03	−0.04
5	32% Polyglycol AU, 67.0% Di-2-ethylhexy azelate.			1.0 BG	−0.99	−1.4	−3.5	−69.9
7	32% Polyglycol AU, 67.5% Di-2-ethylhexyl azelate.	0.5 PHTZ			21.5	Nil	−0.16	−1.0
8	44% Polyglycol BU [5], 54.2% Di-2-ethylhexyl azelate.	0.5 PHTZ	1.0 DBMP	0.3 BG	−0.19	Nil	−0.06	−0.32
9	44% Polyglycol BU, 53.7% Di-2-ethylhexyl azelate.	1.0 PANA [6]	1.0 DBMP	0.3 BG	+4.9	Nil	−0.12	−7.1
10	44% Polyglycol BU, 55.5% Di-2-ethylhexyl azelate.	0.5 PHTZ			−11.7	0.03	−0.58	−7.3
11	44% Polyglycol BU, 55% 2-ethylhexyl azelate.	1.0 PANA			−8.2	0.02	0.18	−9.4
12	30% monobutyl ether of polypropylene glycol (S.U.S. Visc., 1,145 at 100° F.), 70% diethylhexyl sebacate.	1.0 PANA	1.0 MEBP [7]	0.5 BG	−7.2	−0.007	−0.299	−2.38
	Corrosion-Oxidation Test Limits				−5.0 to +12.0	±0.2	±0.4	

[1] 43.4 percent (weight) Ucon LB 625
56.6 percent (weight) Ucon LB 1715.
[2] Phenothiazine.
[3] 2,6-ditertiarybutyl-4-methylphenol.
[4] Benzoguanamine.
[5] Ucon LB 625.
[6] N-phenylalphanaphthylamine.
[7] 2,2′,-methylene-bis (4-ethyl-6-t-butylphenol).

It can be seen that only compositions which contained all three of the critical additives of this process (Examples 3, 4, 8, 9 and 12) met the critical specification for a lubricant as defined by the Allison specification. Other lubricant blends were prepared having varying compositions. The compositions of these lubricant blends are set out in Table 3 below.

TABLE 3

	Blend A, percent	Blend B, percent	Blend C, percent	Blend D, percent	Blend E, percent
Di-2-ethylhexyl azelate	55.5	55.4	55.2	58.8	58.62
Ucon LB-625 [1]	44.0	44.0	44.0	31.36	31.26
Ucon LB-1715 [2]				7.84	7.82
Phenothiazine	0.5	0.5	0.5	0.50	0.5
2,6-Ditertiary-butyl-4-methyl phenol				1.0	1.0
Benzoguanamine				0.5	0.5
Acridine		0.1	0.3		0.3
	100.0	100.0	100.0	100.0	100.0

[1] Monobutyl ether of polypropylene glycol having a viscosity of 625 Saybolt Universal Seconds at 100° F.
[2] Monobutyl ether of polypropylene glycol having a viscosity of 1715 Saybolt Universal Seconds at 100° F.

Tables 4 and 5 which follow, illustrate the results of the British Ministry of Supply thermal stability test and the corrosion-oxidation test made on each of the compositions prepared in Table 1.

TABLE 4

THERMAL STABILITY (282° C.)
[Percent Viscosity Change at 100° F.: Test Time, hours]

	6	12	18	24
Blend A	−6.1	−8.5	−17.9	
Blend B	−2.2	−6.8	−8.5	−11.1
Blend C	−2.9	−1.4	7.7	−8.8
Blend D	−12.6			
Blend E	−0.74	+0.49	−2.4	−5.0
British Specification D. Eng. RD 2487, Issue 3	[1]	[1]	[1]	[1]

[1] Limits: −10 to +20%.

TABLE 5

[Corrosion-oxidation test, 72 hours at 352° F.]

	Percent Viscosity Change	Percent Metal Weight Change mg./cm.²		
		Fe	Cu	Pb
Blend A	−11.7	−0.03	−0.58	−7.3
Blend B	−19.3	Nil	−0.68	−3.4
Blend C	-----	-----	-----	-----
Blend D	−0.2	−0.06	−0.03	−0.06
Blend E	+1.1	−0.06	−0.02	−0.08
352° F. Corrosion Oxidation Test Limits	−5 to 12	±0.2	±0.4	±1.0

From the above tests it may be seen that Blend E containing antioxidants, metal corrosion inhibitor and acridine as a thermolysis inhibitor was the only lubricant to successfully pass both the thermal stability test and the corrosion-oxidation test.

Phosphoramidate, Alkylated Arylamine and Alkylated Phenothiazine

F.G. Oberender, A.W. Godfrey and M.A. Wiley; U.S. Patent 3,476,685; November 4, 1969; assigned to Texaco Inc. describe a lubricating composition which comprises the combination of an ester-base fluid having lubricating properties formed from the reaction of a pentaerythritol or trimethylolpropane and an organic acid having from about 2 to 12 carbon atoms per molecule containing from 0.1 to 4% of a phosphoramidate having the formula:

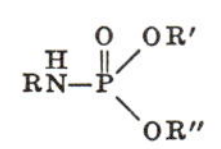

in which R is an aryl or alkaryl radical having from 6 to 20 carbon atoms and R' and R" each represent a hydrocarbyl radical having from 4 to 20 carbon atoms, from about 0.5 to 4% of an alkylated diphenylamine in which the alkyl radicals have from 4 to 20 carbon atoms and from 0.1 to 3% of an alkylated phenothiazine in which the alkyl radicals have from 4 to 20 carbon atoms. The compounded lubricating oil can generally be further improved by adding minor amounts of conventional corrosion inhibitors.

The phosphoramidate, alkylated diphenylamine and alkylated phenothiazine components of the lubricant are readily soluble in the ester-base in the percentage ranges disclosed. The lubricating oil composition can be prepared by adding the various components to the base oil in any order and mixing until a homogenous mixture is obtained.

The lubricating composition was tested for its anticorrosive, oxidation resistance and deposit inhibiting properties in the following tests: the 450°F. Antioxidant Evaluation Test, the 425°F. Oxidation-Corrosion Test for 48 hours, the Oil Mist Deposit Test, the Stainless Steel Corrosion Test and the Erdco Bearing Tests.

The 425°F. Oxidation and Corrosion Test is conducted in accordance with Method 5308.4 of Federal Test Method and Standard No. 791a (issued December 30, 1961) except for the following modifications to conform to Pratt & Whitney Aircraft Specification 521-B (Type 2). The bath temperature is maintained at 425°F. ± 1°F. instead of at 250°F. This test is conducted for a period of 48 hours instead of 168 hours, specified in the original test. The 450°F. Antioxidant Evaluation Test is conducted by passing dry air at a flow rate of 30 l. per hour through the test sample in a glass test cell attached to a head suitable for collecting volatile oxidation products. A steel specimen is immersed in the oil during the test. This test is run for 48 hours with determination of viscosity increase and total acid number taken at 24 and 48 hours.

The Oil Mist Deposit Test is conducted by passing a mist of oil through a glass tube containing a steel liner (shim stock) through the entire length of the tube. The tube is maintained at 600°F. while the oil mist is continuously passed through over a period of 16 hours. At the end of the test period, the steel liner is removed, solvent washed and weighed. The amount of tenacious material adhering to the liner is determined and is reported as grams of deposit per 1,000 g. of oil put through the system. This test is representative of breather tube deposit formation in a jet engine. A description of this test has been published by the American Society of Lubrication Engineers (ASLE Paper No. 66LC-6).

A base oil which was employed in the preparation of lubricating oils tested below was a technical grade pentaerythritol esterified with a mixture of 38% valeric, 13% 2-methylpentanoic, 32% octanoic and 17% pelargonic acid. This base fluid designated Base Fluid A, had the following properties.

Viscosity at 210°F., cs.	4.93	Viscosity index	131
Viscosity at 100°F., cs.	25.6	Flash, °F.	490
Viscosity at -40°F., cs.	7,023	Ryder gear, lb./in.	2,040

The tests set out in Table 1, on the following page, show that there is specificity in the amine component of lubricating

TABLE 1

	Run 1	Run 2	Run 3	Run 4	
Inhibitor Components, wt. percent:					
N-phenyl-O,O′-diphenylphosphoramidate	0.5	0.5	0.5	0.5	
Dioctylphenothiazine	0.7	0.7	0.7	0.7	
Amine	[1] 1.4	[2] 0.6	[3] 0.6	[4] 1.0	
Tests:					
450° F. Anti-Oxidant Test:					
Percent Viscosity Increase:					
At 24 hrs	55.7	93.2	[5] 193	60.0	
At 48 hrs	339	619	----------	503	
TAN Increase:					
At 24 hrs	+2.80	+4.03	+5.24	+4.31	
At 48 hrs	+5.52	+8.17	----------	+8.0	
Deposits (gm.)	0.4	0.8	0.3	0.7	
Sludge (percent)	0.1	0.1	0	[5] 2.0	
425° F./48 Hr. Oxidation Corrosion Test:					(PWA 521-B)
Metal weights Δ in mg./cm.[2]:					
Cu	−0.12	−0.06	−0.02	−0.14	±0.3
Fe	+0.05	0	+0.15	+0.06	±0.3
Al	+0.06	−0.07	0	+0.04	±0.3
Mg	+0.04	−0.06	−0.06	+0.09	±0.3
Ag	0	+0.11	+0.03	0	±0.3
Ti	+0.08	+0.08	+0.02	0	±0.3
Percent Viscosity Increase	20.7	17.0	40.3	23.4	+50 Max.
TAN Increase	+2.04	+1.49	+5.02	2.74	----------
Oil Mist Deposits Test (600° F.), g. deposit/1,000 g. throughput	1.07	[5] 2.5	----------	----------	----------

[1] Dioctyldiphenylamine.
[2] Diphenylamine.
[3] Di-n-hexylamine.
[4] Bis(1,4-dimethylpentyl)paraphenylene diamine.
[5] Fails test.

oil composition of the process. The lubricating oils of Runs 1 through 4 are identical except for the amine component. In addition to the inhibitor components indicated, the lubricating oil compositions of Runs 1 through 4 consisted of 0.1% quinizarin, 0.3% Acryloid Additive and the balance Base Fluid A described above. The Acryloid Additive is a commercially available viscosity index improver consisting of a 25% solution of Acryloid polymers in di-2-ethylhexyl sebacate. The Acryloid polymer is a copolymer of 90% mixed methacrylic esters ranging from butyl to stearyl polymerized with 10% of an unsaturated nitrogenous compound. This VI improver is designated Acryloid A in the examples below. Run 1 is representative of the process. In all of the examples below the additive concentration is in weight percent unless otherwise indicated.

Due to the complex requirements which must be met by a synthetic aircraft turbine oil and the fact that these requirements are constantly becoming more severe with the rapid advances in engine technology, there are no universally accepted criteria for all the properties of such a lubricating oil. It is essential for the lubricant manufacturer to set high performance standards for all the properties of the oil in order to formulate an effective product. In addition to the requirements of the Pratt & Whitney Aircraft corrosion specifications (PWA 521-B), the oils of this process meet high antioxidant and deposits standards as shown by Run 1 above. Run 2 failed on the basis of excessively high Oil Mist Deposits. Run 3 failed because of the excessive viscosity increase of 193% in 24 hours and Run 4 failed because of the excessive 2.0% sludges in the antioxidant test.

Runs 5 through 9 in Table 2 below, show the effectiveness of various base oils and different phosphoramidates. Base Fluid B was a technical grade pentaerythritol esterified with a mixture of 20% 2-methylpentanoic acid, 50% valeric acid and 30% pelargonic acid.

TABLE 2

	Run 5	Run 6	Run 7	Run 8	Run 9	
Composition, wt. percent:						
Base Fluid	B	C	D	A	A	
Quinizarin	0.10	0.10	0.10	0.10	0.10	
Acryloid A	0.30	0.30	0.30	0.30	0.30	
N-phenyl-O,O′-diphenylphosphoramidate	----------	----------	0.50	0.50	----------	
N-phenyl-O,O′bis(nonylphenyl) phosphoramidate	1.00	1.00	----------	----------	1.00	
Dioctylphenothiazine	0.70	0.70	0.70	0.70	0.70	
Dioctyldiphenylamine	1.40	1.40	1.40	1.40	1.40	
Additive A	0.16	0.16	0.16	0.16	0.16	
Dimethyl silicone soln., p.p.m.	50	50	50	50	50	
Tests:						
450° F. AOST:						
Percent Viscosity Increase:						
At 24 hrs	55.6	48.8	11.6	46	61	
At 48 hrs	367	292	44	293	366	
TAN Increase:						
At 24 hrs	2.80	4.88	0.43	2.26	3.6	
At 48 hrs	8.44	9.14	1.37	7.42	7.9	
Deposits, gms	0.5	0.6	0.4	0.2	0.3	
Sludge, percent	Trace	Trace	0.1	0.1	Trace	
425° F./48 Hr. Oxid.-Corr. Test:						(PWA 521-B)
Corr., mg./cm.[2]:						
Cu	−0.25	−0.15	−0.39	−0.14	−0.17	±0.3
Fe	+0.04	+0.02	+0.03	0	0	±0.3
Al	+0.06	+0.03	+0.05	−0.04	−0.04	±0.3
Mg	0	+0.04	+0.03	+0.04	−0.04	±0.3
Ag	0	0	−0.07	0	0	±0.3
Ti	+0.03	+0.03	0	−0.07	0	±0.3
Percent Visconsity Increase	24	25	22	22.3	25.5	+50 Max.
TAN Increase	4.71	6.21	5.35	2.8	3.45	----------
Oil Mist Deposit Test (600° F.), g. deposit/1,000 g. throughput	0.4	1.5	0.9	0.2	0.9	----------

Base Fluid C was a trimethylolpropane esterified with a mixture of 1% valeric, 8% hexanoic, 13% heptanoic, 5% octanoic and 73% pelargonic acids. Base Fluid D was trimethylolpropane tris(neoheptanoate). The Acryloid Additive is the same as that described above. Additive A is a load-carrying additive consisting of a 49 weight percent solution of the monoamide of 2-aminopyridine and trimer acid in the pentaerythritol tetraester described above as Base Fluid A.

The monoamide is prepared by direct condensation of 2-aminopyridine (0.33 equivalent) and the trimer of unsaturated C_{18} fatty acids (1.0 equivalent based on Saponification Number) at 350° to 400°F. for approximately 8 hours followed by removal of the water of reaction by azeotropic distillation. Additive A is effective in a concentration ranging from about 0.05 to 0.5%. The antifoam additive was a 10% solution of dimethyl silicone in kerosene and the solution was employed in a concentration of 50 parts of solution per million of the lubricating oil. The foregoing Runs 5 through 9 show that acceptable synthetic lubricating oil compositions can be prepared from esters of pentaerythritol and trimethylolpropane and that a variety of phosphoramidates are effective.

Runs 10 through 13 in Table 3 below, give comparison data for two turbine oils of the process, a commercial product, and an oil representative of a second commercial product. The oil used in Run 10 is a synthetic aircraft turbine oil in commercial use having the same base fluid as the preferred lubricating oils, namely Base Fluid A described above. The inhibitor package of this oil is based upon phenothiazine and dioctyldiphenylamine. The lubricating oils in Runs 11 and 12 are identical except for the phosphoramidate component. These oils consisted of 1.4% dioctyldiphenylamine, 0.7% dioctylphenothiazine, 0.3% Acryloid A, 0.16% Additive A, 0.1% quinizarin, 50 ppm dimethyl silicone solution plus a phosphoramidate in Base Fluid A. Runs 11 and 12 are representative of the process. The oil used in Run 13 is representative of a second type of synthetic aircraft turbine oil in commercial use. It has the same Base Fluid A but employs an inhibitor package based upon phenyl-α-naphthylamine and dioctyldiphenylamine.

TABLE 3

	Run 10	Run 11	Run 12	Run 13	Test Limits
Composition, wt. percent: Phosphoramidate		(1)	(2)		
Tests:					
410 Stainless Steel Corrosion, 550° F./60 hr., mg	−658	−6	−5	−2	
Anti-Oxidant Test (450° F./48 hr.):					
Percent Vis. Inc	556	293	366	760	
TAN Inc	7.9	7.4	7.9	9.0	
Erdco Bearing Test, Type 1.5:					MIL-L-23699A
Demerit Rating	74	14	16	36	80 Max.
Percent Vis. Inc. (100° F.)	18	18	17	18	−5 to 25
TAN Inc	1.1	1.4	1.4	1.9	2.0 Max.
Erdco Bearing Test, Type 2:					PWA 521-B
Demerit Rating	72	30	33	72	(Pass)
Percent Vis. Inc. (100° F.)	1,436	167	192	181	
TAN Inc	6.7	4.3	5.2	4.6	
Erdco Bearing Test, Type 2.5:					
Demerit Rating	178 (100 Hrs.)	67 (90 Hrs.)			
Hours to Oil Change	43 80 100	90			
Percent Vis. Inc. (100° F.)	291 265 75	352			
TAN Inc	3.4 4.7 2.2	4.4			
Erdco Bearing Test, Type 3:					
Demerit Rating		141 (100 Hrs.)			
Hours to Oil Change		43 88 100			
Percent Vis. Inc. (100° F.)		377 842 44			
TAN Inc		2.5 3.1 1.3			

[1] 0.5% N-phenyl-O,O′-diphenyl phosphoramidate.
[2] 1.0% N-phenyl-O,O′-bis(nonylphenyl) phosphoramidate.

It can be seen from the data that the commercial oil of Run 10 is extremely corrosive to stainless steel as compared to the oils of Runs 11, 12 and 13. The oils of Runs 11 and 12 of the process are far superior to the commercial oils of Runs 10 and 13 in Demerit Rating in the Erdco Bearing Test under Type 1.5 and Type 2 conditions. The Type 2.5 and Type 3 Erdco Bearing Test data compare an oil of this process (Run 11) with a current commercial synthetic lubricant (Run 10) under test conditions which are used to evaluate the operating capability of lubricants under more severe conditions such as will be encountered in the supersonic transport (SST).

The commercial lubricant required three oil changes to complete 100 hours under the Type 2.5 test conditions (550°F. bearing temperature, 450°F. oil in temperature and 490°F. bulk oil temperature), whereas the oil of this process essentially completed this test with one oil change while causing very little deposit formation. The oil of this process successfully completed a Type 3 test (600°F. bearing temperature, 500°F. oil in temperature and 540°F. bulk oil temperature) with three oil changes, which was similar to those of the commercial oil under the more moderate Type 2.5 test conditions. Thus, the oil of this process demonstrates a 50°F. operating temperature improvement over the commercial synthetic turbine oil lubricant.

Trihydroxydiphenyl

J.J. McGrath and H.O. Strange; U.S. Patent 3,326,802; June 20, 1967; assigned to Gulf Research & Development Company have found that a synthetic ester lubricating composition having improved thermal stability for an extended period of time when used to lubricate bearings at temperatures above 300°F. under oxidizing conditions can be obtained by incorporating trihydroxydiphenyl into the synthetic ester lubricant in oxidation-inhibiting proportions.

In order to demonstrate the effectiveness of trihydroxydiphenyl as an antioxidant in a synthetic ester lubricant, comparative oxidation tests have been conducted with a representative synthetic ester lubricant alone and with the same lubricant blended with a commercial trihydroxydiphenyl concentrate and with recrystallized trihydroxydiphenyl. In brief, the comparative tests were conducted by bubbling air at a rate of 1 l. per hour through approximately 20 ml. of test lubricant contained in a glass tube inserted into an aluminum block heated to 347°F. (175°C.). The duration of the test was 48 hours. The viscosity of the test lubricant and its acid number were determined at the beginning and the conclusion of each test. The percent increase in viscosity and the increase in acid number of the lubricant during the test is indicative of the oxidation stability of the test lubricant at a temperature of 347°F. At the end of each test an observation was also made as to whether or not varnish and/or sediment were present.

The results of comparative tests on the lubricant alone and the lubricant containing commercial trihydroxydiphenyl and recrystallized trihydroxydiphenyl are shown in the table below. The commercial trihydroxydiphenyl was a yellowish-red resinous-like material having no sharp melting point but became soft and less viscous as the temperature was increased. The commercial trihydroxydiphenyl which was recrystallized from water comprised yellowish crystals melting at about 55° to 65°C.

	A	B	C
Composition, percent by weight:			
Lubricating Oil: Diisooctyl azelate	100	99	99
Trihydroxydiphenyl:			
Trihydroxydiphenyl (concentrate)	-	1	-
Trihydroxydiphenyl (recrystallized)	-	-	1
Oxidation Test (Air Flow: 1 l./hr.)—			
Viscosity, cs. at 100° F.:			
Initial	12.66	13.68	13.20
After 48-hour test	22.79	13.92	13.39
Increase	10.13	0.24	0.19
Percent Increase in Viscosity	80	1.8	1.4
Viscosity Index:			
Initial	166	176	165
After 48-hour test	141	164	161
Acid Number (ASTM D 974):			
Initial	0.07	0.13	0.01
After 48-hour test	28.68	0.84	0.56
Increase	28.61	0.71	0.55
Visual Observation	(1)	(2)	(2)

[1] Clear, yellow. [2] Clear, red.

The data show that the oxidation stability of diisooctyl azelate at a temperature of 347°F. can be greatly improved by the addition of either a commercially available trihydroxydiphenyl concentrate or recrystallized trihydroxydiphenyl.

9,12-Bis(4-Hydroxyphenyl) Stearic Acid

J.L. Snead, J.F. Messina and H. Gisser; U.S. Patent 3,573,333; March 30, 1971; assigned to the U.S. Secretary of the Army describe the use of inhibitor-generating long-chain fatty acids (and salts) such as hydroxyaryl stearic acids where the aryl group is attached to the long acid chain, at various points, through a tertiary carbon. Examples of the type of structures involved are given below:

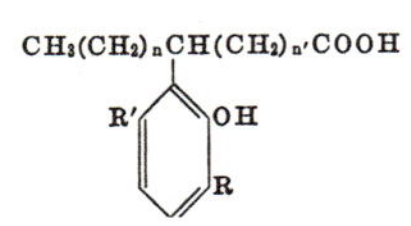

where n+n' = 15
R = isopropyl
R' = CH_3

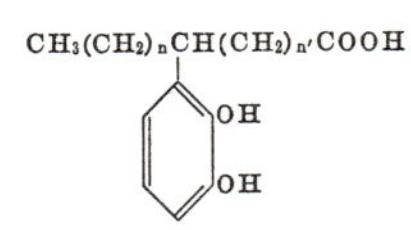

where n+n' = 15

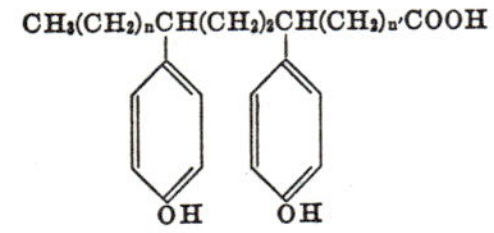

where n+n' = 12

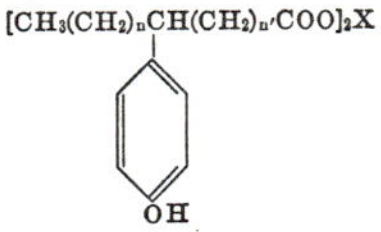

where n+n' = 15
X = calcium, strontium or barium

The above have been found to be compounds effective in inhibiting both oxidation and rust when used in instrument lubricant compositions. The long hydrocarbon chain terminated by the strongly polar acid (or salt) group, adheres to the metal to be lubricated to form a monomolecular layer which prevents rust formation. In addition, hydroxyaryl stearic acids (or salts) are effective as antioxidants because of formation of an inhibitor via autoxidation at the tertiary carbon atom.

In order to determine oxidation stability of the generic group of compounds at different temperatures a dynamic test was employed which is similar to that described in the Federal Specification for "Lubricants, Liquid Fuels, and Related Products: Methods of Inspection, Sampling, and Testing", VV-L-791e, May 21, 1953, Method 5308.3 entitled "Corrosiveness and Oxidation Stability of Light Oils". Two weight percent solutions of the additive compounds in bis(2-ethylhexyl) sebacate, a typical diester lubricant, were oxidized at 120°, 150°, and 175°C. in the presence of copper and steel catalyst strips (1.750 x 0.735 x 0.025 inches). Samples of 10 ml. were withdrawn periodically for viscosity and neutralization number determinations, and the test was continued until after the induction period and considerable increase in acidity and viscosity was obtained. The results are presented in Table 1 below.

TABLE 1: OXIDATION OF ARYL STEARIC ACIDS IN BIS(2-ETHYLHEXYL) SEBACATE[1]

Additive	Neutralization number increase,[2] Eq. mg. KOH/g. oil, hours							Viscosity increase,[3] percent at 37.8° C., hours							Induction period, hours
	24	48	72	96	168	240 at 150° C.	336	24	48	72	96	168	240	336	
None	23							75.0							10
9(or 10)-(2-hydroxy-5-tert-butylphenyl)stearic acid	0.1	7.6						1.8	24						37
9(or 10)-(2,3-dihydroxyphenyl)stearic acid	0.2	0.2	6.8					1.1	1.7	21					62
9(or 10)-(2-hydroxy-3-isopropyl-6-methylphenyl)stearic acid	0.1	12						1.7	61						38
9,12-Bis(4-hydroxyphenyl)stearic acid	0.2	0.3	0.1	—	−0.5	−0.4		0.7	2.5	1.7	—	1.8	0		>240
Calcium 9(or 10)-p-hydroxyphenyl stearate	0	0	7.7					1.6	−2.4	17.7					70
Barium 9 (or 10)-p-hydroxyphenyl stearate	0.3	0.2	0.5					−2	−2	−2.8					>56
Strontium 9 (or 10)-p-hydroxyphenyl stearate	0.2	0.1	0.8	13.3				0.1	−6.2	−5.8	44				81
						240 at 175° C.									
None	3.8							7							4
9(or 10)-(2,3-dihydroxyphenyl)stearic acid	2.1	13.5						7.6	62						26
9,12-Bis(4-hydroxyphenyl)stearic acid	0.3	17.5						1.8	59						33
Barium 9(or 10)-p-hydroxyphenyl stearate	2.3	12.5						5	16.5						28

[1] Two percent of the inhibitor in solution unless otherwise indicated.
[2] Determined in accordance with ASTM D664-54.
[3] Determined in accordance with ASTMD445-53T.

The general effectiveness of a particular additive as an antioxidant may rapidly be determined from Table 1 by noting the duration of the induction period. That is, the longer the induction period, generally, the more effective the antioxidant. Although all of the hydroxyarylstearic acids were effective to some extent as antioxidants, excellent results were obtained through the use of 9(or 10)-(2, 3-dihydroxyphenyl) stearic; 9, 12-bis(4-hydroxyphenyl) stearic acids; and the alkaline earth salts of 9(or 10)-p-hydroxyphenyl stearic acid.

Rust inhibition data were obtained with the use of a cyclic humidity cabinet where test specimens were continuously exposed to alternating cycles of 4 hours at 43.3°C. and 80% relative humidity, followed by 4 hours at 54.4°C. and 95% relative humidity. These conditions are considered severe and give rise to one period of controlled condensation during each 8 hour cycle. Rust inhibition results are presented in Table 2 below.

TABLE 2: HUMIDITY CABINET RUST TEST DATA*

Additive	Hours to Failure
None	24
Oleic acid	84
Linoleic acid	72
9(or 10)-(2-Hydroxy-5-tert-butylphenyl) stearic acid	96
9(or 10)-(2, 3-Dihydroxyphenyl) stearic acid	132
9(or 10)-(2-Hydroxy-3-isopropyl-6-methylphenyl) stearic acid	96
9, 12-Bis(4-hydroxyphenyl) stearic acid	128
Calcium 9(or 10)-p-hydroxyphenyl stearate	162
Barium 9(or 10)-p-hydroxyphenyl stearate	186
Strontium 9(or 10)-p-hydroxyphenyl stearate	132

*Bis(2-ethylhexyl) sebacate containing 2% of the additive.

From the data presented above, it can be seen that most of the compounds exhibit protection against rust. Since 9(or 10)-(2, 3-dihydroxyphenyl) stearic; 9, 12-bis(4-hydroxyphenyl) stearic acids; and the alkaline earth salts of 9(or 10)-p-hydroxyphenyl stearic acid provide better protection that the more highly substituted compounds, it would appear that the bulky alkyl groups may prevent close alignment of the molecules in the absorbed monolayer or decrease the later attractions (Van der Waals forces). The combination oxidation and rust inhibitor compounds were prepared as described below, using 9(or 10)-p-hydroxyphenyl stearic acid, and barium 9(or 10)-p-hydroxyphenyl stearate as examples.

Example: Isomeric Aryl Stearic Acids — The 9(or 10)-aryl stearic acids were prepared by the Friedel-Crafts condensation of oleic acid with phenols in the presence of equimolar quantities of anhydrous aluminum chloride. Phenol (7 parts) was mixed with 99% purity oleic acid (1 part) and anhydrous aluminum chloride (1.3 parts) was gradually added with stirring. After the vigorous reaction had subsided, the mixture was heated to 80°C. for 6 to 8 hours or until the evolution of HCl fumes ceased. The reaction mixture was poured into 20% hydrochloric acid and the organic layer extracted with ethyl ether. The ether extract was washed with water until acid-free and dried over anhydrous Na_2SO_4.

After filtration, the ether was removed by distillation at atmospheric pressure. Excess phenol was then removed by distillation, in vacuo, until the pot temperature of 150°C. at 1 mm. was obtained. The residue, on distillation from a short-path molecular still at 5 to 10 microns pressure, yielded 9(or 10)-p-hydroxyphenyl stearic acid. The 9, 12-diaryl stearic acid was prepared in a similar manner by reaction of linoleic acid with phenol.

The alkaline earth salts of 9(or 10)-p-hydroxyphenyl) stearic acid were prepared by conversion of the sodium salt. To a benzene solution of the sodium salt (2 parts) prepared from the free acid and NaOH in a 1 : 1 ratio was added, dropwise, with stirring, aqueous barium chloride (1 part). The reaction mixture was refluxed for 1/2 hour, cooled, and poured into a separatory funnel. The product was washed free of chloride and then transferred to a clean flask and the remaining water was removed as an azeotrope with benzene. Approximately one-half of the benzene was distilled, in vacuo, in a nitrogen atmosphere and then bis(2-ethylhexyl) sebacate was added and the distillation continued until all the benzene was removed. The final product was a concentrated solution (about 20%) of barium 9(or 10)-p-hydroxyphenyl stearate in bis(2-ethylhexyl) sebacate. The amount of salt was calculated from sulfated ash determinations and the final solution was found to contain less than 0.05% free acid.

Polyalkylated Naphthols

H.J. Andress, Jr., and A.L. Williams; U.S. Patent 3,492,234; January 27, 1970; assigned to Mobil Oil Corporation have found that the polyalkylated naphthols are capable of imparting oxidation resistance particularly under high temperature conditions, from 500°F. and above, to lubricating oil compositions.

In general these polyalkylated naphthols are either alpha or beta-naphthols containing at least two alkyl groups attached to the naphthol nucleus, the alkyl groups being selected so that the sum total of carbon atoms for all the alkyl groups present is at least 12. While the naphthol nucleus must contain a minimum of two alkyl groups, the maximum number of alkyl groups present is seven which is the limit of available positions on the nucleus. A preferred class of polyalkylated naphthols contain from 2 to 4 alkyl groups. As some further examples of the polyalkylated naphthols of the process there may be mentioned: di-(hexyl)-1-naphthol, di-(octadecyl)-1-naphthol, tri-(dodecyl)-1-naphthol, tetra-(hexadecyl)-1-naphthol, tri-(tetrapropyl)-1-naphthol, tri-(hexadecyl)-2-naphthol, di-(dodecyl)-2-ethyl-1-naphthol, tri-(C_{12}-C_{20})-1-naphthol, and the like.

Example 1: A mixture of 96 g. (0.67 mol) 1-naphthol, 168 g. (2 mols) 1-hexene, and 30 g. boron trifluoride-ethyl etherate was stirred at 70°C. for 9 hours. The reaction mixture was washed with hot distilled water until the washings were neutral to litmus. Topping at 225°C. under reduced pressure gave a yield of 236 g. of a product that was chiefly di-(hexyl)-1-naphthol.

Example 2: A mixture of 200 g. (1.39 mols) of 1-naphthol, 700 g. (4.17 mols) propylene tetramer, and 36 g. boron trifluoride-ethyl etherate was stirred at 90°C. for about 12 to 14 hours. The reaction mixture was washed with hot distilled water until the washings were neutral to litmus. The final product, a tri-(tetrapropyl)-1-naphthol, was obtained by topping at 230°C. under reduced pressure.

Example 3: A mixture of 102 g. (0.17 mol) of 1-naphthol, 357 g. (1.42 mols) 1-octadecene, and 45 g. boron trifluoride-ethyl etherate was stirred at 95°C. for about 10 to 12 hours. The reaction mixture was washed with hot distilled water until the washings were neutral to litmus. The final product, a di-(octadecyl)-1-naphthol, was obtained by topping at 240°C. under reduced pressure.

The effectiveness of the polyalkylated naphthol antioxidant is illustrated in the Thin Film Oxidation Test which simulates the conditions to which a lubricating oil is exposed at high temperatures. According to this test each antioxidant is dissolved in a white oil (KV at 210°F. = 8 cs.) at 0.035 mols per kg. of blend. The oil sample is pumped through a preheater where it is heated to the test temperature (575° or 600°F.) into a closed oxidation chamber where the oil is brought into contact with heated pure aluminum disk spinning at 2,500 rpm. The contact time of the oil on the disk is about 1 second.

A pump continuously circulated air through the oxidation chamber where the oil is oxidized in a thin film on the spinning disk. The oxygen consumed is replaced continuously and automatically from an oxygen reservoir. Upon leaving the disk the oxidation of the oil is stopped by passing it through a water jacket held at 90°F. After ten passes over the disk the test is stopped and the deposits on the disk are measured. Test runs were also conducted on oil blends containing di-2-ethylhexyl sebacate oil.

Further details as to the test procedures are described in ASME paper No. 8 presented at the joint symposium of the ASME and ASLE at New York City, 1959. The results are reported in the table on the following page. It will be seen from the forthcoming data that the oil compositions containing the nonalkylated 1- and 2-naphthols (Tests 2 and 6) exhibited high oxidation rates of 350 and 470 (ml. O_2/5 min.) respectively. By contrast, the polyalkylated naphthols of the process exhibit oxidation rates in all instances below 100 and in some cases as low as 20 (Test 7).

TABLE: THIN FILM OXIDATION TEST AT 575°F.

Test No.	Base oil	Antioxidant	Mol per kg. oil	Wt. percent	Oxygen uptake (ml. O_2 per 5 min.)	Lacquer on disk
1	White oil+[1]	None			>500	
2	do [1]	1-naphthol	0.035	0.5	350	Heavy.
3	do	Di-(2-hexyl)-1-naphthol	0.035	1.2	35	Very light.
4	do [1]	Tri-(tertrapropyl)-1-naphthol	0.035	2.25	30	Do.
5	do [1]	Di-(2-octadecyl)-1-naphthol	0.035	2.25	30	Vo.
6	do [1]	2-naphthol	0.035	0.5	470	Medium.
7	do [1]	Tetra-(C_{12}–C_{20})-2-naphthol	0.035	3.7	20	Very light.
8	Di-2-ethylhexyl sebacate.	None			>500	
9	do	Di-(tetrapropyl)-2-ethyl-1-napthol	0.035		30	Do.

[1] Kv. at 210° F.=8 cs.

Catechol and Resorcinol

Q.E. Thompson; U.S. Patent 3,505,230; April 7, 1970; assigned to Monsanto Company describes compositions which exhibit inhibition and control of corrosion of metal surfaces and in particular metal surfaces containing lead and are obtained by the incorporation of certain o- and/or m-hydroxy or hydroxy ester-substituted aromatic compounds into a class of base stocks representative of which are pentaerythritol esters and dipentaerythritol esters.

In order to demonstrate the properties of the compositions of this process, various representative compounds were blended into an ester base stock and the resulting functional fluid composition evaluated for control and inhibition of lead corrosion. One of the major bench scale methods used for evaluating the lead corrosivity of various lubricant formulations is Federal Test Method Standard No. 791, Method No. 5321.1. In this test method, 500 ml. of the test fluid is placed into a tube containing a lead test panel. The tube is inserted in a bath which is maintained at a temperature of 375°F. Air is bubbled through the sample liquid at approximately 2 cubic feet per hour. The weight loss or net weight gain of the test sample is measured after a period of 5 hours. The weight loss is measured in milligrams per square inch.

In Table 1 below, Examples 1 through 12, the base stock that was utilized was a mixture of short-chain pentaerythritol esters and short-chain dipentaerythritol esters having an average chain length of about 6 carbon atoms. In Examples 1 through 9 the ester base stock contained 1% phenyl-α-naphthylamine. In Examples 10 through 12, the ester base stock contained a minor amount of a triaryl phosphate, 2% of a mixture of phenyl-α-naphthylamine and dioctyl diphenylamine and a minor amount of a potassium salt. In Table 1, percent inhibition of lead corrosion was obtained utilizing the following formula.

$$\text{Percent inhibition of lead corrosion} = \frac{\text{mg./in.}^2 \text{ loss without additive} - \text{mg./in.}^2 \text{ loss with additive}}{\text{mg./in.}^2 \text{ loss without additive}} \times 100$$

TABLE 1

Ex. No.	Formulation	Corrosion Inhibitor Compound	Concentration, Weight, Percent	Percent Inhibition of Lead Corrosion
1	I			
2	I	Catechol	0.05	99.8
3	I	1,3-indane dione	0.05	75.0
4	I	Catechol carbonate	0.05	89.0
5	I	Catechol diacetate	0.05	92.5
6	I	Resorcinol	0.05	98.0
7	I	2,4-dihydroxy acetophenone	0.05	96.7
8	I	2,2′,4,4′-tetrahydroxy benzophenone.	0.05	99.2
9	I	1,5-diacetoxy naphthalene	0.05	93.7
10	II	2-hydroxy-3-naphthoylanilide	0.10	94.0
11	II	Resorcinol diacetate	0.05	69
12	II	Diacetyl resorcinol	0.05	90

In order to demonstrate that the compounds did not adversely affect critical fluid properties such as oxidative stability, certain fluid compositions were prepared with and without additives and the corresponding oxidation rate determined. One of the major bench scale methods used for evaluating the oxidative stability of a lubricant is the procedure given in Federal Test Method Standard No. 591, Method No. 5308, according to which the lubricant to be tested is heated at a specified temperature in the presence of certain metals and oxygen and the viscosity increase of the lubricant is determined. Various functional fluid compositions containing the additive were tested according to the above procedure, and the conditions utilized were 125 g. of fluid, temperature of 450°F. and an air rate of 10 l. per hour.

The metal specimens used were as specified in the cited procedure, steel, copper, silver, titanium, magnesium alloy and aluminum alloy. The viscosity of the fluid before and after testing was determined at 100°F. The fluid that was utilized in Table 2 below, was the same formulation as was utilized in Examples 10 through 12 in Table 1.

TABLE 2

Ex. No.	Corrosion Inhibitor Compound	Concentration, Weight Percent	Percent Viscosity Increase	
			48 hrs.	72 hrs.
13	None		27	45
14	Catechol	0.03		49
15	2-hydroxy-3-naphthoylanilide	0.05	27	

As is demonstrated by the data, it is evident that the incorporation of compounds into an ester base stock provides a functional fluid composition which has the ability to inhibit and control lead corrosion and therefore to greatly extend the useful life of a functional fluid composition. Thus, the percent inhibition in most cases was in the order of 90% or better. This is of particular significance since the control of lead corrosion damage extends the useful life of a functional fluid composition. Thus, control of lead corrosion inhibits the formation of deposits, which deposits can adversely affect the lubricating function of a particular fluid. In addition, deposits plug filters as well as critical orifices through which a fluid of necessity has to flow in order to lubricate critical areas in a jet engine.

Anthranilamides

D.S. Bosniack and S.J. Metro; U.S. Patent 3,585,137; June 15, 1971; assigned to Esso Research and Engineering Company have found that synthetic neutral saturated carboxylic acid ester lubricating oils such as the dibasic acid esters, polybasic acid esters, diester-complex esters, and the tri- and tetraesters of polyglycols with mono- or dibasic acids are improved especially as to their supersonic turbine engine performance by incorporating a metal passivator which is an anthranilamide-type compound having the formula:

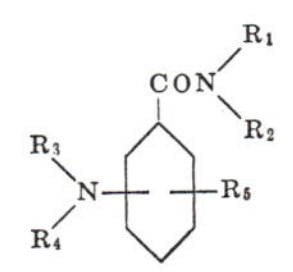

where R_1 and R_2 are each hydrogen, C_1 to C_{12} alkyl, phenyl, phenalkyl or cycloalkyl, R_3 and R_4 are each hydrogen, C_1 to C_9 alkyl, C_1 to C_{13} acyl or phenyl, and R_5 is hydrogen, C_1 to C_9 alkyl or halo C_1 to C_9 alkyl. Preferably, the nuclear amino radical is ortho to the amide group. An arylamine may be also employed, as an antioxidant, in conjuction with the anthranilamide-type compound. An arylthiazine may be used in conjunction with the antioxidant.

In the following examples, the test data have been obtained using standard tests for determining the suitability of the aviation ester lubricants for their intended use. One of the tests which have been employed is the so-called Oxidation-Corrosion Stability Test. This is a standardized test procedure described in Specification FTMSTD 791 Method 5308. Still another test which has become more or less standard, designed to measure the load carrying ability or antiwear ability of the synthetic oil under test is known as the Ryder Gear Test which is also be found and described in FTMSTD 791 Method 6508.

Example 1: The following compounded synthetic aviation oil was prepared: 100 parts constituted the tetraester and the hexaester of a mixture of 90% monopentaerythritol and 10% dipentaerythritol polyesterified with mixed C_5 to C_{10} normal alkanoic acids. Based on the 100 parts of the ester base oil, (this and all subsequent oil compositions are also based on 100 parts by weight of ester base oil) the following additives (also in parts by weight per 100 parts of base oil) were added:

p,p'-Dioctyl diphenylamine	1.6
Phenothiazine	0.4
Sebacic acid	0.02
Benzotriazole	0.05

and 0.1 of a mixture of phosphate esters which amounted to about 3 parts of the dodecylamine salt of dimethyl phosphate and 2 parts of didodecylamine salt of monomethyl phosphate. In the table on the following page, the designation, 1a, in the base oil example column, indicates the 100 parts of base oil but without any of the above additives.

Example 2: The same base ester oil as in Example 1 was used. The following additives were added:

p,p'-Dioctyl diphenylamine	1.0	N-phenyl-beta-naphthylamine	1.0

Example 3: The same base ester oil as in Example 1 was used. The following additives were added:

p,p'-Dioctyl diphenylamine	1.6	Quinizarin	0.1
Sebacic acid	0.02	Amine neutralized phosphate ester mixture of Example 1	0.1

Example 4: The same base ester oil as in Example 1 was used. The following additives were added:

p,p'-Dioctyl diphenylamine	1.6	Sebacic acid	0.02
Phenothiazine	0.4	Amine neutralized phosphate ester mixture of Example 1	0.1

Example 5: The same base composition as in Example 4 was used, but contains in addition, 0.1 quinizarin.

Example 6: 30 parts tetraester of neodecanoic acid monoethoxylate esterified with azeleic acid was added to 70 parts of the triester of trimethylol propane fully esterified with C_7 to C_{10} (average C_8) n-alkanoic acids. Then the following were added:

p,p'-Dioctyl diphenylamine	1.6	Quinizarin	0.1
Phenothiazine	0.4	Amine neutralized phosphate	
Sebacic acid	0.02	ester mixture of Example 1	0.1

Example 7: To 35 parts monopentaerythritol ester of a mixture of C_4 to C_7 normal alkanoic acids were added 50 parts of the base oil (without additives) of Example 1 and 15 parts of dipentaerythritol fully esterified with n-pentanoic acid plus some additives in the same amounts as in Example 6.

Example 8: To 50 parts of base oil of Example 1 (without additives) were added 15 parts of dipentaerythritol fully esterified with n-pentanoic acid and 35 parts of the same triester oil of Example 6 plus the same additives in the same amounts as in Example 6.

Example 9: To 70 parts of the triester of Example 6 were added 30 parts of dipentaerythritol fully esterified with n-pentanoic acid plus the same additives in the same amounts as in Example 6.

Example 10: To 100 parts of the coester of equal parts of 2-ethylhexyl alcohol and hexadecyl alcohol fully esterified with azelaic acid were added:

p,p'-Dioctyl diphenylamine	1.6	Quinizarin	0.1
Phenothiazine	0.4	Amine neutralized phosphate	
Sebacic acid	0.02	ester mixture of Example 1	0.1

The U.S. Navy Specification MIL-L-23699A for 5 centistokes gas turbine lubricants for oxidation-corrosion tests at 400°F. for 72 hours is as follows.

±0.4 copper corrosion in milligrams per square centimeter
±0.2 milligram per square centimeters magnesium, silver, aluminum and iron corrosion
A differential increase or decrease in the used oil as to the viscosity at 100°F. in percentage change of -5 to +25%
A differential increase in used oil total acid number of 3.0 milligrams KOH per gram of oil maximum

Examples 11 through 47:

Example No.	Base oil, Example No.	Additives	Add. amount, pts.	Corrosion, mg./cm.² Cu	Corrosion, mg./cm.² Mg	Used oil: Percent change in viscosity, increase cs. at 100° F.	Used oil: Total acid number, increase-Mg·KOH/gm.
11	1	None		−0.73	+0.09	36.13	4.03
12	1	Anthranilic acid	0.05	−0.56	+0.09	33.96	3.23
13	1	Anthranilic acid N-phenyl β Naphthylamine	0.05 0.25	−0.67	+0.09	35.93	1.84
14	1	Anthranilamide	0.1	−0.26		24.47	1.59
15	1	do	0.2	−0.01	−0.01	24.79	1.41
16	1	Anthranilamide Quinizarin	0.1 0.1	−0.18		24.68	0.96
17	1	Anthranilamide Quinizarin	0.2 0.1	−0.18		24.37	1.14
18	1a	None		−0.99	−23.48	(1)	9.41
19	1a	Anthranilamide	0.1	−0.61	−18.87	11,500.00	7.84
20	1a	do	0.25	−0.10	−15.55	294.00	7.85
21	1a	Tricresylphosphate	1.00	−0.30	−30.59	424.00	9.00
22	1a	Anthranilamide Tricresylphosphate	0.25 1.00	−0.03	−30.09	764.00	8.51

(continued)

Example No.	Base oil, Example No.	Additives	Add. amount, pts.	Corrosion, mg./cm.[2] Cu	Corrosion, mg./cm.[2] Mg	Used oil: Percent change in viscosity, increase cs. at 100° F.	Used oil: Total acid number, increase-Mg·KOH/gm.
23	2	None		−1.29	−1.92	35.40	13.6
24	2	Anthranilamide	0.1	−0.14	−0.02	31.70	1.04
25	2	Anthranilamide Phenothiazine	0.1 0.4	−0.23	−0.01	33.57	1.69
26	2	o-(N-neocaprylyl amino) benzamide	0.1	−0.74	−0.06	33.86	1.59
27	2	o-NH_2, N-octyl benzamide	0.1	−0.53	+0.01	33.63	1.09
28	2	o-NH_2, N-phenyl benzamide	0.1	−0.47	+0.01	33.39	0.96
29	2	o-(CH_3NH) benzamide	0.1	−0.32	+0.01	32.86	0.92
30	2	O-(CH_3NH),N,N diethyl, benzamide	0.1	−0.64	0.00	22.39	0.98
31	2	O-NH_2,N,N diethyl, benzamide	0.1	−0.52	−0.01	20.32	4.72
32	3	None		−0.74		44.24	12.53
33	3	Anthranilamide	0.1	−0.61		41.12	8.01
34	3	Phenothiazine	0.4	−0.52		16.10	1.15
35	3	Anthranilamide Phenothiazine	0.1 0.4	−0.20		18.19	0.92
36	4	None		−1.18	+0.07	24.61	2.65
37	4	Anthranilamide	0.2	−0.16	+0.05	24.17	2.17
38	4	Benzotriazole	0.2	−0.05	−7.28	11.23	6.39
39	5	None		−0.65	+0.06	21.78	2.30
40	5	Anthranilamide	0.1	−0.20	+0.04	22.00	1.65
41	5	Benzotriazole	0.1	−0.28	−5.70	12.10	5.69
42	6	Anthranilamide	0.1	−0.10	+0.11	21.00	2.52
43	7	do	0.1	−0.23	+0.12	24.95	1.04
44	8	do	0.1	−0.18	+0.12	22.00	1.75
45	9	do	0.1	−0.15	+0.11	19.56	1.50
46	10	None		−0.73	+0.20	135.00	22.00
47	10	Anthranilamide	0.1	−0.26	+0.20	51.90	19.24

[1] Too heavy.
NOTE.—All of the data set forth in the table were obtained by carrying out the tests at 400° F. for 72 hours.

From the tabulated data shown in the preceding table, it is apparent that for maximum effectiveness, there is decided improvement in meeting these specifications in the use of both anthranilamide type compounds and aromatic amines and that the presence of benzotriazole, if anything, is not so important and not necessary. The presence of quinizarin although not required to meet Navy specifications, is nevertheless advantageous in improving used oil properties. Anthranilamide and its derivatives are beneficial when used in synthetic ester lubricating oils alone or along with conventional additives in that copper and magnesium corrosion is minimized, and the viscosity increase and acidity increase are also held to a minimum.

N-Guanidinosalicylamide-Tall Oil Acid Products

A process described by R.H. Boehringer and R.E. Vail; U.S. Patent 3,595,793; July 27, 1971; assigned to Emery Industries, Inc. is concerned with compounds that may be used as metal deactivator additives in engine lubricants. These compounds are salts of an amide prepared from a benzoic acid or a substituted benzoic acid and aminoguanidine bicarbonate, and a salt forming aliphatic or aromatic carboxylic acid having from 6 to 40 carbon atoms.

The salts are prepared by first forming intermediate amides. These amides are made from aminoguanidine bicarbonate and a benzoic acid. The benzoic acids which may be used in the preparation of the intermediate amides may have alkyl, alkenyl, halogen or hydroxyl groups substituted for a hydrogen atom on the benzene ring. Specific examples of benzoic acids which may be used in forming the compounds are salicylic acid, orthonitrobenzoic acid, orthochlorobenzoic acid and benzoic acid itself.

The salt forming carboxylic acids which may be used to make the salts may be chosen from aromatic or aliphatic mono- or polycarboxylic acids having from 6 to 40 carbon atoms. Representative carboxylic acids are benzoic acid, oleic acid, stearic acid, isostearic acid (described in U.S. Patent 2,812,342), adipic acid, azelaic acid, dodecanedioic acid, and tall oil fatty acids, or other unsaturated fatty acids, such as soybean oil acids, linseed oil acids, or cottonseed oil acids. These acids may be used either alone or in combination. When the tall oil fatty acids or other unsaturated fatty acids are used, it has been found that they work best when used in combination with rosin acids.

Example 1: N-guanidinosalicylamide, an intermediate compound in the preparation of salts of this process, was prepared by reacting 136.8 g. of aminoguanidine bicarbonate with 138 g. of salicylic acid in 400 ml. of an alcohol mixture (12% isopropanol and 88% methanol). The mixture was stirred for 15 minutes without heat and then for 30 minutes at a constant temperature of 60°C. The reaction mixture was then placed in a three-necked 500 ml. flask and the alcohol was removed using agitation, heat and vacuum, leaving the N-guanidinosalicylamide product.

Example 2: Another intermediate was prepared following the procedure in Example 1 but substituting 122.1 g. of benzoic acid for the salicylic acid. The product, N-guanidinobenzamide, was recovered and retained for use in preparation of a salt.

Example 3: A salt was prepared by adding 50 g. of the N-guanidinosalicylamide prepared in Example 1 to 300 ml. of methanol and 72.3 g. of tall oils acids, a mixture of fatty acids and rosin acids. The fatty acid portion of the tall oil acid was composed of acids having 14 to 18 carbon atoms, about 50% of the acids being unsaturated. The fatty acids comprised 80% by weight of the acid mixture.

The remaining acids are primarily resin acids of the abietic and pimaric types having a phenanthrene nucleus. They are unsaturated and therefore reactive. An unsaponifiable portion of the resin (3 to 10%) contains hydrocarbons and high molecular weight alcohols. After stirring for 15 minutes the alcohol solution of amide and acids was distilled in a three-necked 500 ml. flask with stirring, heat and vacuum. When the temperature reached 120°C., the product was discharged.

Example 4: A salt was formed following the same procedure described in Example 3 except that 42.1 g. of dodecanedioic acid were substituted for the tall oil acids.

Example 5: A salt was formed following the procedure in Example 3 except that 106.0 g. of dimer acid, a polymerized oleic acid, sold under the trademark Empol 1016, were substituted for the tall oil acids.

Example 6: A salt was formed following the procedure in Example 3 except that 55.4 g. of the intermediate prepared in Example 2 were substituted for the N-guanidinosalicylamide.

Example 7: A salt was prepared following the procedure in Example 2 except that 82.7 g. of isostearic acid, a modified fatty acid described in U.S. Patent 2,812,342, were substituted for the tall oil acids.

Example 8: A salt was prepared following the procedure set forth in Example 3 except that 72.2 g. of oleic acid were substituted for the tall oil acids.

Examples 9 through 24: Each of the salts prepared in Examples 3 through 8 was compounded in synthetic ester lubricant base fluids along with other conventional lubricant additives. The compositions of these fluids are shown in Table 1 below. For comparative purposes, two synthetic lubricant compositions shown in Examples 22 and 23 were prepared without the salts of this process. A commercial metal deactivator, benzotriazole, was used in place of a salt of the process in Example 24 for comparative purposes.

TABLE 1

Example	9	10	11	12	13	14	15	16	17	18	19	20	21	22	23	24
Base fluid:																
Pentaerythritol ester [1]	94.93	93.73	93.73	93.73	94.93	94.93								94.93	93.73	94.93
Diisodecyl azelate							90.23									
Ditridecyl azelate								97.45								
Dioctyl azelate									72.90							
PE $(nC_5)_4$									22.00		94.93	22.00				
1276-65R [2]										94.925						
Di-2-ethylhexyl adipate												52.93				
Di-2-ethylhexyl dodecanedioate												20.00				
Emery 3383 [3]													46.87			
TMP $(1210)_3$ [4]													46.87			
Additives, salt of Example:																
3	0.05						0.05	0.05	0.05	0.05	0.05	0.05	0.05			
4		0.05														
5			0.05													
6				0.05												
7					0.05											
8						0.05										
Commercial:																
Metal deactivator (benzotriazole)																0.05
Antioxidant	2.00	2.00	2.00	2.00	2.00	2.00	2.00	2.00	2.00	2.00	2.00	2.00	2.00	2.00	2.00	2.00
Thickener		1.2	1.2	1.2			2.50						1.2		1.2	
Detergent							4.20									
Quinizarin	0.02															
TCP	3.00	3.00	3.00	3.00	3.00	3.00			3.00	3.00	3.00	3.00	3.00	3.00	3.00	3.00
Azelaic										0.005			0.005			0.005

[1] An ester of monopentaerythritol and a mixture of C_5 to C_9 aliphatic monobasic acids.
[2] An ester of technical grade pentaerythritol [12% di and 88% monopentaerythritol] and a mixture of C_5 to C_9 aliphatic monobasic acids.
[3] An ester of trimethylol propane and a mixture of C_6 to C_9 aliphatic monobasic acids.
[4] 1210 is a mixture of C_6, C_7, and C_8 monobasic acids which have an average carbon atom content of 7.

Examples 17 and 19 through 21 in Table 1 were evaluated by using Federal Test 791, Method 5308, at several temperatures. The results of these tests showed the compounded lubricants to have good oxidation stability, slight effect on the metals and low sludge content. Based upon these results, the lubricants are satisfactory for use in diesel engines. The lubricant compositions prepared above were evaluated for oxidation stability using Federal Test 791, Corrosion and Oxidation Stability, Method 5308, as modified by Pratt and Whitney specifications described in PWA 521-B.

The oxidation stability test, described in Federal Test 791, Method 5308, is performed by heating an aerated lubricant with five metal coupons present in the fluid. These metals are copper, steel, aluminum, magnesium and silver. This test is conducted at a temperature of 425°F. instead of the usual 347°F. for 72 hours. At the end of this period each metal weight must not have changed by more than ±0.30 mg./cm.^{2}, and the viscosity of the fluid should not have increased by more than 50%.

The results of the tests performed on the compositions of Table 1 are provided in Table 2. These results show that the salts of Examples 3 through 6 and 8 when used in lubricant compositions 9 through 12 and 14 effectively inhibited metal loss to well within the tolerable limits; whereas in comparative Examples 22 and 23 which did not have a metal deactivator added, the copper and magnesium losses were far in excess of tolerable levels. It may also be seen that in comparison with the commercial metal deactivator used in composition 24, the salts are superior in their ability to inhibit metal loss.

TABLE 2

Example	9	10	11	12	13	14	15	16	18	22	23	24
Viscosity change at 100° F.	71.70	48.9	46.2	48.6	59.1	69.7	56.7	69.7	71.8	80.8	81.3	84.6
Corrosion:												
Copper ($mg./cm.^2$)	−0.220	−0.09	−0.07	−0.08	−0.285	−0.238	−0.161	−0.238	−0.146	−1.070	−1.32	−0.86
Steel ($mg./cm.^2$)	−0.008	+0.05	+0.06	−0.02	+0.015	−0.092	−0.015	−0.092	+0.023	−0	+0.05	+0.02
Silver ($mg./cm.^2$)	−0.254	−0.02	−0.05	+0.02	−0.269	−0.238	−0.208	−0.238	+0.062	−0.238	−0.01	−0.02
Aluminum ($mg./cm.^2$)	−0.023	−0.01	−0	+0.01	−0.061	−0.069	−0.06	−0.068	−0	−0.010	−0.03	−0.02
Magnesium ($mg./cm.^2$)	−0.254	−0.28	−0.07	−0.19	−1.046	−0.246	−0.142	−0.246	−0.020	−1.575	−0.15	−2.26
Sludge (mg.)	60.6	9.5	105	26.0	153.1	49.4	36.8	49.4	39.8	(1)	161	(1)

[1] Unfilterable.

The salt deactivator used in the lubricant composition shown in Example 13 did not perform as effectively as the other salt deactivators; however, it did show marked improvement in copper corrosion over comparative Examples 22 through 24.

The use of the salts reduced the viscosity change of the lubricant fluid as may be seen in Table 2 by comparing Examples 9 through 16 with Examples 22 through 24. In certain instances the viscosity change was not lowered sufficiently to meet the specifications of PWA-521-B; however, the specification of PWA 521-B can be easily met by using a commercial thickener as was done in Examples 10 through 12.

Organometallic Compounds

A.D. Forbes and P. Gould; U.S. Patent 3,413,223; November 26, 1968; U.S. Patent 3,422,014, January 14, 1969 and U.S. Patent 3,399,139; August 27, 1968; all assigned to The British Petroleum Company Limited, England describe a lubricating composition based on a blend of two ester oils, viz:

(1) a liquid aliphatic ester consisting of one or more neutral polyesters that have been prepared by reacting together under esterification conditions and in one or more stages:

 (a) a mono- and/or polyhydric alcohol having from 5 to 15, preferably from 5 to 8 carbon atoms per molecule and having no hydrogen atoms attached to any carbon atom in a β-position with respect to any —OH group and

 (b) a mono- and/or polycarboxylic acid having from 2 to 14, preferably from 6 to 10, carbon atoms per molecule, and

(2) a liquid aromatic ester consisting of one or more diesters, XOOCRCOOY, where R is a saturated hydrocarbon group having from 1 to 14 carbon atoms and X and Y are hydrocarbon or oxygen-containing hydrocarbon groups having from 6 to 20 carbon atoms each of which either is a benzene ring or contains a benzene ring directly attached to the —COO— group.

Preferably the blend contains from 10 to 95% weight of Component (2), more preferably from 40 to 80%. For most purposes, it is desirable that the viscosity of the blend shall be within the range from 1 to 30, preferably from 2 to 9, centistokes at 210°F. and these ranges also represent suitable ranges to achieve in the production of the individual components.

It has also been found that the good high temperature oxidation stability of the resulting blends can be improved still further by the addition of certain organometallic compounds, which are described in U.S. Patent 3,399,139 and U.S. Patent 3,422,014.

According to a preferred form of the process the lubricating composition also contains, dissolved in the base oil blend, an organometallic compound in an amount which gives a metal content of up to 500 parts per million (ppm), preferably from 1 to 15 ppm, based on the total weight of the composition.

The organometallic compounds present in the compositions are those compounds which oxidize or reduce the organic radicals formed during oxidative degradation of esters, that is, salts of aliphatic acids having more than 8 carbon atoms or complexes in which the ligands are composed of any or all of the elements carbon, hydrogen, oxygen and nitrogen. Suitable metals are the transition metals, especially the first transition series (according to the Periodic Table based on that of Mendeleeff), and the nontransition metals such as cerium, which metals can take part in "electron transfer" reactions. It is preferred to use those metals in which the oxidation potential of a couple between a lower and a higher oxidation state is greater than or equal to +0.75 (using the British sign convention and couples in acid solution). The organometallic compound must not be so volatile that it is distilled or evaporated out of the lubricating compositions at high temperatures, for example, above 200°C. It must also be soluble in the base oil.

The preferred transition metals are manganese and cobalt, though chromium or iron could also be used. Copper has been found to be a suitable transition metal. Cerium is a suitable nontransition metal. The preferred complexes are the acetylacetonates, especially the hydrated acetylacetonates, for example, manganous and cobaltous acetylacetonates. Cupric phthalocyanine has also been found to be suitable.

A suitable copper salt is cupric stearate. It has been found that organometallic compounds of this type act as high temperature antioxidants for example, at >200°C. in aliphatic/aromatic ester blends.

Example: High temperature oxidation tests were carried out on a series of aromatic/aliphatic ester blends, with and without organometallic additives and conventional amine antioxidants. The samples were heated for 5 hours at 259°C. Air was bubbled through the sample at a rate of 15 l. per hour. Oxidation stability was determined by measuring the viscosity and acidity increase in the oil during the test and the amount of oxygen consumed. The composition of the blends tested are given in Table 1 below.

TABLE 1

Additive	A	B	C	D	E
p,p'-Dioctyldiphenylamine, percent wt	1.9	1.0	1.9	Nil	Nil
Benzotriazole, percent wt	0.2	0.1	0.2	Nil	Nil
Cobaltous acetylacetonate dihydrate, p.p.m. Co.	10	Nil	Nil	10	Nil

In the case of Blends A and C the composition of the base fluid was 73.4% weight of di-o-t-butylphenyl azelate and 24.5% weight of trimethylol tricaprylate based on the total weight of the composition. In the case of Blend B the composition was 74.2 and 24.7% weight respectively and in the case of Blends D and E, 75.0 and 25.0% weight respectively.

In the above table, p, p'-dioctyldiphenylamine is the conventional amine antioxidant, cobaltous acetylacetonate dihydrate is the organometallic derivative and benzotriazole is a copper passivator. The results obtained for the oxidation of these blends were compared with a reference oil (Blend F), which was a typical aliphatic ester oil of comparable viscosity at 210°F. to that of the aromatic/aliphatic ester blends, i.e., from 7 to 9 centistokes. The composition of the reference oil was p, p'-dioctyldiphenylamine (4.0% weight), benzotriazole (0.25% weight) and a mixture of esters (95.75% weight) synthesized from caprylic acid (11 mol), trimethylolpropane (4.3 mol) and sebacic acid (1 mol). The oxidation tests of the various blends gave the results listed in Table 2 below.

TABLE 2

Blend	Time for 500 g. sample to consume 0.5 mol O_2 (h.)	O_2 uptake per 6 g. sample (ml.)	$KV_{210°}$ c. increase (%)	Acidity increase (mg. KOH/g.)
A	3.2	180	67	2.5
B	0.50	462	242	35.6
C	0.55	540	160	35.9
D	0.45	490	140	36.6
E	0.10	756	383	45.4
F	0.20	812	[1]	[2]

[1] Solidified.
[2] Not determined.

From a study of this table it can be seen that: (a) all the aromatic/aliphatic ester blends were more stable to oxidation than the aliphatic ester oil. (b) The most striking result was obtained for the blend containing both the organometallic compound and the amine antioxidant. The organometallic compound is apparently primarily responsible for the protection of the aromatic ester component and the amine antioxidant is apparently primarily responsible for the protection of the aliphatic ester component. The copper passivator, benzotriazole, is considered to have no effect on the oxidation. A comparison of the results for Blends A and B indicates that it is the presence of the organometallic compound, which is responsible for the great resistance to oxidation of this blend.

(c) The organometallic compound retards oxidation of the blend in the absence of other additives, as can be seen by comparing the results for Blends D and E. (d) The oxidation of the aromatic/aliphatic ester blends is retarded by conventional amine antioxidants, as can be seen for the results for Blends B and C. From these results it can be seen that an increase in the amine concentration from 1 to 2% weight has little effect on the oxidation resistances. It is considered that amine antioxidants are relatively ineffective, compared with organometallic additives, in aromatic ester systems. Therefore, for effective protection of the aromatic/aliphatic ester blends the organometallic additive must be present.

Tests were carried out comparing the properties of Blend A with an inhibited aliphatic ester Blend G. The ester used in Blend G was of similar constitution to that used in Blend F, but from a different batch, and was inhibited to the same degree by p, p'-dioctyldiphenylamine (4.0% weight). Blend G also contained benzotriazole (0.25% weight) as a copper passivator. The tests were carried out in a General Electric Roxana Panel Coker. In this apparatus, a sample of the oil is heated at 400°F. in a sump. A rotating paddle, half immersed in the sample, throws the liquid on to a hot plate held at 200°F. The duration of the test is 8 hours and, at the end of the test, the weight of coke deposited on the hot plate and the changes in viscosity and acidity are measured.

The test is a measure of the oxidative stability of a lubricant and of its coke depositing characteristics. The results are given in Table 3 on the following page.

TABLE 3

Sample	Coke deposit (mg.)	$KV_{100°F.}$ increase (percent)	Acidity increase (mg. KOH/g.)
A	49	74	3.8
G	19	473	4.7

For both samples the coke deposit and acidity increases are well within the General Electric Specification A50T28A requirements (100 mg. and 6 mg. KOH/g. respectively). However, the viscosity change obtained for Blend A is six to seven times less than that obtained for Blend G.

p-Diethylaminophenyl Silanes

H.M. Schiefer and D.R. Weyenberg; U.S. Patent 3,340,286; September 5, 1967; and U.S. Patent 3,344,070; September 26, 1967; both assigned to Dow Corning Corporation have found a method for rendering high temperature lubricants, in particular certain ester lubricants, extremely resistant to oxidation at high temperatures, such as those found in aircraft jet engines.

The best lubricants for jet engines have been found to be esters of alkanoic acids and polyhydric alcohols which have at least two methylol groups bonded to a quaternary carbon atom, such as 1,2,2-trimethylolpropane. Even though these compounds have a high degree of oxidative and heat stability, they are not stable in jet engines when the sump temperatures are in excess of 170°C. and the temperature at some of the bearings is in excess of 315°C.

The composition comprises a mixture of (1) a high temperature organic lubricating fluid and (2) from 0.1 to 5% by weight based upon the weight of (1) of a compound of the formula

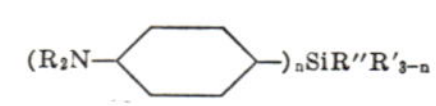

where each R is an alkyl radical of from 1 to 10 inclusive carbon atoms, R' is a radical selected from the group consisting of alkyl radicals of from 1 to 10 inclusive carbon atoms, phenyl, vinyl, hydrogen atoms and —OR''' radicals where R''' is an alkyl radical containing from 1 to 4 carbon atoms, R'' is a radical selected from the group consisting of phenyl and —OR''' radicals where R''' is defined above and n is an integer of from 1 to 3 inclusive.

The antioxidant (2) is present in an amount from 0.1 to 5% by weight based upon the weight of the lubricating fluid. However, the preferred range of the antioxidant is from 0.5 to 3% by weight. There is generally very little improvement in the stability of the lubricating fluid when more than 2% of the antioxidant is added. R in the organosilicon antioxidant is an alkyl radical of from 1 to 10 inclusive carbon atoms. The antioxidants are prepared by first preparing a Grignard reagent and then reacting this reagent with a chlorosilane. The first reaction is represented by the formula:

(1)

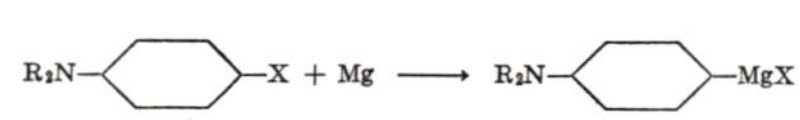

The antioxidants are obtained by reacting the above Grignard reagent with a chlorosilane in accordance with the following formula:

(2)

R, R', R'' and n are defined above. X is any halogen atom. Alternatively, these antioxidants can be prepared by first preparing a lithium reagent by the reaction of a p-halogeno-N,N-dialkylaniline with lithium. The lithium reagent is then reacted with the proper chlorosilane. Tetrahydrofuran is a suitable solvent for both reactions. Reaction (1) can be conducted at a temperature slightly above room temperature. Reaction (2) is best carried out at reflux for a substantial period of time. The product can then be separated from the metal salts produced in Reaction (2). The best antioxidants are those where R is ethyl, R'' is phenyl, R' is either phenyl or methyl, preferably phenyl and n is an integer of from 1 to 2 inclusive. Examples of such compounds are:

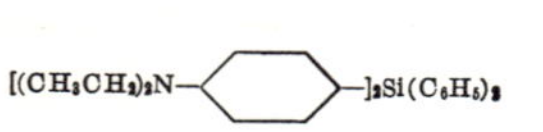

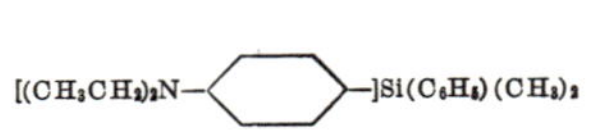

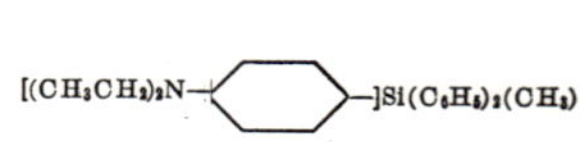

The preferred antioxidant is

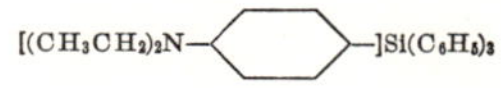

The antioxidants of the process besides proving antioxidant properties and stabilizing the viscosity and the acidity of the high temperature lubricating fluids, particularly the esters of alkanoic acids, also have the property of preventing sludge formation in the high temperature lubricating fluids. Antioxidants which contain only alkyl radicals attached to the silicon atom of the silane other than the dialkylaminophenyl radicals, do not prevent sludge formation.

Example 1: 50 ml. of tetrahydrofuran was added to a 500 ml. three-necked flask equipped with condenser, stirrer and dropping funnel. 10 g. (0.412 g. atom) of magnesium turnings was then added. 68.0 g. (0.298 mol) of p-bromo-N,N-diethylaniline was dissolved in 100 ml. of tetrahydrofuran. The reaction was primed twice by placing a few pieces of magnesium in a test tube with a few milliliters of tetrahydrofuran and p-bromo-N,N-diethylaniline. Once the reaction had started in the test tube, the contents were dumped into the flask. The p-bromo-N,N-diethylaniline in tetrahydrofuran was then added over a 90 minute period. The pot temperature was maintained at 30° to 40°C. during this time. The contents were then stirred for 45 minutes at room temperature. 51.2 g. (0.22 mol) of $(C_6H_5)_2(CH_3)SiCl$ was then dissolved in 50 ml. of tetrahydrofuran and the solution was added to the reaction mixture over a 30 minute period. The pot temperature was maintained at 30° to 35°C. during this time. The mixture was then refluxed for 7 hours and then stirred at room temperature for 11 hours.

A solution of 42.72 g. of NH_4Cl and 16 g. of NaOH in 800 ml. of H_2O was prepared and placed in a beaker containing ice. The reaction mixture was poured into this solution. The organic layer was separated and distilled at a reduced pressure of 0.20 to 0.22 mm. of Hg. 50.4 g. (65.4% yield) of a yellow liquid boiling at 202° to 208°C. was obtained. The yellow liquid crystallized to give a tan solid with a melting point of 68° to 70°C. The tan solid was recrystallized from ethanol and dried. 40.3 g. of (p-diethylaminophenyl)diphenylmethylsilane, a white solid with a melting point of 73°C. was obtained.

Example 2: The following organosilicon compounds were prepared in accordance with the procedure of Example 1. Any modifications of the procedure of Example 1 are set forth in the following paragraphs.

(A) The compound (p-diethylaminophenyl)triphenylsilane was prepared by the reaction of 68.43 g. (0.30 mol) p-bromo-N,N-diethylaniline with 10 g. (0.41 g. atom) of magnesium and 88.35 g. (0.30 mol) of triphenylchlorosilane. The mixture was refluxed for 19 hours and then hydrolyzed. The crude product was recrystallized from diethylether to yield 56.2 g. (45.2% yield) of the pure product, (p-diethylaminophenyl)triphenylsilane.

(B) A compound bis(p-diethylaminophenyl)diphenylsilane was prepared by the reaction of 68 g. (0.298 mol) of p-bromo-N,N-diethylaniline with 6.25 g. (0.90 g. atom) of lithium wire and 28.6 g. (0.113 mol) of diphenyldichlorosilane in diethylether. The ether was replaced by tetrahydrofuran and the reaction refluxed for 8 hours and then hydrolyzed. The crude product was recrystallized from a commercial aliphatic solvent to give 22.65 g. (41.9% yield of theoretical) of the pure product, bis(p-diethylaminophenyl)diphenylsilane.

(C) The compound (p-diethylaminophenyl)phenyldimethylsilane was obtained by the reaction of 68 g. (0.298 mol) of p-bromo-N,N-diethylaniline with 10 g. (0.412 g. atom) of magnesium turnings and 37.55 g. (0.22 mol) of phenyldimethylchlorosilane. The mixture was refluxed for 7 1/2 hours and then hydrolyzed. The crude product was separated by distillation at reduced pressure (0.25 to 0.36 mm. Hg). Redistillation of the crude product at reduced pressure (0.1 mm. Hg) gave 32.2 g. (51.6% yield of theoretical) of the pure product, (p-diethylaminophenyl)phenyldimethylsilane.

Example 3: The antioxidant properties of the compounds prepared in the preceding examples when used alone or in combination with phenyl-α-naphthylamine in a lubricating fluid are measured in the following tables. The lubricating fluid in which these antioxidant properties were tested is the triheptanoate of 1,2,2-trimethylolpropane. The antioxidant to be tested was merely added to the lubricating fluid in the quantity set forth in the following tables. It was necessary in some cases, to heat the antioxidant-containing lubricating fluid in order to obtain a homogeneous solution.

In Table 1 on the following page, the stability of these lubricating fluids was tested by heating the fluid at 218°C. and then measuring the properties at the expiration of certain time periods. Air was bubbled through the lubricating fluid at a rate of 8 l. of dry air per hour per 20 g. of lubricating fluid. The air was bubbled into the fluid through a 1/16 inch (inside diameter) tube. At the end of various time periods the viscosity of the lubricating fluid was tested at 99°C. The acid number of the fluid was also checked at this time in accordance with ASTM D-974-58T (1958). The concentration of antioxidant is expressed in weight percent based upon the weight of the lubricating fluid.

The percent increase in viscosity in each case is based upon the viscosity of the lubricating fluid prior to heating. The increase in acid number is merely the difference between the acid number prior to heating and the acid number after heating for a certain period of time.

TABLE 1

Antioxidant	Time in hours	Visc. in cs. at 99° C.	Percent Increase in Viscosity	Acid No.	Increase in Acid Number
(1) None	0	3.53	----------	0.017	----------
	24	10.05	185	23.1	23.1
	30	15.28	333	25.5	25.5
(2) 1% phenyl-α-naphthylamine	0	3.60	----------	0.014	----------
	24	5.99	66.4	18.6	18.6
	30	7.19	100.0	19.3	19.3
(3) 2% phenyl-α-naphthylamine	0	3.66	----------	0.02	----------
	24	4.85	32.5	7.2	7.2
(4) 1% (p-diethylaminophenyl)-triphenylsilane (Ex. 2A) and 1% phenyl-α-naphthylamine.	0	3.52	----------	0.10	----------
	22	3.72	5.7	0.39	0.3
	30	3.68	4.5	0.45	0.4
(5) 1% (p-diethylaminophenyl)methyldiphenylsilane (Ex. 1) and 1% phenyl-α-naphthylamine.	0	3.65	----------	0.10	----------
	24	----------	----------	0.50	0.40
(6) 0.5% bis(p-diethylaminophenyl)diphenylsilane (Ex. 2B) and 0.5% phenyl-α-naphthylamine.	0	3.52	----------	0.056	----------
	30	4.74	35	12.7	12.6

The lubricating fluids tested in Table 2 were tested in the same manner as those tested in Table 1 except that the air was bubbled into the sample at a rate of 1 l. per hour per 20 g. sample. The viscosity and acid number were determined after 24 hours at 232°C. The lubricating fluid without any additive had, prior to heating, a viscosity of 3.47 at 99°C. and an acid number of 0.28.

TABLE 2

Additive	Visc. in cs. at 99° C.	Percent Viscosity Increase	Acid No.	Acid No. Increase
(1) None	4.39	26.5	22.2	22.0
(2) 0.5% phenyl-α-naphthylamine	4.2	25.0	11.2	11.0
(3) 0.5% (p-diethylaminophenyl) triphenylsilane (Ex. 2A) and 0.5% phenyl-α-naphthylamine.	3.93	13.3	9.10	8.9
(4) 0.5% (p-diethylaminophenyl) triphenylsilane (Ex. 2A) and 0.5% phenyl-α-naphthylamine.	3.82	10.1	3.14	2.96
(5) 0.5% (p-diethylaminophenyl)diphenylmethylsilane (Ex. 1).	4.17	18.1	19.6	19.1
(6) 0.5% (p-diethylaminophenyl) methyldiphenylsilane (Ex. 1) and 0.5% phenyl-α-naphthylamine.	4.18	18.8	9.5	9.0
(7) 0.5% (p-diethylaminophenyl) phenyldimethylsilane (Ex. 2C).	4.11	16.6	16.8	16.4
(8) 0.5% (p-diethylaminophenyl)phenyldimethylsilane (Ex. 2C) and 0.5% phenyl-α-naphthylamine.	3.75	3.3	4.06	3.6

EXTREME PRESSURE ADDITIVES

Triphenyl Phosphorothionate and Chlorinated Polyaryl Compounds

A.C.B. MacPhail, F.T. Barcroft and C.B. Milne; U.S. Patent 3,297,574; January 10, 1967; assigned to Shell Oil Company have found that improved extreme pressure properties are imparted to an aliphatic acid ester lubricating oil by the addition of minor amounts of each of (1) a chlorinated polyaryl compound and (2) a triaryl ester of phosphorothionic acid. More specifically, improved lubricating compositions comprise: (A) an aliphatic carboxylic ester lubricating base; (B) a minor amount of one or more chlorinated polyaryl hydrocarbons represented by the formula

$$\left(\underset{Cl_x}{\overset{|}{Ar}}-\underset{Cl_y}{\overset{|}{Ar}}\right)_z$$

where Ar is a mono- or polynuclear aromatic hydrocarbon such as benzene, naphthalene, or anthracene, preferably at least one of which is benzene, x and y are integers of from 1 to 8 and preferably from 2 to 4, the chlorine atoms being attached directly to the carbon atoms of the aromatic nuclei, and z is an integer from 1 to 10, preferably from 1 to 4; and (C) a minor amount of triaryl ester of phosphorothionic acid, each group containing 6 to 10 carbon atoms.

The chlorinated polyaryl compounds described as additive (B) above, are used in concentrations between 0.5 and 10% by weight and preferably between 1 and 6% by weight of the lubricating oil compositions. Compounds of this type include chlorinated and fluorinated diphenyl, triphenyl, tetraphenyl, pentaphenyl, phenylnaphthyl and mixtures thereof. The halogen content of these compounds varies from 20 to 80% by weight, preferably 40 to 70% by weight. A class of compounds of this type are commercially available under the trade name Aroclor. Particularly preferred are chlorinated (40 to 70%) diphenyls commercially available as Aroclor 1254 and Aroclor 1247, the former being a 40% chlorinated diphenyl. Although these compounds themselves produce desirable extreme pressure characteristics in synthetic lubricating oils, the desirable effects of these compounds do not substantially increase with a corresponding increase of additive concentration. In other words, although the addition of a small amount of these compounds imparts desirable properties to the oil, addition of increased amounts does not produce the expected further increase in extreme pressure properties. The low rate of increase of scuffing load with increasing Arochlor concentration is shown in Tables 1 and 2 (Examples 2 through 4) below.

TABLE 1: COMPOSITION

Composition of Example	1	2	3	4	5	6	7	8
Di(isooctyl azelate), percent w	100.0	94.0	95.0	92.0	95.0	92.0	93.0	90.0
Aroclor 1254 (chlorinated diphenyl), percent w	---	3.0	5.0	8.0	---	3.0	5.0	8.0
Triphenyl phosphorothionate, percent w	---	---	---	---	2.0	2.0	2.0	2.0
3,7-Dioctyl phenothiazine, percent w	---	1.5	---	---	1.5	1.5	---	---
5-Methyl benzotriazole, p.p.m	---	25	---	---	25	25	---	---
D.C. 200/12,000 silicone, p.p.m	---	5	---	---	5	5	---	---
Phenyl-α-naphthylamine, percent w	---	1.5	---	---	1.5	1.5	---	---

TABLE 2: PERFORMANCE DATA

Example	1	2	3	4	5	6	7	8
I.A.E. Scuffing Load, lbs.:								
2,000 r.p.m.—								
60° C	45	55	70	75	90	100	110	>120
110° C	45	74	70	75	80	105	110	>120
6,000 r.p.m.—								
60° C	35	45	47	50	40	40	65	85
110° C	25	34	37	40	45	30	60	85
Sum of scuffing loads, lbs	150	208	224	240	255	275	345	>410
Increase of scuffing loads with increasing Aroclor concentration	---	58	16	16	---	20	70	>65
Rate of increase of scuffing load per 1% increase in Aroclor concentration	---	19	8	5	---	7	35	>22

The extreme pressure effect of the chlorinated polyaryl compounds is considerably enhanced by the addition of 0.5 to 10%, preferably 1 to 6% by weight of the lubricating oil composition, of a triaryl ester of phosphorothionic acid additive (C), above. Addition of the prescribed amount of these esters surprisingly increases the effectiveness of the chlorinated polyaryl additive; i.e., as the concentration of chlorinated polyaryl is increased in the presence of a constant amount of triaryl phosphorothionate, there is a corresponding large increase in antiscuff properties. Thus, the combination of these additives produces excellent extreme pressure properties not attainable with the conventional chlorinated polyaryl additive alone.

In Tables 1 and 2 (Examples 6 through 8), it is shown that the rate of increase of scuffling load with increasing Arochlor concentration in the presence of 2% triphenyl phosphorothionate is considerably higher than in the absence of this compound. The resulting lubricating oil has excellent extreme pressure properties which are only available with the use of the phosphorothionate additive.

Several compositions with varying Arochlor concentration, with and without triphenyl phosphorothionate, were tested for their load-carrying capacity on an I.A.E. Gear Rig as described in the Institute of Petroleum Method 166/60T using BSS-EN34 steel gears and operating at an oil temperature of 60°C. or 110° C. The mean failure loads at 2,000 and 6,000 rpm are shown in Table 2 above.

Examples of triaryl esters of phosphorothionic acid suitable for use in compositions are triphenyl phosphorothionate, tritolyl phosphorothionate, tribenzyl phosphorothionate, and trixylyl phosphorothionate. The preferred ester is triphenyl phosphorothionate.

In related work A.C.B. MacPhail and N.E. Wright; U.S. Patent 3,483,122; December 9, 1969; assigned to Shell Oil Company describe lubricating oil compositions having excellent anticorrosion and antioxidant properties which consist of a major amount of an ester base lubricating oil and minor amounts each of a triaryl phosphorothionate and triaryl phosphate.

Phosphonated Dilinoleate Esters

W.E. Parker and R.E. Koos; U.S. Patent 3,423,319; January 21, 1969; assigned to the U.S. Secretary of Agriculture describe thermally stable compositions for use as extreme pressure, antiwear lubricants.

Effective lubricants for any type of machinery should be capable of performing their function over long periods of time without appreciable changes in their physical and chemical characteristics. Mineral hydrocarbon lubricating oils are not suitable for many lubrication problems of modern industry. Consequently, synthetic lubricants such as bis(2-ethylhexyl) sebacate and dibenzyl sebacate, employed as controls for evaluating the compounds of the process, have been prepared and used as specialty products. There is however, continued need for better and more effective lubricants and lubricant additives capable of sustained performance under more demanding operating conditions. The compounds of the process are represented by the formula shown on the following page.

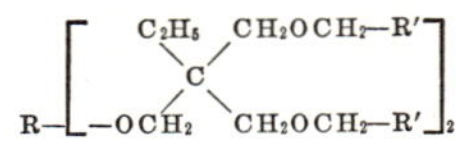

R is the diacyl radical derived from dimerized linoleic acid or the hydrogenated product thereof, R' represents $-CH{=}CH_2$ or

$$-(CH_2)_2-\overset{\overset{\displaystyle O}{\|}}{P}-(OR'')_2$$

and R" is any straight or branched chain aliphatic alkyl radical containing 1 to 8 carbon atoms. These compounds and the lubricant compositions formulated with these compounds as additives are thermally stable and represent a marked improvement in performance characteristics as evaluated by wear test and extreme pressure tests.

Antiwear properties were determined using the Shell Four Ball Wear Tester as described by Peale et al., American Society of Lubrication Engineers Transactions 3, 48 (1960). Approximately 10 ml. of the test oil were placed in the test cup so that the three bottom stationary balls were covered to about a 2 mm. depth. After positioning the cup on its stand in contact with the fourth ball, the oil was heated to 120°C., a 50 kg. load was placed on the weight tray, and the upper ball was allowed to rotate at 600 rpm for 1 hour. The diameters of the circular scars worn on the three stationary balls were measured by means of a low power microscope. The results are shown in the table below. The 5% by weight of additive (lubricants 1, 2 and 3) in the formulation of lubricant mixtures is arbitrary and is not intended to limit the amount used.

Lubrication as Measured by Wear Scar

Lubricant No.	Lubricant Name	Additive (5% by weight)	Wear Scar Dia., mm.
1	Dilinoleate ester	None	0.760
2	53% phosphonated dilinoleate ester.	do	0.483
3	100% phosphonated dilinoleate ester.	do	0.410
4	Bis (2-ethylhexyl) sebacate.	do	0.830
5	100 paraffin oil	do	0.803
6	Bis(2-ethylhexyl) sebacate.	Dilinoleate ester	0.778
7	100 paraffin oil	do	0.597
8	Bis(2-ethylhexyl) sebacate.	53% phosphonated dilinoleate ester.	0.455
9	100 paraffin oil	do	0.460
10	Bis(2-ethylhexyl) sebacate.	100% phosphonated dilinoleate ester.	0.442
11	100 paraffin oil	do	0.460
12	Bis(benzyl) dilinoleate	None	0.653
13	do	100% phosphonated dilinoleate ester.	0.470
14	Hydrogenated bis-(benzyl) dilinoleate.	None	0.748
15	do	100% phosphonated dilinoleate ester.	0.447
16	Bis(p-isopropyl benzyl) dilinoleate.	None	0.843
17	do	100% phosphonated dilinoleate ester.	0.587
18	Dibenzyl sebacate	None	0.950
19	do	100% phosphonated dilinoleate ester.	0.620

In the wear tests, the smaller value denotes the oil with the better lubricating property. The compounds of the process (Nos. 1, 2 and 3) and the lubricant compositions (Nos. 10, 11, 13, 15, 17 and 19) formulated by the addition of the fully phosphonated ester (No. 3) to the reference lubricants (Nos. 4, 5, 12, 14, 16 and 18) show a very marked improvement in antiwear lubricant properties over the reference compounds. For example, when bis(2-ethylhexyl) sebacate (No. 4) was used without additive the diameter of the wear scar measured 0.830 mm. However, by adding 5% by weight of the fully phosphonated ester (No. 3) to the bis(2-ethylhexyl) sebacate, the diameter of the wear scar was reduced 50% to 0.442 mm. (No. 10).

Another example is the paraffin oil (No. 5). Without additive, the wear scar measured 0.803 mm. in diameter. With the fully phosphonated ester (No. 3) added to it, the wear scar was only 0.460 mm. in diameter (No. 11). Other examples showing similar improvement in lubricating properties can be found in the above table.

In a typical procedure for the preparation of the esters, a 2 to 1 molar ratio of alcohol to dimer acid and a small amount of an acid catalyst were dissolved in an inert organic solvent such as benzene and refluxed until the reaction was complete, as indicated by the amount of water collected. The benzene solution was washed until free of acid, dried over calcium sulfate, the benzene removed, and the crude product purified by molecular distillation.

In the preparation of the phosphonated esters, about 1 mol of the above ester was reacted with about 9 mols of a dialkyl phosphonate (dialkyl hydrogen phosphate) in the presence of a peroxide catalyst.

The catalyst was added in 1/4 portions at 4 hour intervals. The excess phosphonate was distilled off under reduced pressure, and the phosphonated ester was purified by molecular distillation. The extent of reaction was determined by infrared spectra, iodine number and percent phosphorus. The alkyl radical of the dialkyl phosphonate can be any straight or branched chain aliphatic alkyl radical containing 1 to 8 carbon atoms such as methyl, ethyl, propyl, butyl, amyl, hexyl, heptyl, octyl and any of their isomers. The di-2-ethylhexyl phosphonate derivatives are used to exemplify this process.

The dimer acid employed in the preparation of the esters contained 95% dilinoleic acid, 4% trimer acid, and 1% monobasic acid and had the following characteristics: acid number, 191.2, neutralization equivalent, 293.4 and molecular weight, 565. The dialkyl phosphonate (dialkyl hydrogen phosphite) was redistilled before use. Best results were obtained with use of naphthalene-2-sulfonic acid as an esterification catalyst. Naphthalene-1-sulfonic acid, methane sulfonic acid and sulfuric acid can also be used as the catalyst, but give poorer results than naphthalene-2-sulfonic acid. The catalyst used for the dialkyl phosphonate addition was t-butyl perbenzoate. Preparation of the esters is illustrated by the following examples.

Example 1: Dimer acid, 141 g., trimethylolpropane diallyl ether, 107 g. and 2.8 g. naphthalene-2-sulfonic acid were dissolved in 450 ml. benzene in a 2 l. three-necked flask. The flask was equipped with a magnetic stirrer, thermometer and a Dean and Stark tube with a water-cooled condenser. The mixture was heated at 90°C. reflux temperature until 7.5 ml. of water was taken off in 7 hours reaction time. Since the amount of water collected indicated that the reaction had not gone to completion, an additional 1.4 g. catalyst was added to the reaction mixture, and the rejuvenated reaction gave off an additional 1.5 ml. water in 2 hours. Total reaction time: 9 hours; total weight of catalyst: 4.2 g. After cooling, the reaction mixture was washed with water until acid-free. The benzene solution was dried over calcium sulfate, filtered, and the benzene evaporated under vacuum. The crude product, a dark amber fluid, weighed 233.5 g.; percent yield, 96.6; acid number, 5.5; n_D^{25}, 1.4776.

The crude product was purified by molecular distillation, the main fraction being collected above 225°C. at 7 micron pressure. This fraction weighed 101 g. The purified product, bis(trimethylolpropane diallyl ether) dilinoleate, a medium amber liquid had the following characteristics: n_D^{25}, 1.4782; iodine number 135.1; saponification number 121.1, calculated, 117.7; saponification equivalent 465.5, calculated 476.8; percent carbon 75.78, calculated 75.88; percent hydrogen, 11.08, calculated 10.99; molecular weight 931, calculated 953.

Example 2: Bis(trimethylolpropane diallyl ether) dilinoleate, 50 g., di-2-ethylhexyl phosphonate, 144 g., and 1/4 of the total amount (4.6 g.) to t-butyl perbenzoate were mixed in a 500 ml. three necked flask. The flask was equipped with a magnetic stirrer, water condenser, thermometer and gas inlet tube. The reaction mixture was blanketed with nitrogen until the reaction was completed. While the reaction mixture was heated at 110°C. for 16 hours, 1/4 portions of the catalysts were added at 4 hour intervals. After each catalyst addition, except the last, the reaction temperature rose five to six degrees. The last addition resulted in only a two degree rise. The reaction mixture was degassed in a molecular still at room temperature under a partial pressure. The excess phosphonate was distilled off at 120° to 150°C. at 15 microns pressure. The main fraction, a viscous amber liquid, was collected above 200°C. at 7 microns. The purified product weighed 65.0 g.; percent carbon, 71.42, calculated 68.34; percent hydrogen, 11.15, calculated 11.29; percent phosphorus, 3.03, calculated 5.68; iodine number, 75.7. The infrared spectra, iodine number and the percent phosphorus indicated that the compound had attained 53% of the maximum amount of dialkyl phosphonate addition.

Polyalkylene Oxide and Phosphate Esters

W. Katzenstein; U.S. Patent 3,567,636; March 2, 1971; assigned to GAF Corporation describes a lubricating composition comprising a major proportion of an ester-based lubricating fluid and a minor amount of a phosphate ester or salt of a nonionic surface active compound. The phosphate ester is selected from the group consisting of monoesters, diesters and mixtures, and the nonionic surface active compound is the condensation produce of an organic hydroxy compound of from 8 to 50 carbon atoms selected from the group consisting of alkyl phenols and alkanols with at least 1 mol of an alkylene oxide having from 2 to 3 carbon atoms. The nonionic surface active compound contains a maximum of about 50% by weight of alkylene oxide based on the weight of the nonionic surface active compound. The following examples illustrate the preparation of useful phosphate esters.

Example 1: 2.7 mols of a nonionic surface active compound derived from dinonylphenol condensed with 4 mols of ethylene oxide is reacted with 1 mol of P_2O_5 in the manner described in the examples of U.S. Patent 3,004,056. The product consists of about equal amounts of mono- and diester with about 10 to 15% unreacted nonionic.

Example 2: The procedure of Example 1 is repeated except that the nonionic is a dodecyl phenol plus 2 mols of ethylene oxide condensate.

Example 3: The procedure of Example 1 is again repeated employing as the nonionic surface active agent a dinonyl phenol plus 5 mols ethylene oxide (40% ethylene oxide content).

Example 4: The procedure of Example 1 is once again repeated employing as the surfactant a dinonyl phenol plus 7 mols ethylene oxide (47% ethylene oxide content). The following examples illustrate the process.

Example 5: To 97.0 g. of di-2-ethylhexyl sebacate are added 3.0 g. of the barium salt (pH=7.0) of the phosphate ester of Example 4. The mixture is heated to 65°C. with stirring until homogeneous in appearance. A similar preparation is made containing 3 g. of tricresyl phosphate. Viscosity measurements are made and antiwear properties ascertained in a Brown modified Shell Four Ball machine using four steel balls SAE 52-100 as described in U.S. Patent 3,036,072. The tests are run at 200°F., 600 rpm at loads of 10 and 40 kg. The following Tables 1 and 2 list the results obtained with the two formulations and the unmodified sebacate.

TABLE 1: VISCOSITY

Composition	KV,[1] 210° F.	KV, 100° F.	KV, −65° F.	VI [2]	Pour point, °F.
1. Di-2-ethyl hexyl sebacate	3.41	12.44	7,868	165	<−65
2. Di-2-ethyl hexyl sebacate plus tricresyl phosphate	3.42	12.42	10,700	166	<−65
3. Di-2-ethyl hexyl sebacate plus barium salt of Example D	3.78	14.48	8,000	178	<−65

[1] Kinematic viscosity.
[2] Viscosity index.

TABLE 2: SHELL FOUR BALL TESTS

Composition	Scar diameter (mm.) at 10 kg.	Scar diameter (mm.) at 40 kg.
1. Di-2-ethyl hexyl sebacate	0.30	0.70
2. Di-2-ethyl hexyl sebacate plus tricresyl phosphate	0.25	0.40
3. Di-2-ethyl hexyl sebacate plus barium salt of Example D	0.20	0.30

Example 6: Example 5 is repeated using however, di-2-ethylhexyl azelate as the ester base. The results are summarized in Tables 3 and 4.

TABLE 3: VISCOSITY

Composition	KV,[1] 210° F.	KV, 100° F.	KV, −65° F.	VI [2]	Pour point, °F.
1. Di-2-ethyl hexyl azelate	2.96	10.68	6,500	152	<−65
2. Di-2-ethyl hexyl azelate plus tricresyl phosphate	3.04	10.68	8,500	155	<−65
3. Di-2-ethyl hexyl azelate plus barium salt of Example D	3.20	11.80	8,180	180	<−65

[1] Kinematic viscosity.
[2] Viscosity index.

TABLE 4: SHELL FOUR BALL TEST

Composition	Scar diameter (mm.) at 10 kg.	Scar diameter (mm.) at 40 kg.
1. Di-2-ethyl hexyl azelate	0.40	0.70
2. Di-2-ethyl hexyl azelate plus tricresyl phosphate	0.30	0.60
3. Di-2-ethyl hexyl azelate plus barium salt of Example D	0.20	0.40

From the above data it is clear that the phosphate ester salt used in the process is superior to the unmodified ester lubricant and the tricresyl phosphate-ester lubricant with respect to viscosity and lubricity characteristics. The compositions are not only superior in these properties but also these compositions meet the viscosity specifications of MIL-L-78080 which are:

Kinematic Viscosity

3 centistokes minimum	210°F.
11 centistokes minimum	100°F.
13,000 centistokes maximum	-65°F.

It should be noted that the azelate base fails to meet these specifications at 210° and 100°F. and the tricesyl phosphate-azelate combinations fails at 100°F. Of further outstanding significance is the fact that the barium salt of Example 4 as used in Examples 5 and 6 is only 50% active because it is prepared by reacting 50 g. of ester of Example 4 with 21 g. of barium hydroxide-octahydrate in 50 g. of ester base. This is reacted by heat and agitation for 2 hours at 85°C.

OTHER ADDITIVES AND APPLICATIONS

Polyesters from Tetraalkylcyclobutanediol as Thickeners

E.W. Wilson and C.W. Hequembourg; U.S. Patent 3,360,547; December 26, 1967; assigned to Eastman Kodak Company describe the use of complex esters as thickeners for synthetic lubricants. In general the esters are complex esters or polyesters formed by reacting a dicarboxylic acid with a tetraalkylcyclobutanediol and with a chain-terminating monohydric alcohol or monocarboxylic acid. The products have the formula:

(1)

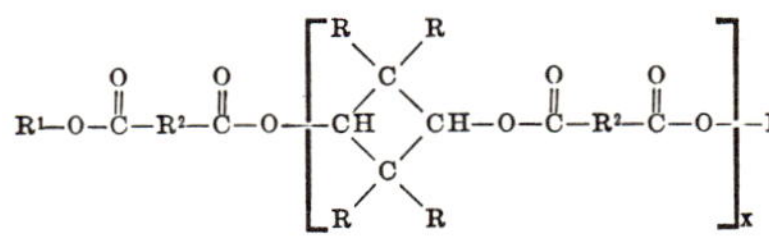

when a monohydric alcohol is used for terminating the polyester chain, or the formula:

(2)

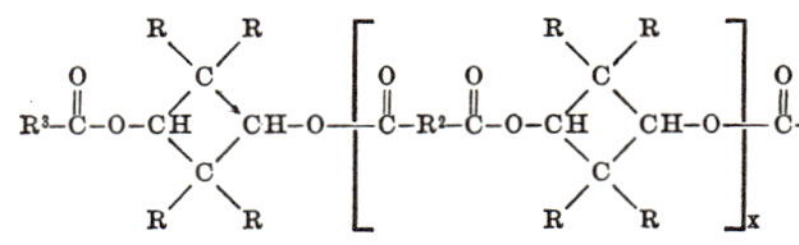

when a monocarboxylic acid is used for terminating the polyester chain. In Formula (1), R^1 is the alkyl residue of the monohydric alcohol, R^2 is the alkylene radical of the dicarboxylic acid, R is a lower alkyl radical and x is an integer from 1 to 10. In Formula (2), R^3 is the alkyl residue of the monocarboxylic acid and the other symbols are as indicated for Formula (1). An essential feature of the complex esters is that they are derived from a tetraalkylcyclobutanediol of the formula:

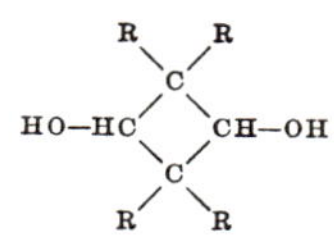

where the substituents, R, are the same or different straight or branched chain lower alkyl groups, i.e., alkyl groups of from about 1 to 4 carbon atoms. Such diols are advantageously prepared by the method described in U.S. Patent 2,936,324.

The complex esters can be prepared by heating a mixture of the tetraalkylcyclobutanediol, the dicarboxylic acid and the monohydroxy alcohol or monocarboxylic acid in the presence of an esterification catalyst. Catalysts providing the most satisfactory results include: titanium esterification catalysts such as described in U.S. Patent 2,727,881, particularly titanium alkoxides such as tetraisopropyl titanate; tin esterification catalysts such as described in U.S. Patent 2,720,507, and especially tin alkoxides such as dibutyl tin oxide and alkyl tin compounds such as tetrabutyl tin.

The complex esters are employed as minor components of lubricating compositions having as a major component a synthetic ester base oil. The complex esters are useful for improving the temperature-viscosity relationship for synthetic ester lubricants. The amount of complex ester added to the base oil should be sufficient to improve the viscosity-temperature relationship of the oil, that is, to form a mixture having a high-temperature viscosity, e.g., at 210°F., higher than that of the base oil while having a low-temperature viscosity, e.g., at -40°F., not excessively higher than that of the base oil. Generally, an amount of complex ester in the range of about 5 to 40 weight percent of the base oil will provide the desired thickening result.

The following example illustrates the preparation of an especially preferred complex ester. This ester is prepared from azelaic acid, 2,2,4,4-tetramethylcyclobutanediol and 2,2,4-trimethylpentanol-1, and is characterized by exceptionally good heat stability and low corrosivity toward lead. The latter property is especially desirable for jet engine lubricants that come in contact with lead or Babbitt metal bearings.

Example: In a 3 l. flask fitted with a stirrer, thermometer, nitrogen inlet tube, steam condenser connected to a Dean-Stark moisture trap which in turn is connected to two water cooled condensers, are charged the following ingredients.

	Grams
Azelaic acid	1,017
2,2,4,4-Tetramethyl-1,3-cyclobutanediol	730
2,2,4-Trimethylpentanol	257
Xylol	100

All of the above ingredients were heated simultaneously under a nitrogen atmosphere to 90°C. and tetraisopropyl titanate (1.9 g.) was added as the catalyst. Water of esterification removed from the reaction at 150° to 200°C. totaled 193 g. or 98.5% of theoretical. The reaction time, including 2.5 hours of vacuum stripping at 200°C. and 1 to 5 mm. of mercury, was 20.5 hours. The final crude product was diluted with heptane, washed with 5% NaOH solution and washed with water until neutral. The crude mixture was filtered through a diatomaceous earth and stripped of low boilers to 90°C. at 150 to 200 mm. The final product was vacuum stripped to 200°C. at 50 to 200 microns. The final acid number was 0.2 and the number-average molecular weight was 1100.

Table 1 below, records physical properties of a complex ester "A" of the process, prepared as described in the above example. It also records comparative data for two complex esters "B" and "C" prepared from diols other than the tetraalkylcyclobutanediol used in preparing the complex esters. Specifically, the two comparative complex esters were prepared by the reaction of azelaic acid and 2,2,4-trimethylpentanol with 2-methyl-1,3-pentanediol or with 2,2-dimethyl-1,3-propanediol.

TABLE 1

	Properties of Complex Esters		
	Viscosity (centistokes), 100° F.	Acid Number	Molecular Weight
Complex Ester "A": Azelaic Acid+2,2,4,4-Tetramethyl-1,3-Cyclobutanediol+2,2,4-Trimethylpentanol	2,589	0.16	1,100
Complex Ester "B": Azelaic Acid+2-Methyl-1,3-Pentanediol+2,2,4-Trimethyl Pentanol	3,538	0.15	1,500
Complex Ester "C": Azelaic Acid+2,2-Dimethyl-1,3-Propanediol+2,2,4-Trimethylpentanol	750	0.66	1,600

Table 2 below, records properties of synthetic ester lubricant compositions employing the complex esters as thickeners. In each of the three compositions the base oil is bis(2,2,4-trimethylpentyl)azelate.

TABLE 2: PROPERTIES OF LUBRICANT COMPOSITIONS

[Lube base, bis(2,2,4-trimethylpentyl)azelate]

Weight Percent Polyester	Viscosity (centistokes)			Foaming, Initial (vol., ml; collapse, min.) [1]			Lead Test [2]		Panel [3] Coking	536° F., Thermal [4] Stability
	100° F.	210° F.	−40° F.	Sequence 1, 75° F.	Sequence 2, 200° F.	Sequence 3, 75° after 200° F.	Initial	Stored 2 Weeks		
21% "A"	38.5	7.70	9,190	Non-Foam	15/0.2	Non-Foam	0.3		2.6	1.0
16.5% "B"	36.54	7.74	8,561	do	10/0.2	do	0.0	0.2	4.7	−52.30
27% "C"	37.24	7.48	10,003	do	25/0.3	do	−12.7	−82	23.7	−6.2

[1] Fed. Test Meth., Standard No. 791, Meth. 3211.1; ASTM D892-46T.
[2] Lead Corrosion Test, Fed. Test Method Standard No. 791, Meth. 5321.
[3] Fed. Test Meth., Standard No. 791, Meth. 3462.
[4] D. E. R. B. 2487, Issue 3, par. 6.7, Thermal Stability.

The table shows that the complex ester "A" prepared from 2,2,4,4-tetramethyl-1,3-cyclobutanediol imparted good high-temperature viscosity to the lubricant composition without having the drawbacks of known thickening agents. The table further shows results of other important lubricant tests. The foaming properties of jet engine lubricants is significant because foaming in the lubricant reservoir system of the jet engine would lead to bearing starvation and ultimate failure. This composition demonstrates satisfactorily low foaming characteristics. The lead corrosion test indicates whether or not the lubricant will be corrosive to the lead-indium alloy in the jet engine bearing. The panel coking test is an indication of the amount of coking that the lubricant will undergo in tubing, bearing, bearing housing and other areas within the hot spots of the engine. Table 2 shows that the lubricant composition of the process containing polyester "A" was markedly superior to the compositions containing polyesters "B" and "C" in the lead corrosion, panel coking and thermal stability tests.

Polyester Thickeners Containing Glycidyl Ester Acid Scavengers

A process described by R.H. Boehringer, R.D. Aylesworth and W.E. Utz; U.S. Patent 3,472,775; October 14, 1969;

assigned to Emery Industries, Inc. involves synthetic lubricant compositions containing a polyester thickening agent that has been refined with a glycidyl ester of an aliphatic acid having from 5 to 22 carbon atoms. The polyester thickeners used in the lubricant compositions may be prepared from long chain dicarboxylic acids and glycols. The preferred acids used in the preparation of the polyesters are polymerized unsaturated fatty acids, such as linoleic and oleic acid. The glycols may be short chain glycols, such as propylene glycol and ethylene glycol, preferably hindered glycols, such as neopentyl glycol, and 2,2,4-trimethylpentanediol-1,3. After the polyester is prepared, it contains excess acid which cannot be introduced into a lubricant composition. Thus, the glycidyl ester is added in excess of the amount needed to neutralize the excess acidity of the polyester and acids as an acid "scavenger". The preferred acid scavenging glycidyl ester is a compound having the general formula:

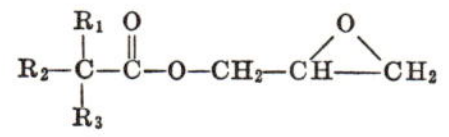

where R_1, R_2 and R_3 are all saturated alkyl groups and contain a total of about 7 to 9 carbon atoms. The glycidyl esters having this structure produce, when reacted with the free acid of the ester thickener, materials which are particularly stable and compatible with the lubricant system. Among the other glycidyl esters which may be used as an acid scavenger in addition to the neo-acid esters are the glycidyl esters of pelargonic acid, heptanoic acid, valeric acid, isostearic acid and oleic acid.

In performing the refining operation on the polyester thickeners used in the lubricant compositions, the excess glycol used in the preparation of the polyester is removed by heating the reaction mixture to about 260°C. with a pressure of about 1 to 2 torr. The polyester is then cooled. The acid value of the polyester is measured and then a 100% excess of the stoichiometric amount of the glycidyl ester needed to neutralize the acid still present is added to the polyester. The reactants are further reacted until the acid value of the polyester is reduced to a value lower than 0.05. The excess glycidyl ester is removed by heating the ester to 245°C. and applying vacuum.

The polyester thickeners may be used in an amount of about 0.5 to 30.0% by weight of the composition and preferably from about 1.0 to 10.0 weight percent based on the total lubricant composition weight. The following examples illustrate the process.

Example 1: A preferred polyester thickener for use in preparing the lubricant compositions was prepared by reacting 2,760 g. of dimer acid, a polymerized oleic acid sold under the trademark Empol 1010, 490 g. of neopentyl glycol, and 57 g. of propylene glycol. The esterification was continued until the acid value was determined to be 11.2, then 159 g. of the glycidyl ester of a neo-acid, sold under the trademark Cardura E, and having a molecular weight of 245, were added. The 159 g. represented a 100% excess over the stoichiometric amount required to react with all of the free acid contained in the polyester thickener. The reaction vessel was heated to a temperature of 210° to 220°C. for a period of 6 hours after which the excess glycidyl ester was stripped off at a temperature of about 230°C. while subjecting the reaction vessel to a pressure of 1 to 2 torr. The resulting polyester had an acid value of 0.034, and a molecular weight of about 3500.

Example 2: A lubricant composition was prepared using 57.2% by weight of diisodecyl azelate, 24.52% by weight of diisodecyl pelargonate, and 12.21% by weight of a commercial thickener, methyl acrylate, sold as Acryloid 710. The remaining portion of the composition was a combination of commercial antioxidants and metal deactivators. This composition was tested for sonic shear stability using ASTM Stand Test Appendix III (1961 Edition). After the test was run, the fluid had a 51.3% viscosity loss at 100°F.

Example 3: A composition was made in accordance with Example 1 except that 2.51% by weight of a methyl acrylate, Acryloid 710 and 4.25% by weight of the glycidyl ester refined polyester prepared in Example 1 was substituted for the thickener used in Example 2. This composition was tested for sonic shear stability as in Example 2. It showed a viscosity loss of 24.3% at 100°F.

Example 4: A lubricant was made using 91.43 parts by weight of diisodecyl azelate, 2.5 parts by weight of the glycidyl ester refined polyester prepared in Example 1, and the composition was tested for sonic shear stability as in Example 2. It showed a 4.0% viscosity loss at 100°F. It can be seen from the sonic shear stability test results in Examples 2 through 4 that the use of a polyester thickener in the lubricant compositions vastly improve the shear stability of the lubricant compositions over the shear stability achievable with the conventional thickeners shown for comparative purposes.

Examples 5 through 7: Lubricant compositions were prepared in accordance with the formulations shown in Table 1 on the following page. These compositions were evaluated using Federal Test 791, Corrosion and Oxidation Stability, Method 5308, run at 425°F. for 72 hours. This test subjects lubricant compositions to heat and aeration in the presence of five metals, copper, silver, steel, magnesium and aluminum. The data in Table 2 on the following page, shows that the change in metal weight was not great, and that the change in metal weight was within the specifications of

Pratt and Whitney PWA 521-B, which allows a maximum change in metal weight of ±0.30 mg./cm.^{2} and a maximum increase in viscosity at 100°F. of 50%. Lubricant compositions containing polyesters which have not been refined have excess acidity and consequent corrosive propensities which render them of marginal utility.

TABLE 1

Example No.	5	6	7
Pentaerythritol ester [1]	93.73		
Pentaerythritol ester [2]			93.73
Trimethylol propane ester [3]		46.50	
Trimethylol propane ester [4]		46.50	
Polyester [5]	1.20	4.48	1.20
Commercial Anti-oxidants	2.00	2.00	2.00
Commercial Metal Deactivator	0.05	0.05	0.05
Quinizarin	0.02	0.02	0.02
TCP	3.00	3.00	3.00
Azelaic Acid	0.005	0.005	0.005

[1] An ester prepared from monopentaerythritol and a mixture of C_5 and C_9 monobasic acids.
[2] An ester prepared from monopentaerythritol and a mixture of C_5 to C_9 monobasic acids.
[3] An ester prepared from trimethylol propane and a mixture of C_6 to C_9 monobasic acids.
[4] An ester prepared from trimethylol propane and a mixture of C_6 to C_8 monobasic acids.
[5] Prepared in Example I.

TABLE 2

Example No.	5	6	7
Viscosity change at 100° F. (Percent)	33.7	38.4	25.5
Sludge (mg.)	19.2	6.8	30.7
Acid value	4.67	1.58	2.7
Copper corrosion (mg./cm.^{2})	−0.08	−0.07	−0.17
Steel corrosion (mg./cm.^{2})	0	0	0
Silver corrosion (mg./cm.^{2})	−0.01	−0.03	−0.02
Magnesium corrosion (mg./cm.^{2})	0	−0.08	0
Aluminum corrosion (mg./cm.^{2})	0	−0.06	−0.04

In related work, J.B. Boylan, R.D. Aylesworth and R.H. Boehringer; U.S. Patent 3,485,754; December 23, 1969; assigned to Emery Industries, Inc. describe a method of refining synthetic ester lubricants which involves adding glycidyl esters of various aliphatic acids, preferably neo-acids to the unrefined ester lubricant. The glycidyl esters, acting as acid scavengers, react with the free acid of the synthetic ester lubricants to reduce the acid value. Additional lubricant materials fully compatible with the ester lubricant are formed in the reaction. The metal corrosion properties of the lubricant are formed in the reaction. The metal corrosion properties of the lubricant are enhanced by either adding additional glycidyl ester to the lubricant or leaving a certain amount of excess glycidyl ester in after refining.

Sodium Perfluoroborate as a Haze Control Agent

A process described by T.S. Chao; U.S. Patent 3,329,611; July 4, 1967; assigned to Sinclair Research, Inc. is concerned with liquid lubricating oil compositions comprising a synthetic ester base oil of lubricating viscosity containing a dimer acid extreme pressure agent, an antioxidant combination of an aromatic amine and a soluble alkali metal compound, and a halo-substituted carboxylic acid as a haze inhibiting agent. In addition, the composition may contain an alkyl ester of an alkylene diamine tetraalkanoic acid as an anticorrosion agent.

The amine and alkali metal additives are incorporated in the base ester oil in small amounts sufficient to retard oxidation of the oil at temperatures in excess of 400°F., e.g., 450°F., and the concentrations employed for optimum results may be dependent on the particular base oil and other additive component selected. The amount of the additives is insufficient to destroy the fluidity of the base ester oil, i.e., the oil remains essentially liquid at normal conditions and the alkali metal and other components are not used in an amount which produces a grease or gel-like structure.

Ordinarily about 0.005 to 5%, preferably about 0.01 to 2%, by weight of the alkali metal derivative provides satisfactory results, and often the alkali metal content of the lubricant is in the range of about 5 to 1,000 ppm, preferably at least about 10 to 40 ppm to 100 to 500 ppm. The amine antioxidant additive component provides satisfactory results when present in amounts of 0.05 to 5% and preferably 0.1 to 2%.

Normally the amount of dimer acid load-carrying agent is between 0.01 to 1 weight percent with the preferred amount being between 0.1 to 0.5 weight percent. The amount of halo-substituted carboxylic acid haze reducing agent is generally between 0.01 and 1 weight percent with the preferred amount being between 0.02 and 0.5 weight percent.

Anticorrosion agents such as the tetraester of alkylene diamine tetraacetic acid when used are normally present in amounts between 0.05 to 5 weight percent and preferably 0.1 to 2 weight percent. The effectiveness of the haze reducing agent is shown by the following examples.

Example 1: A synthetic lubricant was prepared by blending the ingredients shown on the following page, at a temperature of 160° to 170°F. All parts are given in weight percent.

Hercolube A[1]	72.26	Tetrabutyl ester of EDTA[3], containing mono-Na salt of triester (% Na=0.518)	1.50
Hercolube F[2]	24.31	Half amide of hydrogenated dimer acid[4]	0.20
Phenyl-α-naphthylamine	1.00	Dow Corning fluid (200 to 60,000)[5]	0.001
Di-(p-octylphenyl)amine	1.00		

[1] An essentially complete pentaerythritol ester of fatty acids having an average of 6 carbon atoms and with the following approximate inspection data: Acid No. 0.01; saponification No. 420–428; viscosity at 210° F., 4.25–4.35 cs.; viscosity at 100° F., 20.5–21.5 cs.

[2] An essentially complete ester of dipentaerythritol and a mixture of alkanoic acids containing an average of 5 to 6 carbon atoms per molecule and characterized by an acid number of about 0.1, a saponification number of about 391–397 and a kinematic viscosity at 210° F. of 8.7–8.9 cs. and at 100° F. of 56–58 cs.

[3] EDTA is ethylene diamine tetra acetic acid.

[4] The hydrogenated dimer acid was Emery 389R and the half amide analyzes approximately: Acid No. 104.9 and percent N, 2.16.

[5] A Dow Corning antifoam silicone having a viscosity of 200 cs. at 250° C. and a molecular weight of 60,000.

After intimate mixing the lubricant was filtered through paper coated with a layer of Hyflo Super-Cel (a diatomaceous silica) yielding a clear oil of bright reddish purple color. In spite of its superior oxidation resistance, the fluid developed a haze after standing for 3 months in the presence of light. In the absence of light, e.g., as in metal drums and cans, the development of haze was much slower. However, in accelerated oven storage tests the haze developed in a few days. After 7 days in a 221°F. oven, a noticeable amount of precipitate was found at the lower corners of the glass storage bottle.

Example 2: Another synthetic lubricant was prepared from the following components. All parts are in weight percent.

Hercolube A	71.99	Tetrabutyl ester of EDTA	1.50
Hercolube F	24.25	Sodium perfluorobutyrate	0.06
Phenyl-α-naphthylamine	1.00	Half amide of hydrogenated dimer acid of Example 1	0.20
Di-(p-octylphenyl)amine	1.00	Dow Corning fluid (200 to 60,000)	0.001

This formula is the same as that given in Example 1 except that the Na salts of partial esters of EDTA were eliminated and in their place, sodium perfluorobutyrate was added. The resultant lubricant was much more stable toward haze and precipitate formation. No haze was observed during storage for months at room temperature. The oil was bright and clear after a 7 day 221°F. storage test and no precipitate was observed after 3 1/2 months in a 130°F. oven. Table 1 below, shows the superior oxidation resistance, load carrying property, noncorrosiveness to metals, compatibility to seal materials, nonfoaming and nonshearing characteristics and other properties of this lubricant.

TABLE 1

(1)	O_2 Absorption Test — 450°F., 1 ft.^{3} O_2/hr., 75 g. samples*.	
	T_i, min.	>1,145
	T_t, min.	1,145
	V_t, ml.	2,500
(2)	Type 2 Bearing Rig Test —	
	Acid No., final	0.93
	Viscosity rise, final (KV/100), percent	50.0
	WADD demerit rating	58
	100 mesh filter deposit, g.	1.24
(3)	Ryder Gear Scuff Test —	
	Scuff load:	
	A side, lb./in.	2590
	B side, lb./in.	2860
	Relative rating, percent	100
(4)	Gear Fatigue Test — more than 110 hours.	
(5)	Corrosion-Oxidation Test (425°F., 48 hours) —	
	Weight change, mg./cm.^{2}:	
	Cu	-0.30
	Ag	+0.024
	Steel	+0.054
	Al	+0.007
	Mg	-0.77
	Ti	+0.038
	Viscosity rise (KV/100), percent	22.2
	Acid No. increase	1.27
(6)	SOD Pb Corrosion, 325°F. 1 hr., mg./in.^{2}	+6.48, +4.46

(continued)

TABLE 1: (continued)

(7) Carbon Seal Wear — in./min., 0.017×10^{-5}.

(8) Inspection Tests —

KV/210°F., cs.	5.104
KV/100°F., cs.	27.63
KV/ -40°F., cs.	9.824
Flash point, °F.	495
Pour point, °F.	below -80
Acid No. (pH 11)	0.3
Foam test	0-0-0 0-0-0 0-0-0
Percent swell, H rubber	26.1
Percent swell, Viton	18.4
Durometer, Viton	65
Evaporation loss, 6.5 hours:	
400°F., sea level, %	3.41
450°F., 5.5" Hg, %	18.62
Percent sonic shear	none
450°F., 5.5" Hg.	Clear and fluid

*T_i - Time in minutes elapsed before a sharp change in the rate of O_2 absorption occurs.

T_t - Total time of test in minutes

V_t - Volume of O_2 absorbed during the period T_t in ml.

Table 2 below, shows a number of other high temperature synthetic lubricants which are haze-free and remained clear after a 7 day 221°F. oven test.

TABLE 2: COMPOSITION

Component	1	2	3	4	5	6
Hercolube A	72.57	72.91	72.11	72.15	72.18	72.14
Hercolube F	24.17	24.31	24.04	24.05	24.06	24.04
Phenyl-α-naphthylamine	1.00	1.00	1.00	1.00	1.00	1.00
Di(p-octylphenyl)amine	1.00	1.00	1.00	1.00	1.00	1.00
Tetrabutyl ester of EDTA	1.00	1.50	1.50	1.50	-------	-------
Tetra-2-ethylhexyl ester of EDTA	-------	-------	-------	-------	1.50	1.50
C_3F_7 COONa	-------	-------	-------	-------	0.06	0.02
C_3F_7 COOLi	0.06	-------	0.10	0.10	-------	0.10
C_3F_7 COOK	-------	0.03	-------	-------	-------	-------
Half amide of hydrogenated dimer acid as in Example I	0.20	-------	-------	0.20	0.20	0.20
NPG-bis-(dimeracid)[1]	-------	0.25	0.25	-------	-------	-------
Dow Corning Fluid (200–60,000)	0.001	0.001	0.001	0.001	0.001	0.001

[1] NPG-bis (dimer acid) a condensation product of 1 mole neopentylglycol and 2 moles of hydrogenated linoleic dimer acid of Example I.

Diisotridecyl Adipate as Piston Engine Lubricant

R.D. Aylesworth; U.S. Patent 3,481,873; December 2, 1969; assigned to Emery Industries, Inc. describes a method for lubricating a piston engine by employing as the lubricant diisotridecyl adipate prepared from the condensation of adipic acid and an oxo alcohol prepared from propylene tetramer. The diisotridecyl adipate ester lubricant has excellent lubricity, low viscosity at subzero temperatures, low volatility at high temperatures, suitable high viscosity at high temperatures, and a very low order of elastomer swell thus making it well suited for use in piston or Diesel engines.

Example 1: In order to compare the properties of the ester lubricant of this process, diisotridecyl adipate in which the alcohol portion is prepared from propylene tetramer, with other similar adipates, the adipates of 2-ethylhexanol and C_{10} primary alcohol prepared from propylene trimer by the oxo process were prepared and evaluated as lubricant base stocks. The comparative results obtained are shown in Table 1 below.

TABLE 1

	Di-2-ethylhexyl adipate	Diisodecyl adipate	Diisotridecyl adipate
Viscosity at 210° F., cs	2.32	3.60	5.38
Flash Point, ° F	390	420	445
Fire Point, ° F	440	475	520
Volatility, 2 hrs.:			
340° F., percent	15.4	5.2	2.2
360° F., percent	50.0	9.7	3.4
Ford Elastomer swell, percent	36.0	16.5	8.7
Change in Elastomer Hardness Shore A, percent	21.0	10.0	4.0

Generally, elastomer swell should not exceed 10% in piston engines whereas in aircraft, swell of up to 20 to 30% can be tolerated. It may be seen from the foregoing results that the elastomeric swell of di-2-ethylhexyl adipate and di-isodecyl adipate far exceed tolerable elastomer swell limits while the diisotridecyl adipate had excellent elastomer swell properties. The diisotridecyl adipate is also shown to have superior volatility, viscosity, and flash point characteristics.

Example 2: The viscosity properties of the diisotridecyl adipate lubricant were measured and compared with isomeric esters prepared from the reaction of adipic acid and normal tridecyl alcohol and the C_{13} oxo alcohol made from butylene trimer in order to illustrate the difference between the ester used in the process and the isomeric esters. The results are provided in Table 2 below.

TABLE 2

	Adipate of a linear C_{13} alcohol	Adipate of a C_{13} alcohol derived from—	
		Butylene trimer	Propylene tetramer
Viscosity at—			
0° F., cs	Solid	2,500	1,000
−65° F., cs	do	1,000,000	150,000
−40° F., cs	do	58,000	14,000

The above results illustrate the superior viscosity properties possessed by the diisotridecyl adipate which is formed from the C_{13} alcohol derived from propylene tetramer. The substantial difference between the adipate of the C_{13} alcohol derived from butylene trimer by the oxo process and the adipate of this process is most significant. Because of the similarity of structure their properties should be substantially the same.

With a viscosity of 14,000 cs. at -40°F. the diisotridecyl adipate meets or can easily be blended with other ester lubricants with little or no sacrifice in its other properties to meet most low temperature specifications. For example, Military Specification MIL-L-10295A Subzero Internal Combustion Lubricating Oil requires a maximum of 8,500 cs. at -40°F. With but a small amount of diluent such as di-2-ethylhexyl sebacate or any of a number of other ester lubricants, the viscosity of the adipic acid ester of a C_{13} alcohol derived from propylene tetramer via the oxo process can be easily lowered to meet the above noted specification. Considerably more diluent is required to achieve desired viscosity properties for the butylene trimer derived adipic acid ester lubricant with attendant sacrifice in other desired properties.

Spreading Lubricant for Timing Mechanisms

J.F. Messina; U.S. Patent 3,493,509; February 3, 1970; assigned to the U.S. Secretary of the Army describes a spreading lubricant for use on nonjeweled timers. The lubricant has good viscosity properties at -65°F. and good volatility properties at about +180°F., and consists essentially of idodecyl pelargonate and diproplyene glycol dipelargonate. The useful compositions are given below.

TABLE 1

Compound	Preferred Weight, percent	Approximate Effective range, parts per 100
Isodecyl pelargonate	33.0	25-40
Dipropylene glycol dipelargonate	66.5	60-75
Phenyl-α-naphthylamine	0.5	[1] 0.2-0.8

[1] Added.

The properties of the above oil, when tested in accordance with ASTM D-445-60 yielded the values shown below:

Temperature, °F.	Centistokes
100	6.4
0	62.9
-40	341
-65	below 1,500

The volatility of the lubricant was tested in accordance with ASTM D-972-56 and was subjected to 210°F. for 22 hours with two liters of air per minute passing over the cup in which the lubricant was contained, and a value of 2.6% was obtained. Evaluation of the lubricant compositions was determined using two pairs of sequential fuze timers for use with a typical ordnance weapon. The pairs were selected from different lots and were disassembled, cleaned, and then lubricated with the oil in accordance with the following illustrative procedure.

(a) Preset one pair of timers for 100 seconds and the other set of timers for 50 seconds. (b) Soak each of the four preset timers at -65°F. for 12 hours and activate the timers while maintaining the temperature at -55°F. The timers preset for 100 seconds yielded 99.935 and 99.660 seconds and those preset for 50 seconds, 49.881 and 50.064 seconds.

(c) Soak each of the four timers immediately after obtaining the measurements above in a chamber maintained at +125°F. for approximately 3 hours and permit chamber to return to ambient. (d) Remove the timers from the chamber and rewind and seal them. (e) Soak the four rewound and sealed timers in a chamber for 10 days, the chamber being maintained at 185° to 195°F., and then activate at 189°F. The timers preset for 100 seconds yielded 100.039 and 99.735 seconds, whereas the timers preset for 50 seconds yielded 49.957 and 50.001 seconds.

(f) Return the timers to ambient temperature. (g) Rewind and seal timers once again. (h) Soak each of the timers from (g) above at -65°F. for 12 hours and activate them at this temperature. The timers preset for 100 seconds yielded 100.820 seconds and a dud, whereas 51.28 and 50.588 seconds were obtained on the timers preset for 50 seconds. (i) Immediately following the timing test described in (h) above, soak timers in chamber at +125°F. for approximately 3 hours and permit chamber to return to ambient.

(j) With only the fuze cover removed and no further disassembly, the oil was applied to the pallet uprights and the escapement lever. (k) Rewind and seal the timers once again. (l) Soak the timers for 10 hours at -65°F. and activate at -65°F. The timers preset for 100 seconds yielded 100.071 and 100.456 seconds, whereas the timers preset for 50 seconds yielded 50.325 and 50.088 seconds. (m) Immediately following the timing test described in (l) above, soak the timers in a chamber at +125°F. for approximately 3 hours and permit chamber to return to ambient.

(n) Soak the timers for 10 days at +160°F. and then activate at this temperature. The timers preset at 100 seconds yielded 100.120 and 100.026 seconds, whereas the timers preset at 50 seconds yielded 50.066 and 50.018 seconds. (o) Return the timers to ambient. (p) Rewind and seal the timers once again. (q) Soak the timers at -65°F. for 10 hours and activate at -65°F. The timers preset for 100 seconds yielded 101.943 and 101.848 seconds, whereas the timers preset for 50 seconds yielded 50.250 and 50.572 seconds. (r) Immediately following timing test described in (q) above, soak the timers in chamber at +125°F. for approximately 3 hours and permit the chamber to return to ambient. A brief summary of the results are presented in Table 2 below.

TABLE 2: LUBRICANT OIL DATA

	Entire Fuze First Cleaned and Oiled	Fuze Rewound	Fuze Rewound	Fuze Rewound and Escapement Reoiled	Fuze Rewound	Fuze Rewound
Soak time	16 hours	10 days	16 hours	16 hours	10 days	8 hours.
Temperature, ° F	−65	+189	−65	−65	+610	65.
Fuze:						
No. 1—Preset at 100 sec	99.935 sec	100.039 sec	100.820 sec	100.071 sec	100.120 sec	101.943 sec.
No. 2—Preset at 100 sec	99.660 sec	99.735 sec	Dud	100.436 sec	100.026 sec	101.848 sec.
No. 3—Pre et at 50 sec	49.881 sec	49.957 sec	51.280 sec	50.325 sec	50.066 sec	50.250 sec.
No. 4—Preset at 50 sec	50.064 sec	50.001 sec	50.588 sec	50.088 sec	50.018 sec	50.572 sec.

Dilauryl Maleate as Lubricant for Aluminum

R.K. Smith and S.J. Barber; U.S. Patent 3,396,111; August 6, 1968; assigned to E.F. Houghton & Co. describe a method of lubricating two solid surfaces between which there is relative movement, at least one of which surfaces is selected from the group consisting of stainless steel, the structural metals and the alloys of the structural metals of Groups II, III and IV of the Periodic Table, which comprises maintaining between such surfaces a compound of the formula

$$R'-O-\overset{\overset{\displaystyle O}{\|}}{C}-X-\overset{\overset{\displaystyle O}{\|}}{C}-O-R$$

where X is a divalent radical selected from the group consisting of —CH=CH— and

$$-CH_2-\underset{\underset{\displaystyle CH_2}{\|}}{C}-$$

R is a constituent selected from the group consisting of alkyl, methyl alkyl and ethyl alkyl where the alkyl chain contains from 6 to 20 carbon atoms, and R' is a constituent selected from the group consisting of alkyl, methyl alkyl and ethyl alkyl where the alkyl chain contains from 12 to 20 carbon atoms.

Preferred compounds for use according to the process comprise dilauryl maleate and distearyl maleate. This process is particularly useful in lubricating bearing surfaces composed of aluminum and its alloys. In the following examples, the coefficient of friction was determined by the manner described and illustrated in French Patent 1,322,451. The cup or rider was made of steel and the washer of 2 S aluminum. The rider, which had a flat annular surface of 0.393 square inch was rotated at 0.88 rpm to give a surface speed of 0.0461 inch per second against the test washer under a load of 10 kilograms. Both the rider and washer were immersed in a cup containing the test lubricant. The tests were run at room temperature for 20 minutes.

Example 1: 25 g. (0.25 mol) of maleic anhydride and 135 g. (0.5 mol) of stearyl alcohol were dissolved in 350 ml. of xylene along with 0.1 g. of p-toluene sulfonic acid in a 1 l. flask equipped with a stirrer, thermometer and reflux condenser having a water trap. This solution was heated to reflux for 4 hours at the end of which time a total of 5 g. of water was collected. At this stage of the reaction, the acidity was found to be 0.3% calculated as stearic acid. The xylene was stripped at 140°C. and at a pressure of 20 mm. of mercury, a yield of 147 g. (95.5% of theoretical) of distearyl maleate was obtained.

The distearyl maleate was stirred with an oil of 100 SUS to form a lubricating composition containing 10% by weight of maleate ester. This composition was tested for coefficient of friction and wear characteristics under boundary lubrication conditions in the manner described hereinabove. The coefficient of friction between the steel cup and the aluminum washer was determined to be 0.04, as compared with that of more than 0.5 for the 100 SUS oil alone. The lubricating composition containing 10% distearyl ester merely permitted polishing of the aluminum washer wear surface in contrast to gouging and galling caused by the 100 SUS oil.

Example 2: 2,300 g. of xylene were placed in a 5 l. pot equipped with agitator, reflux condenser and dropping funnel. 220 g. of maleic anhydride and 815 g. of lauryl alcohol were added to the xylene and stirred until the material was adequately mixed. 5 g. of p-toluene sulfonic acid were added and the reaction temperature increased to approximately 140°C. The solution was refluxed about 6 hours during which time a total of 44 g. of water were collected and discarded. The free fatty acid of the reaction mixture was found to be 0. The xylene was stripped from the reaction product by heating to 150°C. at a pressure of 20 mm. of mercury. After removal of the xylene, dilauryl maleate was obtained which had a viscosity of 25.53 centistokes at 100°F. and 3.80 centistokes at 210°F. The pour point of the product was 40°F. and the flash point was 310°F.

The lubricity of the dilauryl maleate was measured by evaluation in the friction machine described above. The coefficient of friction between the steel cup and the aluminum washer was found to be 0.12. The washer showed a very fine polish with slight wear track, but showed no evidence of grooving or galling.

Example 3: The same reaction system was used as in Example 2, employing 250 g. of xylene. 159 g. of maleic anhydride were placed in the flask along with 872 g. of stearyl alcohol. 8 g. of p-toluene sulfonic acid were added and the system was heated to reflux and held at that temperature for 6 hours. 30 g. of water were removed and discarded. After stripping the xylene following the procedure of Example 2, the product was distearyl maleate which has a melting point of 120°F. The coefficient of friction between the steel cup and aluminum washer was found to be 0.11. The washer showed a well-polished track with no evidence of appreciable wear.

In a process described by F.A. Hughes; U.S. Patent 3,468,701; September 23, 1969; assigned to Atlas Chemical Industries, Inc. an ester of a saturated fatty acid and a hexitol, a hexitan, or a hexide having not fewer than 2 nor more than 4 fatty acid groups per molecule of the ester is applied as a lubricant coating to aluminum sheet material in order to render the sheet material nonblocking in character.

GREASES

Polyfunctional Esters

A. Harrison; U.S. Patent 3,579,448; May 18, 1971; assigned to Stauffer Chemical Company has found that a lubricating grease composition with exceptional lubricating properties can be manufactured employing a trifunctional ester or tetrafunctional ester as the base fluid. These base fluids can be thickened with a bentonite clay composition along with other secondary gelling agents and dispersing agents.

Example 1: Three different grease formulations were mixed together and studied for oxidative thermal stability. The first composition consisted of 90.9 parts by weight of trimethylol propane triheptanoate, 1.0 part by weight phenyl-α-naphthyl amine, 5.6 parts by weight bentonite clay, 1.0 part by weight pyrogenic silica gelling agent and 1.5 parts by weight propylene carbonate. The second grease formulation had a base stock consisting of 91.0 parts by weight pentaerythritol monobutyrate triheptanoate cross-linked with azelaic acid to a viscosity of 5.6 to 6.0 centistokes measured at 210°F., 6.5 parts by weight bentonite clay, 1.0 part by weight pyrogenic silica gelling agent and 1.5 parts by weight propylene carbonate.

The third grease formulation was provided with 88.4 parts by weight pentaerythritol monobutyrate triheptanoate cross-linked with azelaic acid to a viscosity of 5.6 to 6.0 centistokes measured at 210°F., 2.5 parts by weight phenyl-α-naphthyl amine, 6.6 parts by weight bentonite clay, 1.0 parts by weight pyrogenic silica gelling agent, and 1.5 parts by weight propylene carbonate.

Each of the above grease compositions were tested in accordance with ASTM methods to provide the results shown in Table 1 on the following page.

TABLE 1

	A	B	C
Dropping point	575	600	640
Evaporation (22 hours):			
210° F	1.34		
350° F		3.8	3.4
Separation:			
212° F	2.65		
350° F		4.30	4.36
Oxidation stability, 100 hrs	0	0	0
Thin film test:			
300° F	Pass		
350° F		Pass	
400° F			Pass

Example 2: Several different grease formulations were made and the rust resistance was studied. Thus, stainless steel panels were coated on one side with the greases and suspended in a beaker containing sufficient water to completely cover the grease. The beaker was placed in an oven at a temperature of 100°F. for 48 hours. At the end of the test period, the panel was removed from the beaker, wiped clean of grease and examined for evidence of rusting and pitting. The rusting is classified as none, slight, moderate and extreme. The grease formulations and results of the rust resistance tests are indicated in Table 2 below.

TABLE 2

	A	B	C	D	E	F
Trimethylol propane triheptanoate	87.0	92.0	90.1	86.5	89.0	89.0
Bentonite clay	13.0	5.7	5.6	12.5	3.75	5.0
Pyrogenic silica gelling agent		1.0	1.0		1.0	1.0
Bentonite clay with sodium nitrite in the matrix					3.75	2.5
Propylene carbonate		1.5	1.5	1.5	1.5	1.5
Phenyl-α-naphthyl amine			1.0	1.0	1.0	1.0
Amount of rusting	E	S	S	E	N	N

Example 3: The shear stability of several different grease formulations were studied. Thus, the penetration of each of the formulations was conducted in a conventional manner. The grease formulations and the results of the tests are given in Table 3 below.

TABLE 3

	A	B	C	D	E	F	G
Trimethylol propane triheptanoate	91.5		87.0	92.0	90.9	86.5	91.7
Pentaerythritol monobutyrate triheptanoate cross-linked with azelaic acid to a viscosity of 5.6-6.0 centistokes measured at 210° F		91.5					
Bentonite clay	7.0	8.0	13.0	5.7	5.6	12.5	
Pyrogenic silica gelling agent				1.0	1.0		8.3
Propylene carbonate	1.5	1.5		1.5	1.5		
Phenyl-α-naphthyl amine					1.0	1.0	
Penetration, strokes:							
0	272	279	237	259	267	261	310
60	276	304	261	262	269	270	
10,000	337	387	285	315	330	332	
100,000	400	420	300	356	365	367	

Example 4: The extreme pressure of the greases was tested by formulating two greases and testing the same in accordance with standard procedures. The greases and results of the tests are set out below in Table 4.

TABLE 4

	A	B
Trimethylol propane triheptanoate	90.9	96.5
Bentonite clay	5.6	12.5
Pyrogenic silica gelling agent	1.0	
Propylene carbonate	1.5	
Phenyl-α-naphthyl amine	1.0	1.0
Falex load test, lbs	2,000	2,150

The standard Falex load test was carried out on five commercially available greases to compare with the greases of Example 4. The highest test load was well under 1,800 pounds.

Lithium Soap Thickened Diester Grease

R.L. Crouch and P.A. Cook; U.S. Patent 3,428,562; February 18, 1969; assigned to Texaco Inc. describe a method for preparing a lithium soap thickened grease containing synthetic lubricating oil as the sole lubricating oil component which comprises saponifying a saponifiable fatty material with a lithium saponifying agent in water at a temperature of between 160° and 200°F., adding between 23.0 and 41.0 weight percent synthetic lubricating oil based on the total weight of the synthetic lubricating oil component in the finished grease while the saponification mixture is at a temperature between 150° and 200°F., heating the mixture at a rate of at least 0.7°F. per minute to a top temperature of between 380° and 450°F. while adding additional lubricating oil in an amount between 30.0 and 56.0 weight percent based on the total weight of the synthetic lubricating oil component, maintaining the temperature of the mixture between 380° and 450°F. for between 0 and 30 minutes, cooling the mixture while adding any balance of synthetic lubricating oil and shearing the mixture. The grease is then preferably milled to the proper consistency.

Example 1: 10.9 pounds of distilled water and 1.56 pounds of lithium hydroxide monohydrate were charged into a grease kettle to which was added 9.8 pounds of 12-hydroxystearic acid after the temperature of the mixture of water and lithium hydroxide monohydrate was raised to 160°F. The mixture was then heated and stirred at 160° to 180°F. for 1 hour. It was maintained at this temperature while 20.0 pounds of di-2-ethylhexyl azelate were added through a swing line into the top of the kettle at a rate of 0.35 pound per minute. The first few pounds of oil were added in shots.

The grease mixture was then heated to a top temperature of about 390°F. while there was added at the rate of 0.35 pound per minute 41.08 pounds of di-2-ethylhexyl azelate. When the temperature reached 390°F. all of the di-2-ethylhexyl azelate had been added. The mixture was maintained at a temperature of 390° to 400°F. for 15 minutes. It was then cooled by circulating cold oil through the jacket surrounding the vessel. A minor stream of the grease mixture was withdrawn from the maintained agitated body at a rate of 2 gallons per minute and recirculated via a grease pump and shear value operating with a pressure drop across the valve of 60 psig. When the temperature of the grease mixture reached 325°F. there was added at a rate of 0.35 pound per minute 20.36 pounds of di-2-isooctyl adipate via the swing line and cooling of the mixture continued. There was subsequently added 1.0 pound of N-phenyl-α-naphthylamine, 2.0 pounds pale lead naphthenate and 6.0 pounds of tricresyl phosphate.

Example 2: Into a grease vessel was charged 10.9 pounds distilled water and 1.5 pounds lithium hydroxide monohydrate. The mixture was then stirred to break up any lumps present and was heated to 160°F. at which temperature 9.8 pounds of 12-hydroxystearic acid were added. The mixture was held at 160° to 180°F. for 1 hour with the kettle cover over the vessel. While the mixture was held at this temperature, 20 pounds of di-2-ethylhexyl azelate were added via a swing line into the top of the kettle at a rate of about 0.35 pound per minute with the addition of the oil at first being in shots until soap formation began. The mixture was then heated to 390°F. as quickly as possible while 43.7 pounds of di-2-ethylhexyl azelate were added to the mixture. When the temperature reached about 390°F. all of the di-2-ethylhexyl azelate had been added. The mixture was maintained at a temperature of 390° to 395°F. for 15 minutes.

It was then cooled by passing cold oils through the jacket surrounding the vessel and a minor stream of the mixture was withdrawn and recirculated at a recirculation rate of 2 gallons per minute via a grease pump and a shear valve operating with a pressure drop across the valve of about 60 psig. When the temperature was decreased to 325°F. there was added to the mixture 21.25 pounds of diisooctyl adipate at a rate of 0.35 pound per minute through the swing line. The grease mixture continued to cool until it reached a temperature of about 300°F. When all of the di-2-ethylhexyl adipate had been added 1.0 pound of Manvel Residuum was added. When the temperature decreased to between 180° and 200°F., 1 pound of phenyl-α-naphthylamine, 2 pounds of pale lead naphthenate, 1.0 pound of an extreme pressure antiwear agent and 46 g. of 2,5-bis(octyldithio) thiadiazole copper corrosion inhibitor was added. The mixture was then stirred for 45 minutes.

Refined Esters

A process described by R.H. Boehringer and R.E. Vail; U.S. Patent 3,585,138; June 15, 1971; assigned to Emery Industries, Inc. involves synthetic grease compositions comprising a selected ester refined with a glycidyl ester of an aliphatic acid containing from 5 to 22 carbon atoms and a metallic soap.

Examples of the esters are pentaerythritoltetraisostearate, pentaerythritoltetrapelargonate, di-(2-ethylhexyl) dimerate, and tri-(2-ethylhexyl) trimerate, etc. Examples of complex esters which may be treated by the process are the reaction products of sebacic acid, trimethylpropane, and pelargonic acid; sebacic acid, 1,3,4,7-octanetetrol and pentanoic acid; and ethylene glycol, 1,4-butanediol, adipic acid and azelaic acid.

Prior to this process these esters were not suitable for this use because (1) the excess acidity resulting from the free acid present prevented the ester from thickening to a satisfactory grease consistency when compounded with a metallic soap thickener, and (2) the standard methods for refining either left soap in the ester or too much ester was lost in the process of removing the soap.

The selected ester base fluids are refined in accordance with the method of U.S. Patent 3,470,619. In general, the process is performed by introducing into the unrefined ester system a sufficient amount of a glycidyl ester to react with the entire residual acid content of the unrefined ester. The glycidyl esters react with the free acids in the ester to form a hydroxy ester compound with the basic reaction being as follows:

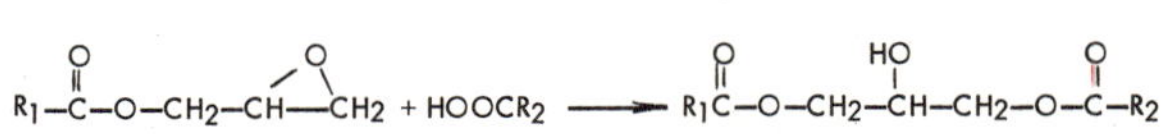

where R_1 is a branched or straight chain alkyl group having from 4 to 21 carbon atoms, preferably having a neo structure and from about 7 to 9 carbon atoms, and R_2 is the residue of a monocarboxylic acid, a dicarboxylic acid, a half ester of a monohydric alcohol, a dicarboxylic acid or carboxyl terminated polyester.

In carrying out the process, the acid value of the unrefined ester is determined and then at least a stoichiometric amount of glycidyl ester is added, but preferably an excess of from 5 to 100 equivalent percent over the stoichiometric amount required for reaction with all of the carboxyl groups. The reaction may be carried out over a wide temperature range of from 150° to 250°C. and preferably 210° to 230°C. The time of the reaction is dependent upon the temperature used and the amount of the free acid which is present. Normally, at a temperature of about 220°C. the reaction may be carried out for a period of from 2 to 4 hours. After the desired level of acid scavenging is completed, the excess glycidyl ester remaining in the ester may be stripped from the product at a temperature of about 250°C. and at a pressure of from 1 to 5 torr.

The hydroxy ester products formed in reducing the excess acid components are fully compatible in the synthetic grease system and possess excellent high temperature qualities, and accordingly, do not detract from the lubricity of the grease. In fact, they supplement the lubricity while alleviating, because of stability, the usual need for removal of the free acid product.

Following the method described above, it has been possible to lower the acid value of those esters suitable for high temperature synthetic greases to the point that no carboxyl content is detectable. This process not only makes available certain esters for synthetic lubricant greases that were not previously practical to use, but it also makes it possible to refine the ester in the same vessel in which it is prepared in the initial esterification reaction alleviating the need for additional handling and equipment. This is possible because of the relatively short time which is required to accomplish the reduction of the acid value to the desired level and the simplicity of the process.

Example 1: 3,500 g. of an ester, tri-(2-ethylhexyl) trimerate, the reaction product of 2-ethylhexyl alcohol and trimeric acid (a tribasic polymerized oleic acid having about 54 carbon atoms), was prepared. The acid value of the tri-(2-ethylhexyl) trimerate was determined to be 2.8. A reaction vessel containing the tri-(2-ethylhexyl) trimerate was charged with 85.5 g. of the glycidyl ester of a neo-acid (Cardura E), having a molecular weight average of 245. The 85.5 g. represented a 100% excess over the stoichiometric amount required to react with all of the free acid contained in the ester. The reaction vessel was heated to a temperature of 220° to 230°C. for 3 hours after which the excess glycidyl ester was stripped off at a temperature of about 245°C. and a pressure of 2 to 3 torr. The ester had an acid value of 0.05.

Example 2: 1,800 g. of an ester, pentaerythritoltetraisostearate, was prepared by the esterification of pentaerythritol with isostearic acid. The acid value of the ester was determined to be 3.50. The unrefined ester was divided into three portions, A, B, and C. The first portion was retained as an unrefined ester to be compounded with a metallic soap. The second portion B, was refined as described in Example 1, with a resulting acid value of 0.05. This portion was also retained to be compounded with a metallic soap.

The third portion C, was refined with sodium hydroxide by the following procedure. The unrefined ester was first diluted with an equal volume of toluene, then sodium hydroxide in an amount of about 300% excess of the stoichiometric requirements for reaction with the free acid was added as a 10% aqueous solution. After vigorous agitation, the excess caustic solution, including the soap heel, was allowed to settle from the organic layer and drawn off. The organic solution was then washed with copious quantities of distilled water until the washwater drained from the solution was neutral to phenolphthalein. The organic layer was then heated in vacuo to distill off the toluene and the last traces of moisture. The alkali refining was accomplished with considerable difficulty and loss of ester. Each of the three portions of ester were then compounded according to the following formulation.

Ester	89.5%
Lithium 12-hydroxystearate	8.0%
Sorbitan monooleate	2.0%
Purified phenothiazine	0.5%

After heating the formulations to 200°C. and stirring until clear, each was quickly cooled and homogenized. The first grease using the unrefined ester did not have a good grease consistency and when tested by ASTM Designation D-217-48 it had a penetration below the lower limit of the test (greater than 380 tenths of millimeters). The second and third greases had a good grease consistency and had a penetration of 245 and 255, respectively.

SILICONES

GENERAL FLUID SYNTHESIS

Aralkylmethyl Polysiloxanes

A.L. Culpepper; U.S. Patent 3,532,730; October 6, 1970; assigned to Dow Corning Corporation describes organopolysiloxane hydraulic fluids. For the last 20 years or longer, various organopolysiloxane fluids have been advocated for use in hydraulic systems. They have found considerable use in such systems, particularly in automotive equipment and in other equipment where the requirements of the fluid are not as severe as is found in aircraft. To date, however, organosiloxane hydraulic fluids have suffered from one or more disadvantage with respect to lubricity, pour point, compressibility and/or cost. The fluids which show the best lubricity have been deficient in compressibility and those which show the lowest compressibility have been deficient with respect to lubricity and/or freezing point.

For example, the best fluids previously employed, are the trifluoropropylmethylsiloxane fluids. However, these fluids have a compressibility of 7.45% at 20,000 psi at 25°C. Other siloxanes, such as the dimethylsiloxane have excellent low temperature properties, but their compressibility is high and their lubricity is poor. Phenylmethylpolysiloxanes have low compressibility, but their lubricity is poor with respect to steel on steel and the freezing point is 0°C. Thus, although all of these fluids have been advocated for use in hydraulic systems they all fail to meet the critical requirements which are now demanded of hydraulic fluids in the advanced type aircraft. This process provides organopolysiloxane copolymers, which meet the criterion of low compressibility, excellent lubricity of steel on steel and low pour point required of such hydraulic fluids. This process involves siloxanes of the average formula

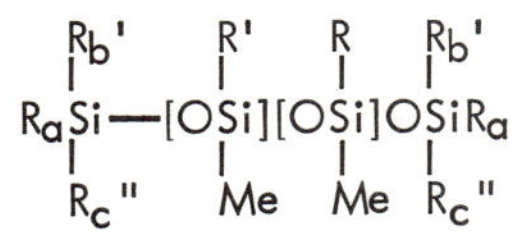

in which R is of the formula

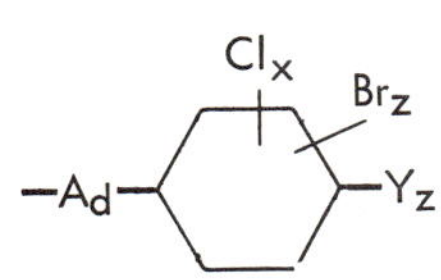

in which A is an alkylene radical of 2 to 3 inclusive carbon atoms, d is 0 or 1, x is 0 to 4, Y is an alkyl radical of 1 to 2 carbon atoms inclusive, and each z is independently 0 to 2; R' is an alkyl radical of 6 to 10 inclusive carbon atoms, R" is an alkyl radical of 1 to 5 carbon atoms, a is 0 to 2, b is 0 to 2, c is 1 to 3, and the sum of a+b+c being 3, the proportions of the various siloxane units being such that the mol ratio of R to R' is from 1:9 to 1:0.67. The above compositions have the required low compressibility together with other desirable properties to make them suitable for hydraulic fluids in high speed aircraft. The lowest compressibility is obtained when a has a value of 1 or 2. The second lowest compressibility is obtained when b has a value of 1 or 2, and a is 0 and the highest compressibility is obtained when c is 3. Any ratio of R to R' in the range 1:9 to 1:0.67 is operative, however, the best range is from 1:5 to 1:2. The following examples are illustrative of the process.

Example 1: 848 g. of a trimethylsiloxy-endblocked methylhydrogen polysiloxane containing 1.18% by weight (10 mols) silicon-bonded hydrogen was heated with stirring to 125°C. under a slow stream of nitrogen. 0.5 ml. of a 0.1 molar solution of chloroplatinic acid in isopropanol was added and then 236 g. (2 mols) of alpha-methylstyrene at a rate to maintain the temperature at 125° to 150°C. The fluid was then heated an additional hour to insure complete reaction and 1,096 g. (8 mols plus 200 g. excess) of octene was added at a rate to maintain the temperature in the same range. The fluid was heated at 125°C. overnight to insure complete reaction of the olefin. The fluid was then stripped by heating to 170°C. at 0.4 mm. and held at that temperature for 1.5 hours. The fluid was then decolorized with activated charcoal and filtered through a 1 to 2 micron filter. The resulting product was the trimethylsiloxy-endblocked copolymer of octylmethylsiloxane and beta-phenylpropylmethylsiloxane in the ratio of 4 octyl groups per beta-phenylpropyl group. The copolymer fluid had the following properties. The viscosity was as follows.

48.1 cs. at 25°C.
31.7 cs. at 38°C.
7.8 cs. at 99°C.
2,180 cs. at -40°C. and
11,200 cs. at -54°C.

The flash point was 290°C., the pour point was -84°C., and the specific gravity was 0.921. The ratio of R to R' was 1:4. The compressibility of the fluid was 6.45% at 20,000 psi at 25°C. The fluid was tested on a Shell 4-Ball test machine at 204°C. at 1,200 rpm under load of 20 kg. for one-half hour to give a scar diameter of 1.38 mm. This shows the lubricity is sufficient for use in an aircraft hydraulic system. The lubricity can be improved by the selection of proper lubrication additives.

Example 2: A solution of 213.5 g. (1 mol) of beta-phenylpropyldimethylchlorosilane and 374 g. (3.25 mols) of methyldichlorosilane in 378 g. of toluene was added continuously to water at 60°C. at a rate of 130 cc per minute. The hydrolyzate was azeotroped to remove water and the solvent was distilled to produce a copolymer of the average formula.

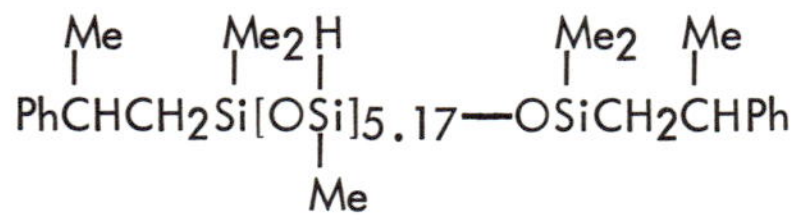

298.3 g. of the above fluid which contained 2.26 equivalents of hydrogen was mixed with 51.45 g. (0.436 mol) of alpha-methylstyrene and 5 drops of a one-tenth molar solution of chloroplatinic acid in isopropanol. The mixture was heated until exothermic reaction increased the temperature at 150°C. and thereafter the heating was continued at 130° to 150°C. for one hour. There was then added an excess of ocetene-1 to the reaction mixture which was heated at 130°C. until silver nitrate gave a negative test for SiH. The product was stripped at 190°C. at less than 1 mm. pressure for 30 minutes. The resulting fluid was a copolymer of the average formula

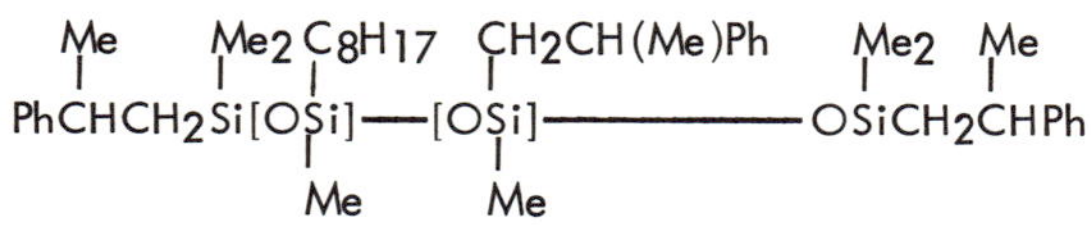

having the following properties. The viscosity was 77.7 cs. at 25°C., 49.9 cs. at 38°C. and 11.7 cs. at 99°C. The pour point was -65°C. and the compressibility was 5.92% at 20,000 psi at 25°C. The fluid was an excellent lubricant for steel on steel surfaces. The ratio of R to R' in this example is 1:1.4.

Silphenylene-Siloxane Copolymers

R.L. Merker; U.S. Patent 3,332,973; July 25, 1967; assigned to Dow Corning Corporation describes fluid silphenylene-siloxane copolymers having the average general formulas

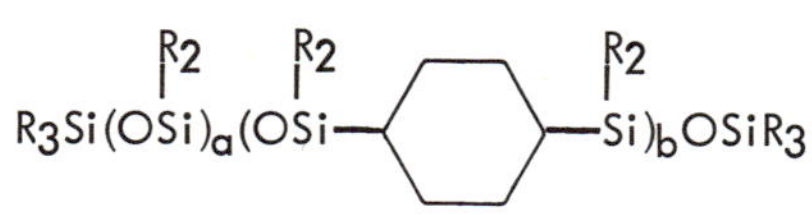

and

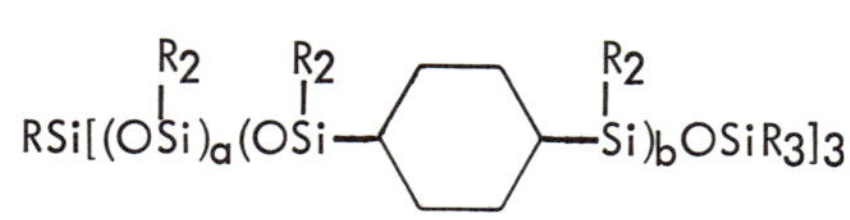

where each R is independently selected from the group consisting of methyl and 3,3,3-trifluoropropyl radicals, a and b are integers such that the ratio of a to b is in the range from 1:1 to 12.5:1. The copolymers have viscosities from 15 to 350,000 cs. at 25°C. The fluid copolymers are useful, for example, as lubricants, dielectric coolants, damping and viscous drive fluids and as viscosity index improvers for silicate and other hydraulic fluids.

The best method for preparing the siphenylene-siloxane copolymer fluids consists of equilibrating the desired molar ratios of para-bis(dimethylhydroxysilyl)benzene, dimethylsiloxane cyclics and a trimethylsilyl-endblocked polydimethylsiloxane fluid having a viscosity of from 0.65 to 300,000 cs. at 25°C. at a temperature of from 100° to 190°C. for 10 to 35 hours in the presence of an alkaline catalyst such as potassium hydroxide. It is preferred that a trimethylsilyl-endblocked polydimethylsiloxane having a viscosity of 1 to 100 cs. at 25°C., a temperature of 140° to 170°C. and a time of 15 to 30 hours be employed. The following examples illustrate the process.

Example 1: 247 g. of para-bis(dimethylhydroxysilyl)benzene, 265.5 g. of polydimethylsiloxane cyclics, 17.8 g. of a trimethylsiloxy-endblocked dimethylsiloxane fluid having a viscosity of 1.5 cs. at 25°C., 530 g. of dry toluene and 1% potassium hydroxide (catalyst) were added to a reaction vessel and refluxed for 16 hours at 110° to 120°C. The catalyst was then neutralized with trimethylchlorosilane. The toluene and other low molecular weight materials were stripped from the fluid to 255°C. at a pressure of 1 mm. of mercury. The 452 g. of product obtained was treated with 2.5 g. of Nuchar (an activated charcoal) for 2 hours at 75°C. and then filtered thru Super-Cel (a diatomaceous silica). The filter product consisted of 397.3 g. of a fluid having a viscosity of 385.5 cs. at 25°C.

Example 2: 219 g. of para-bis(dimethylhydroxysilyl)benzene, 437.5 g. of polydimethylsiloxane cyclics, 20.4 g. of a trimethylsiloxy-endblocked dimethylsiloxane fluid having a viscosity of 1.5 cs. at 25°C., about 677 g. of toluene and 1% of potasium hydroxide were added to a reaction vessel and refluxed for 16 hours at 114° to 115°C. The fluid was then neutralized with 7.0 g. of trimethylchlorosilane. The toluene and low molecular weight materials were stripped from the fluid to 269°C. at a pressure of 1 mm. of mercury. The 546 g. of product obtained was treated with 5 g. of Nuchar for 2 hours at 70° to 75°C. and then filtered thru Super-Cel. The filtered product consisted of 490 g. of a fluid having a viscosity of 209.5 cs. at 25°C.

Example 3: 113 g. of para-bis(dimethylhydroxysilyl)benzene, 200 g. of polydimethylsiloxane cyclics, 124 g. of a trimethylsiloxy-endblocked dimethylsiloxane fluid having a viscosity of 1.5 cs. at 25°C., 426 g. dry toluene and 4.3 g. of potassium hydroxide pellets were added to a reaction vessel and equilibrated for 18 hours at 110° to 120°C. The fluid was then neutralized with 12 g. of trimethylchlorosilane. The toluene and low molecular weight materials were stripped from the fluid to 250°C. at a pressure of 0.275 mm. of mercury. Then the product was treated with Nuchar and filtered. The filtered product consisted of 191.7 g. of a clear liquid having a viscosity of 36.6 cs. at 25°C.

Example 4: 338 g. of para-bis(dimethylhydroxysilyl)benzene, 771 g. of polydimethylsiloxane cyclics, 7.5 g. of a trimethylsiloxy-endblocked dimethylsiloxane fluid having a viscosity of 5 cs. at 25°C. and 10 g. of the potassium salt of polydimethylsilanolate were added to a reaction vessel and heated for 24 hours at 150° to 160°C. The fluid was then diluted with toluene, neutralized with Dry Ice and filtered. The toluene and other volatiles were stripped from the fluid in two stages, the first stage being to 170°C. at a pressure of 28 mm. of mercury and the second stage being to 300°C. at a pressure of 1 mm. of mercury. The 964 g. of product obtained was a fluid having a viscosity of 5,000 (i.e., 4,959) cs. at 25°C.

Example 5: This example illustrates the fact that the fluid silphenylene-siloxane copolymers have better thermal stability at high temperatures than similar fluids containing only methyl groups. The thermal stability test consisted of heating the fluids, under nitrogen, in an oven at 250°C. for 24 hours. The viscosities of the fluids were measured at 25°C. both before and after they were heated in the oven. The fluid samples were prepared for the oven by placing them in a tube, evacuating the air from the tube, purging with nitrogen and then sealing the tube. The silphenylene-siloxane fluid used in the test was that of Example 4. The test results are set forth below and show that the percent change in the viscosity of the polydimethylsiloxane fluid was more than twice as great as the change in the viscosity of the silphenylene-siloxane fluid of this process.

Fluid	Viscosity at 25° C. (cs.)		Percent Change
	Before heating	After heating	
Silphenylene-siloxane	4,959	3,595	27.5
Polydimethylsiloxane	5,070	1,858	63.3

Cyclic Organosilicon Compounds

C.R. Sporck; U.S. Patent 3,452,071; June 24, 1969; assigned to General Electric Company describes a cyclic organosilicon compound having the the structure where R is a monovalent hydrocarbon radical free of aliphatic unsaturation, and n is from 2 to 6. The compounds are very stable.

(1)

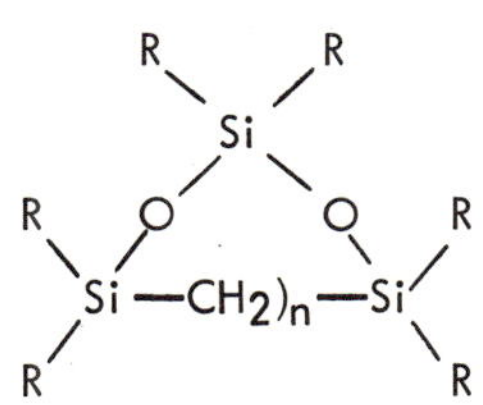

Preferably, the compounds of the process are prepared by reacting a dichlorodisilaalkane compound of formula:

(2) $ClSiR_2{-}(CH_2)_n{-}SiR_2Cl$

with a diorganosilanediol of formula:

(3) $R_2Si(OH)_2$

where R and n are as previously defined. However, the reaction may also be carried out with a dihydroxydisilalkane compound of formula:

(4) $HOSiR_2{-}(CH_2)_n{-}SiR_2OH$

with a diorganodichlorosilane of formula:

(5) R_2SiCl_2

where R and n are, again, as previously defined. In either case, the two compounds are preferably reacted in equimolar amounts, but a 10% excess, based on the stoichiometric ratio of 1:1, of either compound can be employed. The reaction is carried out in a solvent with a total reactant concentration of from 0.1 to 1.0 mol per liter. Preferably, each of the reactants is dissolved in a portion of the solvent, and the two solutions are added to a third portion of the solvent. However, neither the order of addition nor the manner of forming the solutions is critical, so long as the reactants dissolve quickly in the reaction mixture for easy reaction. The formation of the cyclic organosilicon compounds is shown in the example below.

Example: This example illustrates the preparation of the material of Formula (1) where the R substituents on one of the silicon atoms are phenyl, the remaining R substituents are methyl, and n is 2. A quantity of 21.5 parts of 2,5-dichloro-2,5-dimethyl-2,5-disilahexane was dissolved in 140 parts of benzene to form a first solution. A quantity of 21.6 parts of diphenylsilanediol was dissolved in a mixture of 62.5 parts of acetonitrile and 70 parts of benzene to form a second solution. The temperature of the second solution was maintained at 50°C. to assure the retention of the diphenylsilanediol in solution. A third solution was formed containing 32 parts of pyridine dissolved in 350 parts of benzene.

Over a period of 35 minutes, the first and second solutions were added, simultaneously and at equal rates, to the third solution, with stirring. The reaction mixture was stirred for 15 minutes following completion of the addition and the solids were then filtered from the reaction mixture and the solvent evaporated. The residue from the distillation was dissolved in toluene, filtered, and the toluene evaporated from the filtrate. The solid residue was then vacuum distilled at 125°C. at 0.02 mm. of mercury and yielded a product distillate with a melting point of 55° to 56.5°C. This material had the structure:

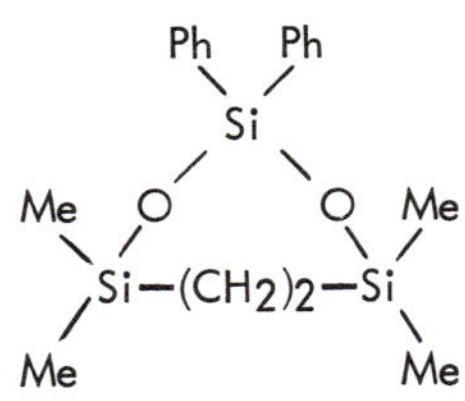

where Ph is the phenyl radical and Me is the methyl radical. This material corresponds to generic Formula (1) as previously described in this example. The structure of the compound was substantiated by an infrared spectrum. After recrystallization twice from ethanol, employing 0.4 part of ethanol for each part of solids, the material melted at 57° to 58°C. Vapor phase chromatography showed a single peak consistent with the expected material. Analysis showed 59.8% C, 7.3% H, and 23.5% Si. Theoretical is 60.4% C, 7.3% H and 23.5% Si.

A quantity of 4 parts of the cyclic organosilicon compound was heated, with stirring, at 125°C. A quantity of 40 parts per million, based upon the cyclic material, of potassium naphthalene was added. No change in viscosity was observable after 5 minutes. A second quantity of 40 parts per million of potassium naphthalene was added and no change was observable after a further 10 minutes. The temperature of the stirred mixture was raised to 150°C. and a third 40 parts per million addition of potassium naphthalene was made. No change could be seen in the viscosity of the mixture after 20 minutes. At that time a fourth 40 parts per million quantity of potassium naphthalene, making a total of 160 parts per million, was added and no change could be seen in the material after a further 20 minutes.

Similarly, 2 parts of the material was stirred at 145°C. and a quantity of 0.044 part of dimethylsulfoxide was added. Following this addition, a quantity of 40 parts per million, based upon the weight of the material, of potassium naphthalene was added and no change in the viscosity or the appearance of the stirred mixture was noted after several minutes. A second addition was made of the same amounts of dimethylsulfoxide and potassium naphthalene, but again no change was noted. A quantity of 80 parts per million of potassium naphthalene was added and heating was continued with stirring for an hour, but there was no detectable change in the appearance of the stirred mixture. An infrared spectrum was run on a quantity of the stirred mixture, but no change was detected from the original structure.

A quantity of 2 parts of the cyclic material was dissolved in 8.5 parts of toluene and the solution was stirred. An 86% solution of sulfuric acid was added to the stirred solution in an amount of 0.11 part. The solution was stirred for several days, at room temperature, with no apparent change in the viscosity of the material. An infrared spectrum was run on this sample and showed no change from the original absorption pattern.

Diaryl Dialkoxy Silanes

B.B. Millward; U.S. Patent 3,291,742; December 13, 1966; assigned to Monsanto Chemicals Ltd., England describes a system where a triaryl aliphaticoxy silane or a diaryl dialiphaticoxy silane is employed as a functional fluid. Where the silane contains a single aliphaticoxy group, the group is one having at least six carbon atoms. However, where the silane contains two aliphaticoxy groups, one of such groups can have as few as two carbon atoms provided that the total number of carbon atoms in the aliphaticoxy groups is at least 12.

For use as functional fluids, the diaryl dialiphaticoxy silanes are preferred since they possess the most desirable properties. It is particularly preferred to employ such a silane where the aryl groups are phenyl, tolyl or xylyl, and where the aliphaticoxy groups are alkoxy groups of 2 to 18 carbon atoms and the two alkoxy groups contain a total of at least 12 carbon atoms. Illustrative members of this group of compounds include diphenyl n-octyloxyl n-decyloxysilane, diphenyl n-decyloxyl n-tetradecyloxysilane, ditolyl 2-ethylhexyloxy n-hexyloxysilane, diphenyl n-octadecyloxy n-hexadecyloxysilane and ditolyl 2,2-dimethylpentyloxy dodecyloxysilane.

In the process for preparing the aryl aliphaticoxy silanes, an arylhalosilane is reacted with an appropriate alcohol. The halogen atoms linked to the starting silane are preferably chlorine. When it is desired to prepare one of the groups of diaryl dialkoxy silanes previously described, the diaryl dihalosilane is first reacted with an alkanol to replace one of the halogen atoms. The resulting diaryl alkoxyhalosilane is then reacted with a different alcohol to replace the other halogen atom and produce the mixed alkoxysilane. Stoichiometric or near stoichiometric quantities of the reactants are usually employed, although a slight excess of the alcohol (for example, five to fifteen percent excess) is sometimes advantageous to the reaction.

As heat transfer media, the products of this process can be used advantageously in the transmission of heat as a transport fluid, i.e., a fluid may be circulated through a boiler or other heating medium to the location where heat is needed and then returned to the boiler for reheating, whether it be high-grade heat for chemical processing, or low-grade space heat used, e.g., in heating homes or offices. The process is illustrated by the following examples.

Example 1: This example describes the process for the production of diphenyl di(n-octyloxy) silane from n-octanol and diphenyldichlorosilane. A solution of 13 cc of pyridine in 25 cc of benzene was added slowly to a solution of 25 cc of n-octanol and 15 cc of diphenyldichlorosilane in 75 cc of ether, and the mixture was heated under reflux for 2 3/4 hours. Excess ammonia gas was then passed in, and the reaction mixture was filtered. The solvents were evaporated from the filtrate, and distillation of the residue gave 25.7 g. of diphenyl di(n-octyloxy) silane having a boiling range of 190° to 205°C. at a pressure of 0.4 mm. of mercury. Purer material having a boiling range of 204° to 206°C. at a pressure of 0.6 mm. of mercury was obtained by redistillation. This was a colorless, mobile oil at room temperature.

Example 2: This example shows certain of the properties, relevant to the process, of a number of aryl alkoxy silanes which can be used as thermally stable functional fluids. The thermal stability of an aryl alkoxy silane was assessed by recording the rate of isothermal vapor pressure increase of a sample in an isoteniscope at a decreasing series of temperatures between 400° and 300°C. The decomposition temperature of the material was recorded as the temperature at which its isothermal vapor pressure increased by 0.84 mm. of mercury per minute. These data illustrate the wide temperature ranges over which the diaryl dialkoxy silanes can be used.

Compound	Boiling Range ° C./mm. of Mercury	Freezing Point (° C.)	"Decomposition Temperature" (° C.)
Diphenyl di(n-hexyloxy) silane	156–159/0.5	Below −36	354
Diphenyl di(n-octyloxy) silane	204–206/0.6	Below −36	344
Diphenyl di(2-ethylhexyloxy) silane	184–188/0.2	Below −36	347
Diphenyl di(3,5,5-trimethyl hexyloxy) silane	176–178/0.1	Below −36	334
Diphenyl di(n-decyloxy) silane	210–212/0.05	−32 to −30	348
Diphenyl di(n-dodecyloxy) silane	245–250/0.3	−6 to −4	356

Example 3: The resistance to hydrolysis of diphenyl di(n-octyloxy) silane was shown by boiling five grams with 30 cc of water under reflux for 17 hours. At the end of this time, only a small quantity of n-octanol had been formed by hydrolysis, and 85% of the diphenyl di(n-octyloxy) silane was recovered unchanged by distillation.

Example 4: This example describes the production of diphenyl di(2,2-dimethylpentyloxy) silane. A solution of 25 cc (0.31 mol) of pyridine in 50 cc of benzene was added slowly to a stirred solution of 36.1 g. (0.14 mol) of diphenyldichlorosilane and 41 g. (0.30 mol) of 2,2-dimethylpentanol in 150 cc of benzene, and the mixture was boiled under reflux for four hours. Excess ammonia gas was then passed in, and the reaction mixture was filtered. After evaporating the solvent from the filtrate, the residue was distilled to give 49.6 g. of diphenyl di(2,2-dimethylpentyloxy) silane having a boiling range of 172° to 176°C. at a pressure of 0.5 mm. of mercury. Its decomposition temperature was 335°C.

m-Phenylphenoxy Substituted Siloxanes

In a process described by H.C. Kaufman; U.S. Patent 3,354,194; November 21, 1967; assigned to Olin Mathieson Chemical Corporation meta-phenoxy phenol is reacted with a dialkyl or diaryl dihalosilane or disiloxane to form an intermediate product which in the second stage of the reaction is reacted with a dihydroxy-substituted aromatic compound, such as resorcinol, to yield a high temperature resistant, silicon-containing fluid composition. The product of the process satisfactorily meets all requirements for high performance aircraft lubricants and hydraulic fluids. The combination of meta-linked phenyl ethers and siloxane or silane results in hydrolytically stable fluids exhibiting pour points down to -40°F. or below. The following examples are illustrative of the process.

Example 1: To a reaction flask equipped with a Dry Ice condenser there was charged 0.15 mol of 1,3-dichlorotetramethyl disiloxane in 100 g. of Decalin at ambient conditions. To the mixture was rapidly added 0.15 mol of meta-phenoxyphenol with stirring. The Dry Ice condenser prevented the escape of the volatile reactant with the evolving hydrogen chloride. The hydrogen chloride evolution began slowly. Heat was applied and hydrogen chloride was evolved vigorously at 60°C. for 3 hours.

Next, the contents of the flask were stirred at 115°C. for 12 hours. After the hydrogen chloride evolution had ceased, the flask containing the yellow-colored solution was cooled to ambient temperature and 0.08 mol of solid resorcinol (insoluble in Decalin) was added and the slurry stirred as heat was applied. At 50°C. the evolution of hydrogen chloride began; at 80°C. hydrogen chloride was vigorously evolved and, after one hour, all the resorcinol had dissolved. The temperature of the mixture was raised to the reflux temperature of the Decalin, viz 194°C., and maintained at that temperature for 12 hours. The resulting amber-colored solution was cooled and distilled to remove the Decalin and unreacted phenols. Further distillation yielded products having the general formula shown below.

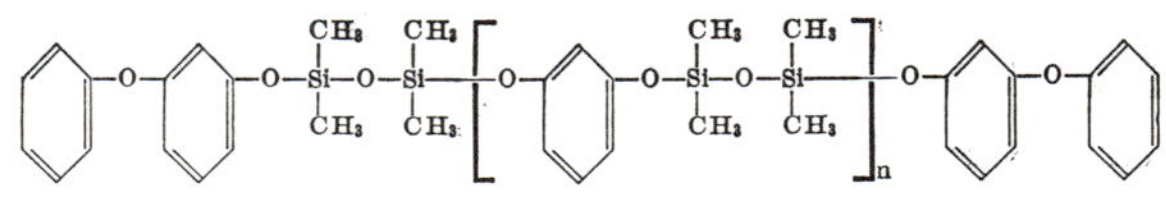

Product A — Sym-tetramethyl bis(meta-phenoxyphenoxy) disiloxane. Formula: see general formula above where n = 0. Weight recovered: 25 g. — distilled as a colorless liquid at 225° to 228°C. (0.7 mm. Hg). Analysis — Calculated for $C_{28}H_{30}Si_2O_5$: Si, 11.16%; Found: Si, 11.1%. Molecular weight — Calculated for $C_{28}H_{30}Si_2O_5$: 502; Found: 485.

Product B — Sym-bis(tetramethyl, meta-phenoxy phenoxydisiloxanyl) meta-benzene. Formula: see general formula above where n = 1. Weight recovered: 10 g. — distilled as a pale yellow liquid at 278° to 280°C. (0.07 mm. Hg). Analysis — Calculated for $C_{38}H_{46}Si_4O_8$: Si, 15.09%; Found: Si, 15.20%. Molecular weight — Calculated for $C_{38}H_{46}Si_4O_8$: 742; Found: 720.

Example 2: To an apparatus similar to that of Example 1 there was charged 0.50 mol of resorcinol and 1.00 mol of water dissolved in 300 ml. of dioxane. The reaction flask was a 5 liter, glass, round bottomed reactor equipped with a conventional heating mantle, a stirrer, and a condenser, cooled by circulating ice water and surmounted by a second condenser cooled by acetone and Dry Ice. The outlet of the final condenser was protected by a drying tube. A jacketed addition funnel, cooled by circulating ice water and protected from the atmosphere by a drying tube, was also connected to the reactor.

A slow stream of nitrogen was introduced over the reactants. 2.00 mols of dimethyldichlorosilane were added to the reaction mixture with no visible sign of reaction. The temperature was raised to 36°C. and hydrogen chloride began to evolve slowly. When a temperature of 71°C. was reached, the dimethyldichlorosilane began to reflux and hydrogen chloride was copiously evolved. After 24 hours the temperature was raised to 103°C. and the reaction mixture was refluxed for another 24 hours. After the contents of the reactor had been cooled, 1.10 mols of meta-phenoxy phenol was added. As the temperature was raised to reflux at 103°C. hydrogen chloride began to be evolved. After 24 hours the evolution of hydrogen chloride had ceased and the contents of the flask were cooled and stripped of dioxane and unreacted phenols. Further distillation yielded the following four products.

Product A — Weight 95 g.; distilled at 180°C./0.08 mm. Hg. By analysis this fraction was shown to be dimethyl bis-(m-phenoxy phenoxy) silane.

Product B — Weight 120 g.; distilled at 223° to 226°C./0.08 mm. Hg. This product was sym-tetramethyl bis(meta-phenoxy phenoxy) disiloxane.

Product C — Weight 120 g.; distilled at 275° to 278°C./0.06 mm. Hg. This product was sym-bis(tetramethyl, meta-phenoxy phenoxydisiloxanyl) meta-benzene.

Product D — Weight 30 g.; could not be distilled. Product D is an exemplification of a product of the formula below where n is greater than 1.

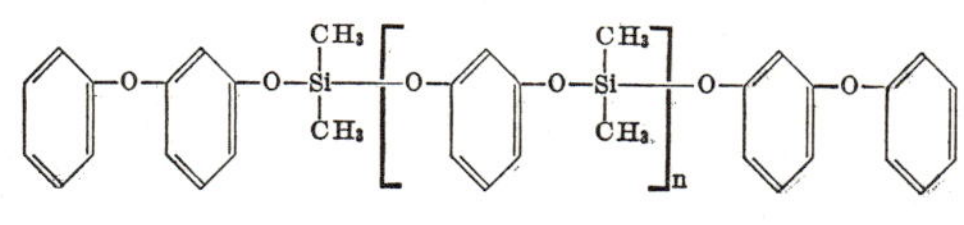

The table below lists the results obtained in certain viscosity tests made with the three products of Example 1.

VISCOSITY MEASUREMENTS

Sample	−40° F.	100° F. cs.	210° F. cs.	500° F. cs.
Product A	Pour point	22.0	4.3	0.86
Product B	do	32.4	5.8	1.13
Product C	do	95.0	23.0	3.92

Siloxane Esters

P.M. Blanchard; U.S. Patent 3,290,248; December 6, 1966; assigned to The British Petroleum Company Ltd., England describes complex esters of siloxane dicarboxylic acids having good viscosity/temperature properties and thermal stability and which are useful as thermally stable fluids, synthetic lubricants and hydraulic fluids. The process involves liquid neutral ester products prepared by reacting together under esterification conditions in one or more stages: (a) a saturated aliphatic monocarboxylic acid having 1 to 18, preferably 3 to 10, carbon atoms per molecule; (b) a saturated polyhydric alcohol having 5 to 15 carbon atoms per molecule and preferably having no hydrogen attached to any carbon atom in a 2 position with respect to any hydroxyl group; and (c) an organo siloxane dicarboxylic acid of the formula shown below where the R_1's are alkylene groups, preferably straight chain having 1 to 10 carbon atoms.

(1)
$$HOOCR_1\overset{CH_3}{\underset{CH_3}{\overset{|}{\underset{|}{Si}}}}—O—\overset{CH_3}{\underset{CH_3}{\overset{|}{\underset{|}{Si}}}}R_1COOH$$

The relative proportions of reactants (a), (b) and (c) are chosen so as to produce an ester product having a viscosity in the range of 2 to 200, especially 2 to 10, centistokes at 99°C. (210°F.). Suitable molar ratios of (a):(b) are 1 to 2.9:1, especially 1.1 to 2.8:1. The proportion of the other ingredient (c) can be calculated so as to give a carboxyl/hydroxyl balance in the reactants though it may be an advantage to use in the reaction mixture an excess of one or two of the reactants over the stoichiometric proportions in order to accelerate the reaction. The proportions of reactants (a) and (c) will, of course, depend on the number of hydroxyl groups in reactant (b). Thus where (b) is a trihydric alcohol, suitable relative molecular proportions for the reactants (a), (b) and (c) are 2 to 2.9:1:0.5 to 0.05, particularly 2.2 to 2.8:1:0.4 to 0.1.

It has been found that the viscosity index of the neutral complex esters described above may be increased by equilibrating them with a cyclic dimethylsiloxane. Accordingly the process also includes neutral liquid complex ester products, prepared by equilibrating a complex ester as hereinbefore described with a cyclic dimethylsiloxane containing 2 to 6 silicon atoms, e.g., octamethyltetrasiloxane. The equilibrating reaction should be carried out in the presence of a catalyst, preferably sulfuric acid. Three neutral complex ester products according to the process are described below. Two of these, identified as Esters P1 and P2, are unequilibrated products and the third, identified as Ester A, is an equilibrated product.

Preparation of Esters P1 and P2: Ester P1 was prepared from the following ingredients: 28.8 g. (0.2 mol) caprylic acid; 10.72 g. (0.08 mol) 1,1,1-trimethylolpropane; and 6.12 g. (0.02 mol) formula (1) acid where R_1 = trimethylene. The components were refluxed with toluene as water entrainer and 0.05 g. tetrabutyl titanate as catalyst for 72 hours when the theoretical amount of water was collected. The toluene was distilled off and the product stripped of light ends to 220°C. at 1 mm. mercury. Ester P2 was prepared from the same ingredients and by the same method except that no catalyst was used.

Preparation of Ester A: Ester A was prepared by equilibrating Ester P2. 0.75 ml. of 98% sulfuric acid was added with stirring to Ester P2. 2.25 g. (0.02 mol) of octamethyltetrasiloxane were added afterwards. The mixture was stirred for 2 hours at room temperature and stood overnight. The solution was taken up in 40/60 petroleum ether and washed free of acid and then stripped to 200°C. at 0.2 mm. mercury. The properties of Esters A, P1 and P2 are given in the table below.

	Ester A	Ester P1	Ester P2
Kinematic at 99° C. (210° F.)	7.38	7.25	7.23
Viscosity at 38° C. (100° F.)	37.6	43.0	42.8
(Centistokes) at —40° C. (—40° F.)	9,540	17,700	17,670
Viscosity Index	142	135	135
ASTM slope (210° F. to —40° F.)	0.642	0.660	0.661
Pour Point ° F	—65	—65	—60
Flash Point, ° F	----------	500	500

Methyl (C_6 to C_{10} Alkyl) Silicone Fluids

J.A. Brennan; U.S. Patent 3,445,390; May 20, 1969; assigned to Mobil Oil Corporation describes a lubricant which comprises a methyl (C6 to C10 alkyl) silicone fluid. The fluid is produced by reacting an aliphatic 1-olefin having from 6 to 10 carbon atoms, i.e., hexene-1, octene-1 or decene-1, with a methyl hydrogen siloxane polymer having a molecular weight in the range of from 800 to 3,000, and preferably from 1,000 to 1,500, having no more than one hydrogen atom per silicon atom, there being from 10 to 40, and preferably from 11 to 20, such Si—H bonds (or reaction sites) available for reaction with the olefin in the methyl hydrogen silicone. The methyl alkyl siloxane polymers of this process have a kinematic viscosity at 210°F. of from 6 to 35 cs. The following example illustrates the process.

Example 1: Preparation of Products — A methyl hydrogen siloxane having a molecular weight of 1,120 was reacted with a stoichiometric excess of (1) hexene-1 and (2) octene-1 using a stoichiometric excess of the olefin in each reaction. The general procedure in each of the reactions is as follows: in a suitable reactor excess olefin and a solution of isopropanol containing chloroplatinic acid (5×10^{-6} mol of platinum per mol of olefin) are mixed together and heated to reflux. Thereafter, the methyl hydrogen siloxane is added and the reaction mixture is maintained at reflux for 4 hours. The excess olefin is distilled off.

The reaction with hexene-1 and the silicone leads to a methyl hexyl silicone having a molecular weight of 2,350, an increase in molecular weight of 1,230. The number of reaction sites along the silicone polymer chain is 14.5 (which also indicates the number of mols of 1-hexene added onto the polymer). The kinematic viscosity of this product at 40°F. is 2,660 cs. The fluid evidences excellent lubricating and hydraulic fluid properties. The reaction of the methyl hydrogen silicone polymer with octene-1 resulted in a product having a molecular weight of 2,700, an increase of 1,580, or 14.0 reaction sites.

Evaluation of Products — These products were evaluated in the autogenous ignition temperature test. This test involves immersing a flask in a heated bath, the temperature is increased at a constant rate. At every 20°F. rise, one or two drops of the test sample are introduced until the temperature increment at which the sample will spontaneously ignite is attained. Once this increment has been established, the autogenous ignition temperature (AIT) is found by lowering and raising the temperature until the minimum which will cause one or two drops of sample to ignite is ascertained.

The products of the above example were also tested for thermal stability at 700°F. Twenty grams of a sample of known viscosity are placed in a glass ampule and purged with nitrogen for two minutes. This container is fitted into an aluminum block at 700°F. and held for 90 minutes. During this time, half a pound of nitrogen is passed over the open tip of the ampule. The evaporation loss, the neutralization number and the kinematic viscosity at 100°F. are thereafter measured, using the usual ASTM methods. The percent loss in kinematic viscosity at 100°F. is deemed a measure of the thermal stability of the sample. Both of the products of Example 1 had an AIT of over 825°F. The percent loss in kinematic viscosity was from 1.3 to 3.7%, and the percent loss in material by evaporation ranged from 0.32 to 0.48.

Example 2: A methyl hydrogen siloxane polymer having a molecular weight of 2,015 is reacted with hexene-1 in the presence of 0.50% by weight of reaction mixture of 1% platinum-on-charcoal. The product has a molecular weight of 3,640. This represents an increase of 1,625 in molecular weight, indicating that 19.1 mols of olefin were added to the silicone. The AIT is 840°F., percent loss in kinematic viscosity of 6.6% and the material loss 0.32%. The results of these examples show that the specific limitations of the methyl hydrogen siloxane polymers described in this process are of critical importance. The number of reaction sites on the polymer chain should be in the range of 10 to 40 and preferably 11 to 20 in order to obtain products having the desired properties. Using silicone polymers having less or more reaction sites either lead to products which have a lower AIT or which suffer either excessive loss in kinematic viscosity or material loss at high temperatures. The products used in the method are capable of performing at extremely high temperatures as lubricants or as hydraulic fluids. They do not form solid materials after long use at such temperatures, but retain their stability and utility over long periods of service.

D.L. Bailey and F.M. O'Connor; U.S. Patent 3,480,583; November 25, 1969; assigned to Union Carbide Corporation describes the preparation of organosilicone block copolymers which are useful as metal to metal lubricants.

1,3-Disilyl-1,3,2,4-Diazadisiletidines

W. Fink; U.S. Patent 3,535,358; October 20, 1970; assigned to Monsanto Company describes a process for preparing 1,3-bis-disilyl-1,3,2,4-diazadisiletidines of the formula

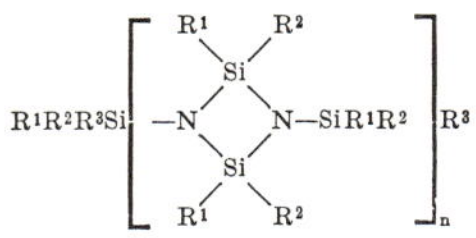

or

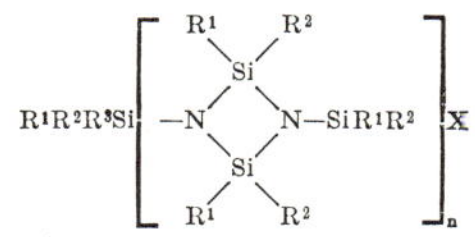

where X is a reactive halogen and n is 1 to 100, by reacting a compound of the formula

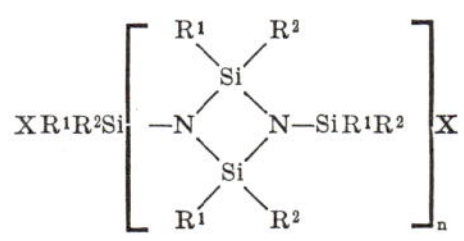

where X is a reactive halogen atom, and NH_2 group or a secondary or tertiary amino group, with a compound of the formula MR^3 where M is an alkali metal atom or MgX in halogen exchange, or a hydrogen linked via oxygen or nitrogen in the amine exchange process, and compounds of the first or second formula above where n is at least 2. These 1,3-bis-disilyl-1,3,2,4-diazadisiletidines are useful as lubricants, oil additives, hydraulic oils and heat exchanger fluids.

Pentafluorophenylethyl Polysiloxanes

G.W. Holbrook and O. R. Pierce; U.S. Patent 3,423,445; January 21, 1969; assigned to Dow Corning Corporation describes the use of monomers, homopolymers and copolymers of organosilicon compounds of the formula

$$[(C_6F_4X)CH_2CH_2]_n\underset{\displaystyle R_m}{\overset{}{Si}}O_{\frac{4-m-n}{2}}$$

Illustrative of this class of compounds is

$$C_6F_4ClCH_2CH_2\underset{\displaystyle CH_3}{\overset{\displaystyle CH_3}{Si}}O_{1/2}$$

The compounds of this process can be made in a number of ways. One method is to react $C_6F_5CH{=}CH_2$, $C_6F_4BrCH{=}CH_2$, or $C_6F_4ClCH{=}CH_2$ with an $\equiv$SiH compound at elevated temperatures in the presence of chloroplatinic acid. Another method is to react pentafluorobenzene, tetrafluorobromobenzene, or tetrafluorochlorobenzene with an $\equiv SiCH{=}CH_2$ compound at elevated temperatures in the presence of $AlCl_3$. Pentafluorobenzene and pentafluorostyrene are commercially available. Tetrafluorochlorobenzene is known to the art, and a method for making tetrafluorobromobenzene. The following examples are illustrative of the process.

Example 1: (A) 38.8 g. of pentafluorostyrene and 0.004 g. of chloroplatinic acid which was dissolved in 0.04 cc of isopropanol were mixed and heated to 80°C. To this 23.6 g. of dimethylchlorosilane were added dropwise, and the mixture was heated at 55° to 60°C. for 2 hours. An additional 0.04 cc of the chloroplatinic acid solution was then added, and the mixture was further heated at 55° to 60°C. for another 8 hours. Fractionation yielded 47.8% liquid $C_6F_5CH_2CH_2Si(CH_3)_2Cl$ (BP 103° to 104°C. at 11 mm.; n_D^{25} 1.4440).

(B) 28.9 g. of this product were dissolved in 50 ml. of diethylether and added with stirring to an equal weight mixture of water and diethylether. The ether phase was then washed with water until neutral, and then it was freed of water and ether. The product was mixed with 0.25 g. of sodium hydroxide, and distilled at reduced pressure. A nearly quantitative yield of $[C_6F_5CH_2CH_2Si(CH_3)_2]_2O$ (BP 150° to 151°C. at 4.0 mm.; n_D^{25} 1.4430) was recovered.

Example 2: 55.4 g. of chlorotetrafluorobenzene were vigorously stirred with a few grams of $AlCl_3$. 24.2 g. of vinyltrichlorosilane were then added in a dropwise manner with stirring, and the mixture was heated with stirring for 4 1/2 hours. Distillation yielded 8 g. of a product with a boiling point of 142° to 145°C. at 18 mm. of pressure. The major component of the distillate was $C_6F_4ClCH_2CH_2SiCl_3$.

Example 3: (A) The experiment of Example 1(A) was repeated, substituting methyldichlorosilane for dimethylchlorosilane. The product was

$$C_6F_5CH_2CH_2\underset{\displaystyle CH_3}{\underset{|}{Si}}Cl_2$$

(B) 15 molar parts of the above were cohydrolyzed with 10 molar parts of trimethylchlorosilane and 75 molar parts of dimethyldichlorosilane to form a fluid have a viscosity of 46.3 cs. at 25°C.

(C) 15 molar parts of the product of (A) above were cohydrolyzed with 10 molar parts of trimethylchlorosilane, and 5 molar parts of dichlorophenyltrichlorosilane to form a fluid having a viscosity of 74.8 cs. at 25°C.

(D) 2 ml. each of fluids (B) and (C) above were placed in separate 20 ml. beakers, 2 ml. of a trimethylsiloxy-endblocked dimethylpolysiloxane having a viscosity of 116 cs. at 25°C. were placed in a third 20 ml. beaker. The beakers were heated at 250°C. and the time required until the fluids gelled was recorded.

Fluid	Gel time, hrs.
Dimethylpolysiloxane	16 to 24
(B)	48 to 56
(C)	64 to 72

(E) Both (A) and (B) showed superiority over both dimethylpolysiloxane and a commercial dichlorophenylphenyl-methylpolysiloxane lubricant as high speed ball bearing lubricants, in that less scarring of the ball bearings occurred when (A) or (B) were used.

Halophenylsiloxane-Fluoroalkylsiloxane Copolymers

G.W. Holbrook and H.M. Schiefer; U.S. Patent 3,334,120; August 1, 1967; assigned to Dow Corning Corporation describe organosiloxane copolymers containing halophenylsiloxane units and fluoroalkylsiloxane units which are particularly useful as lubricants and hydraulic fluids. A major improvement in siloxane lubricants came with the introduction of halophenylsiloxanes such as those described in U.S. Patent 2,599,984. It was found that by incorporating halophenylsiloxane units into hydrocarbon substituted siloxanes one could greatly improve the lubricating quality of siloxanes. However, the improvement was not sufficiently good to make these materials generally satisfactory for the lubrication of steel sliding surfaces.

An even greater advance came with the introduction of fluoroalkyl substituted siloxanes such as trifluoropropylmethylsiloxane. Homopolymers of these materials are excellent lubricants for sliding steel on steel bearing surfaces and in most respects they are equivalent to the best hydrocarbon oils. However, the fluoroalkylsiloxanes suffer from the serious disadvantages that they have a relatively steep temperature viscosity slope and are not useable as lubricants at temperatures below -40°F. or as an aircraft hydraulic fluid below 0°F. Thus, fluoroalkylsiloxane homopolymer fluids are not useable as lubricants in a wide variety of applications which require that the lubricant function at low temperatures. It has been found that copolymers containing both fluoroalkylsiloxane units and the halophenylsiloxane units as defined below have lubricating properties approaching, or better than, those of the fluoroalkyl homopolymers but still will operate at extremely low temperatures. This process relates to copolymers having

(1) units of the group

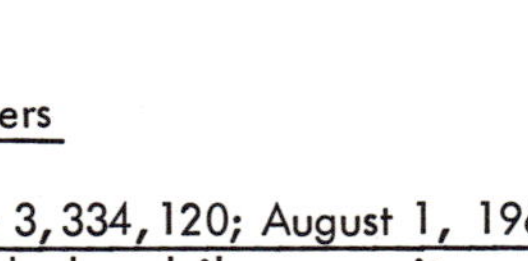

and $R_3SiO_{1/2}$ on the ends of the molecules, the remaining units in the copolymer being substantially all

(2) units of the group consisting of

R'SiO and 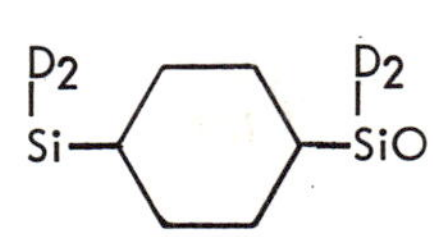

(a) (b)

the total units (2) being present in amount of at least 24 mol percent of the total copolymer and any units (b) being present in amount from 0 to 25 mol percent of the amount of units (a) and in the amount of 0 to 15 mol percent of the total copolymer

(3)

$$\begin{matrix} C_nG_{2n+1} & R''' \\ | & | \\ QCHCH_2 & \text{—} SiO \end{matrix}$$

units in amount of at least 5 mol percent of the total copolymer

(4)

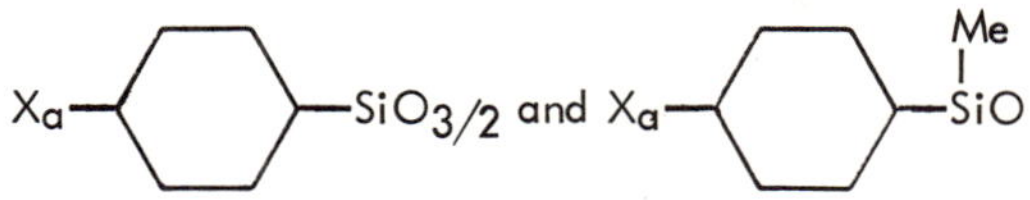

units in amount such that the total number of units of the formula

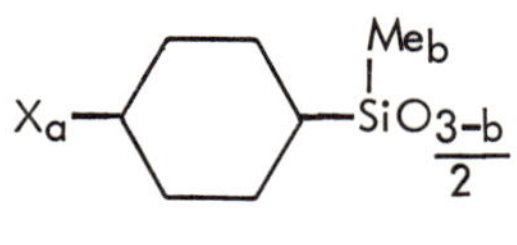

in the copolymer is from 1 to 15 mol percent of the total copolymer, and

(5) $ZSiO_{3/2}$ units in amount of from 0 to 10 inclusive mol percent, the units being interconnected through SiOSi linkages in which R is of the group methyl, phenyl, and

$$\begin{matrix} & Q \\ & | \\ C_nF_{2n+1} & CHCH_2\text{—} \end{matrix}$$

radicals, Q is of the group hydrogen and $C_nF_{2n+1}CH_2CH_2$—, n is an integer from 1 to 8 inclusive, R' and D are each of the group methyl and phenyl radicals, at least one R' and D per silicon being on the average a methyl radical, R''' is of the group phenyl and methyl radicals, Z is of the group phenyl, methyl and

$$\begin{matrix} C_nF_{2n+1} \\ | \\ QCHCH_2\text{—} \end{matrix}$$

radicals, X is of the group chlorine and bromine, a is an integer from 1 to 4 inclusive, and b has a value from 0 to 2 inclusive.

As can be seen, the copolymers are basically three or four component copolymers in which the first component is a triorganosilyl unit which serves as an endblocker. The copolymers are best prepared by cohydrolyzing the corresponding chlorosilanes followed by equilibration of the cohydrolyzate with alkali metal hydroxide catalysts or catalysts comprising the alkali metal salts of silanols. During the equilibration it is often desirable to remove water as it is formed and to maintain the equilibrating system in an anhydrous condition. This is particularly true where the halophenyl substituents contain more than two halogen atoms.

The molecular size of the copolymers is not critical since they are all excellent lubricants regardless of polymer size. However, in many applications it is preferred that the viscosity of the copolymer be in the range of from 25 to 10,000 cs. at 77°F. The copolymers are particularly desirable for use as hydraulic fluids and for this use a particularly desirable form has a viscosity from 50 to 150 cs. at 77°F. The viscosity of the copolymer will be determined by the mol ratio of triorganosiloxy groups to the other siloxy groups. Thus, the viscosity can be varied at will by increasing or decreasing the mol percent of triorganosiloxy groups in the system.

The compositions shown in the following examples were tested for lubricity on the Shell four-ball wear testing machine. In this test, a 1/2 inch steel ball is rotated against three stationary 1/2 inch steel balls at a rate of 1,200 rpm at a temperature of 167°F. under loads of 4, 10 or 40 kilograms for 2 hours. At the end of this time the length and width of the scar formed on each stationary ball is determined and the average of these six measurements is taken as the scar diameter and reported in millimeters. The smaller the scar diameter, the better the lubricant.

Example 1: 1,000 g. of a copolymer of 6.2 mol percent dichlorophenylsiloxane, 75.3 mol percent dimethylsiloxane, 13.5 mol percent trimethylsiloxane and 5 mol percent phenylmethylsiloxane was mixed with 200 g. of

$$\begin{matrix} & Me \\ & | \\ (CF_3CH_2CH_2 & SiO)_3 \end{matrix}$$

and 10 g. of KOH and the mixture was stirred and heated for 18 hours at 160° to 170°C. The resulting fluid was neutralized with Me_3SiCl, filtered and the volatiles were removed by heating to 255°C. at 1 mm. The resulting copolymer had the average composition: 5.6 mol percent dichlorophenyl $SiO_{3/2}$, 67.6 mol percent Me_2SiO, 12.2 mol percent $Me_3SiO_{1/2}$, 4.9 mol percent $C_6H_5(Me)SiO$ and 9.7 mol percent $CF_3CH_2CH_2(Me)SiO$. This copolymer had the following properties.

Temperature, °F.	Viscosity, cs.
-65	4,990
77	82.5
122	49.1

(Pour point -70°F.)

The fluid was tested on the Shell 4-ball wear tester and gave a scar diameter of 0.16 mm. after 2 hours at 1,200 rpm at 167°F. under a load of 4 kg.

Example 2: A series of fluids were made and tested as shown in the table below. In each case a mixture of Me_3SiCl, Me_2SiCl_2, $Cl_2C_6H_3SiCl_3$ and

$$\begin{array}{c} \quad\quad\quad\;\; Me \\ \quad\quad\quad\;\; | \\ CF_3CH_2CH_2SiCl_2 \end{array}$$

in the ratios needed to give the compositions shown below, was added to 2,850 g. of water. Following addition of the chlorosilane mixture, the product was heated for 15 min. and cooled. One liter of toluene was added and the organic phase was separated and washed with sodium bicarbonate solution until the toluene solution was neutral. 5 g. of KOH was added and the toluene was removed by distillation. The residue was heated and stirred for 16 hours at 160° to 170°C. 10 g. of Me_3SiCl was then added to neutralize the alkali and the product was then filtered and heated to 270°C. at 0.25 mm. under a slow N_2 purge and then heated to 275°C. at 2 mm. under a vigorous N_2 purge. The residue was then heated on a steam bath for two hours for 5 g. of Na_2CO_3 and then filtered. The Shell 4-ball wear tests were run at 167°F.

Run No.	Composition of copolymer in mol percent				Viscosity in cs.				Scar Diameter in mm.	
	$Me_3SiO_{1/2}$	$Cl_2C_6H_3SiO_{3/2}$	Me_2SiO	$CF_3CH_2CH_2Si(Me)O$	77° F.	122° F.	210° F.	−65° F.	4 kg.	40 kg.
1	10	5.0	60.0	25	142.6	55.6	25.5	37,000	0.29	0.93
2	10	5.0	65.0	20	101.7	41.8	19.7	8,161	0.20	0.88
3	10	5.0	70.0	15	77.0	34.9	16.9	1,233	0.20	0.81
4	10	5.0	75.0	10	80.8	38.9	17.7	4,179	0.19	0.85
5	10	4.0	71.0	15	72.2	33.2	16.6	4,189	0.23	---------
6	10	4.7	60.3	25	100.8	47.6	21.1	10,340	0.30	1.02
7	10	4.5	65.5	20	71.4	32.5	16.8	4,740	0.25	0.85
8	10	6.0	54.0	30	132.2	48.9	22.3	42,700	0.26	0.97
9	10	3.0	47.0	40	100.3	53.1	18.8	14.840	0.15	0.99
10	10	3.5	46.5	40	93.5	49.9	18.2	9,283	0.13	1.05

Example 3: Employing the procedure of Example 2, a copolymer having the following composition was obtained.

	Mol percent
$CF_3CH_2CH_2Si(Me_2)O_{1/2}$	10
$Cl_2C_6H_3SiO_{3/2}$	5
Me_2SiO	65
$CF_3CH_2CH_2Si(Me)O$	20

The copolymer was a fluid having good low temperature stability and excellent lubricity for steel on steel bearing surfaces.

Tetrachlorophenyl-Containing Polysiloxanes

E.D. Brown, Jr.; U.S. Patent 3,291,735; December 13, 1966; assigned to General Electric Company describes the prepatation and use of a blend comprising a major amount of an organopolysiloxane fluid containing haloaryl radicals attached to silicon and a minor amount of an organopolysiloxane fluid containing chemically combined alkylarylsiloxy units. The following examples illustrate the process.

Example 1: A mixture of chlorosilanes consisting of 9 parts of tetrachlorophenyltrichlorosilane, 83 parts of dimethyldichlorosilane, and 8 parts of trimethylchlorosilane was hydrolyzed at a temperature between 60° and 80°C. for a period of 30 minutes. The hydrolyzate was separated from the aqueous acid layer. It was then treated with a 15% solution of sodium carbonate to neutralize residual hydrolysis acid. The resulting hydrolyzate was then dried with fuller's earth. The hydrolyzate was equilibrated to a temperature of 180°C. for 2 hours in the presence of 0.1% by weight of potassium hydroxide. There was obtained a fluid having 6 mol percent of chemically combined tetrachlorophenyl siloxy units and a viscosity of 65 cp. at 25°C.

A methylphenylpolysiloxane fluid was prepared by initially hydrolyzing a mixture of methylphenyldichlorosilane, dimethyldichlorosilane, and trimethylchlorosilane. There was utilized in the mixture a proportion of 6 mols of dimethyldichlorosilane per mol of methylphenyldichlorosilane, and 9% by weight of trimethylchlorosilane based on the total weight of the respective components of the chlorosilane mixture. The hydrolysis was performed at a temperature of 60°C. for 30 minutes. The hydrolyzate was separated and neutralized as shown above. The hydrolyzate was then dried and equilibrated in the presence of a base catalyst resulting in the production of a methylphenylpolysiloxane fluid having 15 percent of chemically combined methylphenylsiloxy units and viscosity of 90 cp. at 25°C.

A lubricant composition was prepared by mixing together 2 parts of the above described tetrachlorophenylpolysiloxane fluid, per part of the methylphenylpolysiloxane fluid. The load carrying capacity of the lubricant composition was determined by the standard Navy Gear Wear Test, specification MIL-G-3278. In this test a stainless steel gear is turned in a reciprocating motion against a bronze gear. The steel gear is preloaded with either a 5# or 10# weight. The wear of the bronze bearing in terms of weight loss in milligrams per 1,000 cycles is determined after 2,000 cycles. There is shown below the results of the Navy Gear Wear Test utilizing a 5# load after 2,000 cycles. A comparison is given of the wear that resulted in the bronze gear when utilizing the tetrachlorophenylpolysiloxane fluid, tetrachlorophenyl, the methylphenylpolysiloxane fluid, methylphenyl as compared to a blend of two parts of the former with 1 part of the latter.

	Navy Gear Wear Test [Wear Scar (mg.)/1,000 cycles]
Tetrachlorophenyl	8.3
Methylphenyl	6.7
Blend	6.3

In order to measure the advantages achieved under sliding friction conditions with the lubricant composition, as compared to the halogenated arylpolysiloxane fluid, and the alkylarylpolysiloxane fluid, the lubricity of these fluids was measured on a Shell Four-Ball Wear Tester. The table below shows the results obtained with the tetrachlorophenylpolysiloxane fluid, the methylphenylpolysiloxane fluid, and the blend thereof. The test was run at 212°F. utilizing a 20 kg. load at 600 rpm for 1 hour. Wear Scar indicates the average diameter in mm. of the scar of the three balls.

	Four-Ball Wear Tester (Wear Scar, mm.)
Tetrachlorophenyl	0.47
Methylphenyl	2.8
Blend	0.63

Example 2: A lubricant composition was prepared by mixing together the tetrachlorophenylpolysiloxane fluid and the methylphenylpolysiloxane fluid of Example 1 utilizing 2 parts of the tetrachlorophenylpolysiloxane fluid per part of the methylphenylpolysiloxane fluid. There was then added to the resulting blend, naphthenic acid and a fluid methylhydrogen polysiloxane consisting essentially of chemically combined

$$\begin{array}{c} CH_3 \\ | \\ SiO \\ | \\ H \end{array}$$

units. The naphthenic acid and the methyl hydrogen polysiloxane fluid were utilized respectively in amounts sufficient to provide 1% by weight of each in the mixture based on the total weight of the components. The mixture was then tested in accordance with the Navy Gear Wear Test and the Shell Four-Ball Wear Test utilizing the conditions of Example 1. However, with respect to the Navy Gear Wear Test, the lubricity of the mixture was measured under a 5# load, and 10# load. The results of the test are shown below.

	Navy Gear Wear Test (mg./1,000 cycles)	Shell Four Ball Wear Test (mm.)
5#	3	0.5
10#	4.8	----

Iodoethyl-Substituted Organosilicon Compounds

E.D. Brown, Jr.; U.S. Patent 3,515,670; June 2, 1970; assigned to General Electric Company has found that organosilanes and organosiloxanes containing silicon bonded iodoethyl groups are useful per se as lubricants and lubricant additives, particularly for lubricating difficult to lubricate surfaces, such as stainless steel and titanium. A typical lubricating compound is bis(beta-iodoethyl)tetramethyldisiloxane. This compound is useful as an additive for silicone fluids and conventional hydrocarbon lubricants. The following example illustrates the process.

Example: To a reaction vessel was added 46.5 g. of 1,3-divinyltetramethyldisiloxane at 20°C. To this disiloxane was added 80 g. (25% molar excess) of gaseous hydrogen iodide at room temperature. During the beginning of the addition, the initially clear disiloxane became cloudy and then cleared and turned to a light pink color. During the two hours over which the gaseous hydrogen iodide was bubbled into the reaction mixture, the temperature rose 40°C. At the end of the addition, infrared analysis showed the disappearance of the peaks at 6.25 and 7.1 microns, which corresponded to the silicon bonded vinyl group initially present in the disiloxane.

A new peak at 8.2 microns, corresponding to the beta-iodoethyl group, was formed. At the end of this time, the reaction mixture was washed twice with water to remove unreacted hydrogen iodide and then the reaction mixture was dried with sodium sulfate and filtered. The resulting product was bis(beta-iodoethyl)tetramethyldisiloxane which was a low viscosity fluid. The lubricating characteristics of this bisiodoethyltetramethyldisiloxane as an additive in a methylphenyl silicone fluid was compared with the lubricating characteristics of the methylphenyl silicone fluid alone and with a methyl silicone fluid alone, with respect to the lubrication of 52-100 tool steel.

The lubricating characteristics of these fluids were evaluated with a Shell Four-Ball Wear Tester which employed four 52-100 steel balls maintained in a reservoir of the fluid under test at a temperature of 167°F. and employing a speed of 600 rpm for the rotating ball. The test was run for one hour and the wear scar was measured at the end of this time. Tests were run under loads of 10 kg., 15 kg., 20 kg., or 40 kg. In the below are listed the various lubricants evaluated and the wear scar observed in millimeters.

The methylphenyl lubricant listed below was tris(trimethylsiloxy)phenylsilane, which had a viscosity of 5 cs. at 25°C. The methyl lubricant was a trimethylsilyl chain-stopped dimethylpolysiloxane having a viscosity of 50 cs. at 25°C. The three compositions within the scope of the process were solutions of the bisiodoethyltetramethyldisiloxane dissolved in the methylphenylpolysiloxane fluid just described. Three different levels of the additive were employed, 10%, 1%, and 0.5%, based on the weight of the methylphenyl silicone fluid.

Lubricant	Wear scar 10 kg.	15 kg.	20 kg.	40 kg.
Methylphenyl	2.9	3.6		5.8
Methyl				1.8
10% iodoethyl in methylphenyl	0.30			
1.0% iodoethyl in methylphenyl			0.30	0.79
0.5% iodoethyl in methylphenyl	0.30			1.3

As shown above, the compositions of the process are superior to the methylphenyl silicone fluid alone. Stated alternatively, as little as 0.5% by weight of the bisiodoethyltetramethyldisiloxane in the methylphenylsiloxane fluid reduces the wear scar under a 10 kg. load from 2.9 to 0.30 mm. Likewise, at a 40 kg. load, the wear scar under the same circumstances is reduced from 5.8 to 1.3 mm. The compositions are also superior to conventional methyl silicone lubricating fluids in that as little as 0.5% of the bisiodoethyltetramethyldisiloxane in the methylphenyl silicone fluid has a wear scar of only 1.3 mm. under a 40 kg. load as compared with a 1.8 mm. wear scar from a conventional methyl silicone lubricating fluid. Examination of the 52-100 tool steel balls at the end of the test showed that the corrosion of the balls caused by the solution of the disiloxane in the methylphenyl silicone fluid ranged from slight for the 10% concentration down to none for the 0.5% concentration.

ADDITIVES

Diphenylthiourea

F.A. Buehler; U.S. Patent 3,352,781; November 14, 1967; assigned to Mobil Oil Corporation has found that organosilicone lubricants may be stabilized against the gelling effects of high temperature conditions by the incorporation of a minor amount of certain sulfur-containing compounds. These sulfur compounds are: diphenylthiourea, thiophenol, including alkyl thiophenols and mixed alkyl thiophenols, diphenyldisulfide, tetraethyl-thiuramdisulfide and sodium mercapto-benzothiazole. These sulfur compounds by themselves evaporate in a few hours upon exposure to 500°F. However, in combination with the silicone fluids, they are remarkably effective. The silicone fluids which are conventionally formulated to contain a minute amount of a stabilizer, usually iron oxide, or an iron salt of an organic carboxylic acid, such as iron octoate, are rendered more heat stable at high temperatures by the addition of the sulfur compounds of this method as to indicate a synergistic combination. The following examples illustrate the process.

High Temperature Gelling Test: This test determines the ability of a silicone fluid to remain in a fluid state, i.e., resist gelation at 500°F. in the presence of air. The test procedure is as follows: a sample of the fluid is placed in an oven and maintained at 500°F. under forced air circulation. The fluidity of the sample is observed periodically until the sample shows no gravity flow. The elapsed time (from the beginning of the test) to this point is then recorded as the gel time of the sample.

Example 1: The base silicone fluid was methylchlorophenylsilicone which had a molecular weight of 3,000. The silicone had an evaporation loss of 2.1% after 22 hours at 400°F., a pour point of below -99°F., a chlorine content of 7.2%, and kinematic viscosities as follows.

At (°F.)	Centistokes
-65	2024
100	45.2
210	15.1

Into this base fluid was dispersed 0.1% by weight of a finely divided iron oxide (0.001 to 1% of less than 0.5 micron particle size). This composition was subjected to the high temperature gelling test and exhibited a gel time of 70 hours. The base fluid was also tested alone without the finely divided iron oxide; the gel time was less than 70 hours.

Examples 2 through 6: To samples of the composition of Example 1, were added varying amounts of the specific additives of this process. The compositions were also tested in the gelling test at 500°F. The results are tabulated below.

Gel Time of an Iron Oxide-Containing Silicone Lubricant with S-Containing Additives

Examples	Additives	Percent	Gel Time at 500° F., hrs.
	None	0	70
2	Thiophenol	4.1	288
3	Diphenylthiourea	4.7	288
4	Diphenyldisulfide	7.1	237
5	Sodium mercaptobenzo-thiazole	4.4	213
6	Tetraethyl thiuramdisulfide	4.2	213

The base fluid was tested with a representative of the additives of this process, namely, diphenylthiourea with 4.7% by weight without iron oxide; the gel time was less than 70 hours. The tests illustrate the uninhibited silicones have short service life at high temperatures. Silicone fluids containing a conventional stabilizer iron oxide, have slightly extended life. However, the sulfur-containing compositions are valuable and significantly long lived high temperature silicone fluids.

Tetracyanoethylene

J.D. Austin and R.H. Baney; U.S. Patent 3,542,732; November 24, 1970; assigned to Dow Corning Corporation describe a method of stabilizing organopolysiloxane compositions which comprises adding to the siloxane from 0.1 to 5 weight percent, based upon the weight of organopolysiloxane, of tetracyanoethylene. The following examples illustrate the process.

Example 1: Tetracyanoethylene (0.1 g.) was added to 2 g. of trimethylsiloxy-endblocked dimethylpolysiloxane fluids having a viscosity of 6,000 cs. at 25°C. The mixture was heated at 250°C. for 4 hours. The tetracyanoethylene was completely dissolved in the polymer. Thermograms indicated that untreated starting fluids (dimethylopolysiloxane) began to degrade rapidly in air at 320°C., whereas, the sample containing the tetracyanoethylene stabilizer did not begin to decompose until a temperature of 370°C. was attained. It is believed thatpolysiloxanes degrade at higher temperatures by siloxane rearrangement which results in the formation of volatile cyclic species and that the tetracyanoethylene is capable of forming a charge transfer complex which inhibits siloxane rearrangement.

Example 2: The experiment of Example 1 was repeated except that 0.05 g. of tetracyanoethylene were added to 2 g. of the dimethylpolysiloxane and the mixture was heated for 4 hours at 250°C. Thermograms indicated thermal stability equivalent to the mixture of Example 1.

Iron Compounds, Polycarboxylic Acids and Amino Acids

F.A. Buehler; U.S. Patent 3,350,309; October 31, 1967; assigned to Mobil Oil Corporation has found that the high temperature stability of silicone fluids may be improved by adding a minor amount of an iron compound and an aliphatic polycarboxylic acid compound. The polycarboxylic acid compounds which may be employed in this process have the formula

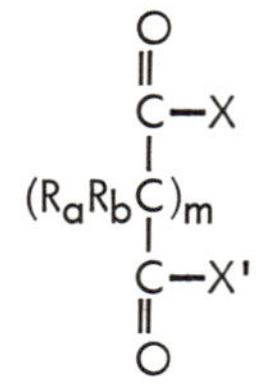

where R_a and R_b each may be hydrogen, hydroxy, carboxy, sulfhydryl, thio, amino, and alkyl, and they may be the same or different, X and X' are either each hydroxy or together a single oxygen atom shared by the two carboxy-carbon atoms, and m is from 0 to 8. The following examples illustrate the process.

The high temperature stability of the silicone compositions is measured by subjecting the fluid composition to 600°F. temperature in the presence of air and by determining the time at which the composition gels or commences to turn solid. The test procedure is as follows: a sample of the silicone fluid composition is added to a Pyrex beaker and is placed in an oven held at 600°F. under forced air circulation. The fluidity of the compositions is checked visually at regular intervals until the sample shows no gravity flow. The time elapsed from the beginning of the test to this point is denoted as the gel time of the sample. The compositions in these examples may be prepared by any simple mixing means using any desired sequence of addition.

Example 1: The silicone fluid used in this example is a methyl phenyl polysiloxane having a methyl to phenol mol ratio of 8.1, a molecular weight of 3,000 and a kinematic viscosity at 210°F. of 20 cs. To three portions of the silicone fluid were added varying quantities of oxalic acid, iron octoate and the two in a combined composition. The high temperature gel test was applied to the uninhibited silicon fluid and to the three compositions. The results are tabulated below.

Additive	Weight percent	Gel Time, Hrs.
1 None	----------	15
2 Oxalic acid	1.0	15
3 Iron octoate	0.20	550
4 Oxalic acid	0.5	1,100
Iron octoate	0.20	

Example 2: The silicone fluid of Example 1 was combined with 0.2% by weight of iron octoate (or 0.017% by weight of iron). To samples of the resulting composition were added varying amounts of polycarboxylic acids. The high temperature stability test was applied to these compositions with the following results tabulated below. In this table, columns headed R_a, R_b and m refer to the aforementioned formula.

Polycarboxylic Additive	Weight Percent	m	R_a, R_b	Gel Time, Hours
None	0	----	----------	550
Succinic Acid	1.5	2	H_4	800
	1.0	2	H_4	1,045
Maleic Anhydride	0.5	2	H_2	1,400
Malic Acid	1.0	2	H_3, OH	1,300
Glutaric Acid	1.0	3	H_6	990
Citric Acid	1.0	3	H_4, OH, COOH	1,155
Adipic Acid	0.5	4	H_8	1,100

It will be noted from the results of the gel test in the above examples that the polycarboxylic acids, when used in combination with iron containing stabilizers, are surprisingly effective in improving the high temperature stability of silicone fluids. Alone, however, these acids appear to have no effect on stability whatever.

F.A. Buehler; U.S. Patent 3,309,312; March 14, 1967; assigned to Mobil Oil Corporation has also found that the high temperature stability of silicone fluids may be increased by adding a minor amount of an iron compound and an amino acid which has the formula

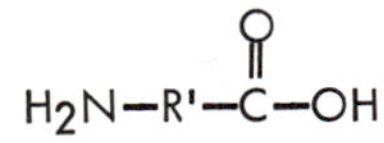

where R' may be alkyl, cycloalkyl, aralkyl, aryl, and alkaryl having from 1 to 20 carbon atoms. R' may also have other substituents, namely hydroxy, sulfhydryl, thio and additional amino groups.

Organophosphorus Compounds

H.M. Schiefer; U.S. Patent 3,386,917; June 4, 1968; assigned to Dow Corning Corporation has found that small amounts of organophosphorus lubricant additives mixed with fluoroalkylmethylpolysiloxane exhibit marked enhancement of antiwear characteristics of metal-on-metal. The following examples illustrate the process.

Example 1: Organophosphorus lubricant additives were added to a trimethylsiloxane-endblocked 3,3,3-trifluoropropylmethylpolysiloxane which had a viscosity at 25°C. of 232 cs., and these mixtures were tested in a Shell Four-Ball Tester at 400°F. for two hours at 1,200 rpm. The nature of the additive, the weight percent added based on the weight of the silicone fluid, the load applied to the four balls in the tester, and the resulting scar diameter on the balls were measured.

Additive	Percent Added	Load (kg.)	Scar Diameter (mm.)
None		4	0.58
Do		10	0.65
Do		40	1.68
Dioctyl hydrogen phosphite	5	4	0.54
Do	5	10	0.61
Do	5	40	0.79
Diisopropyl hydrogen phosphite	5	4	0.35
Do	5	10	0.48
Do	5	40	0.78
Tri-n-butyl phosphate	5	4	0.52
Do	5	10	0.58
Do	5	40	0.62
Tris-2-ethylhexyl phosphate	3	4	0.47
Do	3	10	0.52
Do	3	40	0.65
Tricresyl phosphate	1.5	4	0.51
Do	1.5	10	0.64
Do	1.5	40	0.94
Triethyl phosphate	3	40	0.87
Trioctyl phosphate	3	40	0.65
Do	10	40	0.94
Tris(β-chloroethyl)phosphate	1	40	1.19

The following experiments were performed for comparison: the compositions named below were tested as lubricants in a Shell Four-Ball Tester at 400°F. for two hours at 1,200 rpm. The load applied to the four balls in the tester and the resulting scar diameter on the balls were measured.

Composition	Scar Diameter (mm.) at—	
	10 kg. load	40 kg. load
(a) Dimethylpolysiloxane having a viscosity at 25° C. of 100 cs	0.63	>2.0
(b) 95.8 weight percent of (a) above plus 4.2 percent of trioctyl phosphate	0.85	....
(c) 90 weight percent of (a) above plus 10 percent of trioctyl phosphate	1.39	>2.0
(d) A neutralized fluid hydrolyzate of 84.5 parts by weight of dimethyldichlorosilane, 8.5 parts of tetrachlorophenyltrichlorosilane, and 7.0 parts of trimethylchlorosilane		2.07
(e) 95 weight percent of (d) above plus 5 percent of tricresyl phosphate		2.26
(f) 95 weight percent of (d) above plus 5 percent of trioctyl phosphate		1.90

Example 2: The following lubricant additives were added to a trimethylsiloxane-endblocked 3,3,3-trifluoropropylmethylpolysiloxane having a viscosity at 25°C. of 100 cs., and tested in a Shell Four-Ball Tester at 400°F. for one-half hour with a load of 20 kg. and a rate of 1,200 rpm. The results in each case were as follows.

Additive	Weight Percent Added	Scar Diameter (mm.)
None		0.90
Tris(3,3,3-trifluoropropyl)phosphate	1	0.46
Do	2	0.50
One percent of tricresyl phosphate plus 2 percent of a fatty acid triester of trimethylolpropane as a solubilizing agent	1	0.66

Haloalkyl-Containing Silicone Fluids and Tricresyl Phosphates

E.D. Brown, Jr.; U.S. Patent 3,479,290; November 18, 1969 and U.S. Patent 3,505,378; April 7, 1970; both assigned to General Electric Company describes a lubricating composition which contains both methylchloroalkylsiloxane units and methyl higher alkylsiloxane units, and which also contain tricresyl phosphate. This combination of components provides a lubricating composition which has improved high temperature stainless steel extreme pressure lubricating characteristics. The compositions of the process comprise (A) an organopolysiloxane fluid having the formula shown on the following page.

(1) $$(CH_3)_3SiO\left[\begin{array}{c}Cl\\ |\\ C_nH_{2n}\\ |\\ SiO\\ |\\ CH_3\end{array}\right]_a\left[\begin{array}{c}R\\ |\\ SiO\\ |\\ CH_3\end{array}\right]_b Si(CH_3)_3$$

where R is an alkyl radical containing at least 8 carbon atoms, e.g., from 8 to 18 carbon atoms, n is equal to from 2 to 6, inclusive, a has an average value of at least 1, b has an average value of at least 2.9, the sum of a plus b is equal to from 4 to 40, inclusive, and the ratio of a to the sum of a plus b is equal to from 0.1 to 0.4, inclusive, and (B) from 1 to 15% by weight, based on the weight of the polysiloxane, of tricresyl phosphate.

The fact that tricresyl phosphate will increase the stainless steel lubricating characteristics of the compositions of Formula (1) is unexpected since the addition of tricresyl phosphate to a methyl higher alkyl polysiloxane fluid of the type described in U.S. Patent 3,418,353, or the addition of tricresyl phosphate to a trimethylsilyl chain-stopped methylchloroalkylpolysiloxane fluid fails to increase the lubricating characteristics of such fluids. On the other hand, the addition of the tricresyl phosphate to the compositions of this process has a dramatic effect in increasing the lubricating characteristics of the composition. The organopolysiloxane fluids within the scope of Formula (1) can be copolymers in which the average molecule contains both methylchloroalkylsiloxane units and methyl higher alkylsiloxane units or can be a blend of two fluids, one of which is a trimethylsilyl chain-stopped methylchloroalkylsiloxane fluid having the formula

(2) $$(CH_3)_3SiO\left[\begin{array}{c}Cl\\ |\\ C_nH_{2n}\\ |\\ SiO\\ |\\ CH_3\end{array}\right]_c Si(CH_3)_3$$

when n is as previously defined and c has a value of from 4 to 40, inclusive, and the other of which is a trimethylsilyl chain-stopped methyl higher alkylpolysiloxane having the formula

(3) $$(CH_3)_3SiO\left[\begin{array}{c}R\\ |\\ SiO\\ |\\ CH_3\end{array}\right]_d Si(CH_3)_3$$

where R is as previously defined and d has a value of from 4 to 40, inclusive. The trimethylsilyl chain-stopped methylchloroalkylpolysiloxane fluids of Formula (2) are formed by the relatively straightforward hydrolysis and condensation of trimethylchlorosilane and a methylchloroalkyldichlorosilane having the formula

(4) $$(CH_3)(ClC_nH_{2n})SiCl_2$$

where n is as previously described. Generally, this hydrolysis and condensation is effected by mixing the stoichiometric amounts of the trimethylchlorosilane and the methyl higher alkyldichlorosilane, adding the resulting mixture to a stirred excess of ice water to effect the hydrolysis and condensation, allowing the reaction mixture to separate into an upper silicone phase and a lower aqueous phase, separating the phases, washing with water to remove all acidic by-products and drying the resulting product to form the polysiloxane within the scope of Formula (2). The trimethylsilyl chain-stopped methyl higher alkylsiloxane of Formula (3) is prepared from the corresponding trimethylsilyl chain-stopped methyl hydrogen polysiloxane which has the following formula.

(5) $$(CH_3)_3SiO\left[\begin{array}{c}H\\ |\\ SiO\\ |\\ CH_3\end{array}\right]_d Si(CH_3)_3$$

The organopolysiloxane fluids within the scope of Formula (1) which are copolymers in which the average molecule contains both the methylchloroalkylsiloxane unit and the methyl higher alkylsiloxane unit are prepared by the process described in U.S. Patent 3,505,378. In general, the procedure for preparing the copolymers of Formula (1) is to first form a trimethylsilyl chain-stopped copolymer of methylchloroalkylsiloxane units and methyl hydrogen siloxane units having the formula

(6) $$(CH_3)_3SiO\left[\begin{array}{c}Cl\\ |\\ C_nH_{2n}\\ |\\ SiO\\ |\\ CH_3\end{array}\right]_a\left[\begin{array}{c}H\\ |\\ SiO\\ |\\ CH_3\end{array}\right]_b Si(CH_3)_3$$

where a, b and n are as previously defined. This copolymer of Formula (6) is prepared by the cohydrolysis and cocondensation of a mixture of trimethylchlorosilane, methyl hydrogen dichlorosilane, and a methylchloroalkyldichlorosilane within the scope of Formula (4). The process for forming the composition of Formula (6) is conventional, with the appropriate chlorosilanes being mixed together in the proper ratios, and added to an excess of ice water, with the subsequent isolation of the product. The following examples illustrate the process.

Example 1: A methylchloropropylpolysiloxane fluid within the scope of Formula (2) was prepared by mixing 10 g. (0.92 mol) trimethylchlorosilane with 197 g. (1.0 mol) methylchloropropyldichlorosilane and the mixture was added to 600 g. of water and stirred for one hour. At the end of this time, the two layers were separated and the upper methylpolysiloxane layer was washed with water three times until neutral and then dried over sodium sulfate to produce a polysiloxane of Formula (2) where n is 3, and c has a value of 25.

Following the procedure of U.S. Patent 3,418,353, a methyltetradecylpolysiloxane fluid was prepared by slowly adding 196 g. of tetradecene-1 to 63 g. of a trimethylsilyl chain-stopped methyl hydrogen polysiloxane. This resulted in a trimethylsilyl chain-stopped methyltetradecylpolysiloxane within the scope of Formula (3) in which R is tetradecyl and d is 40. A mixture was prepared of 92 parts of the methyltetradecylpolysiloxane and 8 parts of the methylchloropropylpolysiloxane. This blend was within the scope of Formula (1) when n was equal to 3, R is equal to tetradecyl, and the average value of a was 6 and the average value of b was 32. To 100 parts of this blend was added 5 parts of tricresyl phosphate and the wear scar was determined with type 302 stainless steel balls at 600 rpm, at ambient temperature, and under loads of 10 kg. and 20 kg. The wear scar at 10 kg. was 0.36 mm. and at 20 kg. was 0.43 mm.

In comparison to this, the wear scar of a mixture of 100 parts of the methyltetradecylpolysiloxane and 8 parts of the methylchloropropylpolysiloxane at 10 kg. was 2.4 mm. In contrast also, the wear scar of a solution of 5 parts of tricresyl phosphate in 100 parts of the methyltetradecylpolysiloxane fluid was 2.6 mm. Also in contrast to this, the wear scar of a solution of 5 parts of tricresyl phosphate in 100 parts of the methylchloropropylpolysiloxane fluid was 2.5 mm. With the methyltetradecylpolysiloxane fluid, destructive failure occurred after 15 minutes and the test could not be run for a full hour.

Thus, the data of the above example show that with a polysiloxane fluid containing both the methyltetradecylsiloxane groups and methylchloropropylsiloxane groups and the tricresyl phosphate, satisfactory lubrication occurred under loads of 10 kg. and 20 kg. In contrast to this, when any of the three components was omitted from the reaction mixture, the wear scar increased drastically.

Example 2: Following the procedure of U.S. Patent 3,505,378, a trimethylsilyl chain-stopped copolymer of methylchloropropylsiloxane units and methyl hydrogen polysiloxane units within the scope of Formula (6), when n is 3, a is 8, and b is 32 (80 to 20 copolymer), was prepared by the hydrolysis and condensation of a mixture of trimethylchlorosilane, methylchloropropyldichlorosilane, and methyl hydrogen dichlorosilane. This polymer was then reacted with tetradecene-1 following the precedure of the same application, to produce a copolymer within the scope of Formula (1) in which n is 3, a is 8, b is 32, and R is tetradecyl. To 100 parts of this fluid was added 5 parts of tricresyl phosphate and evaluation of this material in the lubrication of type 302 stainless steel showed a wear scar at 10 kg. of 0.37 mm.

Example 3: A mixture was prepared from 29.7 parts of the methyltetradecylpolysiloxane described in Example 1, and 100 parts of the copolymer containing methyltetradecylsiloxane units and methylchloropropylsiloxane units of Example 2, to produce a polysiloxane blend within the scope of Formula (1) corresponding to the average composition in which n is 3, R is tetradecyl, a is 6 and b is 32, and then 5 parts tricresyl phosphate was added. Evaluation of this material in the Shell Four-Ball Wear Tester gave identical results to the results obtained in Example 1. The wear scar at 10 kg. was 0.36 mm. and the wear scar at 20 kg. was 0.43 mm.

Dihalogenoethyl-Substituted Organosilicon Compounds

In a process described by E.D. Brown, Jr.; U.S. Patent 3,555,062; January 12, 1971; assigned to General Electric Co. mixtures of dihalogenoethyl-substituted organosilicon compositions have been prepared by adding either IBr or ICl to vinyl-substituted silanes and siloxanes. Lubricating compositions of extreme pressure and high temperature characteristics are prepared by adding the mixtures to oils of lubricating viscosity. The lubricating compositions are especially useful for lubricating high temperature alloys. The following examples illustrate the process.

Example 1: To a reaction vessel was added 344 parts of 1,3,5,7-tetramethyl-1,3,5,7-tetravinylcyclotetrasiloxane in 500 parts of carbon tetrachloride and 828 parts of IBr was slowly added over a 2 hour period, with the rate of addition being controlled to keep the autogeneous temperature from rising above 45°C. Upon completion of the addition of the IBr, the reaction mixture was examined by infrared analysis and completion of the analysis was evidenced by the disappearance of the 6.25 and 7.1 micron peaks corresponding to the silicon-bonded vinyl groups in the cyclotetrasiloxanes. After determining that the reaction had been completed, the solvent and unreacted IBr was stripped at a temperature of 90°C. at 20 mm. Chemical analysis of the reaction product showed the presence of 68% total halogen and 42% iodine as compared with theoretical values of 70.6% total halogen and 43.3% iodine. Solubility of the product in white mineral oil was 8% by weight. The reaction product comprised a dihalogenoethyl-substituted cyclotetrasiloxane having the following formula.

Components	Percent Composition of Components in Siloxane
$[(ICH_2CHI)(CH_3)SiO]$	4
$[(BrCH_2CHBr)(CH_3)SiO]$	4
$[(ICH_2CHBr(CH_3)SiO]$	85
$[(BrCH_2CHI)(CH_3)SiO]$	7

The principal component of the reaction mixture was the alpha-bromo-beta-iodoethyl-substituted cyclotetrasiloxane as indicated by the above table which shows the approximate percentages of each component of the reaction mixture. The reaction mixture was a reddish-brown liquid of relatively low viscosity.

Example 2: A lubricating composition was prepared from the mixture of dihalogenoethyl-substituted organosilicon compounds of Example 1 by dissolving varying amounts of the mixture of dihalogenoethyl-substituted organosilicon compounds to white mineral oil. In the table below is listed the wear scar of a control and of lubricating compositions of this application in a conventional Four-Ball Wear test, where a steel ball was rotated at a speed of 600 rpm in contact with three stationary metal balls immersed in the lubricating composition, with the test running for one hour with a load of 30 kg. on the rotating ball. The test was run employing both 52-100 tool steel and 302 stainless steel. In the table is listed the amount of the product of Example 1 (siloxane additive) added to 100 parts of mineral oil, the wear scar in millimeters for tool steel and for stainless steel.

Mineral Oil Parts	Siloxane Additive Parts	Wear Scar, mm. 52-100 Steel	Wear Scar, mm. 302 Stainless 302 Stainless
100	0.00	0.46	NG
100	0.25	0.52	0.41
100	0.50	0.49	0.52

As the table shows, the lubricating characteristics of the compositions of the process are comparable to the lubricating characteristics of mineral oil alone in terms of 52-100 steel lubrication. However, mineral oil alone fails completely to lubricate 302 stainless and it is literally impossible to run the Four-Ball Weat test. On the other hand, with the additive compositions of the process in mineral oil at a level of either 0.25 part or 0.50 part per 100 parts mineral oil, the lubricating characteristics are comparable to lubricating characteristics on simple to lubricate tool steel.

As a further illustration of the special utility of the lubricating compositions, a control was prepared by forming a mixture of 50% white mineral oil, 45% of a commercial mixture consisting of 26.5% diphenyl and 73.5% diphenyloxide and 5% by weight iodine. While this composition was completely effective in lubricating both the 52-100 tool steel and the 302 stainless steel at room temperature, the material was highly volatile at elevated temperatures and, therefore, unsatisfactory for high temperature lubricating applications. Thus, when this material was maintained at a temperature of 60°C. for 10 hours, all of the material evaporated. On the other hand, there is no measurable loss of the lubricating compositions of this example after several weeks at 60°C.

GREASE FORMULATIONS

Methyl Alkyl Silicone Composition

J.H. Wright; U.S. Patent 3,537,997; November 3, 1970; assigned to General Electric Company describes a grease composition which contains a polysiloxane, the organic substituents of which are primarily methyl radicals and C_6 to C_{12} alkyl radicals, a thickener, optionally an amount of a polyether which is sufficient to cause the thickener to disperse into the polysiloxane, but insufficient to deleteriously affect the high temperature properties of the grease, and optionally a sufficient amount of a base to render the grease alkaline. The grease is made by heating together the polysiloxane, the polyether, the thickener and the base to a temperature of 400° to 500°F., after which the grease composition is cooled to room temperature, then milled. The grease composition is used in windshield wiper bearings and drive mechanisms, which are subject to oscillatory motion, often while being flooded with water.

The fluid methylalkylpolysiloxanes employed in the process are known in the art. This type of methylalkylpolysiloxane can be characterized as having the average formula shown on the following page.

$$(CH_3)_n(R)_m[R'(CH_2)_3]_p(R'')_qSiO_{\frac{4-n-m-p-q}{2}} \qquad (1)$$

The sum of (m+n+p+q) has a value of from 2.002 to 3.0; n has a value of from 0.50 to 1.95; m has a value of from 0.50 to 1.00; p has a value of from 0 to 0.5; q has a value of from 0 to 1/4 (m+n+p); R is an alkyl radical containing from 6 to 12 carbon atoms, e.g., hexyl, octyl, decyl, dodecyl, etc. radicals; and R' is a t-butyl-substituted hydroxyaryl radical and has the formula

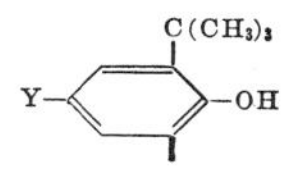

where Y is a member selected from the class consisting of hydrogen, monovalent hydrocarbon radicals, hydroxyaryl radicals, hydroxyaryl-substituted monovalent hydrocarbon radicals, hydroxyaryl ethers joined to the t-butyl-substituted hydroxyaryl radical through the ether linkage, hydroxyarylthioethers joined to the t-butyl-substituted hydroxyaryl radical through the thioether linkage and hydroxyarylmethylene ethers joined to the t-butyl-substituted hydroxyaryl radical through the methylene ether linkage. R" is selected from the class consisting of lower alkyl radicals having one to 5 carbon atoms.

Example 1: An allylated product of the formula

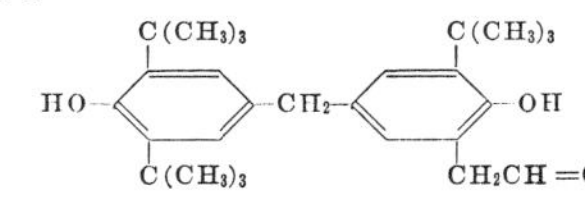

was prepared by dissolving 424 g. (1 mol) of 4,4'-methylenebis(2,6-ditertiarybutylphenol) in an equal weight of toluene and an equal weight of ethyl alcohol. One thousand grams of a solution containing 112 g. of potassium hydroxide in ethyl alcohol was made and slowly added to the phenol to provide the stoichiometric equivalent of the phenolic hydroxyl groups. A brilliant purple solution resulted which, when dried, showed no evidence of phenol and tested completely for complete conversion to the potassium phenylate.

An additional equivalent amount of ethyl alcohol was added and 1.5 mols of allyl chloride was slowly introduced to the reaction mixture, which was refluxed for 2 hours at 70°C. All solids were filtered from the reaction mixture and the product was washed and stripped. Infrared analysis showed that the phenolate had been converted to phenol and that the allyl group was in place. Nuclear magnetic resonance evidence pointed to a replacement of one tertiary butyl group on one of the two aryl radicals by an allyl radical. To a reaction vessel was added 300 g. of a liquid trimethylsilyl chain-stopped methylhydrogenpolysiloxane of the formula $(CH_3)_{1.5}(H)_{0.75}SiO_{0.875}$.

To this mixture was added 0.00125 g. of chloroplatinic acid hexahydrate and 28.8 g. of the allylated product over a period of 0.5 hour, while the temperature of the reaction mixture was maintained at 110°C. External heating was discontinued. Then 500 g. of decene-1 was added slowly to the reaction mixture over a one hour period, during which time the temperature was maintained at 110°C. by the exothermic reaction resulting from the addition. After complete addition of the decene-1, heat was applied to the flask to maintain temperature at 110°C. for an additional 30 minutes to insure that all ≡SiH is totally reacted and then the reaction product was vacuum stripped at 282°C. and 10 mm. Hg using a nitrogen purge. This resulted in a base oil within the scope of Formula (1) in which R is decyl, R' is a radical of the formula

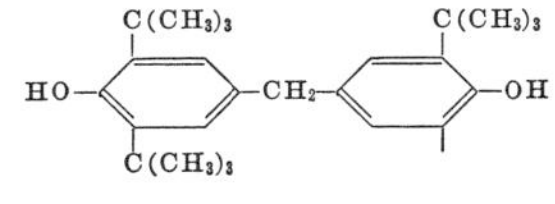

n has a value of 1.5, m has a value of 0.734 and p has a value of 0.016. Since there are fewer than one of the t-butyl hydroxyarylpropyl-substituted silicon atoms per 8 silicon atoms and the polysiloxane contains 8 silicon atoms per molecule, it is apparent that the compositions comprise a blend of products within the scope of Formula (1), some of which contain the internal antioxidant radical and some of which do not. To 36 g. of the base oil and 18 g. of lithium myristate in a grease kettle were added 0.75 g. of a polyether of the formula

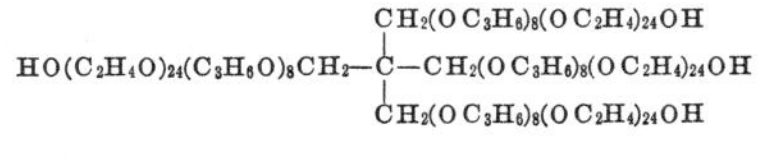

and 0.01 g. of finely divided lithium hydroxide. The mixture was stirred and heated to 240°C. The mixture was maintained at 240°C. for 10 minutes with stirring. The mixture was then slowly cooled at a rate of 1.4°C. per minute to 150°C. at which temperature 0.45 g. of N-phenyl-alpha-naphthylamine, 0.1 g. of finely divided lithium hydroxide, 36 g. of the base oil and 7 g. of lithium myristate were added. The slow cooling was continued to room temperature. The mixture was then milled three times through a Morehouse colloid mill set at 3 mils clearance. The resulting material

was a light grade (290-330 Penetration) grease with the following properties.

Penetration — 296
Bleed 24 hours at 150°C. — 7.8%
Evaporation 24 hours at 150°C. — 1.4%
Appearance — Smooth grease, purple in color

The grease was used to lubricate the oscillating recording mechanism in a chart recorder and was still functioning perfectly after 6 months of continuous use. Under identical conditions a commercial petroleum grease failed after 2 weeks. A windshield wiper mechanism consisting of a cam operating in a housing was lubricated with the grease of this example and operated for 6 months with no difficulty. The same mechanism lubricated with petroleum grease failed after 200 hours. A result similar to that achieved with the petroleum grease was achieved using a conventional silicone grease.

Example 2: To a reaction vessel was added 300 g. of a liquid trimethylsilyl chain-stopped methylhydrogenpolysiloxane of the formula

$$(CH_3)_3SiO\left(\begin{array}{c}CH_3\\|\\SiO\\|\\H\end{array}\right)_{60}Si(CH_3)_3$$

To this mixture was added 0.00125 g. of chloroplatinic acid hexahydrate and 28.8 g. of the allylated product described in Example 1 over a period of 0.5 hour, while the temperature of the reaction mixture was maintained at 110°C. Heating was discontinued. Then 683 g. of decene-1 was added slowly to the reaction mixture over a one hour period, during which time the temperature was maintained at 110°C. by the exothermic reaction resulting from the addition. After complete addition of the decene-1, heat was applied to the flask to maintain the temperature at 110°C. for an additional 30 minutes to insure that all ≡SiH was totally reacted and then the reaction product was vacuum stripped at 282°C. and 10 mm. Hg using a nitrogen purge. This resulted in a base oil within the scope of Formula (1) where R is decyl, R' is a radical of the formula

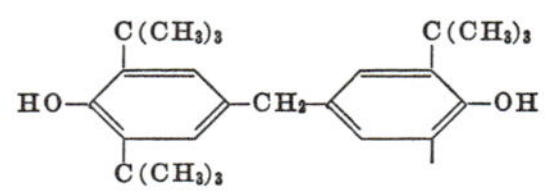

n has a value of 1.065, m has a value of 0.953 and p has a value of 0.0148. Since p, the ratio of the t-butylhydroxyaryl radicals to silicon atoms is less than one in 62, and there are 62 silicon atoms per polysiloxane molecule, it is apparent that the compositions of the method comprise a blend of products within the scope of Formula (1) in which the majority of the polysiloxane molecules contain one R' radical and a minor of the polysiloxane molecules which do not contain an R' radical. To 36 g. of the base oil and 18 g. of lithium myristate in a grease kettle were added 0.75 g. of a polyether of the formula

$$HO(C_2H_4O)_{24}(C_3H_6O)_8CH_2-\begin{array}{c}CH_2(OC_3H_6)_8(OC_2H_4)_{24}OH\\|\\C\\|\\CH_2(OC_3H_6)_8(OC_2H_4)_{24}OH\end{array}-CH_2(OC_3H_6)_8(OC_2H_4)_{24}OH$$

and 0.01 g. of finely divided lithium hydroxide. The mixture was stirred and heated to 240°C. The mixture was maintained at 240°C. for 10 minutes with stirring. The mixture was then slowly cooled at a rate of 1.4°C. per minute to 150°C. at which temperature 0.45 g. of N-phenyl-alpha-naphthylamine, 0.1 g. of finely divided lithium hydroxide, 36 g. of the base oil and 7 g. of lithium myristate were added. The slow cooling was continued to room temperature. The mixture was then milled three times through a Morehouse colloid mill set at 3 mils clearance. The resulting material was a light grade (290-330 Penetration) grease.

In related work, E.D. Brown; U.S. Patent 3,579,467; May 18, 1971, assigned to General Electric Co. describes the basic synthesis of the hydroxyaryl substituted polysiloxanes. These formulations exhibit improved oxidative stability at elevated temperatures and as shown above (U.S. Patent 3,537,997) can be used in a variety of grease compositions.

Dimethyldiphenylpolysiloxanes

A.L. Hajjar; U.S. Patent 3,544,464; December 1, 1970; assigned to General Electric Co. describes a grease composition containing a polysiloxane, the organic substituents of which include phenyl and tetrachlorophenyl radicals in amounts of from 5 to 50%, a higher fatty acid soap of lithium, an amount of a particular polyether sufficient to cause the lithium soap of the fatty acid to disperse into the polysiloxane, and a sufficient amount of a base to render the grease alkaline.

The grease is made by heating the polysiloxane, the polyether, the lithium soap of the higher fatty acid, and the base to a temperature of 400° to 500°F., after which the grease composition is cooled to room temperature and milled. The grease composition is used in the journal bearings of starter motors having high current capacity and which run hot but are subjected to extremely low temperatures when not running.

Hydrophobic Silica

H.M. Schiefer, J.W. Vandyke and J.S. Delphenich; U.S. Patent 3,344,066; September 26, 1967; assigned to Dow Corning Corporation have found that the addition of a fluid, hydroxylated organopolysiloxane plus a strongly hydrophobic, treated silica to a silicone fluid results in a grease having a significant increase in mechanical and thermal stability when compared with previously known silicone-containing greases. Many of the greases of this process are particularly useful as lubricants for high speed bearings.

More specifically, this process relates to a grease consisting essentially of (a) 100 parts by weight of an organopolysiloxane fluid of the average unit formula $R_nSiO_{4-n/2}$ where R is selected from the group consisting of monovalent hydrocarbon and haloaryl radicals of no more than 7 carbon atoms, free of aliphatic unsaturation, n has an average value of 1.9 to 2.1, the viscosity of (a) being from 20 to 30,000 cs. at 25°C.; (b) from 5 to 25 parts of a powdered silica filler having a surface area of at least 100 square meters per gram, which filler is rendered hydrophobic by treatment with (c) from 2 to 150 weight percent, based on the weight of (b), of a compound selected from the group consisting of cyclic polysiloxanes of the formula $(R'_2SiO)_3$, at least 50 mol percent of the R' groups being methyl; silazanes of the formula $(R'_3Si)_aNH_{3-a}$ where a has a value of 1 to 2; polysilazanes of the unit formula

$$(R'_2\underset{|}{\text{Si}}\underset{\underset{\text{H}}{|}}{\text{N}}-)$$

which have a viscosity of no more than 200 cs. at 25°C., and silanes of the formula R'_bSiX_{4-b}, where X is chlorine, bromine, lower alkoxy, lower acyloxy, or hydroxyl, b has a value of 2 to 3, and R' is in all cases as defined below; and (d) from 5 to 100 weight percent, based on the weight of (b), of a hydroxylated organopolysiloxane fluid containing an average of 1.8 to 2.1 R' groups per silicon atom where R' is selected from the group consisting of monovalent hydrocarbon radicals of no more than 7 carbon atoms, the fluid having a silicon-bonded hydroxyl content of at least 1 weight percent. It is preferred for R to be both methyl and phenyl. Preferred ingredients (c) are sym-trimethyltriphenylcyclotrisiloxane and hexamethyldisilazane, and a preferred ingredient (d) is a hydroxylated fluid consisting primarily of phenylmethylsiloxane units. The following examples illustrate the process.

Example 1: To 10 parts by weight of a powdered silica having a surface area of 200 square meters per gram there was added 200 parts of Chlorothene as a dispersing agent, 3.2 parts of a phenylmethylsiloxane fluid having approximately a 6 weight percent hydroxyl content, 0.5 part of sym-trimethyltriphenylcyclotrisiloxane, 0.3 part of other cyclic and linear phenylmethylsiloxanes which were impurities in the ingredients, and two parts of ammonium carbonate. This was heated at reflux to 150°C. for 5 to 10 minutes. Then 100 parts by weight of a trimethylsiloxy-endblocked copolymer consisting of 92.5 mol percent of dimethylsiloxane units and 7.5 mol percent of phenylmethylsiloxane units, having a viscosity of 100 cs. at 25°C., were added, and the chlorothene was removed under vacuum.

The product was a grease. It was tested for consistency by the test described in ASTM 1403-56T, converted as therein described to the full scale cone penetration of ASTM D217-52T. Briefly, this test measures the depth that a cone shaped weight sinks into the grease. Low penetration values therefore indicate a thick grease, and high values indicate a thin grease. A penetration of over 400 indicates that the material is more soup-like than grease-like, whereas a penetration of less than 100 indicates that the grease is too thick to be normally used. Penetrations were measured before and after working the grease by forcing it through a metal plate with one millimeter holes, performed by moving the plate through the grease. The working test consists of 60 one second strokes of the plate. The bleed and evaporation rates of the grease were measured in terms of weight loss of the grease. The results were as follows.

Grease	Penetration		Penetration after 24 hrs. at 400° F.		Bleed after 30 hrs. at 400° F. (percent)	Evaporation after 22 hrs. at 400° F. (percent)
	Unworked	Worked	Unworked	Worked		
(a) The above grease	279	297	249	286	9.22	2.
(b) A commercial soap-thickened silicone grease	283	297	95	219	5-10	5-18 10

The above two greases were tested in an MRC S-17, size 204 bearing in accordance with Federal Standard 791, Method 333. The bearing was run at 10,000 rpm under a five pound load and a temperature of 450°F. The bearing with grease (a) ran for over 1,000 hours without failing, while the bearing with grease (b) failed in less than 24 hours.

Example 2: (a) Another batch of grease was made by the method of Example 1, except that hexane was used in place of Chlorothene as the dispersing agent. (b) A grease was made in the following manner: to 100 parts by weight of a powdered silica having a surface area of 400 square meters per gram, there was added one part of ammonium carbonate and 27.5

parts of sym-trimethyltriphenylcyclotrisiloxane. This was rolled at room temperature for 5 days, and then heated at 150°C. under vacuum to remove all remaining ammonia. A hydrophobic silica was recovered. To 18.1 parts by weight of the above treated silica there was added 81.9 parts of a trimethylsiloxy-endblocked polysiloxane copolymer consisting of 92.5 mol percent of dimethylsiloxane units and 7.5 mol percent of phenylmethylsiloxane units, having a viscosity of 1,000 cs. at 25°C.

(c) To 9 parts by weight of a powdered silica having a surface area of 200 square meters per gram there was added 100 parts of a blend of dimethylpolysiloxane fluids, the blend having a viscosity of 6,000 cs. at 25°C. This was milled until a grease was formed. These greases were tested as in Example 1.

Grease	Pentration		Penetration after 24 hrs. at 400° F.		Bleed after 30 hrs. at 400° F. (percent)	Evaporation after 22 hrs. at 400° F. (percent)
	Unworked	Worked	Unworked	Worked		
Sample (a)	279	290	272	286	12.7	1.25
Sample (b)	256	>400				
Sample (c)	212	>400				

A fourth grease (d) was prepared from 86.5 weight percent of a dimethylpolysiloxane fluid with a viscosity of 500 cs. at 25°C., 10.3% of powdered silica filler with a surface area of over 100 square meters per gram, 2.7% of a hydroxylated methylpolysiloxane fluid, having a hydroxyl content of 4 weight percent, 0.36 percent of boric acid and 0.14% of pentaerythritol, the latter two ingredients being known silicone grease stabilizers. Grease (d), and grease (a), were both tested in the bearing test of Example 1. The bearing with grease (a) operated for over 1,040 hours without failing, while the bearing with grease (d) failed in less than 24 hours. Grease (a) is within the scope of the process. Greases (b), (c), and (d) are outside the scope of the process, being included to show the superiority of grease (a).

J.B. Christian; U.S. Patent 3,518,189; June 30, 1970; assigned to U.S. Secretary of the Air Force describes grease compositions which consist of a mixture with from 58.5 to 64.5 parts by weight of the Dow Corning Corporation QF6-7039 inhibited silicone as a base fluid, of from 35 to 40 parts by weight of the powdered ammeline, and of from 0.5 to 1.5 parts by weight of the Cabosil H-5 silica powder. The structural formula of the silicone is

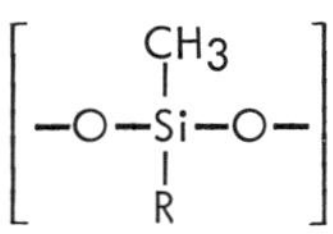

where R is a phenyl group, n is an integer from 1 to 3 and $-Si(CH)_3$ groups are chain stoppers. Greases formulated of the ingredients and according to the preferred proportions demonstrated a dropping point, when tested according to Federal Test Method Standard 791, Method Number 1421, of higher than 650°F. Penetration testing according to ASTM Method D217 gives values of from 270 to 340 in both the unworked and worked condition demonstrating the maintenance of the stability of the grease through working conditions. Further testing of these greases according to Military Specification MIL-G-38277 finds oil separation at 600°F. of only from 16 to 20%; while Federal Test Method Standard 791, Method 350 testing finds evaporation at 600°F. only from 25 to 28%.

Testing according to ASTM Method D-1478 shows low temperature torque at 25°F. of from 760 to 1,120 gram-centimeters at starting and of from 530 to 800 gram-centimeters after running. On the Pope Spindle Test, under a five pound load at 20,000 rpm and 600°F., the grease operated satisfactorily for from 150 to 200 hours. Testing according to Federal Test Method Standard 791, Method Number 3453 for oxidation stability of these greases disclosed a drop of two pounds. Moreover, actual use of these compounds as all purpose greases for high speed, high altitude aircraft has proven that they continue to perform satisfactorily in the low pressure conditions normally encountered at altitudes on the order of 100,000 feet above sea level, under both light and heavy loads, at both low and high temperatures up to beyond 600°F., at both low and high rotational speeds, under various types of motion and in various types of bearings.

Silica Surface Treated with Hexamethyldisilazane

S. Meghir; U.S. Patent 3,526,594; September 1, 1970; assigned to Standard Oil Company describes lubricant greases especially suitable for use at elevated temperatures of 400° to 600°F., and higher, are made by thickening a normally liquid lubricant with a finely divided silica which has been surface treated with a hexa(C_{1-4} alkyl) disilazane, e.g., hexamethyldisilazane, or with a fluorosilane derivative prepared by reacting dimethyldichlorosilane with a perfluoroalcohol, e.g., a C_{11}-fluoroalcohol.

In preparing the lubricant grease, the surface-treated silica, in sufficient quantity, e.g., 2 to 40% of the final grease composition, is added to the oleaginous lubricant vehicle and stirred at room temperature until the mixture is uniform and then milled. The milled mixture is heated to 400° to 500°F. and maintained at such temperature for a period of 0.1 to

10 hours; and thereafter cooled to room or ambient temperature to form a homogenous grease product. Illustrative of hexaalkyldisilazanes which may be employed are hexamethyldisilazane, hexaethyldisilazane, hexabutyldisilazane, trimethyltripropyldisilazane and triethyltrimethyldisilazane. The perfluoroalcohol reacted with the dimethyldichlorosilane has the general formula $H(CF_2CF_2)_xCH_2OH$, where x is an integer of 1 to 8. In preparing the fluorosilane derivative, the perfluoroalcohol is reacted at room temperature with the dimethyldichlorosilane in equimolar amounts.

In the preparation of the surface treated silica, the hexaalkyldisilazane or the fluorosilane derivative is reacted with the finely divided silica for 2 hours in a reflux system in toluene suspension at the toluene boiling point temperature under a slow stream of nitrogen. The toluene is subsequently distilled off and the surface treated silica dried at 400°F. under slight vacuum. Some boric acid has been found to be an effective stabilizer for silica thickened greases, it is therefore advantageous to add 2%, based on the silica, to the mixture of silica and the hexaalkyldisilizane or the fluorosilane derivative.

The finely divided silicas, which are surface treated with either the hexaalkyldisilazane or with the above described fluorosilane, have particle sizes predominantly in the 0.005 to 0.020 micron range, with surface areas (BET nitrogen adsorption isotherm) of 100 to 350 square meters per gram. Such silicas are prepared by various known techniques, including low temperature precipitation from aqueous solutions of soluble silicates, high temperature, i.e., 100°C., vapor phase hydrolysis of silicone tetrachloride, etc. A particularly suitable silica is a product known as "Cab-O-Sil H5," a pyrogenic silica prepared by the high temperature hydrolysis of silicone tetrachloride. The following examples illustrate the process.

Example 1: A finely divided silica, Cab-O-Sil H5, was treated with a toluene solution of an amount of hexamethyldisilazane in excess of that calculated to completely cover the surface of the silica. The coating technique used was adapted from the technique used for coating chromatography supports as described in Technical Bulletin No. 12 dated March 1, 1966, issued by Applied Science Laboratories, Inc. A grease was prepared by thickening a silicone polymer oil (a phenylmethyl silicone polymer fluid known as QF6-7024 silicone fluid) with 13% of the above surface treated silica.

For the purpose of comparison the same finely divided silica was treated with n-butyl (C_4), isooctyl (iC_8), dodecyl (C_{12}) and cetyl (C_{16}) alcohols to form the corresponding estersils, and greases were prepared by thickening QF6-7024 silicone fluid with such estersils. The above greases were subjected to a bearing performance test at 600°F., run at 10,000 rpm in the atmosphere, using a 25 lb. axial and 50 lb. radial load. The test is continued to bearing failure. The results of this test are tabulated below.

Type of thickener	Percent thickener	Time to failure, hr.
(1) Cab-O-Sil—hexamethyldisilazane	13.0	126
(2) Cab-O-Sil—"Estersil PLI 71" [1]	7.5	37
(3) Cab-O-Sil—iC^8alcohol	14.0	2.5
(4) Cab-O-Sil—C^{12} alcohol	11.1	24
(5) Cab-O-Sil—C^{16} alcohol	13.6	17
(6) Cab-O-Sil—untreated	4.3	40

[1] n-Butanol coated silica marketed by Du Pont.

The above data demonstrate the superior thermal stability of greases thickened with silica treated hexamethyldisilazanes.

Example 2: A C_{11} perfluoroalcohol [$H(CF_2—CF_2)_5CH_2OH$] was reacted at room temperature with dimethyldichlorosilane in equimolar amounts, to give a fluorosilane, perfluoroalkoxydimethylchlorosilane. Cab-O-Sil H5 was treated for two hours with the above fluorosilane in a reflux system in toluene suspension at the toluene boiling point temperature under a slow stream of nitrogen. Two percent boric acid, based on the silica was added to the mixture of silica and fluorosilane. The relative amounts used were such that one-half of the estimated surface hydroxyls of silica should be recovered. Toluene was then distilled off and the thickener dried at 400°F. under slight vacuum. The following greases were prepared with the above fluorosilane surface treated silica.

	Percent thickener	Vehicle fluid
Grease A	20	Silicone polymer.[1]
Grease B	15.4	Perfluoro polymer.[2]
Grease C	8.0	Do.

[1] QF6-7024, marketed by Dow Corning.
[2] PR 143 a perfluoro-polymeric fluid prepared by Du Pont.

Greases A and B were subjected to static aging tests at 400° and 600°F. for 18 hours. The results are shown below.

Grease	Aging, 18 hours: 400° F.	Aging, 18 hours: 600° F.
A	Stable	Liquid.
B	do	Stable.

The 400° and 600°F. aged Grease B samples were combined and additional PR143 fluid added to give a grease containing 6.2% of the thickener. This grease product was stable after aging at 600°F. for an additional 140 hours, for a total aging period of 158 hours. Grease C, after being subjected to two high speed bearing performance tests, at 10,000 rpm at 50 lb. radial and 25 lb. axial loading, one at 450°F. and the other at 550°F. At 450°F. the bearing was still running at 1,250 hours of operation; at 550°F. the bearing failed after operating 270 hours.

N.H. Kasten and H. Teicher; U.S. Patent 3,374,169; March 19, 1968; assigned to Monsanto Company describe a thickening agent comprising silica material such as fume silica or a silica aerogel and a silicone glycol copolymer having the general formula

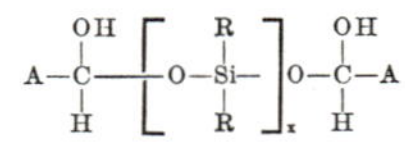

where R is an alkyl group having 1 to 4 carbon atoms; A is either hydrogen, an alkyl group having 1 to 3 carbon atoms; or a hydroxy-substituted alkyl group having 1 to 3 carbon atoms; and x is an integer representing the number of units in the molecular chain and has a value of 1 to 25. The thickening agent is utilized for increasing the viscosity and thixotropy of a variety of organic liquids, such as an oleaginous liquid, or a liquid organic resin.

Silica and Polyphenyls

P.R. McCarthy and C.S. Tempalski; U.S. Patent 3,291,733; Dec. 13, 1966; assigned to Gulf Research & Development Company have found a lubricating composition having improved lubricating characteristics for an extended period of time when used to lubricate bearings operating at temperatures up to 600°F. and speeds up to 20,000 rpm can be obtained by incorporating into a synthetic lubricating oil in oil thickening proportions a mixture of p-polyphenyl and finely divided silica.

The amount of the p-polyphenyl which is employed in the lubricating composition is a small amount which in the absence of the finely divided silica would not be sufficient to thicken the lubricating oil to the consistency of a grease. For example, if p-polyphenyl is employed by itself to thicken a silicone oil to the consistency of a grease, it is used in amounts above 10% by weight, preferably at least 15% by weight. The amount of the p-polyphenyl employed in the composition of the process comprises 0.5 to 2% by weight of the total composition. If more than 2% of p-polyphenyl is employed in combination with the finely divided silica, the performance life of the composition is adversely affected. The lubricating oil used in preparing the lubricating composition shown below was a synthetic oil known as QF-6-7024 Fluid.

The Cab-O-Sil employed in preparing the compositions is a finely divided amorphous silica. This silica is a submicroscopic particulate silica prepared in a hot gaseous environment (1100°C.) by the vapor phase hydrolysis of a silicon compound. On a moisture-free basis, this silica is 99.0 to 99.7% silicon dioxide which is practically free from contaminating metallic salts. In preparing the illustrative lubricating compositions, the oil, the p-polyphenyl and the silica were mixed at room temperature for a period of 10 to 30 minutes. The slurry thus formed was passed twice through a Premier Colloid Mill set at a stator-rotor clearance of 0.0015 inch. The makeup and properties of the thickened lubricating compositions thus prepared are set forth below.

TABLE 1

Composition, Percent By Weight	A	B	C	D
Lubrication Oil QF-6-7024	95.5	95.0	94.5	94.0
p-Polyphenyl	0.5	1.0	1.5	2.0
Finely-divided silica [1]	4.0	4.0	4.0	4.0
Total Thickener	4.5	5.0	5.5	6.0
Ratio of p-polyphenyl to finely-divided silica	1:8	1:4	1:2.7	1:2
INSPECTION				
Penetration D-1403:				
Unworked (Conv. to D217)	321	309	302	295
Worked (Conv. to D217)	339	328	321	302
Dropping Point, °F	[2] 686	[2] 651	[2] 655	[2] 690
Performance Life, Hours: Average results of two tests at 20,000 r.p.m. and 600° F	149	117	158	161

[1] Cab-O-Sil.
[2] Drop of fluid.

The long performance life of compositions at a high rotational speed and a high temperature is self-evident from the above data. In order to illustrate the importance of maintaining the p-polyphenyl content at a maximum of 2% by weight, the performance life of a composition of the process (Composition D) was compared with other similar compositions containing from 3 to 6% by weight of p-polyphenyl. The results of the comparative tests are set forth below.

TABLE 2

Composition, Percent By Weight	D	E	F	G	H	I
Lubricating Oil QF-6-7024	94.0	93.0	92.0	91.0	90.0	92.0
p-Polyphenyl	2.0	3.0	4.0	5.0	6.0	4.8
Finely-divided silica	4.0	4.0	4.0	4.0	4.0	3.2
Total Thickener	6.0	7.0	8.0	9.0	10.0	8.0
Ratio of p-polyphenyl to finely-divided silica	1:2	1:1.3	1:1	1.25:1	1.5:1	1.5:1
Performance Life, Hours: Average results of two tests at 20,000 r.p.m. and 600° F	161	47	38	72	50	66

The long performance life of Composition D as compared with the performance lives of Compositions E through I which contain more than 2% of p-polyphenyl is indicative of the critical amount of p-polyphenyl in compositions of the process. Other lubricating compositions within the scope of the process are illustrated below.

TABLE 3

Composition, Percent By Weight	J	K	L	M	N
Lubricating Oil QF-6-7024	95.5	95.4	95.0	94.6	94.3
p-Polyphenyl	1.5	1.15	1.0	0.9	0.7
Finely-divided silica	3.0	3.45	4.0	4.5	5.0
Total Thickener	4.5	4.6	5.0	5.4	5.7
Ratio of p-polyphenyl to finely-divided silica	1:2	1:3	1:4	1:5	1:7

E. Pirson and S. Nitzsche; U.S. Patent 3,518,188; June 30, 1970; assigned to Wacker-Chemie GmbH, Germany have found that compounds and greases based on organosiloxane fluids filled with silica or other pulverulent material are stabilized on storage and against the effects of heat by incorporating an organosiloxane resin.

I.J. Spilners and C.S. Tempalski; U.S. Patent 3,291,732; December 13, 1966; assigned to Gulf Research & Development Company also describe a lubricating composition which comprises a dispersion in a synthetic lubricating oil of a sufficient amount of p-polyphenyl to thicken the lubricating oil to a grease consistency.

Clay-Silicone Fluid Reaction Product

A.E. Axworthy; U.S. Patent 3,293,180; December 20, 1966; assigned to Shell Oil Company describes an extremely high temperature clay grease which is not merely a lubricating base oil physically thickened with clay, but which is an actual reaction product of the clay, the lubricating base oil, and a metal having an atomic number of from 26 through 29. More particularly, the grease composition is the thermal reaction product of a polysiloxane oil, a sulfate of a metal having an atomic number of 26 through 29, and clay.

The clays to be used in the process are those having a substantial base exchange capacity, preferably between 60 and 100 milliequivalents of exchangeable base per 100 g. of clay. These include particularly the montmorillonites, especially the sodium, potassium, lithium and other bentonites such as Wyoming bentonite, magnesium bentonite (hectorite) and saponite. The clays are preferably treated to remove the substantial amount of gangue which is normally associated therewith. This can be done by dispersal of the crude clay in water and allowing the product to separate into fractions which may be separated from one another either by gravitational means or more preferably by high speed centrifuging. The clay forms a hydrous sol (hydrosol) with water, a preferred concentration being 1 to 5% by weight of clay in order to have a system which is relatively fluid and easy to transport, such as by pumping.

In order to prepare the clay in a more suitable manner for full reception of the waterproofing compound, the clay while in hydrosol form may be treated with a mineral acid, preferably a phosphorus acid such as phosphoric acid in an amount at least equal to the basic exchange capacity of the clay. When utilizing the phosphoric acid, this will ordinarily be in an amount between 4 and 12% by weight based on the clay. The colloidal nature of clay is such that it is an effective thickening agent only when the particles are in a swollen or expanded state. Though the clay is expanded when in the form of an aqueous slurry or hydrosol, it is necessary that it remain expanded in the presence of the coreactants after removal of the water. To accomplish this isopropyl alcohol is added to the clay slurry in an amount equivalent to at least 7 parts by weight alcohol per weight of water and the polysiloxane oil reactant is added before removal of water and alcohol from the clay is completed. The polysiloxane oils which can be used are the methyl-phenyl silicone fluids of lubricating oil viscosity which will generally be of the following types.

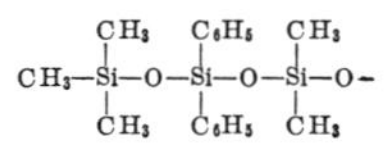

and/or

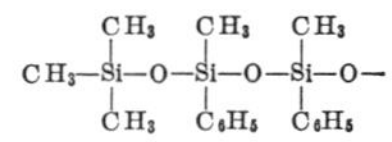

The above types of methyl-phenyl silicone fluids, in addition to being the most thermally stable, are also readily available in commercial quantities. The high temperature grease is prepared by the following procedure.

(1) To an aqueous slurry (hydrosol) of clay is admixed a sulfate of a metal having an atomic number of 26 through 29, isopropyl alcohol in an amount equal to at least seven times the weight of the water contained in the slurry and a methyl-phenyl polysiloxane oil.

(2) The mixture is heated to a temperature of at least 212°F., and preferably at least 250°F., for a period such that essentially all the water and isopropyl alcohol are removed by evaporation.

(3) The dehydrated mixture is then reacted by heat treatment at a temperature of 400° to 550°F. for a period of at least one hour.

(4) The reaction mixture is cooled and milled to form a grease.

Example 1: An aqueous solution of 1.2 g. of copper sulfate was added to 800 g. of a 2% by weight slurry of Hectorite clay in water at 160°F. and well mixed. Two gallons if isopropyl alcohol were admixed with the heated slurry and the admixture was boiled gently to remove greater than 50% by weight of the alcohol and water. A DC-550 silicone oil (382.8 g.) was then added to and mixed with the partially dehydrated slurry and the oil-containing mixture was dried, with agitation, at 270°F., and, after essentially complete removal of all the water and alcohol, heated at 450°F. for a period of three hours. The product, which was very dark brown in color, was milled three times to form a stable grease structure. The relative amounts of the reactive ingredients of the grease were as follows.

	Percent by Weight
Clay	4.0
$CuSO_4$	0.3
Silicone oil	95.7

However, during the high temperature heating or baking step, at least 10% by weight of the silicone (polysiloxane) was volatilized by decomposition and reaction in the presence of the clay and copper sulfate. Thus the relative amounts of clay and sulfate are somewhat higher in the finished grease. As is implied above, the heat treatment and reaction of the clay, polysiloxane oil and metal sulfate is accompanied by the evolution of vapors from reaction of the polysiloxane oil. In general, from 10 to 30 by weight of oil is volatilized in this manner. Concomitantly, the reaction mixture turns to a dark brown or black color.

Because of the relatively high thermal stability of silicone fluids, the reaction must be carried out at a temperature of at least 400°F. and preferably at least 450°F. On the other hand, it has been found if excessively high temperatures are used, the degree of decomposition of the silicone fluid is too great and that the reaction product is too hard and may contain entrapped volatile decomposition products. It is therefore preferred to carry out the reaction at not over 550°F. and still more preferably from 450° to 500°F.

It will be noted that the dehydration step in which the water and alcohol are removed from the clay and the reaction step can be carried out in one continuous heating cycle. However, it is important that all of the water and alcohol be removed before the reaction temperature is attained. In other words, the dehydration step should be conducted at temperatures below that at which chemical (including thermal) reaction occurs. The dehydration step is preferably carried out with agitation in order to facilitate heat transfer, however the reaction step is preferably carried out in a quiescent state. In carrying out the reaction in a quiescent state, it is preferred to spread the reactant mixture into relatively thin (1/2 inch) layers to facilitate even heating. The following tests were performed to examine the effects of compositional and process variables on the character of the grease produced in this manner.

Example 2: Using the same general procedure as Example 1, a number of high temperature grease samples were prepared in which a wide range of copper sulfate and clay concentrations and various intervals of heating were observed with regard to their effect on mechanical and thermal stability and water resistance of the grease. The following tests were used to assess thermal stability and water resistance.

Thermal Stability; Thin Film Oxidation (TFO) Test — In this test a thin film of the grease is placed on a smooth flat plate which is heated in air at specified conditions of time and temperature. At selected intervals, the thermal stability is determined by measuring the weight loss of the sample and visually observing the texture of the grease.

Thermal Stability; Bomb Oxidation — A fixed quantity of the grease is heated to 250°F. in the presence of oxygen in a sealed bomb. Oxidative stability is measured by observation of the pressure drop within the bomb at various intervals. Pressure drops of less than 5 psi at 100 hours or less than 15 psi at 500 hours are considered to be excellent.

Water Resistance; Qualitative Test — A small amount of grease is smeared on the inside of a dry, clear glass beaker. The beaker is filled with water and then heated to boiling. Poor resistance to water is indicated by flaking off of the grease from the side of the beaker within less than 10 minutes.

Water Resistance; Water Washout (ASTM Test Procedure D1264) — In this test the grease is packed in a ball bearing which is inserted in a housing and rotated at 600 rpm with water of specified temperature impinging on the housed bearing. Water washout is measured by the amount of grease washed from the bearing in one hour.

TABLE 1: EFFECT OF COMPOSITION ON GREASE PROPERTIES

	A	B	C	D	E	F	G	H	I	J
Composition of Ingredients, Percent by wt.:										
$CuSO_4$	0.05	0.2	0.3	0.3	0.3	0.5	1.0	1.0	5.1	16.7
Clay	4.0	4.5	3.5	4.0	3.8	4.0	7.0	8.5	8.5	6.7
DC-550 Silicone Oil	95.95	95.3	96.2	95.7	95.9	95.5	92.0	90.5	86.4	76.6
Heat Treatment:										
Time, hrs	15	2	2	22	22	22	3	4	4	4
Temperature, °F	450	450	500	450	450	450	450	450	450	450
Mechanical Stability:										
Penetration, dmm:										
Unworked	205	246	382	243	312	392	347		336	(1)
Worked		263				400	353		362	
Visual Rating	(2)					(3)	(3)		(4)	(4)
Thermal Stability:										
Thin Film Oxidation Test:										
Temperature, °F	450		450	450	450	450		450		
Time, hrs	803		93	880	400	400		400		
Weight Loss, Percent	29		16	18.9	15.5	16		12.7		
Appearance, Consistency	(5)			(3)	(6)	(7)		(6)		
Bomb Oxidation Test:										
Temperature, °F				250						
Time, hrs				100/500/870						
Pressure Drop, p.s.i				3/4/4						
Water Resistance:										
Beaker Test	(8)		(9)	(9)	(9)		(9)	(9)	(3)	
ASTM D 1264, Washout, Percent wt				4.3						

(1) Too soft. (2) Putty-like. (3) Good. (4) Thixotropic. (5) Plastic. (6) Still plastic. (7) No change. (8) Fair. (9) Excellent.

The above data show that the amount of time for which the grease is heat treated in excess of one hour is not important with regard to the properties of the resultant grease. Consequently, heat treatment for any time in excess of one hour is sufficient. However, in order to assure more complete removal of volatiles from the grease, a heat treatment period of at least two hours is preferred. The date in Table 1 also show that the amount of metal sulfate in the grease has a profound effect on the grease properties. As shown by sample A, to obtain adequate thermal stability and water resistance, at least 0.1% by weight metal sulfate must be employed.

On the other hand, as is shown by greases I and J in comparison with grease G, if the amount of metal sulfate exceeds 4% by weight, the grease has poor mechanical stability and is thixotropic. Superior properties are obtained when the amount of sulfate is within the preferred range of from 0.2 to 1.0% by weight. Particularly noteworthy about all of the above greases, which are within the 0.1 to 4% metal sulfate range, is their outstanding oxidation stability, thermal stability and resistance to water, and especially the second of these. It has also been found that the silicone oil plays an important part in achieving the desired properties of the greases, as is shown below.

Example 3: Using the same procedure as Example 1, three greases were prepared and tested from silicone oils of different methyl-to-phenyl ratio. Each was prepared from 0.3% weight $CuSO_4$ and 4.0% weight clay and was heat treated at 450°F. for 20 to 22 hours. The results were as shown on the following page. From the data shown in Table 2, it appears that the high (6.0:1) methyl-phenyl silicone oil was thermally too unstable to be practicable for making the grease composition of the process and gave unsatisfactory water resistance. On the other hand, the low (0.8:1) methyl-phenyl silicone oil yielded a finished grease having only fair thermal stability and was of too hard consistency. It is therefore preferred to employ methyl-phenyl silicone blends in which the mol ratio of methyl-to-phenyl groups is from 1.5 to 3.5.

TABLE 2: EFFECT OF METHYL/PHENYL MOL RATIO IN SILICONE OIL

Methyl/ Phenyl Mole Ratio	Water Resistance	Thermal Stability	
0.8:1	Excellent	Fair	Finished grease too hard.
2.8:1	do	Excellent	Good grease consistency.
6.0:1	Poor		Excessive volatilization during heat treatment before satisfactory grease consistency obtained.

J.D. Stansfield and C.B. Lancaster; U.S. Patent 3,294,683; Dec. 27, 1966; assigned to Shell Oil Company have found that clay based greases may advantageously be prepared by using an alkyl ester of an inorganic acid, the ester having a boiling point above 100°C. as a dispersant together with a codispersant which is a hydroxyl compound also having a boiling point above 100°C. By using the dispersant and codispersant in certain relative proportions it is possible to produce a clay based greased having superior properties with respect to yield, storage stability and mechanical stability.

Polytetrafluoroethylene Thickener

A process described by J.B. Christian; U.S. Patent 3,314,889; April 18, 1967; assigned to the U.S. Secretary of the Air Force provides a grease composition where the thickener will not separate prematurely from the lubricating base fluid. The process involves the combination of a high temperature organosilicon lubricating base fluid with a high temperature polymeric thickener in a particular range of proportions. The specific base fluid employed is a fluorosilicone fluid known as Dow Corning FS 1265.

The specific fluorosilicone fluid (FS 1265) designation is of a viscosity grade of from 250 to 1,000 cs. at the lower limit and up to 10,000 cs. at the upper limit, a specific gravity within the range of from 1.24 to 1.30, a freezing point of from -60° to -25°F., a flash point of from 440° to 530°F., and a trace of acidity. More specifically FS 1265 can be designated as a polymeric trifluoropropylmethylsiloxane of the composition and structure

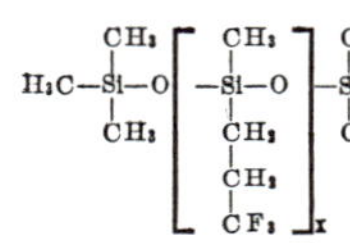

as listed on page 1 of Preprint No. 63LC-12, "Boundary Lubricating Properties of Fluoroalkyl Silicones in Bench and Pump Tests" by H.M. Schiefer and J. Van Dyke, published by American Society of Lubrication Engineers, Chicago, Illinois, 1963. The organic or polymeric thickener to be employed in combination with the fluorosilicone base fluid is a fluoroethylene polymer in the form of a powder, particularly that with a particle size such that it will pass through from a 100 to a 200 mesh screen. The fluoroethylene polymers thus involved are of the class consisting of tetrafluoroethylene and chlorotrifluoroethylene and are of the type known as Teflon.

The fluoroethylene powders of the sizes indicated are mixed with the fluorosilicone base fluid by conventional mixing devices and techniques in the weight ratios of from 30 to 35 parts of the fluoroethylene powder to from 65 to 70 parts of the fluorosilicone base fluid, as a result of which the total grease will be from 30 to 35% fluoroethylene powder by weight and from 65 to 70% fluorosilicone base fluid by weight. The improvements in a multipurpose grease compound which render it particularly adaptable for use in a wide variety of applications, particularly involving high speeds, high temperatures, and high loads such as are encountered in high speed aircraft, rocket-powered space vehicles and the like, are demonstrated by the following examples.

Example 1: 30 parts by weight of a polytetrafluoroethylene powder of a size which would pass through a 200 mesh screen were thoroughly admixed with 70 parts by weight of the before described Dow Corning FS 1265 fluorosilicone base fluid by first stirring in a conventional internal mixer and then blending by three passes through a 3 roll paint mill with the roller opening set at 0.002 inch, all at room temperature. Penetration of this grease, measured by means of the standard ASTM method, was from 293 to 297; and, on the Pope spindle bearing test under five pound load, the following results were achieved.

10,000 rpm at 450°F.	1,000 hrs.
10,000 rpm at 500°F.	221 hrs.
20,000 rpm at 400°F.	1,000 hrs.

Example 2: 20 parts by weight of polytetrafluoroethylene powder of the 200 mesh screen size and 80 parts by weight of the Dow Corning FS 1265 fluid were admixed as above, and the resulting compound showed an ASTM penetration of from 350 to 364. On the Pope spindle bearing test under the same five pounds loading, the following results were achieved.

10,000 rpm at 450°F.	300 hrs.
10,000 rpm at 500°F.	120 hrs.
20,000 rpm at 400°F.	225 hrs.

Example 3: 40 parts by weight of the polytetrafluoroethylene powder of a 200 mesh screen size and 60 parts by weight of the Dow Corning FS 1265 fluorosilicone fluid were admixed as set forth in Example 1. The ASTM penetration of this composition was 227 to 232 and the Pope spindle bearing test under five pounds loading gave the following results.

10,000 rpm at 450°F.	214 hrs.
10,000 rpm at 500°F.	100 hrs.
20,000 rpm at 400°F.	140 hrs.

As a standard of comparison for the above, one of the more advanced high temperature greases available today is that conforming to the published military specification known as MIL-G-27343 which is designated as a "grease, ball and roller bearing for temperatures from -100° to +400°F." The greases meeting this specification are generally of the class consisting of arylurea-thickened silicone type fluids. The requirements of greases according to this specification are that on the Pope spindle test under five pound load, the grease should be operative for 500 hours at 400°F. at 10,000 rpm.

Phenylsiloxanes and Trifluoropropylphenylsiloxanes as Thickeners

F.O. Stark; U.S. Patent 3,288,712; November 29, 1966; assigned to Dow Corning Corporation describes heat resistant greases which contain phenylsiloxanes as thickeners. The phenylsiloxanes of this process are polyorganosiloxanes where at least 75 mol percent of their organic substituents are phenyl groups. They are hard, intractable materials, most of which are infusible. The thickener can be made in several ways. Hexaphenylcyclotrisiloxane and octaphenylcyclotetrasiloxane are both well-known, and are useful as the thickener. Hexaphenylcyclotrisiloxane can be polymerized to higher polymeric siloxanes by means of such alkaline catalysts as n-butyl lithium, sodium hydroxide, sodium phenoxide, or potassium silanolate.

The best temperature range of this polymerization reaction runs from 150° to 250°C., but one must be careful not to overheat the reaction or run it for too long a time, as the higher polymeric siloxanes may degrade to the cyclic tetramer. Four to 24 hours of reaction time at 205°C. generally gives good results when potassium silanolate catalyst is used in a proportion of one potassium atom to 1,000 silicon atoms. The higher polymeric siloxanes made by the polymerization of hexaphenylcyclotrisiloxane are the preferred thickeners of this process.

Polysiloxane fluids are the preferred fluid ingredients. Specifically, $(C_6H_5)_2CH_3SiO[(C_6H_5)(CH_3)SiO]_xSiCH_3(C_6H_5)_2$ fluid is the preferred species as the fluid ingredient for the greases where x is a positive integer that gives a material with a viscosity at 25°C. of from 300 to 10,000 cs. The finely divided solid thickener is mixed with the oil ingredient until a grease-like composition is formed. A method of mixing that imparts shearing action is preferred. The particle size of the thickener is not critical, but the particles should be as small as possible, since it is known that better greases are made when the filler is very finely divided. The following examples illustrate the process.

Example 1: 26.4 weight percent of a diphenylpolysiloxane polymer which had been made by the alkaline equilibration of hexaphenylcyclotrisiloxane was milled into 73.6 weight percent of a phenylmethylsiloxane fluid with diphenylmethylsiloxy endblocks and having a viscosity of 700 cs. The product was a grease, having an unworked penetration of 256 and a worked penetration of 279. A sample of the grease was exposed at 550°F. temperature for 1 1/2 hours, and showed no change in color or bleeding. Only a slight amount of graininess was observed after the test. The grease was tested in a continuous high temperature bearing test using a 204 bearing, a speed of 10,000 rpm, a load of 5 pounds, and a temperature of 600°F. The grease gave a bearing life of 186 hours. A good aryl-substituted urea thickened silicone grease using the same fluid as above was tested in the same bearing test, and gave a life of 160 hours.

Example 2: 16 weight percent of a polymerizate made by heating hexaphenylcyclotrisiloxane for 12 hours at 205°C. in a nitrogen atmosphere with sufficient n-butyl lithium to give one lithium atom per 1,000 silicon atoms was milled with 84 weight percent of a phenylmethylsiloxane fluid with diphenylmethylsiloxy endblocks and having a viscosity of 700 cs. The product was a grease, having an initial penetration of 279, an unworked penetration of 293, and a worked penetration of 283, indicating good work stability. The grease was heated for 24 hours at 600°F. in a cone made from metal screening in order to test the bleed of the grease. 23.02 percent by weight of the grease was lost through bleeding. Under similar conditions it was found that the grease lost 19.19% of its weight through evaporation. A bearing test similar to the test of Example 1 was performed. The bearing life at a 5 pound load was 201 hours. At a 50 pound load, the bearing life was 194 hours, indicating good lubricating life under loaded conditions on a high speed ball bearing.

Example 3: 18 weight percent of a polymerizate made by heating hexaphenylcyclotrisiloxane for 24 hours at 205°C. in a nitrogen atmosphere with sufficient n-butyl lithium to give one lithium atom per 1,000 silicon atoms was milled with 82 weight percent of a phenylmethylsiloxane fluid with diphenylmethylsiloxy endblocks, and having a viscosity of 700 cs. A grease was formed having an initial penetration of 280. The grease was heated for four hours at 400°F. in a cone made from metal screening in order to test the bleed of the grease. 10.5% by weight of the grease was lost through bleeding. Under similar conditions it was found that the grease lost 0.30% of its weight through evaporation. The grease was subjected to the bearing test of Example 1. The bearing failed after 244 hours of continuous operation.

In a process described by C.-L. Lee, G.J. Quaal and O.W. Marko; U.S. Patent 3,575,860; April 20, 1971; assigned to Dow Corning Corporation a solid homopolymer of trifluoropropylphenyl siloxane units is added to a lubricating oil vehicle in an amount sufficient to thicken to a grease.

Organosilicone Fillers and Metal Chromate Addition

J.H. Wright; U.S. Patent 3,304,259; February 14, 1967; assigned to General Electric Company describes an organopolysiloxane grease which comprises at least one organopolysiloxane base fluid selected from the class consisting of a first organopolysiloxane fluid having the formula

$$R_eSiO_{4-e/2} \qquad (1)$$

a second organopolysiloxane fluid having the formula

$$R_nH_mSiO_{4-n-m/2} \qquad (2)$$

and a third organopolysiloxane fluid having the formula

$$R_pVi_qSiO_{4-p-q/2} \qquad (3)$$

and a filler formed by the addition of the SiH groups of the second organopolysiloxane fluid across the silicon-bonded vinyl groups of the third organopolysiloxane fluid, in the presence of a platinum catalyst at an elevated temperature, the filler being an organopolysiloxane material having silethylene linkages of the structure $\equiv Si{-}CH_2CH_2{-}Si\equiv$. In the above formula R is selected from the class consisting of alkyl radicals, substituted alkyl radicals, aryl radicals, and substituted aryl radicals; m is from 0.015 to 1.0, the sum of n and m is from 2.0 to 2.35; q is from 0.015 to 1.0 and the sum of p and q is from 2.0 to 2.3; e from 1.98 to 2.5; and the symbol Vi represents a vinyl group having the structure $-CH{=}CH_2$. By proper selection of the values of m, q, and the chain length of the polymers of the Formulas (2) and (3), the size and structure of the filler material can be accurately controlled. The organopolysiloxane materials of the Formula (2) may range from a cyclotetrasiloxane of the formula

$$(RHSiO)_4 \qquad (4)$$

where R is as previously defined, to an organopolysiloxane fluid of the formula

$$R_2HSi{-}O{-}(R_2SiO)_x{-}SiR_2H \qquad (5)$$

where R is as previously defined and x is an integral number of up to 100. Of course, the hydrogen containing material may be a fluid similar to that just described, but containing hydrogen substitutents on more than 2 of the silicon atoms in the chain, such as one having the formula

$$R_3SiO{-}(R'_2SiO)_y{-}SiR_3 \qquad (6)$$

where R is as previously defined, each R' is independently selected from the class consisting of R and hydrogen and y is an integral number of up to 5,000 which is so selected that m in Formula (2) is from 0.015 to 1.0. Preferably, the SiH containing material is one corresponding to Formula (2), regardless of form, where m is from 0.037 to 0.333. The organopolysiloxane compound of the Formula (3) can similarly vary from a cyclotetrasiloxane of the formula

$$(RViSiO)_4 \qquad (7)$$

where R is as previously defined, to an organopolysiloxane fluid of the formula

$$ViR_2SiO{-}(R_2SiO)_a{-}RViSiO{-}SiR_2Vi \qquad (8)$$

where R is as previously defined and a is an integral number of up to 150. Again, the vinyl-containing material may be a fluid similar to that just described, but having vinyl substituents on more than 3 of the silicon atoms in the chain.

A compound such as that described on the previous page has the following formula

$$ViR_2SiO—(R''_2SiO)_z—SiR_2Vi \qquad (9)$$

where R is as previously defined, each R" is independently selected from the class consisting of R and Vi, and at least two of the R" groups are vinyl and z is an integral number of up to 5,000 which is so selected that q in Formula (2) is from 0.015 to 1.0. Preferably, the Si-vinyl compound is one of Formula (3), regardless of form, where q is from 0.037 to 0.430.

Example 1: A copolymer was formed by equilibrating (a) 2.9 parts of divinyltetramethyldisiloxane, (b) 1.1 parts of methylvinylcyclotetrasiloxane, and (c) 41.0 parts of octamethylcyclotetrasiloxane with 0.05% potassium hydroxide, based on (a) and (b), as a catalyst. This resulted in a fluid equivalent to that of Formula (3) where R is methyl, q is 0.073, and the sum of p and q is 2.05. The filler was formed by reacting the above fluid with 5 parts of a methyl hydrogen polysiloxane fluid equivalent to Formula (6), where y was from 10 to 40, averaging 25, one-half of the R' groups were H and the other half of the R' groups were methyl, and R was methyl.

The catalyst consisted of sufficient chloroplatinic acid to provide 25 ppm of platinum, based on the total weight of SiH and SiVi containing materials. The ingredients were heated at 150°C. for 45 minutes and resulted in a hard gel. The gel was cooled and ground on a three-roll paint mill using two tight passes and one loose pass to give a fine, fluffy powder. This powder was milled with 50 parts of a dimethylpolysiloxane fluid having a viscosity of 500 cs. to give a stable grease.

Example 2: A powder equivalent to that formed in Example 1 was blended with equal parts of a phenylmethylpolysiloxane fluid, equivalent to Formula (1) where R is both methyl and phenyl, and e is 2.0, having a viscosity of 150 cs. The resulting grease was extremely stable.

Example 3: A mixture was prepared containing 1.5 parts of the SiH fluid of Formula (6) as defined in Example 1, 30 parts of a SiVi fluid of Formula (8) where a was 100 and each R group was methyl, 68.5 parts of a dimethylpolysiloxane fluid having a viscosity of 500 cs., and sufficient chloroplatinic acid to give 200 ppm of platinum, based on the SiVi-SiH mixture. The ingredients were mixed and put in an oven at 200°C. for 20 minutes. After the resulting gel was cooled, it was milled on a three-roll paint mill using three passes, two at a tight setting and one at a loose setting. The resultant grease was tested according to MIL S8660B, a beaker and cone test. The grease showed no bleed, 1.6% evaporation at 200°C. for 24 hours, and worked penetration at 60 cycles of 230. By comparison MIL S8660B calls for a maximum of 8% bleed, a maximum of 2% evaporation, and a maximum penetration of 310. These specifications are based on available, standard organopolysiloxane greases.

Examples 4 through 10: A series of preferred greases formed according to the process are shown below. The table shows the parts by weight of SiH fluid, SiVi fluid and organopolysiloxane fluid of Formula (1), and indicates whether the added fluid was present during the cure of the filler or whether it was milled in following cure. Additionally, the amount of catalyst, in parts per million as platinum, based on the total of SiH and SiVi containing materials, and the length of cure time, in minutes, are shown. In each case the SiH fluid was that described in Example 1 and the SiVi fluid was that described in Example 3. The added organopolysiloxane fluid in Examples 4 and 5 was a cyanopropylmethylpolysiloxane fluid having a viscosity at 25°C. of 10,000 cs. and having the average formula

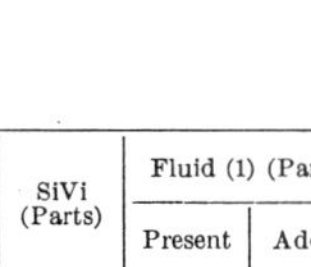

In each of the other examples, the added organopolysiloxane fluid was a dimethylpolysiloxane fluid, equivalent to Formula (1), where R was methyl, and e was approximately 2.0, having a viscosity at 25°C. of 500 cs. The filler was cured in each case at 200°C. using chloroplatinic acid as the catalyst.

TABLE 1

Ex.	SiH (Parts)	SiVi (Parts)	Fluid (1) (Parts)		Catalyst (p.p.m. as Pt)	Cure Time (Min.)
			Present	Added		
4	10	35	55	-	35	30
5	10	35	-	55	35	30
6	3.3	30	33.3	33.4	17.5	12
7	5.0	28.3	33.3	33.4	17.5	12
8	1.4	32.5	33.9	32.2	17.5	30
9	1.0	32.5	66.5	-	17.5	20
10	0.8	39.2	40	20	55	30

The materials of each of the above examples was tested according to MIL S8660B. The results of those tests for bleed, evaporation, and worked penetration at 60 cycles are shown below.

TABLE 2

Example	Bleed	Evaporation	Worked Penetration, 60 Cycles
4	1.61	0.85	238
5	1.48	0.69	252
6	5.3	1.2	219
7	4.0	0.8	215
8	2.36	1.0	260
9	0	1.6	215
10			294

J.H. Wright; U.S. Patent 3,537,995; November 3, 1970; assigned to General Electric Company has found that the addition of a metal chromate or dichromate to a silicone lubricating grease reduces galvanic corrosion during the lubrication of dissimilar metal surfaces which are separated by a lubricating film of the grease.

Inorganic Chromium-Phosphinic Acid Polymeric Thickeners

N.D. Peschko, B.P. Block, G.H. Dahl and L.R. Ocone; U.S. Patent 3,522,178; July 28, 1970; assigned to Pennwalt Corporation have found that greases having unusually good properties, particularly with respect to drop point, are provided by compositions comprising a liquid lubricant and a small, but thickening amount of an inorganic polymer consisting of bridging groups and a triply bridged chromium atom where the bridging groups are the anions of phosphinic acids of structure $R_2P(X)XH$ where R is selected from the group consisting of alkyl and aryl and X is oxygen or sulfur.

The phosphinic acids preferred in the process are those where R is an alkyl group of from one to 18 carbon atoms and where R is an aryl group such as phenyl and naphthyl. The liquid lubricants which may be converted to greases include any of the numerous, well known lubricants such as hydrocarbon fluids, chlorinated polyphenyls, ester based fluids such as sebacates, adipates, azelates and the like, phosphate esters and silicone fluids. Two general methods of grease preparation may be followed. In both techniques a chromium containing inorganic doubly bridged polymer such as that described in U.S. Patent 3,275,574 where one unidentified ligand is water and a second ligand is the hydroxyl anion (that is, a hydroxy-aquo polymer) is first mixed with the phosphinic acid $R_2P(X)XH$ in an organic solvent system such as hydrocarbons and halogenated hydrocarbons exemplified by benzene, chloroform, orthodichlorobenzene, toluene, and solvents such as tetrahydrofuran.

In one technique the solvent is removed to form a thickening composition which is simply added to the liquid lubricant and the mixture stirred with heating until grease consistency is obtained. In the second preparative technique the thickening composition is formed by allowing the hydroxy-aquo polymer-phosphinic acid-solvent system to form a gel by letting it stand at room temperature. Gel formation will generally occur in 24 hours or more and in lesser time at elevated temperatures. The gel thus formed in the solvent system is then dispersed at room temperature in the fluid lubricant and the mixture heated (generally from 100° to 225°C.) with stirring to evaporate any residual solvent, thereby leaving the grease product.

In general, it may be said that when the phosphinic acid used contains R groups which contain four or more carbon atoms, the first method described above is preferred. Alternatively, when the phosphinic acid contains only aryl and/or lower alkyl groups then the second method is the preferred procedure. In both procedures the triply bridged inorganic polymer is formed in the final stages of the process. The hydroxy and water ligands which appear in the double bridged polymer from which the triply bridged polymer is obtained are removed as water during the processing steps and it is believed that their absence in the final grease product contributes to an improved grease. Thus, for example, the absence of any water eliminates any corrosion problems which a water-containing grease might have.

The ratio of phosphinic acid and chromium containing doubly bridged polymers which will be used in making the greases of the process may vary considerably. In general the molar ratio of phosphinic acid to chromium in the polymer will be from 0.4 to 1.3. The preferred ratio range is from 0.6 to 1.1 and most preferably equimolar amounts will be employed. The greases have excellent properties with respect to consistency, wear properties, and drop points, and lend themselves to easy handling. They show excellent drop points, being on the order of from 400° to 700°F. and they have excellent extreme pressure lubricity properties. The following examples illustrate the process.

Example 1: Preparation of Chromium Methylphenylphosphinatedioctylphosphinate Thickener — (1A) A solution of 15.0 g. (38 mmol.) of $Cr(H_2O)(OH)[OP(Me)(Ph)O]_2$ polymer and of 10.95 g. (38 mmol.) of $(C_8H_{17})_2P(O)OH$ in 150 ml. of tetrahydrofuran (THF) is allowed to stand at room temperature for 30 minutes. The thickener product is then recovered by evaporation of the solvent. (1B) A tetrahydrofuran solution is prepared containing 1.65 g. (5.7 mmol.) $(C_8H_{17})_2P(O)OH$ and 3.0 g. (7.6 mmol.) $Cr(H_2O)(OH)[OP(Me)(Ph)O]_2$ polymer. The solution it refluxed for 2 1/2 hours and the thickener reaction product is recovered by evaporation of the solvent. (1C) The polymer $Cr(H_2O)(OH)[OP(Me)(Ph)O]_2$ and the acid $(C_8H_{17})_2P(O)OH$ (1:1 molar ratio) are dissolved in benzene and the solution is heated at 60°C. for 3 hours.

Preparation of Silicone Fluid Greases — (a) An amount of 5 g. of the chromium phosphinate thickener, prepared as under (1A), is introduced into 45 ml. of a poly(methylphenylsiloxane) (Dow Corning 550 Lubricant). The mixture is heated to 140° to 160°C. and stirred vigorously. Grease formation is completed within 2 hours. To produce further setting of the grease it is heated an additional 5 hours at 170° to 180°C. The grease has a dropping point of 340°C. and a penetration of 273 (ASTM D217-60T). Extreme pressure properties measured with a Shell Four Ball Tester show the following typical wear scar diameters.

40 kg. load	0.86 mm.
80 kg. load	1.53 mm.
120 kg. load	2.21 mm.
160 kg. load	weld

(b) A mixture of 0.2 g. of the thickener as prepared under (1A) and 1.8 ml. of poly(methylphenylsiloxane) liquid (Dow Corning 510 silicone fluid) is vigorously stirred and heated to 150°C. for 15.5 hours to give a soft grease. (c) An amount of 0.2 g. thickener as prepared under (1A) is placed into 1.8 ml. poly(methylphenylsiloxane) fluid (Dow Corning 703 silicone fluid). The mixture is stirred and heated at 150°C. for 30 minutes at which time a firm, gel-like grease forms. (d) A grease is obtained from the thickener, prepared as described under (1B), by placing 0.5 g. into 4.5 ml. of Dow Corning 550 silicone fluid and heating with rapid stirring at 110°C. for 16 hours. (e) The benzene solution of the thickener described in (1C) is added to DC 550 lubricant to make a 4% by weight concentration of thickener. Subsequently more thickener solution is added until the final concentration is 10%. During the addition the mixture is heated to 150° to 160°C. until all the solvent is flashed off thereby leaving a soft grease.

Preparation of Greases with Ester Fluids — (a) To 45 ml. of a pentaerythritol ester (Hercolube A-PE) is added 4 g. of thickener, the preparation of which is described in (1A). The mixture is agitated at 100°C. for 15 minutes and then one additional gram of thickener is added. Stirring is continued for a period of 4 hours, while the temperature is maintained at 125° to 130°C. to yield a smooth, homogeneous grease. (b) An amount of 5 g. of thickener, prepared as described under (1A), is added to 45 ml. of di(2-ethylhexyl) sebacate. The reaction mixture is stirred and heated to 105° to 115°C. for 15.5 hours to yield a soft, homogeneous grease.

Example 2: Preparation of Chromium Methylphenylphosphinatedibutylphosphinate Thickener — (2A) Amounts of 3.97 g. (10 mmol.) of $Cr(H_2O)(OH)[OP(Me)(Ph)O]_2$ polymer and 1.78 g. (10 mmol.) of $(C_4H_9)_2P(O)OH$ are dissolved in 19 ml. benzene. The solution is kept at 60°C. for two days and the thickener product recovered by evaporation of the solvent.

Preparation of a Tricresyl Phosphate Grease — A mixture of 0.2 g. of thickener of (2A) and 2.3 ml. of tricresyl phosphate is stirred and heated at 130°C. for 4 hours to yield a soft grease.

Preparation of a Chlorinated Polyphenyl Grease — An amount of 2.3 ml. of a chlorinated polyphenyl (Aroclor 1254, Monsanto Chemical Company) and 0.2 g. of thickener (2A) are thoroughly mixed. The reation mixture is heated to 150°C. for 3 hours while stirring is maintained thereby yielding a firm, smooth grease.

Example 3: Preparation of Chromium Methylphenylphosphinatedihexylphosphinate Thickener — (3A) A solution of 8.0 g. (20.2 mmol.) $Cr(H_2O)(OH)[OP(Me)(Ph)O]_2$ polymer and 4.72 g. (20.2 mmol.) $(C_6H_{13})_2P(O)OH$ in 100 ml. tetrahydrofuran is heated to reflux for 2 hours. The solvent is then evaporated and the thickener product recovered.

Preparation of a Silicone Grease — 4.5 g. of thickener prepared as described in (3A) is placed into 45 ml. of silicone fluid (Dow Corning 550 fluid). The mixture is vigorously stirred and heated at 130° to 160°C. for 4 hours. The resultant grease is further heat treated at 210°C. for 45 minutes. The drop point of this grease is in excess of 260°C.

Diurea and Polyarylurea Compounds as Thickeners

I. Bauer, S. Nitzshe and R. Riedle; U.S. Patent 3,423,318; January 21, 1969; assigned to Wacker-Chemie GmbH, Germany describe the use of certain diurea compounds as thickening agents in polysiloxane oils to form greases exhibiting excellent lubricant and temperature stability properties. The preferred diurea is the one which is obtained by reacting two molecules of cyclohexylamine with one molecule of naphthylenediisocyanate. The following examples illustrate the process.

Example 1: 315 g. naphthylene-1,5-diisocyanate was dissolved in 3 liters of perchloroethylene at 70°C. A solution of 315 g. cyclohexylamine and 1 liter perchloroethylene was allowed to run into the solution slowly and with concurrent stirring. After the reaction product had been stirred for two hours at 70°C. the urea formed was filtered off employing a suction filter. The product was washed clean with more solvent and was dried at 100° to 120°C. The dry urea was pulverized and mixed with 3,100 g. of a trimethylsilyl-endblocked phenylmethylpolysiloxane having a viscosity of 400 cs. and a refractive index of 1.505 measured at 25°C. After processing several times on a three-roll mill, a homogeneous grease having the following properties was obtained.

Stationary penetration, 250 (according to DIN 51804).
Penetration after 60,000 up and down strokes, 275 (according to DIN 51804).
Weld load in a four-roll apparatus, 230/250 kg. (according to DIN 53654).
Dropping point, higher than 320°C.

If a compound is prepared under the same conditions using an equimolar quantity of naphthylamine in place of cyclohexylamine, a much less stable compound will be obtained with the following properties.

Stationary penetration, 220.
Penetration after 60,000 up and down strokes, 350.
Weld load in a four-roll apparatus, 120/140 kg.
Dropping point, higher than 320°C.

Example 2: 500 cc of acetone and 50 g. of cyclohexylamine (0.5 mol) were placed in a 2 liter flask equipped with stirrer, reflux cooler and dropping funnel. 62.5 g. (0.25 mol) of commercially pure diphenylmethane-4,4'-diisocyanate was added dropwise while stirring. This increases the temperature from 21° to 49°C. The principal portion of the acetone was distilled from the precipitated diurea and the slurried residue is mixed with 235 g. of a trimethylsilyl endblocked chlorophenylmethylpoly siloxane having a viscosity of 200 cs. and a refractive index of 1.455 measured at 25°C. The product was processed on a three-roll mill until it was homogeneous and the residual acetone was driven off in a dry chest at 150°C. A grease with the following properties was obtained.

Stationary penetration, 280.
Penetration increases after 60,000 up and down strokes, 28.
Dropping point, more than 320°C.
Bleeding after 24 hours at 150°C., 1.2%.
Maximum pressing pressures in the Almen-Wieland testing machine, 350 kg.

If the same compound is prepared according to the same method, with an equimolar quantity of aniline, a compound with the following properties will be obtained.

Stationary penetration, 270.
Penetration increases after 60,000 up and down strokes, 72.
Dropping point, 270°C.
Bleeding after 24 hours at 150°C., 4.1%.
Maximum pressing pressure in the Almen-Wieland testing machine, 250 kg.

J.F. Hedenburg and C.S. Tempalski; U.S. Patent 3,374,170; March 19, 1968; assigned to Gulf Research & Development Company describe a lubricating composition which comprises a dispersion in a synthetic lubricating oil of a sufficient amount to thicken the lubricating oil to a grease consistency of a heat treated polyaryl urea. The polyaryl urea which is heat treated prior to use in the composition is the reaction product of an aryl isocyanate having the formula $R(NCO)_n$ where R is selected from the group consisting of arylene (e.g., phenylene, biphenylene and naphthylene), diphenylmethane and C_1 to C_4 alkyl substituted arylene radicals containing not more than 12 cyclic carbon atoms and n is an integer of 2 to 3 and an aryl amine having the formula $R_1(NH_2)_m$ where R_1 is selected from the group consisting of aryl, alkaryl, aminoaryl, hydroxyaryl, cyanoaryl, haloaryl, alkoxyaryl, aryloxyaryl and aminoaryloxyaryl radicals where the alkyl portions of the alkaryl and alkoxyaryl radicals contain from 1 to 4 carbon atoms and m is an integer of 1 to 3.

The reaction product is prepared by reacting the aryl isocyanate and the aryl amine in amounts such that the ratio of the isocyanate groups to the amine groups is within the range of 1:1 to 1:2. An especially outstanding polyaryl urea for subsequent heat treatment is obtained by reacting 3,3'-dimethylbiphenylene-4,4'-diisocyanate with a mixture of p,p'-oxydianiline and p-chloroaniline in a mol ratio of 4:4:1. In order to stabilize the polyaryl urea for use in compositions, the polyaryl urea, prior to being incorporated in the oil, is heated to a temperature of 450° to 600°F. for a time sufficient to reduce its weight by 20 to 40%. The heat treated product thus obtained has a melting point above the polyaryl urea prior to being heat treated, now being above 400°C. The heat treated product is a dark brown powder. It has been found that a grease composition prepared from a heat treated polyaryl urea has excellent lubricating characteristics when used to lubricate a bearing operating at a temperature of 600°F. at a speed of 20,000 rpm.

Fluorinated Polysiloxane and Arylurea Thickener

H.M. Schiefer and J.S. Delphenich; U.S. Patent 3,377,284; April 9, 1968; assigned to Dow Corning Corporation describe thermally stable lubricant greases exhibiting superior high speed bearing life and organic solvent-resistance consisting of (a) liquid fluorinated polysiloxane and (b) arylurea thickener. An illustrative example is a mixture of

(a) 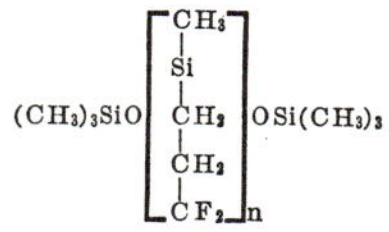 and (b) 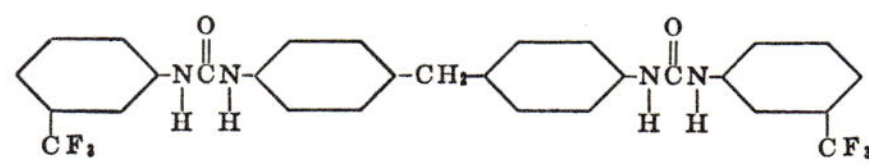

Example 1: (A) To 70 parts by weight of a polymer of the formula

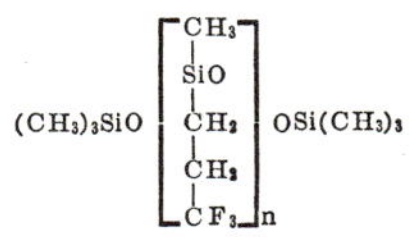

having a viscosity of 1,460 cs. at 25°C., there was added 258 parts of 1,1,1-trichloroethane (Dow Chlorothene), 12.9 parts of

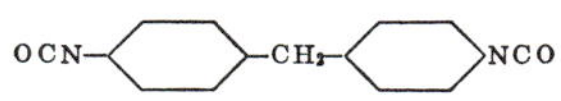

and 17.1 parts of meta-aminobenzotrifluoride

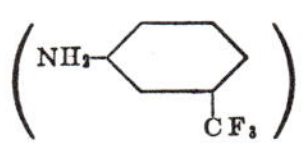

The mixture was heated at 125°C. and stirred until the reaction between the latter two ingredients was complete, forming

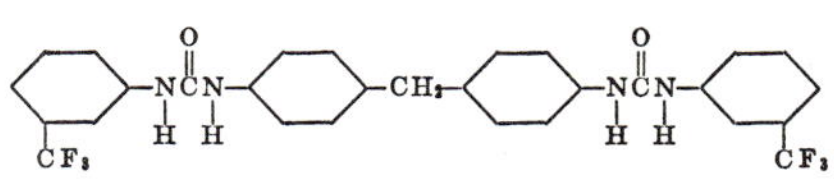

The 1,1,1-trichloroethane was removed by heating the mixture in a vacuum at 150°C. The resulting product was a grease having excellent physical stability, and showing little change in consistency after vigorous working. (B) A film of the grease was placed on an aluminum panel, and the panel was partly immersed in a mixture of 60 weight percent isooctane 5% benzene, 20% toluene, and 15% xylene. The grease film remained intact on the panel for over 8 hours. (C) The grease was used to lubricate a 204 size bearing which was operated continuously at 10,000 rpm, 400°F., and a 5 pound load. The bearing operated for 941 hours before failing. The above bearing test was repeated, using a 197 pound load. The bearing had not failed after 675 hours of operation.

Example 2: To 83.4 parts by weight of a polymer of the formula

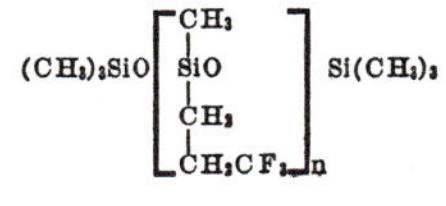

having a viscosity of 1,460 cs. at 25°C., there was added 80.2 parts of 1,1,1-trichloroethane, 8.02 parts of

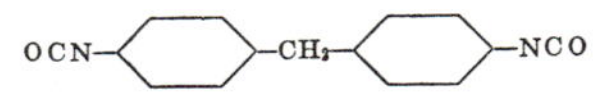

5.16 parts of meta-aminobenzotrifluoride, and 3.42 parts of paratoluidine. The mixture was heated at 125°C. and stirred until the reaction between the latter three ingredients was complete, yielding a compound of the average formula

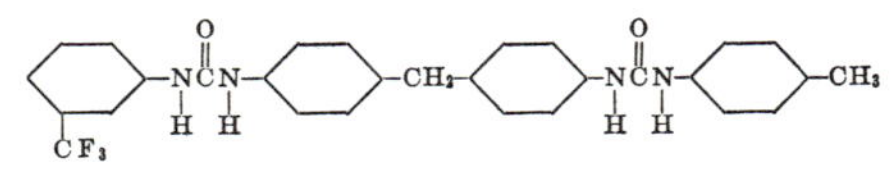

The 1,1,1-trichloroethane was then removed by heating the composition at 150°C. in a vacuum. The resulting product was a grease having excellent physical stability, and showing little change in consistency after vigorous working. A film of the grease was placed on a steel panel and partly immersed in the solvent mixture of Example 1. The grease film remained intact on the panel for over 8 hours. The grease was subjected to the bearing test of Example 1 at a 197 pound load. The bearing operated in excess of 1,082 hours.

Bis(Triphenylsilyl) Perfluorodicarboxylic Acid Ester

In a process described by F.K. Kawahara; U.S. Patent 3,347,794; October 17, 1967; assigned to Standard Oil Company grease compositions comprising bis(triphenylsilyl) perfluorodicarboxylic acid esters are provided for use at elevated temperatures. It has been found that a lubricant grease suitable for use in applications at elevated temperatures in the range of 600°F. and higher is formulated from conventional thickening agents, and a lubricant vehicle made from silicone polymer oils mixed with a bis(triphenylsilyl) perfluorodicarboxylic acid ester. Additionally, it is to be expected that such a grease composition will be effective at temperatures much higher than 600°F. in particular applications where pressure is reduced, such as, for example, in lightly loaded bearings used to control motors for spacecraft operating outside the earth's atmosphere under conditions of essential vacuum. Bis(triphenylsilyl) perfluorodicarboxylic acid esters of the process are characterized by the following structural formula

$$\phi_3\text{—Si—O—}\overset{\text{O}}{\overset{\|}{\text{C}}}\text{—}(\text{CF}_2)_n\text{—}\overset{\text{O}}{\overset{\|}{\text{C}}}\text{—O—Si—}\phi_3$$

where n is 1 to 8, and ϕ is phenyl. Of compounds of this type, bis(triphenylsilyl) perfluoroglutarate having the following structural formula

$$\phi_3\text{—Si—O—}\overset{\text{O}}{\overset{\|}{\text{C}}}\text{—}(\text{CF}_2)_3\text{—}\overset{\text{O}}{\overset{\|}{\text{C}}}\text{—O—Si—}\phi_3$$

is preferred. These compounds are chemically and thermally stable. For example, these compounds will typically melt at elevated temperatures in the range of 300° to 550°F. Specifically, the above described compound, bis(triphenylsilyl) perfluoroglutarate melts at from 446° to 453°F. This high melting point characteristic makes these compounds excellent for use in high temperature grease formulations both in small amounts as additives and in larger amounts together with silicone polymer oil as the lubricant vehicle. The bis(triphenylsilyl) perfluorodicarboxylic acid esters typically are prepared by reacting triphenylsilanol with a dicarboxylic acid chloride of the following general formula

$$\text{Cl—}\overset{\text{O}}{\overset{\|}{\text{C}}}\text{—}(\text{CF}_2)_n\text{—}\overset{\text{O}}{\overset{\|}{\text{C}}}\text{—Cl}$$

where n is 1 to 8, in the ratio of two mols of triphenylsilanol to one mol of dicarboxylic acid chloride. The reaction is carried out in solvents at room temperature and atmospheric pressure. Solvents are then removed by reducing the pressure and the residue remaining is worked with an alcohol. The following examples illustrate the process.

Example 1: A lubricant grease was prepared to contain 35% by weight ammeline and 65% by weight Dow Corning silicone polymer oil (QF-6-7024). This grease was used as a control. It was tested in a lightly loaded antifriction bearing operated at a temperature of 600°F. under a 50 pound radial and a 25 pound axial load at a rate of 10,000 rpm. The apparatus and test procedures used were in cooperation with the Coordinators Research Council Test L-35-59 technique. This grease was operated for 113 hours before failure.

Example 2: A grease composition of the process was prepared by adding 27.6 g. of triphenylsilanol in 58 g. of pyridine to 13.8 g. of perfluoroglutaroyl chloride in 20 ml. of benzene at room temperature. The solvent was removed by reducing pressure and the residue was worked with 170 ml. of absolute ethanol. 16 g. of colorless solid product melting at 446° to 453°F. was obtained. 3% by weight of this product was then added to the grease composition as prepared in Example 1 and the formulation was tested in accordance with the procedure described above. This grease was operated for 183 hours before failure for a gain of 70 hours in effective operating time.

Example 3: A grease composition was prepared as in Example 2 and tested in accordance with the L-35-59 technique. The grease was successfully operated for 207 hours before failure.

Example 4: A grease was prepared as described in Example 2 and tested in accordance with the L-35-59 technique. This grease was successfully operated for 172 hours before failure.

Example 5: A grease formulation was prepared to contain the following: 82.5% by weight of Dow Corning 510 silicone fluid; 9% by weight bitolylene diisocyanate; 3.7% by weight para-toluidine; and 4.5% by weight para-chloroaniline compound. This grease was also used as a control. It was tested in a lightly loaded antifriction bearing operated at a temperature of 450°F. under a 5 pound radial and a 25 pound axial load at a rate of 10,000 rpm in accordance with L-35-59 technique. The control grease was operated for 450 hours to failure.

Example 6: A grease as described in Example 5 was prepared to contain 3% bis(triphenylsilyl) perfluoroglutarate and was tested in the same fashion as described in Example 5. This grease was successfully operated for 790 hours without failure.

Fluorosilicone Grease Composition

J.B. Christian; U.S. Patent 3,563,894; February 16, 1971; assigned to the U.S. Secretary of the Air Force describes a grease formulation capable of maintaining its dimensional stability under heavy load and high temperature conditions of (1) a major proportion of trifluoropropylmethylsiloxane having a viscosity of 300 cs. and consisting essentially of $CF_3CH_3CH_2Si(CH_3)O$ units and endblocking units of the formula $(CH_3)_2SiO_{0.5}$; (2) a minor proportion by weight of a composition consisting of (a) 50 to 55% by weight of bitolylene diisocyanate, (b) 20 to 30% by weight of p-chloraniline, and (c) 20 to 25% by weight of p-toluidine; and (3) 10% by weight of molybdenum disulfide. In one specific formulation, the following recipe is used.

	Parts by Weight
Dow Corning FS 1265 fluorosilicone (300 cs.)	72
Bitolylene diisocyanate	9.54
p-Chloroaniline	4.68
p-Toluidine	3.78
Molybdenum disulfide (technical grade)	10

In the preparation of the above recipe, all of the bitolylene diisocyanate and two-thirds of the fluorosilicone or organopolysiloxane oil are placed in a mixer and heated to from 250° to 300°F. while stirring. All of the p-chloroaniline and one-sixth of the fluorosilicone are placed in a second container and heated to from 250° to 300°F. while stirring; and all of the p-toluidine and the remaining one-sixth of the oil are placed in a third mixer and heated to the same 250° to 300°F. while stirring.

After the solute contents of the three separate mixtures have been completely dissolved in the oil, the chloroaniline and toluidine mixtures are rapidly and simultaneously added to the bitolylene diisocyanate solution while it is being stirred. All of the molybdenum disulfide is then added to mixture and mixed with a spatula. After the mixture is cooled to from 200° to 250°F. the grease is milled on a three roll mill for from five to ten passes. As a demonstration of the improvements in both load carrying characteristics and lubricity of the formulation according to the above recipe and processing procedure, the results shown in Table 1 on the following page are obtained. In addition, the excellent properties and performance of the grease composition (A) of the process compared to a known grease formulation (B) which is MIL-G-27459, is evident from the tabulation shown in Table 2. The grease of Military Specification MIL-G-27549 consists of:

Dow Corning DC-4039 silicone fluid	75.2%
Arylurea thickener	18.8%
Pentachlorophenyl mercapto acetic acid	6.0%

TABLE 1

Test	Method	Result
Unworked penetration	ASTM D217-60T	330.
Worked penetration (60 strokes)	ASTM D217-60T	350.
Dropping point	ASTM D566-42	460° F.
Oil separation at 450° F	Fed test method std. 791 Method 321	10.3%.
Evaporation, 22 hrs. at 450° F	ASTM D972-56	10.2%.
Copper strip corrosion, 72 hrs. 212° F	Fed test method Std. 791 Method 5309	No staining.
Water resistance	ASTM D1264-53T	7.58%.
Rust preventive properties	ASTM D1743-60T	No staining.
Mean hertz value	Fed test method Std. 791 Method 6503	58.9.
Oscillation bearing test, 425° F., 4,000 lbs	Military specification MIL-G-27549	170,886 cycles.
Low temperature torque, −60° F	ASTM D1478 (starting)	8,000 g. cm.
	(running)	2,000 g. cm.
Low temperature torque, −40° F	ASTM D1478 (starting)	3,600 g. cm.
	(running)	332 g. cm.
Shell 4-ball wear test, 2 hrs. 167° F., 1,200 r.p.m., 40 kg., 52-100 balls	Millitary specification MIL-G-25760	0.6 mm. scar.
Shell 4-ball wear test, 2 hrs., 400° F., 600 r.p.m., 40 kg., M-10 balls	do	0.6 mm. scar.
Shell 4-ball wear test, 2 hrs., 400° F., 1,200 r.p.m., 40 kg., M-10 balls	do	0.7 mm. scar.
Shell 4-ball wear test, 2 hrs., 400° F., 600 r.p.m., 50 kg., M-10 balls	do	0.6 mm. scar.
Shell 4-ball wear test, 2 hrs., 400° F., 1,200 r.p.m., 50 kg., M-10 balls	do	0.8 mm. scar.
Worked stability (100,000 strokes)	Fed Test Method Std. 791 Method 313.1	358.

TABLE 2

Test	Method	Results	
		B	A
(1) Unworked penetration	ASTM-D 217	290	330.
(2) Worked penetration	ASTM-D 217	301	350.
(3) Dropping point	ASTM-D 566-42	473° F	460° F.
(4) Mean hertz value	Fed. test method Std. 791, Method 6503	60	58.9.
(5) Oscillation bearing test, 425° F., 4,000 lbs	Military specification MIL-G-27549	68, 952 cycles	170,886 cycles.
(6) Shell 4-ball test, 2 hrs., 10 kg., 1,200 r.p.m., 167° F. (wear scar diameter).	Military specification	1.24 mm	0.40 mm.
(7) Shell 4-ball test, 2 hrs., 40 kg., 1,200 r.p.m., 400° F (wear scar diameter).	Military specification	3.478 mm	0.70 mm.

Siloxanylferrocenes as Stabilizers

P.R. McCarthy and C.S. Tempalski; U.S. Patent 3,306,851; February 28, 1967; assigned to Gulf Research & Development Company have found that the tendency of a synthetic oil to volatilize and to solidify when heated at 600°F. in the presence of air can be decreased by incorporating in the oil a small amount of siloxanylferrocene. It has also been found that when a siloxanylferrocene is incorporated in a polyorganosiloxane or a polyaryl ether together with an oil thickening proportion of a 1,3,5-triazine compound melting above about 600°F. (315°C.), a lubricating composition is obtained having an improved performance life. Specific examples of the symmetrical siloxanylferrocenes are shown below.

1,1'-bis(3-phenyltetramethyldisiloxanyl)ferrocene
1,1'-bis(5-phenylhexamethyltrisiloxanyl)ferrocene
1,1'-bis(7-phenyloctamethyltetrasiloxanyl)ferrocene
1,1'-bis(7,7,7-triphenyl-1,1,3,3,5,5-hexamethyltetrasiloxanyl)ferrocene
1,1'-bis(heptamethyltrisiloxanyl)ferrocene
1,1'-bis(3-phenyltetramethyldisiloxanyl)dimethylferrocene
1,1'-bis(5-phenylhexamethyltrisiloxanyl)dimethylferrocene

Specific examples of the unsymmetrical siloxanylferrocenes are

tris[(1'-phenyldimethylsilylferrocenyl)dimethylsiloxyl]-methylsilane
1,1'-bis[3-(1'-phenyldimethylsilylferrocenyl)-1,1,3,3-tetramethyldisiloxanyl]-ferrocene
1-trimethylsilyl-1'-(3-phenyl-1,1,3,3-tetramethyldisiloxanyl)-ferrocene, and
1,5-bis(1'-trimethylsilylferrocenyl)-1,1,3,3,5,5-hexamethyltrisiloxane.

The symmetrical siloxanylferrocenes can be prepared by the method described in U.S. Patent 3,010,982. In order to illustrate the characteristics of lubricating compositions, various synthetic lubricating oils were compared with the same base oils containing a siloxanylferrocene. The test used in making the comparisons comprises placing ten grams of the test lubricant into 50 ml. beakers cut to a height of 1 5/8 inches. The beakers are covered with stainless steel lids containing one 9/64 inch hole. The beakers are placed in an oven where they are heated at a temperature of 600°F. in the presence of air. The beakers are withdrawn periodically and weighed to determine the extent of evaporation. The mobility of the fluids at 600°F. and after cooling to 80°F. is also observed. In the tests summarized in the table shown below, the lubricating oils employed were synthetic lubricating oils known as DC-550 Fluid and QF-6-7024 Fluid. DC-550 Fluid is considered to be a methylphenylsiloxane polymer wherein the end silicon atoms are substituted to a high degree by methyl groups. QF-6-7024 Fluid is considered to be a methylphenylsiloxane polymer where the end silicon atoms are substituted to a high degree by two phenyl groups and one methyl group.

Composition, Percent By Weight	A	B	C	D	E	F	G	H
Lubricating Oil:								
DC-550 Fluid	100	99.9						
QF-6-7024 Fluid			100	99.99	99.9	99.7	99.5	
Siloxanylferrocene, 1,1'-bis(5-phenylhexamethyltrisiloxanyl)ferrocene		0.1		0.01	0.1	0.3	0.5	100
Inspection:								
Evaporation, percent after the following number of hours at 600° F.:								
48	[c, d]18.5	[a]17.2	[a]7.7	[a]6.5	[a]6.6	[a]7.1	[a]7.7	[e]31
96		[a]28.3	[a]13.1	[a]13.9	[a]13.3	[a]14.0	[a]14.4	
120		[b, d]33.7						
144		[c]37.4	[b, d]20.6	[a]19.5	[a]20.0	[a]19.7	[a]20.0	[f]31
168			[b, d]24.0	[a]22.6	[a]23.4	[a]22.2	[a]23.4	
192			[c]26.2	[b, d]25.2	[a]26.3	[a]24.4	[a]26.8	
216				[b, d]27.6	[a]29.2	[a]27.0	[b, d]30.1	
240				[b, d]29.3	[a]31.3	[a]29.1	[c]31.8	
264				[c]30.8	[a]34.4	[a]31.9		
288					[b, d]37.7	[b, d]35.0		
336					[c]41.5	[c]38.8		

[a] Fluid at 80° F.
[b] Fluid at 600° F.
[c] Solid at 600° F.
[d] Solid at 89° F.
[e] Solid at 600° F. in 24 hours.
[f] Solid at 400° F. after heating at 400° F. for 144 hours.

The improved characteristics of synthetic lubricating compositions containing a siloxanylferrocene is evidenced by the above data showing that the siloxanylferrocene gives an increase in the time over which the lubricating compositions remains fluid at 80°F. and at 600°F. The improved stability obtained by adding the siloxanylferrocene to the organosiloxanes was surprising inasmuch as the siloxanylferrocene alone when heated to 600°F. in the presence of air was solid within 24 hours with a 31% loss in weight. Even when heated to only 400°F. in the presence of air, the siloxanylferrocene solidified within 144 hours with a 31% loss in weight.

The lubricating oils used in preparing the lubricating compositions shown below were m-bis(m-phenoxyphenoxy)phenyl ether and QF-6-7024 Fluid. In preparing the illustrative lubricating compositions, the oil, the ammeline and the siloxanylferrocene were mixed at room temperature for a period of 10 to 30 minutes. The slurry thus formed was passed twice through a Premier Colloid Mill set at a stator-rotor clearance of 0.002 inch. The approximate makeup and properties of the thickened lubricating compositions thus prepared are set forth below.

Composition, percent by weight	A	B	C	D
Synthetic Lubricating Oil:				
QF-6-7024 Fluid	65	64		
m-Bis(m-phenoxyphenoxy)phenyl ether			70	69.65
Ammeline	35	35	30	30
1,1'-bis(5-phenylhexamethyltrisiloxanyl)ferrocene		1		0.35
Inspection, Dropping Point, ° F. (ASTM D566-42)	700+	700+	800+	850+
Penetration (ASTM D217-52T):				
Unworked	291	300	324	257
Worked	283	310	302	242
Performance Life, hrs. 600° F., 20,000 r.p.m.	[1] 104-113	177	66	[1] 146-214

[1] Results of two tests.

The long performance life of the compositions of the process (Compositions B and D) at a high rotational speed and a high temperature is self-evident from the above data. By comparing Composition B and D with Compositions A and C, respectively, it will be noted that the addition of 1,1'-bis(5-phenylhexamethyltrisiloxanyl)ferrocene improves the performance life of the composition.

Alkyl Succinic Partial Esters for Corrosion Inhibition

F.A. Buehler and J.J. Giammaria; U.S. Patent 3,291,736; December 13, 1966; assignee to Mobil Oil Corporation describe improved grease compositions exhibiting the properties of rust inhibition and improved structure, which contain small amounts of a partial ester having the general formula

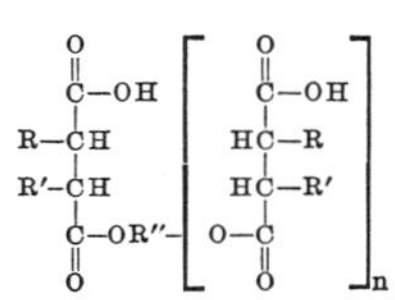

where of R and R' one is hydrogen, and the other is selected from the group consisting of alkyl and alkenyl radicals containing from 8 to 18 carbon atoms and preferably from 9 to 15 carbon atoms, R" is a residue of a polyhydric alcohol and preferably a residue of a polyhydric alcohol containing from 2 to 6 carbon atoms, and n is an integer from 1 to 4.

These corrosion inhibitors, as represented by the above indicated general formula, are partial esters of an alkyl or alkenylsuccinic anhydride and, preferably, are the products obtained by the reaction of one molar equivalent of a polyhydric alcohol with two molar equivalents of the anhydride. The residue of the polyhydric alcohol, as represented by R" may, generically, contain oxygen, sulfur or nitrogen in the chain as well as other substituents. The integer n represents a whole number from 1 to 4 to indicate that within the scope of the inhibitor compound are included the derivatives of diols, triols and other polyhydroxy compounds.

Illustrative of a typical procedure for the preparation of the inhibitor compounds employed in the grease formulations, is the procedure in which the desired substituted succinic anhydride and alcohol charges are weighed into a suitable jacketed acid-resisting vessel equipped with agitation, temperature recording means and a reflux condenser. This charge is heated at a temperature from 95° to 100°C., while agitating, and held at this temperature, while the progress of the reaction is observed periodically by withdrawing a sample and determining the neutralization number thereof. Heating is continued until the desired neutralization number is obtained, or until it remains substantially constant. The resulting reaction is represented as follows.

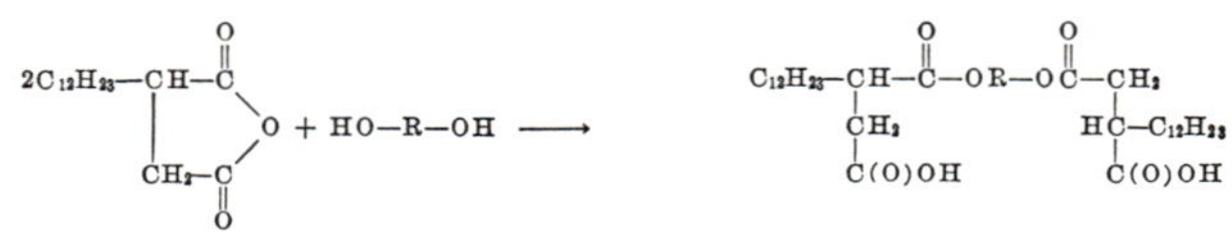

A more detailed description of the preparation and properties of the above described partial esters comprising the corrosion inhibitors present in the grease formulations are described in U.S. Patent 3,117,091 issued January 7, 1964. A particularly outstanding application of the use of the partial esters, is their incorporation in silicone-containing greases as corrosion inhibitors. The following examples illustrate the improved grease compositions of the process and their properties, as evidenced by respective bearing protection and 60 stroke worked penetration tests for the grease formulations described therein. The bearing protection against rust tests employed is identified as D1743-60T, in ASTM Standards on Petroleum Products, 1961. The penetration test employed is identified as D217-52, in ASTM Standards of Petroleum Products, 1961.

Example 1: A lubricating grease composition was prepared by mixing 235 parts by weight of p-methoxyphenyl perylimid pigment, 60 parts by weight of denatured ethanol and 765 parts by weight of a polymethylphenyl siloxane. This polysiloxane had a molar ratio of methyl to phenyl groups of 20:1, molecular weight of approximately 4,000 (as determined by vapor osmometry) and the following kinematic viscosity.

Temperature, °F.	Centistokes
210	30
100	80
-65	2,000

The components of the mixture were paddled in a grease kettle, with the temperature being raised to 230°F. in three hours. The resulting grease was then homogenized in a colloid mill through a clearance of 0.002 inch. This resulting grease composition was found to have a soft appearance, and a 60 stroke worked penetration according to the aforementioned ASTM method D217-52, of 350. The grease was found to fail the bearing protection test against rusting, in accordance with ASTM D1743-60T.

Example 2: A second lubricating grease composition was prepared in a manner similar to the procedure employed in Example 1, except that 783.5 parts by weight of the polysiloxane of Example 1 were employed and 26.5 parts by weight of a partial ester inhibitor, were added to the mixture. This inhibitor comprised the reaction product of two mols of dodecenylsuccinic anhydride and one mol of 2-methyl-2,4-pentane diol. This inhibitor was added to the polysiloxane as a solution in 60 parts by weight of denatured ethanol; also, 190 parts by weight of the perylimid pigment thickening agent of Example 1 were employed.

After processing the resulting grease formulation in the same manner as described in Example 1, a final grease formulation was obtained, which presented a smooth buttery appearance, and having a 60 stroke worked penetration, according to the aforementioned ASTM D217-52 penetration test, of 285, and wheel bearing rust protection sufficient to pass the ASTM bearing protection test D1743-60T.

SOLID LUBRICANTS

Perfluoroalkylsiloxane-Containing Lubricant

J.C. Saylor; U.S. Patent 3,503,926; March 31, 1970; assigned to Dow Corning Corporation has found that copolymers of certain fluoroalkylsiloxane and diphenylsiloxane units in certain proportions are solids. These solid copolymers are excellent lubricants and can be molded into various articles or utilized as coating compositions. According to the process there is provided a solid organopolysiloxane copolymer consisting essentially of (1) from 10 to 70 mol percent of siloxane units of the formula

$$RCH_2CH_2\overset{\displaystyle R'}{\overset{|}{Si}}O$$

where R is a perfluoroalkyl radical of from 1 to 10 carbon atoms and R' is a monovalent hydrocarbon radical, and (2) from 30 to 90 mol percent of diphenylsiloxane units. The copolymers can also include small amounts, up to 10 mol percent, of siloxane units containing cross-linking groups, such as, methylvinylsiloxane, diphenylvinylsiloxane, methylphenylvinylsiloxane and the like. Copolymers of perfluoroalkylsiloxanes and diphenylsiloxane which contain less than 30 mol percent diphenylsiloxane units are not solid, for example, the copolymer of 75 mol percent trifluoropropylmethylsiloxane and 25 mol percent diphenylsiloxane units is a gum. The above described copolymers can be prepared by the alkaline polymerization of cyclic organosiloxanes of the formula

$$(RCH_2CH_2\overset{\displaystyle R'}{\overset{|}{Si}}O)_3$$

with $(Ph_2SiO)_3$ in the desired proportions or by cohydrolyzing the corresponding chlorosilanes followed by equilibration of the cohydrolyzate with alkali metal hydroxide catalysts or catalysts of the alkali metal salts of silanols. The preparation of cyclic trisiloxanes is described in detail in U.S. Patent 2,979,519. The cyclic trisiloxanes are conventionally polymerized by heating in the presence of an alkaline catalyst such as alkali metal hydroxide, quaternary ammonium hydroxide salts and siloxane salts of such hydroxides. Generally, temperatures in the range of 190° to 225°C. are utilized. The following examples are illustrative of the process.

Example 1: A hydroxy-endblocked copolymer of 50.5 mol percent Ph_2SiO units and 49.5 mol percent $CF_3CH_2CH_2(Me)SiO$ units was prepared by mixing $(Ph_2SiO)_3$ and $[CF_3CH_2CH_2(Me)SiO]_3$ vigorously during the addition of catalyst. During catalyst addition the mixture was maintained at 205°C. Me_3SiONa was used as the catalyst and was added in an amount equal to one sodium atom per 4,895 silicon atoms. The mixture was maintained at 205°C. for 11 minutes. Upon cooling a strong white copolymer was obtained. The copolymer was soluble in hot toluene. Upon evaporation of the toluene solution, a white opaque film of considerable strength was obtained.

Example 2: A copolymer of 60 mol percent Ph_2SiO units and 40 mol percent $CF_3CH_2CH_2(Me)SiO$ units was prepared by the method of Example 1. The Me_3SiONa catalyst was added in amount equal to one sodium atom per 5,000 silicon atoms, polymerization temperature was 200°C. After a few minutes the polymerization was complete.

Example 3: (1) A hydroxy-endblocked copolymer of 39.8 mol percent Ph_2SiO units and 60.2 mol percent $CF_3CH_2CH_2(Me)SiO$ units was prepared by heating and mixing the respective cyclic trimers while adding n-butyl lithium catalyst in hexane. The catalyst was added in an amount sufficient to provide one lithium atom per 2,900 silicon atoms. Polymerization was complete after 12 minutes at 225°C.

(2) A copolymer of 49.4 mol percent Ph_2SiO units and 50.6 mol percent and $CF_3CH_2CH_2(Me)SiO$ units was prepared in the same manner as copolymer (1) utilizing the same amount of n-butyllithium catalyst. The polymerization was complete after 12 minutes at 225°C.

(3) A copolymer of 69.7 mol percent Ph_2SiO units and 30.3 mol percent $CF_3CH_2CH_2(Me)SiO$ units was prepared in the same manner as above with the same amount of catalyst. The polymerization was continued for 20 minutes at 210°C.

(4) A copolymer of 54.3 mol percent Ph_2SiO units and 45.7 mol percent $CF_3CH_2CH_2(Me)SiO$ units was prepared as described above by polymerization for 16 minutes at 210°C.

To determine their suitability as solid lubricants, copolymers of the process were dissolved in hot toluene and coated onto a 0.751 ± 0.0001 inch diameter, 2 inch long steel plug. Evaporation of the toluene left a coating of copolymer a few microns thick on the plug. The coated plug was then forced into an annular sleeve having an internal diameter of 0.750±0.0001. The percent of entry of the plug length into the sleeve at the various loadings was measured. The coefficient of friction was also determined. Of course, the greater the entry and the smaller the load to obtain entry the better the solid lubricant. This test is described in detail in ASTM-TCL-1-2. For purposes of comparison, graphite, a known solid lubricant, was also tested. The results are set forth below.

Solid lubricant	Percent entry	Load (lbs.)	Coefficient of friction
Copolymer (1)	100	2,500	.063
Copolymer (2)	100	5,750	.128
Copolymer (3)	100	10,750	.236
Copolymer (4)	100	7,350	.159
Graphite	32	15,000	Fail

Polysiloxanes Containing Fluorinated Alkyl and Phenyl Groups

A process described by C.-L. Lee; U.S. Patent 3,575,921; April 20, 1971; assigned to Dow Corning Corporation provides a linear siloxane polymer consisting essentially of units of the formula

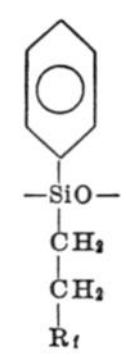

in which R_f is a perfluoroalkyl radical of from 1 to 10 carbon atoms inclusive. The R_f substituent in the above described homopolymer can be $-CF_3$, $-C_2F_5$, $-C_3F_7$ and so on, through to the C_9F_{19} and $C_{10}F_{21}$ radicals. The perfluoroalkyl groups can be straight or branched-chain structures. These homopolymers are characterized as having a very narrow molecular weight distribution. They can be fluids, gums or solids depending upon the degree of polymerization. Those polymers containing at least 150 of the above defined siloxane units are semicrystalline solids and in certain instances, as shown in the examples, can have stereo specific structure. The polymer is readily soluble in aromatic solvents such as benzene, toluene and xylene.

The siloxane polymers are prepared by polymerization of the corresponding cyclic trimer in the presence of lithium catalyst, such as n-butyl lithium or other alkaline catalysts, for example an alkali metal hydroxide, quaternary ammonium salts of such hydroxides or the alkali metal salt of a siloxane. The polymerization proceeds at temperatures of 100°C. and above. The following examples are illustrative of the process.

Example 1: 400 g. of cyclic $[(CF_3CH_2CH_2)(C_6H_5)SiO]_3$ were introduced into a one liter flask which had been purged with dry nitrogen. The trimer was a mixture of the

cis-isomer, 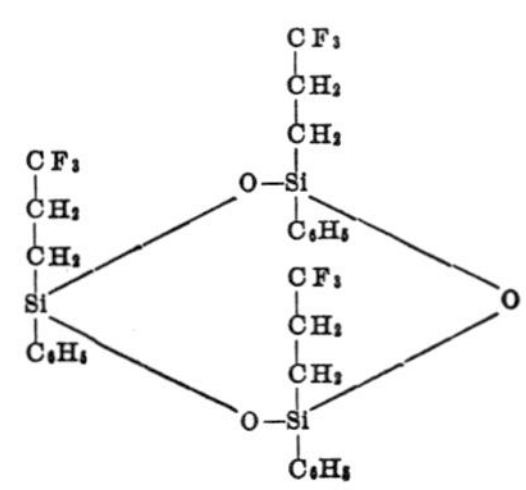 and the trans-isomer, 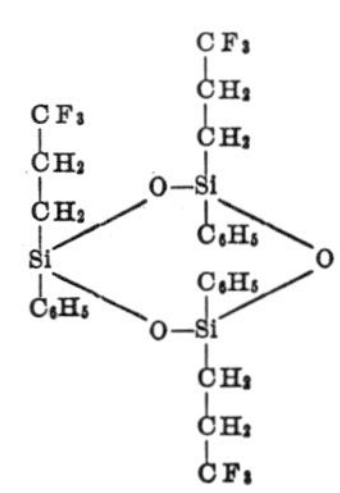

The trans-isomer content of the mixture was about twice that of cis-isomer content. 4 ml. of n-butyl lithium in heptane (1.1 mol of n-butyl lithium/liter) was added to the trimer. The catalyst-containing trimer was then heated to a temperature of 150°C. and stirred vigorously. After four minutes, polymerization was complete. The product was a solid polymer of the general formula

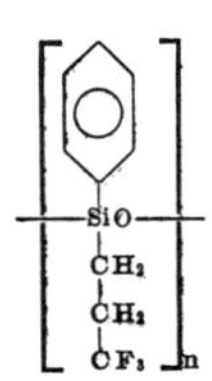

and had a number average molecular weight (M_n) of 100,000 as determined by gel permeation chromatography.

Example 2: 109 g. of the cyclic trimer utilized in Example 1 were heated to 150°C. and polymerized by the addition of 1 ml. of the above described n-butyl lithium solution. After 2 minutes the reaction was complete, yielding a solid polymer of the following units:

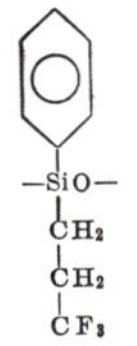

The product was dissolved in toluene and then precipitated with acetic acid to yield

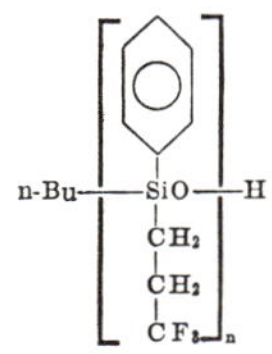

n being approximately 270. The precipitated polymer was vacuum dried at 60°C. to remove the solvent. The number average molecular weight (M_n) was 60,000, while the weight average molecular weight (M_w) was determined to 70,000. Thus the heterogeneity ratio is 1.18, a very narrow molecular weight distribution in the polymer product. A sample of the product was tested to determine its thermal stability by maintaining the material at 450°F. and measuring weight loss at specific time intervals. The results are as follows.

Time Maintained at 450°F., days	Percent Weight Loss
3	0.39
7	0.66
14	0.92

These results demonstrate the superior thermal stability of the polymers.

Example 3: 20 g. of the cyclic trimer utilized in Example 1 were heated to 150°C. and polymerized by the addition of 0.75 ml. of the previously described n-butyl lithium solution. The trimer polymerized to a high viscosity fluid after 2 minutes. The number average (M_n) molecular weight was determined to be 18,800; the weight average molecular weight M_w was 22,200. The heterogeneity ratio, M_w/M_n, of 1.18 shows that this polymer also has a very narrow molecular weight distribution.

Example 4: To demonstrate the effect of water addition to the polymerization, 50 g. of the cyclic trimer utilized in Example 1 were polymerized by the addition of 1 ml. of the previously described n-butyl lithium solution and 0.1 ml. of distilled water. This polymerization was complete in 2 minutes to yield an oily fluid polymer of the formula

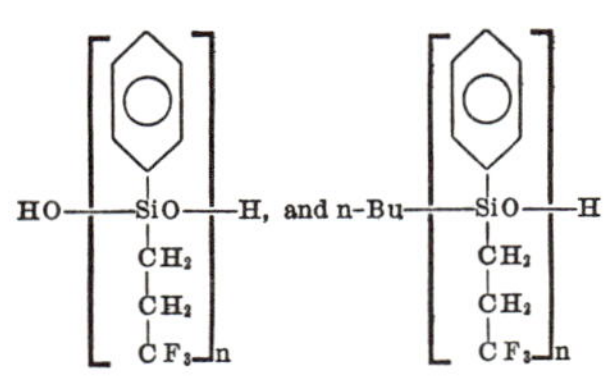

where n was equal to 32. The product had a number average molecular weight of 7,000 and a weight average molecular weight of 9,800. The M_w/M_n ratio was 1.40 thus the addition of water not only lowers the molecular weight, but also broadens the molecular weight distribution of the polymer. The M_w/M_n ratio for known polytrifluoropropylmethylsiloxane is approximately 4. As shown by the above examples, the homopolymers of the process generally have a heterogeneity ratio of less than 2, even when water is added during polymerization.

Example 5: To demonstrate the suitability of the linear homopolymers as solid lubricants, the polymers prepared in

Examples 1 through 4 were each dissolved in toluene and coated onto a 0.751 ± 0.0001 inch diameter, 2 inch long steel plug. Evaporation of the toluene left a coating of the polymer on the steel plug. This coating was a few microns thick. The coated plug was then forced into an annular sleeve which had an internal diameter of 0.750 ± 0.0001 inch. The percent of entry of the plug length into the sleeve was measured at various loadings. Of course, the greater the entry and the smaller the load to obtain entry, the better the solid lubricant. The coefficient of friction was also measured for the polymer of Example 1. This test is described in detail in ASTM-TCL-1-2. For purposes of comparison, graphite, a known solid lubricant, was also tested in the manner described above. The results are shown below.

Solid lubricant	Percent entry	Load (lbs.)	Coefficient of friction
Polymer of Example 1	100	5,750	0.12
Polymer of Example 2	100	5,700	
Polymer of Example 3	100	9,100	
Polymer of Example 4	100	9,300	
Graphite	32	15,000	Fail

These results demonstrate that the polymers of the process are highly effective solid lubricants. The material can be used to impregnate porous bearings, deposited as a dry film on bearing surfaces or it can be bonded to other materials to form a bearing composite.

Silicone-Epoxy Dry Film Lubricant

A process described by R.F. Hollitz; U.S. Patent 3,288,710; November 29, 1966; assigned to Dow Corning Silicones Limited, Canada involves dry film lubricants for metal that have the capability of withstanding heavy loads without rupture of the film, which causes loss of lubrication and final disruption of the film. The compositions of this process are described in the following manner. A lubricating composition consisting essentially of (1) from 40 to 85 weight percent of a dry lubricant, and (2) from 15 to 60 weight percent of a resinous composition consisting of an epoxy resin which is a condensation product (a) from 40 to 100% by weight of an aromatic epoxy resin; and, (b) from 0 to 60% by weight of an organosilicon compound of the formula

$$R_nSi(R'_m)O_{\frac{4-m-n}{2}}$$

where R is selected from the group consisting of monovalent aromatic hydrocarbon radicals and lower alkyl radicals, R' is a hydrolyzable radical, n has an average value of 0.8 to 3, and m has an average value of 0.1 to 3.2, the sum of n+m being not greater than 4. The composition of this process is operative as a dry film lubricant without any silicone ingredient, but it is generally preferred for ingredient (2) to contain from 5 to 40% by weight of silicone. The silicone appears to enhance the load bearing ability and the heat stability of the film.

Example 1: A composition was prepared consisting of 70% by weight MoS_2, 7% graphite, and 23% of a composition consisting of the following: the condensation product of (a) 15 parts of the hydrolyzate of 70 mol percent monophenyltrichlorosilane and 30 mol percent monopropyltrichlorosilane, hydrolyzed in one-half of the molar weight of the silanes of methyl isobutyl Cellosolve and excess toluene; and (b) 45 parts of

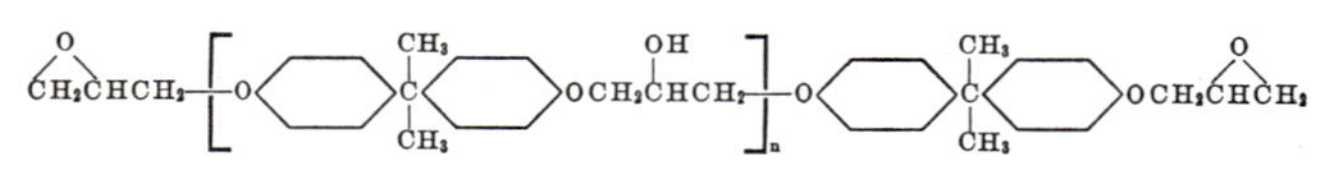

where n equals 2, mixed with 20 parts of xylene, 20 parts of beta-ethoxyethyl acetate, and 1 part of 2-ethylhexylic acid. This composition was milled in a colloid mill, and diluted with toluene until it was sprayable. This composition was sprayed on two vapor-blasted steel V blocks and a pin that fits into the V block, and cured for one hour at 300°F. The pins and blocks used for the tests were one-quarter inch diameter and liquid honed using 50% No. 625 and 50% No. 1250 abrasives, to provide consistent surface finishes with maximum surface area.

Prior to spraying the dry films, pins and blocks were washed in acetone and were not touched with the hands until the curinging cycles and testing were completed. The pins are of SAE 3135 steel, with a Rockwell B hardness of 87-90, and the V blocks of A151-1137 steel with a Rockwell C hardness of 20 and a block angle of 96°. The frictional characteristics of the pin rotating in the blocks were determined by means of a Falex lubricant testing machine, using a step test. By this technique the load on the pin and block is increased stepwise at periodic intervals, the torque of the rotating pin being constantly measured. The results are shown on the following page.

Total Load (lb.)	Torque at beginning of period (lb./in.)	Torque at end of period (lb./in.)	Duration of period of load (min.)
250	5	1	3
500	4.5	4.5	1
750	7	7	1
1,000	8	8	1
1,250	10	9	1
1,500	11	10.5	1
1,750	11.5	11.5	1
2,000	12	12	1
2,250	12	12	1
2,500	12	12	1
2,750	12	12.5	1
3,000	13	13.5	1
3,250	13	13	1
3,500	13.5	13.5	1
3,750	13.5	14.0	1
4,000	14	14	1
4,250	14.5	14.5	1
4,500	15	30	(1)

[1] Failure after 55 seconds.

Example 2: The composition of Example 1 was reformulated with the graphite ingredient being replaced with 7% of calcium fluoride. The composition was run through a colloid mill, and to it was added enough of a mixture of 50% ethylene glycol and 50% xylene to make the composition sprayable. The composition was sprayed on a block and pin set, cured, and step tested in the manner of Example 1. The results are shown below.

Total Load (lb.)	Torque at beginning of period (lb./in.)	Torque at end of period (lb./in.)	Duration of period of load (min.)
250	3	1.5	3
500	4	4	1
750	5.5	5.5	1
1,000	7.5	7.5	1
1,250	8.5	8	1
1,500	9	9	1
1,750	10.5	9.5	1
2,000	11	10	1
2,250	11.5	11	1
2,500	12	11	1
2,750	11	11	1
3,000	12	11	1
3,250	12	11.5	1
3,500	12.5	11.5	1
3,750	12	12	1
4,000	12.5	13	1
4,250	13	13.5	1
4,500	15.5	17.5	[1] 2

[1] Not maintaining load.

Perarylated Silanes for Solid Lubricant Powders

A.M. Adair and L. Spialter; U.S. Patent 3,575,858; April 20, 1971; assigned to the U.S. Secretary of the Air Force describe high temperature lubricants which comprise homogeneous mixtures of 10 to 90% by weight of perarylated silanes with 90 to 10% by weight of solid MoS_2, graphite, BN, aluminum powder, or low melting glass. The lubricants all have fire points of greater than 800°F. and do not burn or decompose during use at temperatures up to 800°F. The lubricants, since they do not decompose at temperatures up to 800°F., may be used under continued service conditions, which is not the case with the best of the prior art materials. The lubricants may be prepared by a number of different methods.

(1) A perarylated silane and a solid lubricant powder may be ground together.
(2) A perarylated silane may be melted to a liquid and solid lubricant powder stirred in.
(3) Molten perarylated silane may be added to and stirred into solid lubricant powder.
(4) Perarylated silane may be dissolved in a solvent which is more volatile than the silane. Then solid lubricant powder may be added to the solution and the solvent evaporated.

Specific examples of perarylated silanes which may be used are: phenyl-tris-(4-biphenylyl)silane, hexaphenyldisilane, tetrakis-(4-biphenylyl)silane, tetra-2-thienylsilane, phenyltri-2-thiensylsilane, and 3-pyridyltriphenylsilane. The term "perarylated silanes", as used in this specification, may be defined as that class of molecules constructed from silicon atoms and aromatic or substituted aromatic moieties. The aromatic moieties may be selected from carbocyclic or heterocyclic groups either alone or mixed. Specific examples are, of course, given above. The following are examples of lubricant preparations according to the process.

Example 1: Powdered graphite was blended with phenyl-tris-(4-biphenylyl)silane. For this purpose, a measured amount of the silane was heated in a container to a temperature in the range from 320° to 500°F. Phenyl-tris-(4-biphenylyl)-silane melts at 320°F. and is a low viscosity liquid. The temperature of the molten silane was maintained and a measured amount of powdered graphite, 150 to 325 mesh, stirred in. Complete blending resulted with ease. The consistency of mixtures prepared in this manner varied from stiff putty-like to thick soup-like depending on the ratio of graphite to silane. The mixtures solidified to glass-like texture upon cooling. They softened to their original consistencies when reheated. Graphite did not tend to settle out except when high weight percentages of silane were mixed with coarser graphite powder. The mixtures could be cooled and heated many times with no apparent change in characteristics. Compositions prepared in this way were 40 weight percent phenyl-tris-(4-biphenylyl)silane to 60 weight percent graphite, 50 weight percent of the silane to 50 weight percent graphite, and 60 weight percent of the silane to 40 weight percent of the graphite.

Example 2: Technical fine molybdenum disulfide powder was mixed with phenyl-tris-(4-biphenylyl)silane. The mixing procedure of Example 1 was used. Consistencies and characteristics of the silane-MoS_2 mixtures were similar to those of the silane-graphite mixtures of Example 1 except MoS_2 did not tend to settle out even when high weight percentages of silane were used. Lubricants consisting of the following percentages by weight were prepared: (1) 30% of the silane to 70% MoS_2, (2) 50% of the silane to 50% MoS_2, and (3) 90% of the silane to 10% MoS_2.

Example 3: Technical fine molybdenum disulfide powder was added to tetrakis-(4-biphenylyl)silane. To prepare such compositions, the silane was placed in a container and the container heated to a temperature in the range from 540° to 700°F. Tetrakis-(4-biphenyl)silane melts at 540°F. and, like the silane of Examples 1 and 2, is, when melted, a low viscosity liquid. After the silane was melted, powdered MoS_2 was added to and stirred into the silane. A 50-50 by weight mixture was prepared. Upon cooling, the composition solidified to a crystalline form rather than to the amorphous (glass-like) form of the mixtures of Examples 1 and 2. However, upon reheating and recooling, the material solidified into an amorphous form. The mixture exhibited no visible change in consistency at working temperatures.

The lubricating properties of compositions were evaluated during extrusion tests and compared with those of the best prior art lubricants. The 50-50 weight percent molybdenum disulfide-phenyl-tris-(4-biphenylyl)silane mixture gave the lowest values of container liner friction at 800°F. that have been obtained to date for extrusion operations under such conditions.

OTHER

Bag Lubricant for Tire Molding

J. Cekada, Jr.; U.S. Patent 3,532,624; October 6, 1970; assigned to Dow Corning Corporation describes a bag lubricant for tire molding. The use of methylpolysiloxanes as release agents for tires and other plastic articles is well-known. In the manufacture of a tire, the methylpolysiloxane fluid is sprayed on the mold for the release of the tire from the metal surface. In such applications dimethylpolysiloxane alone is an excellent release agent. However, in the manufacture of a tire the actual shaping of the tire is caused by inflating a rubber bag inside the green carcass which forces the tire into shape in the mold. Subsequently, the bag must release from the inside of the vulcanized tire without forming defects. The material used as a parting agent between the tire carcass and the bag is commonly called a bag or band ply lubricant. This process is particularly designed to enhance the effectiveness of these lubricants. This process relates to a composition consisting essentially of percent by weight:

(1) 2 to 10% of a dimethylpolysiloxane fluid of at least 30,000 cs. of viscosity at 25°C.
(2) 5 to 20% by weight of an ethylene or propylene polyglycol having a molecular weight of at least 1,000
(3) 25 to 50% by weight mica having a particle size of from 160 to 600 mesh
(4) 0.5 to 5% by weight bentonite clay and sufficient highly polar solvent to cause the bentonite to form a gel structure and thicken the composition
(5) 0.25 to 5% lecithin
(6) 25 to 55% by weight hydrocarbon solvent boiling below 200°C.

The composition can be prepared by mixing the ingredients in any suitable manner, but the best way of producing the composition is to add the hydrocarbon solvent to a mixing vessel and then add the other ingredients in the following order. Throughout the addition thorough agitation should be employed. The bentonite clay is first added followed by the necessary amount of highly polar solvent. Then the soya lecithin followed by the mica. The mica should be added slowly at a rate sufficient that thorough wetting and dispersion of the mica is obtained at all times.

Finally, the silicon and the glycol can be added in any order or simultaneously, particularly if they are compatible. After addition of all the ingredients the mixture is then thoroughly sheared on a colloid mill or other equivalent device. The resulting product is a thixotropic material, which when under pressure will flow quite easily, yet it does not separate when subjected to loading operations spraying operations, pumping operations, or the agitation encountered during shipment. The following example is illustrative of the process.

Example: The composition shown on the following page was prepared using the order of addition shown above.

37.5 % by weight mica, 325 mesh (wet ground)
7.5 % by weight of 500,000 cs. dimethylpolysiloxane fluid
12 % by weight of a copolymer of ethylene and propylene glycol in amount of 50% of each of the copolymer having a molecular weight of 2,700
1 % by weight soya lecithin
2 % by weight bentonite clay known as Bentone 38
0.67% of 95% methanol
39.33% of an aliphatic hydrocarbon solvent, boiling range of 117 to 142°C.

The above formulation has been stored for over one year without any indication of a separation. It has been shipped to various parts of the United States and foreign countries without any separation. This formulation was employed in a tire manufacturing plant over a 4 day period during which 13,685 tires were molded. The composition was sprayed on the green tires at an average nozzle pressure of 80 psi and under a pump pressure of 25 psi. During this period 550 lbs. of material was employed. Of the 13,685 tires molded no defects attributed to the bag lubricant was found. One defective tire was found due to faulty construction. In addition, the bag life obtained from this composition was better than that previously obtained with commercial bag lubricants.

Hydrocarbon-Silicone Copolymer as Fiber Lubricants

A process described by R.L. McKellar; U.S. Patent 3,445,389; May 20, 1969; assigned to Dow Corning Corporation involves a hydrocarbon silicone copolymer consisting essentially of a copolymer selected from the group consisting of (A) a copolymer of the average formula

$$(C_aH_{2a+1})\overset{(CH_3)_2}{\overset{|}{Si}}O{-}[(CH_3)_2SiO]_b{-}\overset{(CH_3)_2}{\overset{|}{Si}}(C_aH_{2a+1})$$

in which a has an average value from 18 to 75 inclusive and b has an average value from 0 to 73 inclusive, (B) a copolymer of the average formula

$$(C_aH_{2a+1})\overset{(CH_3)_2}{\overset{|}{Si}}O{-}[(CH_3)_2SiO]_f\overset{(CH_3)_2}{\overset{|}{Si}}{-\!-\!-}Z{-\!-\!-}\overset{(CH_3)_2}{\overset{|}{Si}}O[(CH_3)_2SiO]_f{-}\overset{(CH_3)_2}{\overset{|}{Si}}(C_aH_{2a+1})$$

in which a is defined above, f has an average value from 0 to 10 inclusive and Z is a divalent n-alkylene radical having from 1 to 10 carbon atoms, (C) a copolymer consisting essentially of dimethylsiloxane units, trimethylsiloxane units and units of the formula

$$(C_aH_{2a+1})\overset{CH_3}{\overset{|}{Si}}O$$

in which a is defined above, there being an average of two trimethylsiloxane units per molecule, from 2 to 75 dimethylsiloxane units per molecule and an average of at least 2

$$(C_aH_{2a+1})\overset{CH_3}{\overset{|}{Si}}O$$

units per molecule and no more than

$$(C_aH_{2a+1})\overset{CH_3}{\overset{|}{Si}}O$$

units per 10 siloxane units based on the total number of siloxane units in (C). The hydrocarbon-silicone copolymers useful as lubricants and are particularly useful as lubricants for fibers. These lubricants are compatible with organic lubricants and are also compatible with dimethylpolysiloxane fluids. The general compatability of these hydrocarbon-silicone copolymers makes them particularly useful in that they can be used with conventional organic lubricants and provide better and longer lasting lubrication on fibers. The hydrocarbon-silicone copolymers are waxy materials and some are waxy solids at room temperature. These hydrocarbon-silicone copolymers can be applied to fibers by conventional techniques, such as passing the fiber through a solution, passing the fiber through a liquid hydrocarbon-silicone copolymer bath, or running the fiber over a hydrocarbon-silicone copolymer which is a solid. Any fiber can be so lubricated, such as yarn or thread made of wool, rayon, hemp, silk, polyesters, polyamides and polypropylene. The hydrocarbon-silicone copolymers reduce the friction between the fibers and the spindles and also between fiber and fiber. The reduction in friction is greater than most conventional lubricants. The following examples illustrate the process.

Example 1: A hydrocarbon-silicone copolymer was prepared by mixing three drops of a solution of one weight percent chloroplatinic acid in ethanol and 188.5 g. of an alpha-olefin mixture consisting of alkenes having 20 carbon atoms or more and alkanes having 20 carbon atoms or more where the alkenes represented about 70 weight percent of the alpha-olefin mixture. This mixture was heated to 110°C. in a one liter, three necked flask equipped with a reflux condenser, thermometer and a mechanical stirrer. To the heated solution, 30.1 g. of a mixture of toluene and 20.5 g. of a dimethylpolysiloxane having two terminal silicon-bonded hydrogen atoms and having an average of 8 silicon atoms per molecule was added dropwise. The addition required 0.5 hours. The mixture was then heated to 145°C. for 24 hours. The mixture was then stripped to 200°C. at 1.0 mm. Hg. The product was a waxy solid melting at 48°C. and was a mixture of a siloxane of the average formula

$$(C_aH_{2a+1})\overset{(CH_3)_2}{\overset{|}{Si}}O-[(CH_3)_2SiO]_6-\overset{(CH_3)_2}{\overset{|}{Si}}(C_aH_{2a+1})$$

plus alkanes in which the sum of the carbon atoms of the C_aH_{2a+1} radicals and the alkanes per C_aH_{2a+1} radical was 61.

Example 2: A hydrocarbon-silicone copolymer was prepared by melting 124.6 g. of an alpha-olefin mixture consisting of alkenes having 20 carbon atoms or more and alkanes having 20 carbon atoms or more where the alkenes represented 70 weight percent of the alpha-olefin mixture. Three drops of a solution of one weight percent chloroplatinic acid in ethanol was added. The mixture was heated to 110°C. To the heated mixture, 50.2 g. of the dimethylpolysiloxane mixture of Example 1 was added dropwise at a rate sufficient to maintain the temperature of the mixture at 110°C. The mixture was then heated to 144°C. for 24 hours with stirring. The mixture was then stripped to 210°C. at 1.0 mm. Hg. The product was a waxy solid at room temperature with a melting point of 54°C. and was a mixture of a siloxane of the average formula of

$$(C_aH_{2a+1})\overset{(CH_3)_2}{\overset{|}{Si}}O-[(CH_3)_2SiO]_6-\overset{(CH_3)_2}{\overset{|}{Si}}(C_aH_{2a+1})$$

plus alkanes in which the sum of the carbon atoms of the C_aH_{2a+1} radicals and the alkanes per C_aH_{2a+1} radical was 40.

Example 3: A hydrocarbon-silicone copolymer was prepared by mixing three drops of a solution of one weight percent chloroplatinic acid in ethanol and 188.5 g. of an alpha-olefin mixture consisting of alkenes and alkanes both having 20 or more carbon atoms per molecule where the alkenes represented 70 weight percent of the alpha-olefin mixture. This mixture was heated to 110°C. and then tetramethyldihydrogendisiloxane was added at a sufficient rate to maintain the temperature at 110°C. After the addition the mixture was heated at 150°C. for 2 hours. The resulting mixture was then stripped at 180°C. at 1.0 mm. Hg. The product was a waxy solid at room temperautre and melted at 55°C. and was a mixture of a siloxane of the average formula of

$$(C_aH_{2a+1})\overset{(CH_3)_2}{\overset{|}{Si}}O \text{———} \overset{(CH_3)_2}{\overset{|}{Si}}(C_aH_{2a+1})$$

plus alkanes in which the sum of the carbon atoms of the C_aH_{2a+1} radicals and the alkanes per C_aH_{2a+1} radical was 61.

Quaternary Ammonium Compounds as Glass Lubricants

In a process described by C.A. Roth; U.S. Patent 3,560,385; February 2, 1971; assigned to Dow Corning Corporation an organosilicon quaternary ammonium compound such as

$$[(CH_3O)_3Si(CH_2)_3\overset{\oplus}{N}(CH_3)_2(C_{18}H_{37})]\overset{\ominus}{Cl}$$

is applied to the surfaces of solid siliceous material, for example, glass fibers, as a lubricant. The following examples illustrate the process.

Example 1: Three liter, three necked flask equipped with a condenser, stirrer, thermometer and a Dean-Stark trap, was charged with 610 g. (2.0 mols) of octadecyldimethylamine and 440 g. (2.2 mols) of chloropropyltrimethoxysilane. To this solution, there was added 1,050 g. of methyl Cellosolve. The mixture was heated for 20 hours at reflux temperature during which time 201 g. of material, chiefly methanol, boiling below 118°C. was removed. This material was replaced with an equal weight of methyl Cellosolve. After 20 hours the reaction was complete. Titration of a sample of the product

$$[(CH_3O)_3Si(CH_2)_3\overset{\oplus}{N}(CH_3)_2C_{18}H_{37}]\overset{\ominus}{Cl}$$

showed the chloride ion concentration to be 3.42%. The calculated chloride ion concentration for the product is 3.38%. The product was miscible with water in all proportions. To determine the glass lubrication properties, 8 ounce glass bottles were dipped into a 0.5% by weight aqueous solution of the above product. The bottles were allowed to air dry. Three bottles were stacked on their sides in a pyramid on a hinged level board. The free end of the board was raised until the top bottle began to slide over the two base bottles. The angle at which sliding of the top bottle occurred was recorded as the angle of repose. Of course, the smaller the angle of repose the greater the lubricity of the glass surface.

In this manner the angle of repose for surfaces treated with the formula shown below

$$[(CH_3O)_3Si(CH_2)_3\overset{\oplus}{N}(CH_3)_2C_{18}H_{37}]\overset{\ominus}{Cl}$$

was determined to be from 3° to 4°, while untreated bottles had an angle of repose of from 24° to 28°.

<u>Example 2:</u>

$$[(CH_3O)_3Si(CH_2)_3\overset{\oplus}{N}(CH_3)_2C_{18}H_{37}]\overset{\ominus}{Cl}$$

was also prepared by heating a mixture of 60 g. of of octadecyldimethylamine and 40 g. of chloropropyltrimethoxysilane in a stainless steel autoclave. The reactants were maintained at 100°C. in the autoclave. After about 70 hours, 89.5 g. of an orange liquid product was obtained. Titration for the chloride ion indicated that the reaction was 72% complete. The product was soluble in water. The glass lubricating properties of the product were determined by the test procedure described in Example 1. The treated glass bottles had an angle of repose of 4°, showing the product to be an excellent lubricant for glass surfaces.

POLYGLYCOLS, PHOSPHATES AND SILICATES

POLYGLYCOLS

Allyl Glycidyl Ether-Propylene Oxide Copolymers

H.H. Fall; U.S. Patent 3,350,461; October 31, 1967; assigned to The General Tire & Rubber Company describes a process for the production of allyl glycidyl ether-propylene oxide copolymeric diols having molecular weights in the range of 2,500 to 700, and having little or no terminal unsaturation. According to the process, polyoxypropylene diols can be prepared by the polymerization of propylene oxide and allyl glycidyl ether in a heterogeneous system employing finely divided solid potassium hydroxide as the catalyst. The potassium hydroxide should have a purity of at least 85% and preferably should have a purity of at least 90%.

It has been found that unless allyl glycidyl ether is employed as an initiator in this copolymerization, the amount of terminal unsaturation in the molecules is not decreased but is within the range produced by the processes previously known in the art. The amount of allyl glycidyl ether employed in the process can range from about 0.15 to 4.3 mols for each 4.3 mols of propylene oxide. When the amounts of the allyl glycidyl ether below 0.15 for each 4.3 mols of propylene oxide are employed, terminal unsaturation results.

It has also been found that the amount and the form of the potassium hydroxide employed in the process has an effect on the product produced. Potassium hydroxide should be employed in a finely powdered state and should have a purity of at least 85%, as stated above, in order to obtain polymers having decreased amounts of terminal unsaturation. The amount of potassium hydroxide employed can be from about 1.5 to 7.3 weight percent of potassium hydroxide based on the total weight of propylene oxide and allyl glycidyl ether employed. The process is preferably conducted at room temperature, i.e., from about 15° to 30°C. The reaction being exothermic, the temperature will rise to about 100°C. for a short time during the polymerization.

According to the process, propylene oxide, allyl glycidyl ether, and solid powdered potassium hydroxide are charged into a glass reaction vessel forming a heterogeneous mixture while being maintained at ambient temperatures with stirring, whereby the propylene oxide and allyl glycidyl ether copolymerize through the epoxide groups under the influence of the potassium hydroxide to produce the essentially linear hydroxy terminated polyoxypropylene-allyl glycidyl ether copolymers.

The process is preferably conducted at atmospheric pressure. Although higher or lower pressures can be employed, no commensurate advantages are obtained thereby. The polyoxypropylene-allyl glycidyl ether copolymers produced according to the process are useful as lubricants. The following examples illustrate the process.

Example 1: Into a glass reaction vessel fitted with a stirrer and water-cooled reflux condenser were charged propylene oxide (250 g., 4.3 mols), allyl glycidyl ether (22.8 g., 0.2 mol) and powdered potassium hydroxide (16.7 g., 0.27 mol, 90% by weight KOH). The contents of the vessel were stirred at room temperature for 28 hours. There was obtained 255 g. of a polymeric product having 0.788 millimol of unsaturation per gram of polymer and a hydroxyl number of 20.4. The amount of unsaturation found in the polymeric product is approximately equal to the amount of unsaturation introduced into the reaction as the allyl group of the allyl glycidyl ether.

Another polymeric product obtained in the same manner as above had 0.801 millimol of unsaturation per gram of polymer and a hydroxyl number of 21.6. The molecular weight of the product calculated from the hydroxyl number was 5,190. Determined by the isopiestic method, the molecular weight was found to be, on an average, 4,880 which substantiates that the polymer obtained is a copolymer of allyl glycidyl ether and propylene oxide having terminal hydroxyl groups, i.e., that it is a polyoxyalkylene glycol.

Example 2: A series of experiments was conducted employing the procedure as described in Example 1 where the amount of the powdered 90% potassium hydroxide catalyst was varied. The amounts of the monomer, i.e., allyl glycidyl ether and propylene oxide were the same as employed in Example 1. It was found that as the amount of potassium hydroxide catalyst was decreased, there was a corresponding decrease in the yield of the copolymer produced (see table below). It was found that the amount of unsaturation present in the polymers isolated corresponded within the limits of experimental error to the amount of the allyl glycidyl ether introduced into the system and, consequently, the polymers had terminal hydroxyl groups.

		Yield	
Run No.	90% KOH (g.)	Grams	Percent
C	4.2	17	6.2
D	8.4	149	54.6
E	10.5	185	67.7
F	12.6	234	85.7
A	16.7	257	94.2

Polyoxyalkylene Adducts of Phenols

A.W. Sawyer; U.S. Patent 3,377,288; April 9, 1968; assigned to Olin Mathieson Chemical Corporation has found that hydraulic fluids which contain as a base or a lubricant for the system, a polyoxyalkylene adduct of a phenol are of low cost, possess a high boiling point, are odorless, colorless and have a high degree of water tolerance. They also possess a high degree of compatibility with other fluids and exhibit a very low rate of corrosivity. Another feature of this process is the highly satisfactory rubber swelling performance of the fluids as shown in tests carried out according to SAE Standard J-70-B. Fluids previously employed have utilized expensive materials such as 2-ethylhexanol, heptanols, butyl ethers of glycols, or diethers of glycols in an amount from about 10 to 30% by weight of the total composition to achieve the desired rubber swelling characteristics. The rubber swelling properties of the fluids of this process are derived from the lubricant or base portion.

The polyoxyalkylene adduct of a suitable phenol which comprises the base or lubricant position of the composition will form from about 20 to 50% by weight of the hydraulic fluid composition. The polyoxyalkylene adducts are formed by condensing aromatic compounds having at least one hydroxyl group with an alkylene oxide. A preferred type of adduct is that formed from an alkyl phenol. The alkylene oxides which can be used in preparing the polyoxyalkylene adducts of the starting phenols include ethylene oxide, propylene oxide and butylene oxide. The following examples illustrate the process.

Example 1:

	Percent by Weight
Adduct of nonyl phenol plus 4 1/2 mols of ethylene oxide	24.4
Diethylene glycol	37.3
Ethylene glycol monomethyl ether	20.4
Diethylene glycol monomethyl ether	12.9
Triethylene glycol monomethyl ether	4.0
Potassium soap of rosin	1.0

Example 2:

	Percent by Weight
Adduct of nonyl phenol plus 4 1/2 mols of ethylene oxide	24.5
Diethylene glycol	9.8
Diethylene glycol monomethyl ether	13.75
Triethylene glycol monomethyl ether	51.25
Glycol-borax condensate*	0.7

*This product prepared by condensing 10.42 mols of ethylene glycol with 1.52 mols of potassium tetraborate.

Example 3:

	Percent by Weight
Adduct of nonyl phenol plus 3 mols ethylene oxide	25.0
Diethylene glycol monomethyl ether	18.7
Triethylene glycol monomethyl ether	56.1
Potassium tetraborate	0.20

Example 4:

	Percent by Weight
Nonyl phenol plus 13 mols of a mixture of 60 weight percent ethylene oxide and 40 weight percent polypropylene oxide (random addition product)	11.20
Nonyl phenol plus 4 1/2 mols of ethylene oxide	11.20
Diethylene glycol	9.80
Ethylene glycol monomethyl ether	7.70
Diethylene glycol monomethyl ether	56.85
Dodecyl benzene	1.50
Water	0.60
Potassium tetraborate	0.35
Dibutylated 2,2-di-(4-hydroxyphenyl)propane	0.30
Butynediol	0.50

Example 5:

	Percent by Weight
Nonyl phenol plus 13 mols of a mixture of 60 weight percent ethylene oxide and 40 weight percent propylene oxide (random addition product)	12.25
Nonyl phenol plus 4 1/2 mols of ethylene oxide	12.25
Diethylene glycol	9.80
Ethylene glycol monomethyl ether	4.50
Diethylene glycol monomethyl ether	58.55
Dodecyl benzene	1.50
Potassium tetraborate	0.35
Dibutylated 2,2-di-(4-hydroxyphenyl)propane	0.30
Butynediol	0.50

Example 6:

	Percent by Weight
Nonyl phenol plus 13 mols of a mixture of 60 weight percent ethylene oxide and 40 weight percent propylene oxide (random addition product)	10.50
Nonyl phenol plus 4 1/2 mols of ethylene oxide	10.50
Diethylene glycol	43.00
Ethylene glycol monomethyl ether	19.00
Diethylene glycol monomethyl ether	14.35
Dodecyl benzene	1.50
Potassium tetraborate	0.35
Dibutylated 2,2-di-(4-hydroxyphenyl)propane	0.30
Butynediol	0.50

Example 7:

	Percent by Weight
Nonyl phenol plus 13 mols of a mixture of 60 weight percent ethylene oxide and 40 weight percent propylene oxide (random addition product)	12.00
Nonyl phenol plus 4 1/2 mols of ethylene oxide	12.00
Diethylene glycol	9.80
Diethylene glycol monomethyl ether	36.39
Triethylene glycol monomethyl ether	27.16
Dodecyl benzene	1.50
Potassium tetraborate	0.35
Dibutylated 2,2-di-(4-hydroxyphenyl)propane	0.30
Butynediol	0.50

Example 8:

	Percent by Weight
Adduct of nonyl phenol plus 2 mols ethylene oxide	25.0
Diethylene glycol monomethyl ether	18.75
Triethylene glycol monomethyl ether	56.05
Potassium tetraborate	0.20

Example 9: Percent by Weight

Adduct of nonyl phenol plus 6.5 mols of a mixture of 50% by weight of ethylene oxide and 50% by weight of propylene oxide (random addition product)*	20.0
Diethylene glycol monomethyl ether	20.0
Triethylene glycol monomethyl ether	59.8
Potassium tetraborate	0.2

*The molecular weight of the mixture is calculated at 44 + 58 ÷ 2 or 51.

The fluid compositions of Examples 4 through 7 were tested according to the procedures set forth in Hydraulic Brake Fluid — SAE Standard J-70-B. Pertinent data relating to these tests, which illustrate the outstanding properties of these fluids, is shown in the table below. All of the fluids tested (i.e., the fluids of Examples 4 through 7) were found to satisfy completely the requirements for Heavy Duty 70R1 Type Hydraulic Brake Fluid.

Testing of Hydraulic Brake Fluid Compositions

Test	SAE 70R Requirement	Example Number			
		4	5	6	7
Boiling Point, ° F	302° F. min	319	350	320	398
Flash Point, ° F	145.4° F. min	195	195	180	220
Viscosity, c.s.:					
−40° F	1,800 cs. max	465	542	1,529	1,035
+122° F	3.5 cs. min	4.1	4.7	6.7	5.73
+212° F	1.3 cs. min	1.7	1.8	2.3	2.12
pH	7.0–11.5	9.8	10.1	11.0	11.3
Stability at High Temperature:					
Boiling Point, ° F	294.8° F. min	326	354	323	398
Boiling Point Change, ° F	±9° F. max	+7	+4	+3	0
Corrosion weight change, mg./sq. cm.:					
Tinned Iron	0.2 max	0.0	0.0	0.0	0.0
Steel	0.2 max	0.0	0.0	0.0	0.0
Aluminum	0.1 max	0.0	0.0	0.0	0.0
Cast Iron	0.2 max	0.0	+0.05	0.0	0.0
Brass	0.4 max	0.0	0.0	−0.06	0.0
Copper	0.4 max	0.0	−0.05	−0.06	−0.08
Pitting or roughening of metal strips	None	None	None	None	None
Jelling of fluid-water mixture	do	None	None	None	None
Crystalline deposit on glass jar walls or metal strips	do	None	None	None	None
Sedimentation, percent by volume	0.10 percent max	<0.1	<0.1	<0.1	<0.05
pH fluid-water mixture	7.0–11.5	9.5	9.6	7.3	10.9
Disintegration rubber cups evidenced by excessive tackiness or blisters.	None	None	None	None	None
Sloughing rubber cups evidenced by carbon black separation	do	None	None	None	None
Decrease hardness rubber cups	15 deg. max	−4	−5	−6	−5
Increase base diameter rubber cups, in	0.055″ max	0.033	0.033	0.053	0.031
Fluidity and Appearance Low Temperatures:					
−40° F.:					
Discernibility black contrast lines	Clearly discernible	Clear	Clear	Clear	Clear
Stratification or sedimentation	None	None	None	None	None
Time air bubble travel to top, sec	10 sec. max	2	2	3	2
−58° F.:					
Discernibility black contrast lines	Clearly discernible	Clear	Clear	Clear	Clear
Stratification or sedimentation	None	None	None	None	None
Time air bubble travel to top, sec	35 sec. max	2	2	4	3
Evaporation:					
Percent Weight Loss	80 percent max	77.2	74.2	78.5	75.0
Gritty or abrasive ppt. in residue	None	None	None	None	None
Pour point residue, ° F	23° F. max	5	5	5	−5
Water Tolerance:					
−40° F.:					
Discernibility black contrast lines	Clearly discernible	Clear	Clear	Clear	Clear
Stratification or sedimentation	None	None	None	None	None
Time air bubble travel to top, sec	10 sec. max	2	2	3	1
+140° F.:					
Stratification	None	None	None	None	None
Sedimentation, percent	0.05 percent by vol. max	None	None	None	0.00
Compatibility:					
−40° F.:					
Discernibility black contrast lines	Clearly discernible	Clear	Clear	Clear	Clear
Stratification or sedimentation	None	None	None	None	None
+140° F.:					
Stratification	do	None	None	None	None
Sedimentation, percent	0.05 percent by volume max	None	None	None	None
Resistance to Oxidation:					
Pitting or roughening metal strips	None	None	None	None	None
Gum deposit metal strips	Trace	None	None	None	None
Weight loss, mg./sq. cm.:					
Aluminum	0.05 max	0.00	0.00	0.01	+0.01
Cast Iron	0.30 max	0.01	0.01	0.00	+0.01
Effect on Rubber, +158° F.:					
Hardness decrease:					
Natural	10 deg. max	4	4	3	4
SBR	10 deg. max	3	4	4	4
Base diameter increase, in.:					
Natural	0.006–0.055″	0.029	0.029	0.036	0.027
SBR	0.006–0.055″	0.024	0.026	0.024	0.022
Disintegration evidenced by excessive tackiness or blisters:					
Natural	None	None	None	None	None
SBR	do	None	None	None	None
Sloughing indicated by carbon black separation:					
Natural	do	None	None	None	None
SBR	do	None	None	None	None
Specific Gravity, 60/60° F	No requirement	1.036	1.038	1.060	1.048
Pounds per gallon, 60° F	do	8.63	8.65	8.83	8.73

Polyglycol-Based Acetals

In a process described by G.L. Doelling, J.E. Peskar and J.A. Siefker; U.S. Patent 3,563,893; February 16, 1971; assigned to Wager Electric Corporation, acetals are formed by reacting methoxyacetaldehyde with tetrahydrofurfuryl alcohol or with mono lower alkyl ethers of mono, di or tri-ethylene or propylene glycols. These acetals are, in general, clear liquids which boil at temperatures well over 400°F. at atmospheric pressures and have structures represented by the formulas shown on the following page.

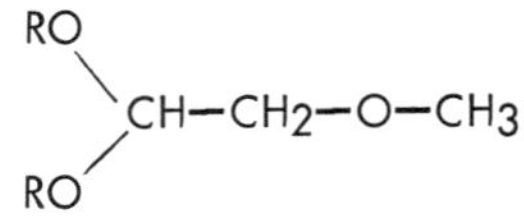

R is one of:

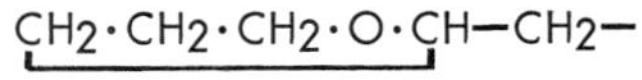

$$R_1O-\left[C_2H_4O\right]_n-C_2H_4-$$

$$R_2O-\left[C_3H_6O\right]_n-C_3H_6-$$

R_1 and R_2 are aliphatic hydrocarbon radicals having up to four carbon atoms.
n is an integer from 0 to 2 inclusive.

In forming the described acetals of methoxyacetaldehyde, conventional preparation methods can be used. Thus, each 1 mol of methoxyacetaldehyde is reacted with at least 2 mols, and preferably with a small excess, of the particular alcohol ingredient selected from the total useful group comprising tetrahydrofurfuryl alcohol and mono lower alkyl esters of mono, di or tri-ethylene or propylene glycols. The reaction is carried out at elevated temperatures in the presence of a catalyst which may be an acid such as hydrochloric or p-toluene sulfonic acid or the catalyst may be boron trifluoride-ethyl ether complex as is known in the art. In general, the reaction temperatures may vary from about 190° to 250°F. in order to form the acetal product at a reasonably rapid rate. The reactants may be dissolved in a suitable solvent, for example, benzene or other organic solvents, and both the solvent and the water formed during the course of the reaction, as the same evaporate from the reaction mixture, may be trapped and collected by a reflux condenser. The process is illustrated by the following examples.

Example 1: Methoxyacetaldehyde-di(methoxydiethoxyethyl)acetal was made by reacting each 1 mol of methoxyacetaldehyde with a slightly more than 2 mols of the monomethyl ether of triethylene glycol. The reactants were dissolved in benzene together with a catalyst comprising boron fluoride-ethyl ether complex and reaction was carried out under reflux at temperatures from 200° to 240°F. The resulting acetal was a liquid with a specific gravity at 20°/20°C. of 1.08, a viscosity at 20°C. of about 16 centistokes, a viscosity at -40°F. of about 1,700 centistokes, a reflux boiling point of 579°F., and it did not freeze up after 16 hours at -58°F. The rubber swelling was about 0.021 inch gain in base diameter of a standard 1 1/8 inch natural rubber wheel cylinder cup after four days at 158°F. (mixed with an equal volume of castor oil). The acetal had the following formula:

$$(CH_3OC_2H_4OC_2H_4OC_2H_4O-)_2CH-CH_2-O-CH_3$$

Example 2: Methoxyacetaldehyde-di(triethoxyethyl)acetal was made, as in Example 1, by reacting methoxyacetaldehyde with the monoethyl ether of triethylene glycol. This acetal was a liquid with a specific gravity at 20°/20°C. of about 1.05, a viscosity at 20°C. of about 15 centistokes, a boiling point at about 2 mm. pressure of about 390°F. It did not freeze up after two days at -40°F., and the rubber swelling at 158°F. on a natural rubber 1 1/8 inch cup was 0.028 inch gain in base diameter (50-50 castor oil). The acetal had the following formula:

$$(C_2H_4OC_2H_4OC_2H_4OC_2H_4O-)_2CH-CH_2-O-CH_3$$

Example 3: Methoxyacetaldehyde-di(methoxyethoxyethyl)acetal was made in the manner described in Example 1 by reacting methoxyacetaldehyde and the monomethyl ether of diethylene glycol. This acetal was a liquid that did not freeze up after 16 hours at -58°F. It had a boiling point at about 2 mm. pressure of about 310°F., a specific gravity at 20°/20°C. of about 1.06 and a viscosity at 20°C. of about 7 centistokes. The acetal had the following formula:

$$(CH_3OC_2H_4OC_2H_4O-)_2CH-CH_2-O-CH_3$$

Example 4: Methoxyacetaldehyde-di(tetrahydrofurfuryl)acetal was made in the manner described in Example 1 by reacting methoxyacetaldehyde with tetrahydrofurfuryl alcohol. This acetal was a liquid with a specific gravity at 20°/20°C. of about 1.09, a viscosity at 20°C. of about 14 centistokes, and a boiling point at atmospheric pressure of about 590°F. It swells at 1 1/8 inch natural rubber cup about 0.050 inch after 24 hours at 150°F. (50-50 with castor oil). It did not freeze up after several hours at -58°F. The acetal had the following formula:

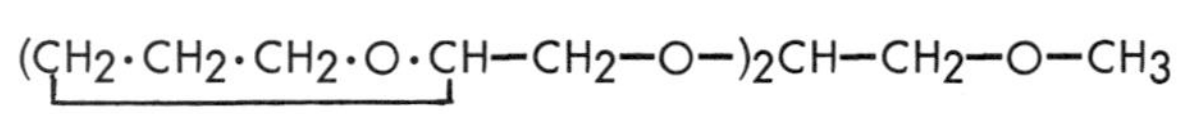

Example 5: A hydraulic fluid was made using the acetal of Example 1 as an ingredient. This fluid had the following composition for approximately 100 cc of fluid: 40 cc of methoxyacetaldehyde-di(methoxydiethoxyethyl)acetal; 34 cc

of triethylene glycol monoethyl ether; 5 cc of dipropylene glycol; 20 cc of polypropylene glycol (molecular weight of 1,025); 0.3 gram of tetramethyl diaminodiphenyl methane; 0.7 gram of potassium ricinoleate. This hydraulic fluid had properties approximately as follows:

Appearance	A clear liquid
Specific gravity at 20°/20°C.	1.043
Viscosity at 20°C.	18.10 centistokes
Viscosity at -40°F.	1,705 centistokes
Rubber swelling, 1 1/8 inch natural rubber cup, 3 days at 158°F.	0.018 inch

Example 6: Another hydraulic fluid was made using the acetal of Example 3 as an ingredient as follows: 60 cc of methoxyacetaldehyde-di(methoxyethoxyethyl)acetal; 21.6 cc of triethylene glycol monomethyl ether; 8.0 cc of dipropylene glycol; 10.0 cc of polypropylene glycol (molecular weight of 2,000); 0.2 cc of triethanolamine; 0.2 grams of trimethyl dihydroquinoline polymer. This hydraulic fluid had properties approximately as follows:

Appearance	A clear liquid
Specific gravity at 20°/20°C.	1.051
Viscosity at 20°C.	12.6 centistokes
Viscosity at -40°F.	707 centistokes
Boiling point, SAE	510 °F.
Rubber swelling, 1 1/8 inch natural rubber cup, 3 days at 158°F.	0.022 inch

Alkanolamine, Organic Acid and Polyoxyalkylene Glycol

R.H. Davis; U.S. Patent 3,374,171; March 19, 1968; assigned to Mobil Oil Corporation has found that a balanced formulation can be obtained in a lubricant composition which is effective as a cutting fluid having improved corrosion-resistant lubricating and cooling efficiency properties and which, in the course of machining operations, maintains its transparency in hard water solutions, avoids the formation of tacky water-insoluble residues, is free from excessive foam formation and is resistant to rancidity.

The composition comprises a balanced formulation and containing as essential components a major proportion of water and, in individual minor proportions, the combination of a saturated organic acid, either straight-chain or branched-chain having from about 6 to 9 carbon atoms per molecule, a water-soluble alkanolamine and a water-soluble polyoxyalkylene glycol. In this respect, it is found that the alkanolamine component, in addition to functioning as a corrosion-inhibiting agent in which the organic acids are soluble, also functions as a plasticizer in that, in the course of use, evaporation of water from the cutting fluid leaves a residue upon machinery parts which is desirably soft or semiliquid. The function of the organic acid component in combination with the alkanolamine is primarily that of a corrosion inhibitor in that it forms the corresponding amine acid salt. The water-soluble polyoxyalkylene glycol in the composition functions as a load-support agent.

The water-soluble alkanolamine should be liquid at room temperature. The lower molecular weight compounds are generally preferred and, for this purpose, it has been found that such alkanolamines as mono, di or tri-ethanolamine are most effective.

The lubricant compositions are preferably prepared by a blending procedure which comprises reacting the alkanolamine and organic acid components with about 10 to 20 parts of water, which is to be present in the finished formulation. This blending procedure may be carried out at room temperature. However, heating to 120°F. with agitation is most satisfactory. The remaining quantity of water required to be present in the finished product is added, together with the polyoxyalkylene glycol, and any of the aforementioned other desired components. It should be noted, however, that if an alkali metal nitrite is to be present in the finished product, such material is added last for the reason that addition of acids such as caprylic acid to a solution of sodium nitrite could result in decomposition of the nitrite, thus reducing or nullifying its rust-inhibiting effects. If a germicidal agent is to be incorporated in the formulation, it is preferably blended with the alkanolamine and organic acid. The compositions were subjected for the purpose of evaluation to the following corrosion, hard water stability and machining efficiency tests.

Corrosion Test — One to three grams of malleable iron turnings are placed in a glass container, washed with the test solution which is to be evaluated, drained and allowed to stand in an atmosphere of approximately 90% relative humidity and at a temperature of about 70°F. for a period of about 72 hours. After this period, the sample is then checked for the appearance of rust. The samples are graded from 0 (no rust) to 3 (severe rusting) in accordance with the standard chart on the following page. This test, and a more detailed general method for carrying it out, is described in U.S. Patent 3,071,545.

Description	Rating
No rusting, or less than 1% of the metal surface coated with yellow iron oxide	0
Light rusting, rusting confined to not more than six spots, each of which is 1 mm. or less in diameter	1
Moderate rusting, rusting in excess of the above, but confined to less than 5% of the surface of the specimen	2
Severe rusting, rusting covering more than 5% of the surface of the specimen	3

Hard Water Stability Test — In this test, 1 part of the aqueous metal working fluid is mixed with 5 parts of the ASTM 1478-57T, paragraph 4a, hard water and stored in a 100-ml. graduated cylinder for a period of 72 hours, at a temperature between 70° and 75°F. At the end of this period, there is recorded the volume of sediment formed as a result of interaction between the additives in the cutting fluid and the polyvalent ions present in the water.

Machining Efficiency Test — The machining efficiency of the improved lubricant compositions is evaluated with respect to a series of holes, drilled in a test bar made from hot rolled steel, which are to be tapped. Prior to testing, samples of the formulations are diluted by employing 1 part, by weight, of the additive concentrate to 5 parts, by weight, of distilled water. The fluid is applied to the tap with a polyethylene squeeze bottle. Four unknown samples are bracketed by two standards. The tapping efficiency is calculated by dividing the torque obtained with the unknown, into the average standard torque, and multiplying by 100. This test, and a more detailed general method for carrying it out is described in U.S. Patent 3,071,545.

In Table 1 below, samples containing triethanolamine, a polyether polyol, in the form of a copolymer, produced by reacting ethylene oxide and propylene oxide, and having a hydroxyl number from about 22 to 38, and a saturated organic acid having from 4 to 9 carbon atoms per molecule, were subjected to the corrosion test, in which the composition of the hard water employed was derived from that described in ASTM Test 1478-57T, paragraph 4a, which shows a preponderance of magnesium and sulfate ions. This water was modified in two ways, viz., the active ingredients were increased by 50% in order to increase total hardness from about 230 to 345 ppm and to increase the sulfate content to about 270 ppm. Sodium hydroxide was used to neutralize excess carbon dioxide employed for stabilization to attain a neutral pH.

As will be seen from Table 1, the samples containing triethanolamine, the ethylene oxide-propylene oxide copolymer, and either hexanoic, heptanoic, caprylic or pelargonic acids (Examples 3, 4, 5 and 6) effectively inhibited rusting of the malleable iron turnings, with a rating in each instance of 0. On the other hand, where the acid component comprised valeric or isobutyric acids (Examples 1 and 2), severe rusting occurred with a rating of 3, clearly indicative of the fact that when the acid component contains less than 6 carbon atoms per molecule, it is unsatisfactory for use in metalworking operations.

TABLE 1: AQUEOUS SYSTEMS CONTAINING AN ALKANOLAMINE SALT OF SATURATED C_4-C_9 ORGANIC ACID AND ETHYLENE OXIDE-PROPYLENE OXIDE COPOLYMER

Ex. No.	Base Formula: Triethanolamine (Percent Wt.)	Isobutyric Acid (Percent Wt.)	Valeric Acid (Percent Wt.)	Hexanoic Acid (Percent Wt.)	Heptanoic Acid (Percent Wt.)	Caprylic Acid (Percent Wt.)	Pelargonic Acid (Percent Wt.)	Copolymer (Percent Wt.)	Distilled Water (Percent Wt.)	Malleable Iron Rust-Test: 1 part base formula to 5 parts ASTM Hard Water
1	20	3	-	-	-	-	-	7	70	3
2	20	-	3	-	-	-	-	7	70	3
3	20	-	-	3	-	-	-	7	70	0
4	20	-	-	-	3	-	-	7	70	0
5	20	-	-	-	-	3	-	7	70	0
6	20	-	-	-	-	-	3	7	70	0

In Table 2 on the following page, samples containing triethanolamine, a copolymer of ethylene oxide and propylene oxide, as noted in Table 1 above, and a saturated organic acid having from 6 to 10 carbon atoms per molecule, were subjected to a hard water stability test which was carried out in the following manner. One part of the specific aqueous metalworking fluid was mixed with five parts of the ASTM hard water and stored in a 100-ml. graduated cylinder for a period of 72 hours and at a temperature from about 70° to 75°F. At the end of this period, the volume of sediment formed as a result of interaction between the additives in the cutting fluid and polyvalent ions present in the water were recorded.

As will be seen from Table 2, the samples containing triethanolamine, the ethylene oxide-propylene oxide copolymer, and either hexanoic, heptanoic, caprylic or pelargonic acids (Examples 1, 2, 3 and 4), revealed no separation after 72 hours. On the other hand, where the acid component comprised capric acid (Example 5), a white insoluble soap (10% by volume) appeared at the top of the graduate. This outstanding difference, with reference to saturated organic acids having more than 9 carbon atoms per molecule, is indicative of the fact that the use of such higher organic acids can result in clogged filters, poor rust protection, and reduced tool life in areas where hard water is encountered.

TABLE 2: AQUEOUS SYSTEMS CONTAINING AN ALKANOLAMINE SALT OF SATURATED C_6-C_{10} ORGANIC ACID AND ETHYLENE OXIDE-PROPYLENE OXIDE COPOLYMER

Ex. No.	Base Formula: Triethanolamine (Percent Wt.)	Hexanoic Acid (Percent Wt.)	Heptanoic Acid (Percent Wt.)	Caprylic Acid (Percent Wt.)	Pelargonic Acid (Percent Wt.)	Capric Acid (Percent Wt.)	Copolymer (Percent Wt.)	Distilled Water (percent Wt.)	Hard Water Stability Test: Percent Volume Separation 1 part base formula to 5 parts ASTM Hard Water
1	20	3					7	70	No separation.
2	20		3				7	70	Do.
3	20			3			7	70	Do.
4	20				3		7	70	Do.
5	20					3	7	70	10% white insoluble residue.

In order to illustrate the importance of the polyoxyalkylene glycol as an effective load-support agent in the aqueous lubricant compositions, it will be seen from Table 3 below that neither the triethanolamine, nor the ethylene oxide-propylene oxide copolymer, when used alone or in mixtures (as in Examples 1, 2 and 3) are effective load-support agents, as indicated by the respective tapping efficiency, which in each case is below 79%. Furthermore, as will be seen from Table 3, the amine salts of the C_6 to C_9 saturated organic acids, as represented by hepranoic, caprylic, and the branched chain 2-ethyl hexoic acid, are more effective, but the tapping efficiency results are still below 89%. It will be noted, however, that a combination of the ethylene oxide-propylene oxide copolymer and the amine salts produces tapping efficiency results which range from 93 to 96% (as in Examples 5, 7 and 9). This represents a significant increase in tapping efficiency over the amine salts when used alone (as in Examples 4, 6 and 8). It will, therefore, be apparent that there is a synergistic extreme pressure effect between the polyoxyalkylene glycol and the amine salts.

TABLE 3: AQUEOUS SYSTEMS CONTAINING AN ALKANOLAMINE, OR SATURATED ORGANIC ACID OR ETHYLENE OXIDE-PROPYLENE OXIDE COPOLYMER, OR MIXTURES

Example No.	Base Formula: Triethanolamine (Percent Wt.)	Heptanoic Acid (Percent Wt.)	Caprylic Acid (Percent Wt.)	2-Ethyl Hexoic Acid (Percent Wt.)	Copolymer (Percent Wt.)	Distilled Water (Percent Wt.)	Tapping Efficiency, Percent 1-5 Dilution Base Formula-Dist. Water
1	30					70	75.7
2					30	70	78.7
3	20				10	70	75.5
4	20	10				70	81.0
5	20	3			7	70	95.5
6	20		10			70	81.7
7	20		3		7	70	96.0
8	20			10		70	87.5
9	20			3	7	70	93.0

Fatty Acid, Morpholine and Alkylene Polyol Oil Emulsions

J.E. Berger; U.S. Patent 3,444,080; May 13, 1969; assigned to Shell Oil Company describes soluble oil compositions comprising lubricating oil containing a fatty acid, a morpholine, and an alkylene polyol which are useful in hot rolling of metals. Illustrative examples of soluble oils of this process comprise:

Composition A	Weight Percent
Emery 3305 fatty acid	23.0
Morpholine	2.0
Hexylene glycol	10.0
HVI neutral (VI 94, 105 SUS at 100°F.)	65.0

Composition B	Weight Percent
Oleic acid	20.0
Morpholine	1.6
Hexylene glycol	11.0
Mineral lubricating oil (same as Composition A)	67.4

Composition C	Weight Percent
Oleic acid	17.0
Morpholine	1.5
Hexylene glycol	8.0
Mineral lubricating oil (same as Composition A)	73.0
Germicidal agent (0.25% wt. Dowicide C.O. + 0.25% wt. Dowicide 6)	0.5

In preparing the soluble oil, the components are merely blended together using conventional mixing means, such as mechanical agitation. Heating may be used, if desired, to facilitate blending, e.g. temperatures of from 80° to 120°F. The preparation of the morpholine-containing soluble oils is relatively simple when compared to the involved cooking steps required in preparing the alkanolamine-containing soluble oils.

Finished emulsions are prepared by mixing, on a volume basis, from about 2 to 10%, preferably 3 to 6%, soluble oil with water. In general, the soluble oils can be used in concentrations lower than those used with conventional soluble oils which is advantageous from a cost standpoint. The emulsions form quite readily, which is an important factor since the mixing of the oil and water is done in the field by means such as air agitation or propeller mixing. The composition of the emulsion yields a proper balance among the forces of attraction operating between the oil and water phase, the oil phase and the metal, and the water phase and the metal. The composition of the emulsion is such that an oil-in-water emulsion and a water-in-oil emulsion are almost equally likely to form spontaneously, yet the resulting emulsion is quite stable. Typical properties of the compositions are given below.

Property	Value
Gravity, °API at 60°F.	29.5
Color, ASTM	2.0
Pour point, °F.	0
Cloud point, °F.	28
Flash, °F.	300
Viscosity, SSU at 100°F.	113
SSU at 210°F.	40.6
VI (ASTM D-567)	100
Initial pH	7.5
TBN-E, mg. KOH/g. of oil	12
TAN-E, mg. KOH/g. of oil	44
Ratio TBN-E/TAN-E	0.27
Saponification No.	45
Sulfated ash, percent weight	0.002

Biodegradable Composition

In a process described by C.L. Zuraw; U.S. Patent 3,507,792; April 21, 1970; assigned to Sinclair Research, Inc. a lubricating composition which is essentially biodegradable and dispersible in water is made from about 0.5 to 30 weight percent of an unsaturated ester of a normal aliphatic monocarboxylic acid and a normal aliphatic alcohol; about 0.1 to 30 weight percent of at least one nonionic emulsifier having an HLB (hydrophile-lipophile balance) range of about 8 to 15 which can be a condensate of ethylene oxide with an ester of a normal aliphatic monocarboxylic acid and a polyhydric normal alcohol, the condensate containing about 2 to 20 mols of ethylene oxide per mol of condensate, a condensate of an ethylene oxide with a normal aliphatic alcohol the condensate containing about 1 to 15 mols of ethylene oxide per mol of condensate, a condensate of ethylene oxide with a normal aliphatic monocarboxylic acid, the condensate containing about 2 to 20 mols of ethylene oxide per mol condensate; and about 1 to 99 weight percent of water. The process is illustrated by the following examples.

Example 1: A batch of the fluid composition had the listed components and was prepared in the following manner.

Component	Weight Percent
45° N.W. sperm oil	25.0
Polyoxyethylene (20 mol) sorbitan monopalmitate	2.8*
Polyoxyethylene (20 mol) sorbitan trioleate	2.1*
Polyoxyethylene (6 mol) oleyl ether	2.1*
Monoethanolamine	5.0
Benzotriazole	0.5
Ethylene glycol	25.0
Water	37.5

*12.5 HLB

The benzotriazole was dissolved in the three polyoxyethylene-containing emulsifiers. One-fifth of the sperm oil was added to the emulsifiers and the mixture was stirred for about 1/2 to 1 hour. The monoethanolamine was then added and the resulting mixture was stirred for about 1/2 to 1 hour. About one-seventh of the water was added and the resulting mixture was stirred for about 1/2 to 1 hour. The remainder of the sperm oil and the water was then added and the resulting mixture was stirred for about 1 1/2 hours. Finally, the ethylene glycol was added and the composition was stirred to a uniform consistency.

Example 2: A batch of the fluid composition was prepared in the same manner as in Example 1 (there being one polyoxyethylene emulsifier instead of three) and had the following composition.

Component	Weight Percent
45° N.W. sperm oil	25.0
Brij 96 [polyoxyethylene (10 mol) oleylether]	8.0*
AT666 (triethanol amine salt of a hydrocarbon sulfamido acetic acid)	2.0
Monoethanolamine	7.5
Ethylene glycol	18.0
Water	39.5

*12.4 HLB

Example 3: The fluid composition prepared in Example 1 was subjected to a biodegradability test, using a simulated activated sludge system. A 14-liter New Brunswick Micro-Ferm fermentor was used to simulate an activated sludge sewage treatment system. Activated sludge and raw sewage were obtained from a local disposal plant. The synthetic sewage used in all tests was prepared as follows:

Synthetic Sewage*	
Bacto peptone	3.5 g.
NaCl	3.0 g.
K_2HPO_4	0.5 g.
$MgSO_4 \cdot 7H_2O$	0.2 g.
Raw sewage	1 liter
Deionized water	11 liters

*Huddleston, R.L. and Allred, R.C., "Evaluation of Detergent Degradation Using Activated Sludge." J. Am. Oil Chemists, 41 (11) 732-735, November 1964.

Ten liters of synthetic sewage containing between 2,000 to 3,000 mg./liter (dry weight) activated sludge as recommended for normal sewage plant operation were added to the fermentor. The temperature was held at 25°C. and the air flow controlled at approximately one liter per minute. Slight mixing (100 rpm) was used to keep the sludge in suspension during operation. Retention time in the system was limited to six hours. At the end of this period, air flow and mixing were discontinued and the sludge allowed to settle. The clear water was siphoned off and fresh synthetic sewage added. Approximately 1.5 liters of sludge remained in the fermentor after the clear effluent was removed. Samples were collected initially, at three and at six hours. Chemical and Biochemical Oxygen Demand determinations were made on clear supernate from each sample, employing the procedures outlined in "Standard Methods for the Examination of Water and Wastewater," American Public Health Association, Inc., 1965.

Three such runs were made. The first and third runs were made without the addition of the fluid composition of this process in order to obtain information on sludge activity. Results presented below demonstrate the complete degradation of the fluid composition as indicated by COD and BOD (Chemical and Biochemical Oxygen Demand) determinations. These procedures are routinely used by disposal plant operators to determine the quality of their effluent. Runs 1 and 3 demonstrate the effectiveness of the system in handling synthetic sewage. The fluid composition added to the system in Run 2 increased the initial COD and BOD, but after six hours exposure the levels had been reduced to close to that of Runs 1 and 3. The Chemical and Biochemical Oxygen Demand determinations obtained therefore indicate that the fluid composition was degraded with no apparent adverse effect on the activated sludge.

TABLE 1: DISPOSAL OF THE FLUID COMPOSITION IN A SIMULATED ACTIVATED SLUDGE SYSTEM

	Chemical Oxygen Demand (COD), p.p.m.[1]			Five-day Biochemical Oxygen Demand (BOD), p.p.m.[2]		
Exposure Time (Hours)	0	3	6	0	3	6
Run Number:						
1. No fluid composition	197	96	48	108	70	33
2. Fluid composition added[3]	426	161	60	300	27	3
3. No fluid composition	211	47	40	65	13	5

[1] Chemical Oxygen Demand is the amount of dichromate oxidizable material in a water or sewage sample. It is reported as mg./liter (p.p.m.) of dichromate oxygen utilized.

[2] Bichemical Oxygen Demand is an estimate of the quantity of oxygen required by bacteria (during a five-day incubation period) to oxidize, or render stable, the more easily decomposable organic substances in a water or sewage sample. It is reported as mg./liter (p.p.m.) of oxygen utilized.

[3] 3.3 percent solution in tap water.

The fluid composition prepared as described on the previous page was subjected to rust and wear tests. For comparison, a soluble oil composition containing petroleum oil which is representative of the fluids often used in industry was subjected to the identical tests. The results listed in Table 2 demonstrate the performance of the fluid composition in inhibiting rust and wear.

TABLE 2

Product	Dilution	Cast iron chip rust test(1)	Shell four ball wear test (scar diameter mm.)(2)
Fluid composition	1:50	No rust	0.47
	1:150	do	0.50
	1:300	Very light rust	0.47
Soluble oil	1:50	Rust	0.65
	1:150	do	0.63
	1:300	do	0.70

[1] This is a standard test often used for testing rust inhibiting properties of soluble oil emulsions. It consists of immersing cast iron chips in test emulsion in a beaker; covering with a watch glass; inverting and allowing emulsion to drain while chips remain covered on the watch glass by the beaker. Chips are observed for rust after 24 hours.

[2] Test conditions: 1,800 r.p.m., 5 kg., 1 hour, at 130° F.

Phosphine Oxide and Sulfide Additives

P. Geymayer, S. Billenstein and W. Renz; U.S. Patent 3,580,849; May 25, 1971; assigned to Farbwerke Hoechst AG, Germany have found that the protection against wear of actuating fluids for hydraulically operated mechanisms consisting of mixtures of glycols and polyglycols and/or their ethers or the aqueous solutions can be improved by adding 0.15 to 20% by weight, preferably 1 to 8%, of tris(hydroxymethyl)phosphine oxide and/or phosphine sulfide as well as the related oxalkylation products with up to 5 mols of ethylene oxide and/or propylene oxide per mol of phosphine oxide or phosphine sulfide.

To test the lubricating properties of the actuating fluids according to the process, the four-ball apparatus of G.D. Boerlage was used. In this test the protection against wear is defined by two characteristic values, i.e. corrected average load according to Hertz and welding load. These values were determined from the third report about the requirements and tests of difficultly inflammable liquids for hydraulic power transmission and steering (Committee of the European Communities, permanent committee for operating security in coal mining, Luxembourg, Oct. 10, 1967). In both cases high values indicate a good lubricating effect.

The abrasion was additionally tested on an abrasion balance according to Reichart (Study group for the exploitation of literature about lubricants, Report No. 267, pages 1-3, Kadmer, 1957). In this test the area of the abrasion mark on the test specimen is taken as measurement for the abrasion. The larger the area, the poorer the lubricating effect. The following examples illustrate the process.

Example 1: Polyethylene glycol having a molecular weight of 200 (PEG 200) was mixed with varying amounts of tris(hydroxymethyl)phosphine oxide (TOMP oxide). The welding loads measured in the four-ball apparatus increased with increasing amounts of additive.

Added Amount of TOMP Oxide to PEG 200, %	Welding Load, kg.
0	140/160
1	160/180
3	180/200
5	240/260
7	260/280

Example 2: Increasing amounts of TOMP oxide were added to a 50% solution of polyethylene glycol having a molecular weight of 4,000 in water. The addition considerably improved the welding loads measured in the four-ball apparatus.

Added Amount of TOMP Oxide to PEG 4000 + H_2O 1:1, %	Welding Load, kg.
0	130/150
1	180/200
3	240/260
5	280/300
7	320/340

Example 3: By adding 2% of TOMP oxide to a mixture of 65% of PEG 200, 34% of PEG 300, 0.1% of $NaNO_3$, and

0.9% of borax, the welding load measured in the four ball apparatus could be raised from 230/250 kg. to 500/550 kg. and the average load according to Hertz could be improved from 40.0 to 86.0 kg.

Example 4: By adding TOMP oxide to difficultly inflammable hydraulic fluids on the basis of glycol and water, the lubricating properties of the fluids, measured in the four-ball apparatus and in the abrasion balance according to Reichert, could be considerably improved. The hydraulic fluids used had different viscosities and consisted of:

(a) 15 to 30% of liquid, high molecular weight, water-soluble oxalkylation mixture, for example polyglycols having a molecular weight of from 10,000 to 20,000.
(b) 25 to 35% of low molecular weight alkylene glycols and/or alkylene glycol ethers, for example monoethylene glycol, and/or ethylpolyglycol.
(c) 40 to 50% of water.
(d) Various additives, for example borax, triethanol amine or morpholine to improve the properties such as protection against corrosion and aging.

According to the third Luxembourg report, difficultly inflammable fluids for hydraulic power transmissions and steering are designated by the initials HS. The letter C indicates the class of the aqueous solutions containing 40 to 50% of water. The number placed behind defines the visocisity, for example:

HS C-2: 11 to 14 centistokes at 50°C.
HS C-4: 20 to 40 centistokes at 50°C.
HS C-8: 50 to 70 centistokes at 50°C.

The fluids used in this example essentially correspond to the standards defined in the above report. By adding the claimed lubricants the properties of the actuating fluids for hydraulically operated mechanisms are modified to a negligible degree except for the lubricating properties.

Fluid	Addition of TOMP Oxide, %	Welding Load, kg.	Av. Load According to Hertz, kg.
HS C-2	Without	150/160	40.7
HS C-2	3	280/300	55.5
HS C-4	Without	140/160	36.4
HS C-4	3	260/280	53.4
HS C-8	Without	140/160	36.0
HS C-8	3	240/260	58.0

Ascorbic Acid Stabilizers

In a process described by T. Turumaru, M. Yotsuzuka, and K. Ogino; U.S. Patent 3,415,891; December 10, 1968; assigned to Takeda Chemical Industries, Ltd., Japan, a method is provided for the stabilization of polyoxyalkylene polyols which comprises mixing the polyoxalkylene polyol with a stabilizer which is from the group consisting of ascorbic acid, isoascorbic acid and alkali metal and alkaline earth metal salts of the acid, alkanoic and alkenoic acid esters of the acid, as well as mixtures of the acids, salts and esters.

As the metal salts of ascorbic acid or isoascorbic acid, there are used alkali metal salts such as sodium salt, potassium salt, or alkaline earth metal salts such as calcium salt. The number of carbon atoms of the aliphatic carboxylic acid constituting the esters is preferably from 4 to 18, inclusive. The aliphatic carboxylic acid may be saturated or unsaturated. For example, the acid may be butyric acid, caproic acid, caprylic acid, capric acid, lauric acid, myristic acid, palmitic acid, stearic acid, linolic acid, linolenic acid, etc.

The addition of the stabilizer to the polyoxyalkylene polyol may conveniently be carried out in a variety of ways. The addition of the stabilizer may be made in the solid state, or as a solution in which the stabilizer is dissolved in a suitable solvent such as methyl alcohol, ethyl alcohol, benzene, etc. The addition may be carried out during or after the process of refining of polyoxyalkylene polyol, though in the former case care should be taken that the addition is carried out at least after the neutralization of alkaline catalyst employed for the polymerization. When the stabilizer is employed in a state of solution, the solvent may be removed under heating in an atmosphere of an inert gas, after mixing the stabilizer sufficiently with the polyoxyalkylene polyol.

When the stabilizer is added to polyoxyalkylene polyol directly, i.e., in a solid form and not in solution in a solvent, the stabilizer is mixed thoroughly by heating in an atmosphere of an inert gas to dissolve the stabilizer uniformly in the polyoxyalkylene polyol. The following results show the effect on stabilizing polyoxyalkylene polyol (polyoxypropylene, av. mol. wt. = 2,000).

(1) Solution of 0.1 part by weight of ascorbic acid in 10 parts by volume of methanol.
(2) Solution of 0.1 part by weight of ascorbic acid caprylic acid ester in 10 parts by volume of methanol.
(3) Solution of 0.1 part by weight of dibutyl-hydroxy-toluene in 10 parts by volume of methanol.
(4) Solution of 0.06 part by weight of dibutyl-hydroxy-toluene and of 0.04 part by weight of ascorbic acid caprylic acid ester in 10 parts by volume of methanol.
(5) Solution of 0.04 part by weight of dibutyl-hydroxy-toluene and of 0.04 part by weight of ascorbic acid in 10 parts by volume of methanol.
(6) Solution of 0.1 part by weight of butyl-hydroxy-anisol in 10 parts by volume of methanol.
(7) Solution of 0.06 part by weight of butyl-hydroxy-anisol and of 0.04 part by weight of ascorbic acid in 10 parts by volume of methanol.
(8) Solution of 0.04 part by weight of butyl-hydroxy-anisol and of 0.06 part by weight of ascorbic acid calcium salt in 10 parts by volume of methanol.

Method of Measurement — 100 parts each by weight of polyoxypropylene glycol whose average molecular weight is 2,000 is mixed well with stabilizer solution, followed by removal of the methanol in an atmosphere of nitrogen gas. Then, the compositions thus stabilized are kept at 120°C. with the aid of an infrared lamp. Decrease in the weight of the compositions are then measured. For purposes of comparison, the stability of polyoxyalkylene polyol is similarly observed without the addition of any stabilizer (control).

Stabilizer	Percent of decrease in the weight of +polyoxypropylene glycol composition				
	2 hrs.	4 hrs.	6 hrs.	8 hrs.	10 hrs.
(1) Ascorbic acid, 0.10%	0.2	0.2	0.4	2.0	
(2) Ascorbic acid caprylic acid ester, 0.10%	0.2	0.4	1.5		
(3) Dibutyl-hydroxy-toluene, 0.1%	0.2	0.6	3.0		
(4) Dibutyl-hydroxy-toluene, 0.06% + Ascorbic acid caprylic acid ester, 0.04%	0.2	0.3	0.4	0.9	3.0
(5) Dibutyl-hydroxy-toluene, 0.04% + Ascorbic acid, 0.04%	0.2	0.3	0.5	0.8	2.5
(6) Butyl-hydroxy-anisol, 0.1%	0.2	0.3	0.4	0.6	1.8
(7) Butyl-hydroxy-anisol, 0.06% + Ascorbic acid, 0.04%	0.1	0.2	0.2	0.3	0.4
(8) Butyl-hydroxy-anisol, 0.04% + Ascorbic acid calcium salt, 0.06%	0.2	0.2	0.3	0.4	0.6
Control	2.0	4.7			

From the above it can be seen that the stabilizer has an excellent effect in the stabilization of polyoxyalkylene polyol. Especially, when the stabilizer is employed together with known stabilizer(s), the effect on stabilization increases synergistically.

PHOSPHATES

Chlorinated Diphenyl Ethers

D.H. Nail; U.S. Patent 3,432,437; March 11, 1969; U.S. Patent 3,432,436; March 11, 1969; and U.S. Patent 3,352,784; November 14, 1967; all assigned to McDonnell Douglas Corporation has found that mixtures of phosphate esters, particularly triaryl phosphates such as tricresyl phosphate, with the combination of a chlorinated diphenyl ether and a chlorinated biphenyl, preferably in certain proportions noted below, provide highly improved fire resistant compositions particularly useful as hydraulic fluids and lubricants, which have good lubricity over a wide range of applied loads and good pump life when employed in a hydraulic system. These fluids also have good physical and chemical stability, good thermal and hydrolytic stability and noncorrosiveness, and particularly, have improved low temperature characteristics including low viscosity at low temperatures of the order of 0°F. and below, and low pour point and cloud point, permitting the composition to remain liquid at such low temperatures and substantially without crystallization of solids at such temperatures.

To such mixture of phosphate ester, chlorinated diphenyl ether and chlorinated biphenyl, there can be added optionally and in some instances preferably, minor amounts of suitable viscosity improvers such as polyalkyl methacrylates, and also inhibitors, as well as additives to enhance lubricity. One of the essential components of the three component fluid compositions are the phosphates. Phosphate esters suitable for use in the process are the trialkyl phosphates, triaryl phosphates, diarylalkyl phosphates and dialkylaryl phosphates. Such phosphate esters have the general formula:

$$\begin{array}{c} OR \\ | \\ R'O-P{=}O \\ | \\ OR'' \end{array}$$

where R, R' and R" are each a radical selected from the group consisting of phenyl, cresyl, xylyl, and an alkyl radical of from about 4 to 12 carbon atoms, preferably from about 4 to 10 carbon atoms. Suitable phosphate esters are, for example, tri-n-butyl phosphate, tri(2-ethylhexyl)phosphate, cresyl diphenyl phosphate and triphenyl phosphate, dihexyl phenyl phosphate. Triaryl phosphates such as cresyl diphenyl phosphate are preferred phosphates.

Suitable chlorinated diphenyl ethers which can be employed as the second essential component of the composition or fluid can include from about 1 to 10 chlorine atoms in the molecule, generally from about 1 to 6 chlorine atoms per molecule. In each instance, the mono-, di-, tri-, tetra-, penta-, and hexa-chloro derivatives of the diphenyl ether according to the process consist primarily or substantially entirely of substitution chlorine on the aromatic rings, rather than addition chlorine, and in each case the ether used can be either a particularly selected isomer or compound, or mixture of isomers, or an unresolved mixture of isomers or mixture of chlorinated ethers of varying numbers of chlorine atoms in the molecule. Such mixtures of chlorinated diphenyl ethers are obtained from the reaction of chlorine with diphenyl ether, e.g. as described in U.S. Patents 2,022,632 and 2,165,813.

The following are blends or mixtures of triaryl phosphates, chlorinated diphenyl ethers, and chlorinated biphenyls according to the process.

		Percent by Weight
(1)	Monochlorodiphenyl ether	25
	Pentachlorobiphenyl	25
	Tricresyl phosphate	50
(2)	Tetrachlorodiphenyl ether	25
	Tetrachlorobiphenyl	25
	Tricresyl phosphate	50
(3)	Pentachlorodiphenyl ether	25
	Pentachlorobiphenyl	25
	Tricresyl phosphate	50
(4)	Tetrachlorodiphenyl ether	25
	Pentachlorobiphenyl	25
	Tricresyl phosphate	50
(5)	Hexachlorodiphenyl ether	25
	Tetrachlorobiphenyl	25
	Tricresyl phosphate	50
(6)	Trichlorodiphenyl ether	15
	Pentachlorobiphenyl	35
	Tricresyl phosphate	50
(7)	Trichlorodiphenyl ether	15
	Tetrachlorobiphenyl	35
	Tricresyl phosphate	50
(8)	Tetrachlorodiphenyl ether	17
	Pentachlorobiphenyl	17
	Pentachlorodiphenyl ether	17
	Tricresyl phosphate	49

Blends (1) to (8) above are illustrative of the large number of blends or forumulations of fluid compositions which can be formulated. The formulations of higher chlorine content are particularly highly fire resistant and the compositions of lower chlorine content tend toward lower viscosity at decreased temperatures. Composition (8) illustrates formulations which contain more than one component of the class of chlorinated diphenyl ethers. Blends of this type are desirable to provide greater versatility in achieving the liquid range and viscosity required in a particular case. Fluid formulations corresponding to compositions (1) to (8) above can be used as hydraulic fluids or lubricants in industrial air compressors, hydraulic presses and hydroelectric turbines.

G.G. Groslambert; U.S. Patent 3,352,780; November 14, 1967; assigned to Labofina, SA, Belgium describes a fire resistant lubricant which comprises a polychlorinated diphenyl as the main component, containing from about 0.5 to 5%, preferably about 2 to 4%, of a phosphoric ester. The composition further usually includes additives for imparting improved nonfoaming, viscosity index improving, antirust and antioxidant characteristics.

Monoamides of Succinic Acids as Corrosion Inhibitors

A process described by B.W. Hotten; U.S. Patent 3,427,245; February 11, 1969; assigned to Chevron Research Co. utilizes mixtures of aliphatic hydrocarbon substituted succinamic acid amine salts, where the amine is a secondary amine.

These are used under extreme corrosion conditions, particularly in phosphate lubricants. The following is a list of succinamic acids which may be used for the acid-amino salt mixtures of this process.

N,N-dimethyl n-eicosenylsuccinamic acid
N,N-diethyl n-tetradecenylsuccinamic acid
N,N-diethyl n-pentadecenylsuccinamic acid
N,N-diethyl n-octadecenylsuccinamic acid
N,N-diethyl n-eicosenylsuccinamic acid
N,N-dibutyl n-tetradecenylsuccinamic acid
N,N-dihexyl n-tetradecenylsuccinamic acid
N,N-diphenyl n-hexadecenylsuccinamic acid
N,N-diphenyl n-heptadecenylsuccinamic acid
N,N-diphenyl n-octadecenylsuccinamic acid
N,N-ditolyl n-hexadecenylsuccinamic acid

The alkenyl succinic acids used in the formation of the succinamic acids are prepared by the reaction of an olefin, preferably an alpha-olefin, with maleic anhydride to form the alkenyl succinic anhydride. This product may then be hydrogenated to form the alkyl succinic anhydride. However, in most instances there would be little if any advantage in the alkyl over the alkenyl succinic anhydride and, to that extent, the hydrogenation step to the alkyl succinic anhydride is considered superfluous. The methods of reacting an olefin with maleic anhydride are known in the art.

When preparing the compounds of this process the desired alkenylsuccinic anhydride is reacted with a secondary amine having the formula: HNR^3R^4, where R^3 and R^4 are hydrocarbyl groups of from 1 to 10 carbon atoms, preferably alkyl or carbocyclic aryl. The alkyl groups will generally be of from 2 to 6 carbon atoms, while the aryl groups will generally be of from 6 to 10 carbon atoms, more usually of from 6 to 8 carbon atoms. Illustrative of various amines are:

Dimethylamine
Diethylamine
Dipropylamine
Dibutylamine
Dipentylamine
Dihexylamine
Diheptylamine
Dioctylamine
Dinonylamine
Didecylamine
Methylhexylamine

The following examples illustrate the process.

Example A: Into a reaction flask in a nitrogen atmosphere was introduced 352 g. (1.5 mols) of olefins, derived from cracking wax, in the range of from C_{10} to C_{20}, having an average molecular weight of 230, and 147 g. (1.5 mols) of maleic anhydride and the mixture heated for 10 hours at a temperature in the range of about 180° to 235°C. At the end of this time, the product was cooled and isolated yielding 355 g. Neutralization equivalent was 216 (calculated 160) indicating about 70% reaction with the remaining portion being unreacted olefin.

Example B: Into a reaction flask in a nitrogen atmosphere was introduced 362 g. (1.5 mols) of olefins, derived from cracking wax, having from 15 to 20 carbon atoms (242 average molecular weight) and 147 g. of maleic anhydride (1.5 mols) and the mixture heated at a temperature of about 180° to 250°C. for a period of 3 hours. At the end of this time, 383 g. of a product was isolated. Neutralization equivalent was 250 (calculated 170) indicating about 70% reaction to form the alkenyl succinic anhydride.

Example 1: Into a reaction flask was introduced 33 g. of the product obtained in Example A and 7.3 g. (0.1 mol)(30 mol percent excess) of diethylamine and the mixture heated at 100°C. for 2 hours. At the end of this time, the reaction mixture was cooled and the product isolated. Neutralization equivalent was 450.

Example 2: Into a reaction flask was introduced 34 g. of the product of Example B and 7.3 g. of diethylamine (0.1 mol) (47 mol percent excess) and the mixture heated at 50°C. for one hour. At the end of this time, the reaction product was cooled and the product isolated. Neutralization equivalent was 343.

Example 3: Into a reaction flask was introduced 9.2 g. (0.019 mol) of the product obtained in Example B and 4.68 g. (10% excess) of diphenyl amine and the mixture heated at 150°C. The reaction mixture was allowed to cool and the product isolated.

Example 4: Into a reaction flask was introduced 9.2 g. (0.019 mol) of the product obtained in Example B and 5.10 g. (20% excess) of diphenyl amine and the mixture heated at 150°C. The reaction mixture was allowed to cool and the product isolated.

The products were tested under a variety of severe conditions to demonstrate their excellence in inhibiting rust formation. The first tests reported are the turbine oil test ASTM D 665-60 and the Emulsibility Test ASTM D 1401-56T, the latter demonstrating the ability of oil and water to separate from each other. Since it is essential that the oil and water separate at a reasonable rate in a turbine, the effect of any additives upon the separation is essential to a determination of their efficacy. The rust test uses both distilled and synthetic sea water on a standard carbon steel specimen.

The table on the following page indicates results obtained both as to emulsibility and rusting in distilled water as well as synthetic sea water. The sample used 0.1 weight percent of the succinamic acid in a turbine oil comprising 92%

heavy and 8% light turbine oil base, o,o'-di-t-butyl-p-cresol antioxidant, 0.3 weight percent and poly(di-methyl-siloxane) defoamer, 0.005 weight percent.

Succinamic Acid, Example	Emulsibility		Rust Observed	
	Vol., ml.	Time, min.	Distilled Water	Synthetic Sea Water
1	3	24	No rust	No rust
2	3	10	No rust	No rust
3	3	21	No rust	No rust
4	3	18	No rust	No rust

The materials of this process permit rapid separation between the oil and water, while providing excellent rust protection. Using tricresyl phosphate in place of the turbine oil in the turbine oil rust test described above, the following results were obtained:

Succinamic Acid, Example	Weight Percent	Rust Observed	
		Distilled Water	Synthetic Sea Water
	0	Heavy rust	Heavy rust
1	1	No rust	Few specks
1	4	No rust	No rust
4	1	No rust	No rust

Using 2 weight percent, the composition of Example 1 was tested in a grease rusting test (ASTM D 1743-60T). The test is described as a method for determining the corrosion preventative properties of greases using grease-lubricated tapered roller bearings stored under wet conditions. This method shows correlation between laboratory results and service for grease lubricated aircraft wheel bearings. The grease used was a 15% lithium stearate-bis(2-ethylhexyl)sebacate grease. At the end of the test, no rust was observed, giving the grease the high rating of 1.

Finally, the exemplary composition of Example 4, the partial salt of N,N-diphenylalkenylsuccinamic acid was tested with a variety of commercial phosphate ester hydraulic fluids in the Humidity Cabinet Test (ASTM D 1748). The test employs 100% relative humidity and 120°F. temperature. The following table indicates the results:

Phosphate Ester Hydraulic Fluid	Rust Observed in 24 Hours	
	Without Additive, %	With 1 Wt. Percent of Ex. 4
Skydrol	100	Trace
Skydrol 500A	100	Trace
Skydrol 7000	100	Trace
Pydraul AC	100	Trace
Pydraul F-9	100	Trace

Skydrols are nonflammable aviation hydraulic fluids; 500A is based on dibutyl phenyl phosphate and 7000 is based on octyl diphenyl phosphate. Pydrauls are nonflammable industrial hydraulic fluids, based on tricresyl phosphate and chlorinated biphenyl.

SILICATES

Trimethylolalkane-Diorthosilicic Acid Esters

H. Göthel and H. Feichtinger; U.S. Patent 3,514,402; May 26, 1970; assigned to Ruhrchemie AG, Germany describe the use of liquid materials comprising diorthosilicic acid esters of the formula: $(Si—O—A—O—Si)—(OR)_{6-n}(OR')_n$ where A represents a radical of an ester derived from a trimethylolalkane containing a quaternary carbon atom and having up to about 10 carbon atoms, preferably a trimethylolalkane, and a saturated carboxylic acid having about 3 to 20 carbon atoms corresponding to the structural formula:

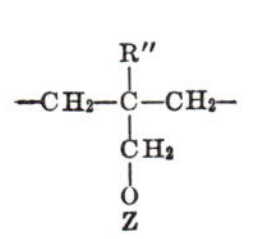

where Z is the ester residue of the carboxylic acid and R" is hydrogen or a lower alkyl of about 1 to 6 carbon atoms, R represents an alkyl group having about 4 to 20 carbon atoms, R' represents a n-butoxy-diethylene glycol ether radical, n has a value of 0 to 3 and R and R' may be the same or different, as lubricants in jet propulsion aircrafts.

The diorthosilicic acid esters where the A-groups are radicals of the monoester of trimethylolethane, trimethylolpropane or trimethylolbutane with pivalic acid, isobutyric acid or β,β-dimethylbutyric acid constitute broadly preferred diorthosilicic acid esters. The monoesters of polyvalent alcohols, characterized by A in the formula given on the previous page can be obtained by reaction of one mol of the polyvalent alcohol and one mol of the carboxylic acid, at a temperature below about 170°C. followed by distillative recovery of the ester. The starting materials for the R groups having about 7 to 20 carbon atoms may be derived from alcohols obtained by hydroformylation of straight chain olefins followed by hydrogenation.

The diorthosilicic acid esters may be prepared by the conventional methods employed for the production of diorthosilicic acid esters as known in the art, as for instance by reacting stoichiometric amounts of corresponding water-free alcohols, polyoxyalkylene glycols and monocarboxylic acid esters of the polyvalent alcohols with gaseous silicon tetrachloride as described in German Patent 1,142,855. The advantages and properties of the diorthosilicic acid esters of the process as lubricants are shown in Table 1, where a comparison of certain of the esters of the process with certain known silicic acid esters has been set out. The esters of the process used in the comparison and listed in Table 1 are the following:

(I) trimethylolethane-monopivalate-diorthosilicate-hexa-n-nonyl ester
(II) trimethylolpropane-monopivalate-diorthosilicate-hexa-n-nonyl ester
(III) trimethylolpropane-monoisobutyrate-diorthosilicate-hexa-n-nonyl ester
(IV) trimethylolpropane-monopelargonate-diorthosilicate-hexa-n-nonyl ester
(V) trimethylolethane-monopivalate-diorthosilicate-hexa-n-octyl ester
(VI) trimethylolpropane-monopivalate-diorthosilicate-hexa-n-octyl ester
(VII) trimethylolpropane-monoisobutyrate-diorthosilicate-di-n-nonyl-di-n-decyl-di-n-undecyl ester
(VIII) trimethylolethane-monopivalate-diorthosilicate-hexa-n-decyl ester
(IX) trimethylolpropane-monopivalate-diorthosilicate-hexa-n-decyl ester
(X) trimethylolpropane-mono-β,β-dimethylbutyrate-diorthosilicate-hexa-n-decyl ester
(XI) trimethylolpropane-monocaproate-diorthosilicate-hexa-n-decyl ester
(XII) trimethylolpropane-mono-3,5,5-trimethylhexanoate-diorthosilicate-hexa-n-decyl ester
(XIII) trimethylolpropane-monopropionate-diorthosilicate-hexa-n-decyl ester
(XIV) trimethylolpropane-monopivalate-diorthosilicate-hexa-iso-decyl ester
(XV) trimethylolpropane-monopivalate-diorthosilicate-tri-n-undecyl-tri(diethyleneglycol-n-butyl ether) ester

Esters used for comparison are the following dibasic esters and orthosilicic acid esters known in the art:

(A) bis-(2-ethylhexyl)-sebacate
(B) trimethylolpropane-tripelargonate
(C) trimethylolpropane-tricaproate
(D) diethylpropanediol-1,3-di-3,5,5-trimethylhexanate
(1) tetra-n-butyl-orthosilicic acid ester
(2) tetra-3,5,5-trimethylhexyl-orthosilicic acid ester
(3) tetraisotridecyl-orthosilicic acid ester

TABLE 1

Ester	Molecular weight	Density viscosity in cst. at ° C. d_4^{20}	98.9	37.8	−40	−54	Viscosity index	Setting point (° C.)	Flame point (° C.)
I	1,116	0.933	5.49	24.0	3,300	15,900	166	−70	255
II	1,130	0.936	5.6	24.2	3,300	15,800	168	−69	258
III	1,116	0.918	9.9	49.0	11,000	61,100	154	−63	257
IV	1,186	0.954	10.4	55.0	11,900	66,000	150	−61	260
V	1,032	0.948	5.71	23.6	2,520	11,050	173	−70	242
VI	1,046	0.946	5.05	21.9	2,715	13,250	168	−70	246
VII	1,200	0.934	8.9	46.0	7,680	35,800	151	−65	262
VIII	1,200	0.925	5.85	26.2	3,040	35,000	163	−63	260
IX	1,214	0.923	5.93	27.0	4,360	25,000	161	−67	267
X	1,228	0.934	7.8	38.0	6,600	36,300	154	−64	270
XI	1,228	0.961	18.0	100.0	50,500	----------	147	−52	265
XII	1,270	0.943	11.3	57.5	15,100	88,000	152	−58	275
XIII	1,186	0.974	10.7	56.0	14,150	----------	150	−54	262
XIV	1,214	0.910	5.25	24.7	6,380	72,000	153	−65	268
XV	1,268	0.968	6.1	26.5	4,800	24,000	167	−65	272
A	426	0.917	3.28	12.5	1,372	10,150	152	−71	232
B	554	0.943	4.6	21.8	6,010	----------	145	−59	278
C	428	0.974	3.0	12.3	2,085	15,400	110	−65	232
D	412	0.914	3.88	20.0	18,800	----------	92	−57	220
1	320	0.896	1.04	1.85	12.0	51.0	230	−127	78
2	600	0.879	3.4	12.78	850	5,400	161	−76	219
3	824	0.875	6.6	36.0	21,400	----------	135	−52	262

Table 2 on the following page gives the results of various tests run on the same compounds as set forth above with respect to their utility as lubricants. The esters having molecular weights above 1,000 boil at temperatures above 450°C. at 760 torr, above 250°C. at 1 torr and above 180°C. at 0.01 torr. Their flame points range above 260°C.

TABLE 2

Ester	Boiling behavior DIN 51751, 51567 boiling temperature (° C.) at torr			Oxidation test		Lubricating action: Attrition values in the four ball apparatus at 200° C. and a load of 40 kg., scratch diameter in mm. after—		
	1	0.01	760	Evaporation value, percent by weight	Numerical characteristic	1 h.	2 h.	3 h.
I	255	199	468	1.30	240	1.36	1.54	1.72
II	262	204	475	1.12	217	1.35	1.56	1.70
III	257	200	470	1.2	238	1.33	1.48	1.63
IV	267	211	483	1.54	288	1.35	1.51	1.63
V	238	181	447	2.02	282	1.40	1.59	1.85
VI	244	186	454	2.0	262	1.44	1.63	1.83
VII	272	216	488	0.5	225	1.31	1.51	1.64
VIII	270	213	485	0.8	207	1.26	1.42	1.65
IX	274	218	490	0.77	181	1.25	1.43	1.58
X	279	223	495	0.9	219	1.25	1.45	1.60
XI	277	221	493	1.1	245	1.27	1.44	1.67
XII	288	231	505	1.17	287	1.25	1.41	1.62
XIII	264	208	480	0.7	210	1.26	1.41	1.61
XIV	274	218	490	1.3	233	1.32	1.52	1.65
XV	284	226	500	1.1	275	1.26	1.58	1.70
A	222	165	428	2.1	536	1.6	1.88	2.1
B	258	194	462	1.6	320	1.51	1.72	1.85
C	220	163	425	1.95	420	1.6	1.78	1.9
D	205	150	410	2.06	468	1.65	1.85	2.0
1	102	----------	283	12.2	1,363	2.27	[1]	----------
2	223	166	430	6.5	507	1.95	2.3	[1]
3	300	242	518	3.5	328	1.58	1.83	2.05

[1] Oil evaporated.

The viscosity-temperature responsivity of the esters, determined according to the viscosity index of Dean and Davis (DIN 51 563, see G. Vögtle, Lexikon der Schmierungstechnik, Stuttgart, 1964) is low. Diorthosilicic acid esters where A consists of a radical of a trimethylolethane- or trimethylolpropane-pivalic acid monoester distinguish themselves from esters with equal R-groups, but other acid components in the A-group by their lowest viscosity and their highest viscosity index. Due to their viscosity range, the esters exhibit outstanding low temperature characteristics. The setting points set out in Table 1 depend on their viscosity and are not caused by solid deposits.

The thermal stability of the esters was determined by the following test: 300 ml. of the ester to be evaluated (containing 1% by weight phenothiazine as oxidation inhibitor), were contacted with 5.1 air/hour at 200°C. (±1°) for 32 hours. The vaporization value is equal to the amount of liquid products split off expressed in percent by weight. The numerical characteristic indicates the amount and the nature of the volatile oxidation products and renders a differentiated characteristic of the oxidation tendency of the esters possible. It corresponds to the sum of the weights in mg. of the carbon dioxide evolved during the oxidation and of the organic acids in the condensed, liquid oxidation products, respectively.

The oxidative stability of the diorthosilicic acid esters was evaluated in comparison with that of dibasic esters and orthosilicic acid esters known in the art. Ester A is illustrative of lubricants which can be employed in the subsonic range. Those lubricants are distinguished by outstanding low temperature characteristics and favorable viscosity indices, but possess only low oxidative stability. Ester B and Ester D are typical for lubricants to be used in the supersonic range up to Mach 2.2. Their oxidative stability is good but their viscosity indices and low temperature characteristics are unsatisfactory. It is seen that the diorthosilicic acid esters according to the process possess the excellent low temperature characteristics as well as viscosity-temperature responsivity as the Ester A combination with a good or even better oxidative stability as the illustrative Esters B to D. As can further be seen from Table 2, orthosilicic acid esters (Esters 1 to 3 given as examples) yield too high evaporation values at the oxidation test.

The esters according to the process exhibit further good lubricating action. Differentiated numerical characteristics were determined by the attrition test with the four-ball apparatus of Boerlage (Lexikon der Schmierungstechnik, G. Vögtle, Franckh'sche Verlagshandlung, Stuttgart, 1964) at 200°C. under a constant load of 40 kg. dependent upon the testing time (after 60, 120 and 180 minutes). The four-ball apparatus was heated in an electrically heated metal bath. The oils to be tested were inhibited against oxidation by addition of 1% by weight of phenothiazine. The diameters of the scratches were determined by microscopic inspection. In comparison to the Esters A to D the process esters yielded the lowest attrition values. The small growth rates of these values show the efficiency of the process esters as lubricants.

In related work, H. Göthel and H. Feichtinger; U.S. Patent 3,444,081; May 13, 1969; assigned to Ruhrchemie AG, Germany describe a diorthosilicic acid ester having a group comprising a pentaerythrite-dicarboxylic acid ester or dipentaerythrite-tetracarboxylic acid ester radical suitable for use as a lubricant for extremely high performance aircraft jet engines.

Polyalkylene Glycol-Diorthosilicic Acid Esters

H. Göthel, H. Feichtinger and H. Noeske; U.S. Patent 3,336,227; August 15, 1969; U.S. Patent 3,383,315; May 14, 1968; and U.S. Patent 3,538,001; November 3, 1970; all assigned to Ruhrchemie AG, Germany have found that liquid material comprising diorthosilicic acid esters of the formula: $A(OSi)_2(OR)_{6-n}(OR')_n$ where A represents a polyoxyalkylene radical having 1 to 6 and preferably 2 to 4 ether-oxygen atoms, R represents a polyoxyalkyleneglycol ether radical having 2 to 8 ether-oxygen atoms, R' represents an alkyl group having 9 to 20 carbon atoms, n is of the class consisting of from 1 to 4 and 0, and where R' and R may be the same or different, provides liquid or fluid compositions which are suitable for use as lubricants, brake fluids, hydraulic fluids, etc. A preferred group of diorthosilicic acid esters has the following formulae:

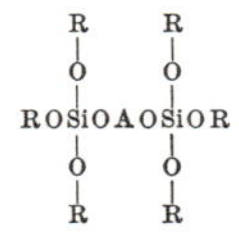

$$A(OSi)_2(OR)_3(OR')_3$$

where A, R and R' are as previously defined. The following examples illustrate the process.

Example 1: A reaction vessel (a so-called bubble-column reactor) consisting of a vertical glass tube with an internal diameter of 69 mm. and a height of 550 mm. provided with a glass frit (Jena apparatus glass, grain coarseness 2) in its bottom was charged with 121 g. diethyleneglycol and 1,122 g. diethyleneglycol-n-butyl ether, which had previously been dried to a water content below 0.1%. Silicon tetrachloride was applied dropwise from a dosing vessel into an evaporator coil heated at 80°C., which was gas-tight connected with the reaction vessel, by a conduit joining the reaction vessel below the glass frit. The silicon tetrachloride vapors evolved in the evaporator coil were sucked through the frit under a vacuum of 180 torr. The ascending $SiCl_4$ vapors together with the hydrogen chloride evolved during the reaction formed a bubble-column of 420 mm. height with the liquid in the reactor. The liquid mixture of alcohol and ether, which had been preheated to 40°C., was heated up to 52°C. by the heat of reaction.

In order to compensate for the decreasing evolution of heat during the reaction, the reaction mixture was heated at about 50°C. during the course of the reaction. 357 g. $SiCl_4$ were introduced and converted during 100 minutes. After termination of the reaction, the reaction vessel was heated at 110°C. for 2 hours, whereby dissolved hydrogen chloride was removed from the reaction mixture under introduction of nitrogen at a vacuum of 20 torr. (After the treatment the Beilstein reaction was negative.) Thereafter the nonreacted excess amounts of the reactants (125 g.) were distilled off from the reaction product under a vacuum of 0.01 torr up to a temperature of 196°C. As distillation residue, 1,150 g. of diethyleneglycol-diorthosilicic acid hexa-diethyleneglycol-n-butylether ester (density d_4^{20} = 1.038, index of refraction n_{20}^D = 1.4452) were obtained. The silicon content of the product ester was determined as 5.02% (calculated 4.98%); the viscosity of the ester at 98.9°C. was 3.9 cs., at 37.8°C. 13.5 cs., at -40°C. 1,273 cs., at -54°C. 7,490 cs.; its viscosity index amounted to 203, its flow point (or setting point) was 71°C.

Example 2: 50 parts by volume of a dibasic ester known in the art as bis-3,5,5-trimethylhexyl-adipate were admixed with 50 parts by volume diethyleneglycol-diorthosilicic acid-tri-(diethyleneglycol-n-butylether)-tri-(iso-decyl) ester. The advantages of the mixture of the ester of the process to the conventional ester lubricant are shown in the following table, where the properties of the two single esters are shown in comparison with those of the ester mixture, hereafter designated admixture 50/50% by volume.

	d_4^{20}	Viscosity in cst. at ° C.			Viscosity index	Flame point,° C.	Welding value load in kg.
		98.9	37.8	−54			
Bis-3,5,5-trimethylhexyl adipate	0.916	3.15	11.86	15,720	147	222	150/160
Diethyleneglycol-diorthosilicic acid-tri-(diethyleneglycol-n-butylether)-tri-(isodecyl)-ester	0.969	3.9	14.43	10,100	188	264	170/180
Admixture, 50/50 percent by vol	0.944	3.53	13.01	12,700	169	238	170/180

Hexaalkoxydisiloxane Mixtures

M.A. Pino; U.S. Patent 3,320,297; May 16, 1967; assigned to Chevron Research Company describes a process for producing mixtures of polysiloxanes by reacting together in one single reaction step a silicon tetrahalide, an alkanol from the group of primary and secondary alcohols containing from 1 to 16 carbon atoms, water, and a tetraalkyl orthosilicate, the alkyl substituents of which correspond to the alkyl portion of the alkanol. The application of tetraalkyl orthosilicate in substantial proportions in this reaction is found to be critical and permits moderating the process so as to influence in a desired direction the distribution of the several components in the final polysiloxane reaction product mixture. It dispenses with the need for additional processing agent, such as "acid acceptors".

The orthsilicate moderator, whose alkyl portion should correspond to that of the alkanol reactant, may be either supplied from the preceding process cycle or yet supplied from a source extraneous to the process. When introduced into the reaction zone, the orthosilicate moderates the reaction so as to contribute to the formulation of polyalkoxypolysiloxane mixtures meeting the particular specifications of the industry.

An important advantage resulting from the use of this coreactant orthosilicate moderator in the preparation of predominantly disiloxane product mixtures is that at the termination of the process the final product mixture contains less orthosilicate than one did originally start with at the beginning of the process. By the same token, in the preparation of predominantly polysiloxane mixtures containing a preponderance of higher polysiloxanes over disiloxanes, the application of orthosilicate as a reaction moderator likewise prevents excessive formation and accumulation of orthosilicate as a by-product in the final product mixture. Thus, whether one strives to obtain a predominantly disiloxane or a predominantly polysiloxane product mixture, the burdensome task of separating from the ultimate product mixture the by-product orthosilicate and finding for it a suitable outlet (use or purchaser) is either greatly minimized or completely eliminated. The following examples illustrate the process.

Example 1: This example describes a series of comparison tests effected to bring out the advantages obtained from the moderating function of orthosilicate in the production of predominantly hexaalkoxydisiloxane mixtures. In the first run of this test series (Run 1), wet alkanol, namely 2-ethyl butanol, and a silicon tetrahalide ($SiCl_4$) were added in substantially stoichiometric amounts into an intially dry flask. The temperature was approximately 100°F. Upon distillation (topping), the composition of the reaction product mixture was determined. The disiloxane-polysiloxane ratio was found to be much lower than the acceptable minimum of 1.5 specified by the industry.

In the next run of the same series (Run 2), the reaction was carried out under the same conditions, except that the wet alkanol contained about 67% of the theoretical amount of water required for the formation of disiloxane. On completion of the reaction, the ratio of disiloxane to polysiloxane was again found not to come up to the acceptable minimum of 1.5 specified by the industry. In Run 3, the reaction was carried out again under similar conditions, but the water in the wet alkanol was present in excess of the theoretical amount required for the formation of disiloxane. The disiloxane to polysiloxane volume ratio was extremely low, and the product obviously was unsuitable for industrial uses.

In Run 4, the reaction flask initially contained 200 ml. of orthosilicate from a previous reaction batch. In this case, the reaction was again carried out at 100°F. and yielded a product mixture with the disiloxane-polysiloxane ratio equal to 1.80 which fully satisfied the specifications. The results of the described four runs are tabulated in the table below which shows the ratio of disiloxane to polysiloxane for each run and the composition of the corresponding reaction product mixtures after having been distilled.

Preparation of Disiloxane Mixtures

Run No.[1]	Ratio of Disiloxane to Polysiloxane	Composition of Product, in Vol. Percent [2]			
		Alkanol	Ortho-Silicate	Disiloxane	Polysi-loxane
1	0.77	8	39	23	30
2	1.33	10	62	16	12
3	0.42	14	12	22	52
4 [3]	1.80	0	2	63	35

[1] All batches made with one-half mol of $SiCl_4$.
[2] By distillation after removal of reaction moderator where used.
[3] About 10% of the reaction mixture lost in blowing with nitrogen.

Example 2: In this test series, carried out to illustrate the preferred operation of the process with recycle of the orthosilicate moderator, stoichiometric proportions of silicon tetrachloride as a typical silicon tetrahalide, 2-ethyl butanol as a typical alkanol, and water (14.4 mol percent, based on the total wet alkanol) were added simultaneously into a glass-lined reactor which initially contained enough tetra-2-ethylbutyl orthosilicate (about one-half of the total volume of all reactants) to assure the desired distribution of the hexaalkoxydisiloxane and higher polyalkoxypolysiloxanes. The product, upon distillation, disclosed a volume ratio of disiloxane to polysiloxane of about 3.8:1.

At the time of starting, the orthosilicate was present in a volume somewhat greater (1.3:1 volume ratio) than half of the total reaction mixture. It reacted with wet alkanol to form disiloxane, thereby desirably moderating the disiloxane to polysiloxane ratio. The main reaction followed the equation:

$$2SiCl_4 + 6C_6H_{13}OH + H_2O \xrightarrow{\text{diluent}} (C_6H_{13}O)_3SiOSi(OC_6H_{13})_3 + 8HCl$$

The temperature was kept at less than about 105°F., and the reaction mixture (a 30-gallon batch) was stirred vigorously at 60 to 100 rpm. When the reaction was completed, the mixture was withdrawn from the reactor and purged of dissolved hydrogen chloride by blowing with an inert gas (nitrogen). The orthosilicate which did not convert to polysiloxanes and the unreacted 2-ethyl butanol, both of which remained in the crude unseparated product mixture, were substantially completely removed from the reactor by topping distillation. The results of this series of test runs are tabulated in the table on the following page.

At the conclusion of the test series, it was observed that invariably less orthosilicate was present in the final reaction mixture than at the start of the reaction. In other tests, which use lesser volumes of orthosilicate at the beginning of the reaction, it was observed that the less orthosilicate was fed in, the more of it was found present in the crude reaction product.

	Run No.				
	1	2	3	4	5
Time of addition in hours	4.5	4.7	5.0	10.8	10.8
Water in alcohol, mol percent	14.4	20	30	20	20
Composition of reaction product mixture in vol. percent.					
2-ethyl butanol	15	15	15	[1] 25	[2] 37
Orthosilicate	45	36	30	20	6
Disiloxane	30	36	35	34	31
Polysiloxane	8	12	17	18	22
Disiloxane/polysiloxane ratio	3.8	3.0	2.1	1.8	1.4

[1] Neutralized by blowing with N_2.
[2] Neutralized with $NaHCO_3$.

On substituting 2-ethyl hexanol in another series of tests for 2-ethyl butanol and on repeating the procedure described under substantially identical conditions, the resulting disiloxane-polysiloxane mixture is found to have a similar disiloxane to polysiloxane ratio of 3:1, and the quantity of orthosilicate contained in the final product mixture is invariably less than the orthosilicate available in the reactor at the beginning of the reaction.

Similar observations are made with ethanol, n-hexanol and n-dodecanol as the alkanol reactant, in accordance with the process.

POLYPHENYL ETHERS

GENERAL SYNTHESIS

Bis(m-Phenoxyphenyl) Ether

E.S. Blake and W.C. Hammann; U.S. Patent 3,451,061; June 17, 1969; assigned to Monsanto Company have found that a limited class of meta-linked unsubstituted polyaromatic polyethers containing at least four benzene rings in the polyether chain possess the desired properties for functional fluid applications. This class of functional fluid compositions can be illustrated by the structural formula

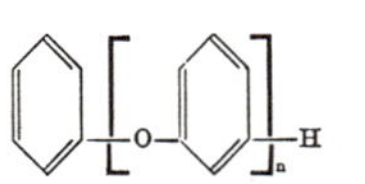

where n is an integer of at least 3 and preferably is an integer from 3 to 6 inclusive, i.e., the series of meta-linked unsubstituted polyethers containing from 4 to 7 benzene rings, respectively illustrated by the compounds:

bis(m-phenoxyphenyl) ether
m-bis(m-phenoxyphenoxy) benzene
bis[m-(m-phenoxyphenoxy) phenyl] ether
m-bis[m-(m-phenoxyphenoxy) phenoxy] benzene

One of the most widely employed commercial functional fluids intended as having high thermal stability is di(2-ethylhexyl) sebacate and this compound has a thermal stability or thermal decomposition point of 525°F., which will be noted to be about 300°F. below the thermal decomposition points of the poly(m-oxyphenylene) benzene compounds of the process. Whereas this is a very substantial numerical difference this difference is far greater than apparent from the comparison of the thermal decomposition points since the practical use of the functional fluids in service is greatly extended even by relatively small differences in thermal decomposition points. Thus in determining the useful life of the functional fluid at about 600°F., i.e., the time in hours at about 600°F. for 10% of the functional fluid to decompose, it is found that the di(2-ethylhexyl) sebacate has a useful life of about 10 hours, whereas the compounds of the process have a useful life of the order of about 6 x 10^6 hours. The following examples are illustrative of the process.

Example 1: A mixture of 55.7 g. (0.3 mol) of m-phenoxyphenol, 15.7 g. (0.28 mol) of potassium hydroxide, and 50 ml. of xylene was heated at reflux under a nitrogen atmosphere until 6 ml. of water had been collected in a Dean and Stark trap. The xylene remaining in the reaction mixture was distilled therefrom and 60 g. (0.24 mol) of m-bromophenyl phenyl ether and 1 g. of copper powder were introduced into the reaction vessel. Then the reaction mixture was heated at 240° to 250°C. overnight.

The reaction mixture was poured into 500 ml. of a 5% aqueous potassium hydroxide solution and extracted with four 250-ml. portions of benzene. The combined benzene extract was then washed with two 250-ml. portions of 5% potassium hydroxide solution and then with three 250-ml. portions of water, dried over anhydrous sodium sulfate, filtered, and the benzene removed by distillation up to 150°C. at 15 mm. mercury to obtain 92 g. of a dark red oil. The reaction products were then distilled through an 11-inch Vigreux column to obtain an 86% yield of crude product, the bulk of which distilled at 222° to 224°C. at 1 mm. of mercury.

This fraction was taken up in 100 ml. of benzene, treated with 5 g. of charcoal, filtered through Attapulgus earth and then through alumina to obtain a colorless solution. The benzene was removed by distillation up to 170°C. at 15 mm. of mercury pressure and the residue distilled through an 11-inch Vigreux column to recover 61 g. of product identified as bis(m-phenoxyphenyl) ether. This product had a distillation range of 214° to 217°C. at 0.7 mm. of mercury, a boiling point of about 860°F., a refractive index of about n_D^{25} 1.6195, a thermal decomposition point of 813°F., and was still fluid down to about 22°F. Analysis — Calculated for $C_{24}H_{18}O_3$: C, 81.34%; H, 5.12%. Found: C, 81.18%; H, 5.20%.

Example 2: A mixture of 87.5 g. (0.47 mol) of m-phenoxyphenol and 24.6 g. (0.44 mol) of potassium hydroxide was introduced into a reaction vessel and heated until reaction occurred over the temperature range of about 80° to 110°C. Then 40 ml. of toluene was introduced and the reaction mixture refluxed until 10 ml. of water had been collected in a Dean and Stark trap. The toluene was then distilled from the reaction mixture while maintaining the reaction mixture above about 135° to 140°C. to avoid the precipitation of the potassium m-phenoxyphenoxide.

Thereafter 50 g. (0.21 mol) of m-dibromobenzene and 1 g. of copper powder were added to the reaction mixture and after the initial reaction the mixture was held at 220° to 230°C. for 4 hours with stirring, then cooled to 80°C. and 150 ml. of benzene added thereto. The reaction mixture was poured into 200 ml. of water, the organic phase recovered and filtered, diluted to about 300 ml. benzene and washed with five 100-ml. portions of 5% aqueous sodium hydroxide solution, dried with anhydrous calcium chloride, filtered, and the benzene distilled therefrom at a temperature up to 120°C. at 15 mm. mercury to obtain 86 g. of a dark red oil.

The oil reaction products were then distilled through an 11-inch Vigreux column and the fraction having a distillation range of about 280° to 294°C. at 2.5 mm. mercury recovered. This fraction was taken up in 200 ml. of benzene, treated with 5 g. of charcoal, filtered, and then passed through alumina to obtain a substantially colorless solution. The benzene was stripped off and the residue distilled through an 11-inch Vigreux column to recover the product identified as m-bis(m-phenoxyphenoxy) benzene. This product had a boiling point of 292° to 294°C. at 2.5 mm. of mercury, a refractive index of about n_D^{25} 1.6304, a thermal decomposition point of 848°F., and was still fluid down to about 44°F. Analysis — Calculated for $C_{30}H_{22}O_4$: C, 80.70%; H, 4.97%. Found: C, 80.71%; H, 4.98%.

Diphenoxybiphenyl Fluid Isomers

R.M. Schisla and W.C. Hammann; U.S. Patent 3,406,207; October 15, 1968; assigned to Monsanto Company have found that specific members of the diphenoxybiphenyls, i.e. compositions having the structural formula

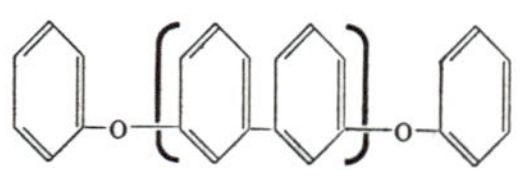

where at least one of the phenoxy substituents is in the meta configuration with respect to the biphenyl nucleus, are particularly useful as thermally stable functional units. Of the twelve possible diphenoxybiphenyls, only four isomers have suitable physical properties to provide the desired wide temperature fluid range necessary for use as suitable functional fluid compositions.

The suitable isomers of diphenoxybiphenyl having fluid properties at room temperature are 2,3'-diphenoxybiphenyl, 3,3'-diphenoxybiphenyl, 3,4'-diphenoxybiphenyl, and 3,4-diphenoxybiphenyl and this restricted class of diphenoxybiphenyls will be referred to as the "fluid diphenoxybiphenyls." All of the other possible isomers of diphenoxybiphenyl are crystalline solids at room temperature having melting points ranging from about 153° to 311°F., i.e. the crystalline solid diphenoxybiphenyls have a melting point of the order of at least about 100°F. above that of the specific class of fluid diphenoxybiphenyls. The melting points of the diphenoxybiphenyls which are not fluids at normal room temperatures are given below:

Isomer	Melting Point, °F.
4,4'-diphenoxybiphenyl	311
2,2'-diphenoxybiphenyl	213
2,4'-diphenoxybiphenyl	153
2,3-diphenoxybiphenyl	197
2,4-diphenoxybiphenyl	172
2,5-diphenoxybiphenyl	202
2,6-diphenoxybiphenyl	179
3,5-diphenoxybiphenyl	198

The use of individual isomers of the dihalogenated biphenyl starting materials can become somewhat expensive and in general the final compositions preferably are mixtures rather than individual compounds per se. Accordingly, one can use chlorinated biphenyl compositions containing about 32% chlorine, corresponding to an average composition containing 2 atoms of chlorine per molecule of biphenyl, i.e. this composition is a mixture of dichlorobiphenyls. Such mixtures of dichlorobiphenyls provide a cheap reactant and an excellent starting material for the preparation of mixed fluid diphenoxybiphenyl compositions. One can prepare such dichlorobiphenyl mixed compositions by the chlorination of biphenyl in the presence of an iron catalyst, such as a ferric chloride catalyst, at reaction temperatures preferably between about 40° and about 150°C. and more preferably still between about 100° and about 140°C. until the chlorinated biphenyl product contains up to about 32% chlorine.

The mixed dichlorobiphenyls preferably are then isomerized in the presence of anhydrous aluminum chloride and anhydrous hydrogen chloride at temperatures from about 50° to about 250°C. and preferably from about 100° to about 210°C., and more preferably still from about 150° to about 200°C. for a time of from about 15 minutes to about 10 hours or more and preferably from about 1 hour to about 5 hours. The amount of anhydrous aluminum chloride isomerization catalyst employed in the isomerization reaction can vary from about 0.1 to about 10 or more, and preferably from about 0.5 to about 2 weight percent, based on the halobiphenyl composition.

Sufficient dry potassium phenate to react with the halogen atoms of the dihalobiphenyl isomerization mixture is added to a reaction vessel together with an effective amount of the copper catalyst, e.g. cupric chloride, and the temperature raised to from about 225° to 245°C. The crude isomerized dihalobiphenyl composition is then added to the dry reaction mixture at such a rate as to maintain the temperature below about 250°C. after all of the dihalobiphenyl composition has been added to the reaction vessel. The reaction is continued at reflux temperatures of from about 200° to 240°C. for several hours, e.g., of from about 2 to about 10 hours or longer.

The diphenoxybiphenyl mixed composition can then be treated in one of several ways. The reaction mixture can be cooled, taken up in an inert organic solvent and washed with water to remove any residual potassium phenate and the potassium halide salt formed during the reaction. The organic phase is then separated from the aqueous phase and dried. Thereafter the inert solvent employed, for example toluene, is removed by distillation. It has been found that even the presence of about 2% of the 4,4'-isomer will cause a mixed composition of the diphenoxybiphenyls to sludge at room temperatures.

It was also found that this deficiency could be readily overcome by the substantial removal of the 4,4'-isomer by treatment of the diphenoxybiphenyl composition with an alkane solvent, e.g. hexane, heptane, nonane, and commercial mixtures of alkanes. The alkane solvent and reaction mixture are warmed together preferably at about 40° to 60°C., agitated, and then cooled, whereby the 4,4'-isomer is precipitated therefrom and removed by filtration. The solvent is then stripped from the reaction mixture which may be used as such as a functional fluid or may be further fractionated to provide any particularly desired cut. The recovered 4,4'-dichlorobiphenyl fraction can be separately isomerized to provide a mixture containing the major components 3,3'-diphenoxybiphenyl (about 41%) and 3,4'-diphenoxybiphenyl (about 45%). The following examples are illustrative of the process.

Example 1: A one-liter, four-necked flask was fitted with a stirring means, thermometer, addition funnel, and Dean-Stark trap and charged with 250 g. phenol (2.66 mols), 56 g. potassium hydroxide (1 mol), and 50 ml. of toluene. The substantial excess of phenol employed in this example was to prevent the potassium phenate from "caking-out" at temperatures around 125°C. The water liberated during the formation of potassium phenate was distilled off at a pot temperature of 125° to 135°C. After removal of the water, the reaction mixture was heated to 170°C. to remove the toluene.

Thereafter, 5 g. of cupric chloride catalyst (0.04 mol) was introduced into the reaction vessel followed by the dropwise addition of 29 g. of 3,3'-dichlorobiphenyl (0.13 mol) over a period of from 20 to 30 minutes. After the final addition of the 3,3'-dichlorobiphenyl, the reaction mixture was refluxed for 12 hours at 195° to 200°C. with agitation. The reaction mixture was allowed to cool to about 75°C. and taken up with benzene, washed with an aqueous potassium hydroxide solution, the organic and aqueous phases separated, and filtered.

The organic phase was further washed with potassium hydroxide solution followed by water, the aqueous phases being removed, and the organic phase dried over anhydrous magnesium sulfate. The benzene was then distilled from the reaction mixture under reduced pressure and the residue fractionated through an 18-inch Vigreux column. Only trace amounts of materials were obtained over the range of 95° to 190°C. at 0.35 mm. of mercury and the 3,3'-diphenoxybiphenyl byproduct was recovered in 87% yield over the range of 190° to 214°C. at 0.35 mm. of mercury. The product fraction was redistilled and the fraction having a boiling point range of 211° to 214°C. at 0.35 mm. of mercury was recovered. The product was identified as 3,3'-diphenoxybiphenyl.

Calculated for $C_{24}H_{18}O_2$: C, 85.18; H, 5.36. Found: C, 84.97, 84.99; H, 5.38, 5.29. The 3,3'-diphenoxybiphenyl compound was found to have a refractive index at 25°C. of 1.6428, a boiling point of about 854°F. and a thermal decomposition point of about 838°F.

Example 2: A glass reaction vessel was placed in an ice bath and 252 ml. concentrated hydrochloric acid together with 252 ml. of water were introduced thereto and chilled, after which 185 g. of 3-aminodiphenyl ether (1 mol) was added thereto. A solution of 64.2 g. of sodium nitrite (0.93 mol) dissolved in 225 ml. of water was added dropwise thereto controlling the temperature below 10°C. so that the diazonium compound formed would not be decomposed and to preclude the loss of nitrous acid. After a positive starch-iodide test to show a slight excess of nitrous acid present, a solution of 166 g. potassium iodide (1 mol) dissolved in 175 ml. of water was added dropwise to the reaction mixture keeping the temperature below 10°C.

During the addition of the potassium iodide solution there was a continuous evolution of nitrogen. After completing the addition of the potassium iodide solution, the reaction mixture was gradually warmed to about 50°C. to evolve all the nitrogen and insure complete reaction. The reaction mixture was then taken up in hexane, and the organic phase washed with a solution of potassium hydroxide, followed by a solution of dilute hydrochloric acid, cnd the organic phase dried over anhydrous magnesium sulfate. The hexane was then removed from the reaction mixture under reduced pressure and the reaction mixture distilled under vacuum. The crude product boiling over the range of 110° to 120°C. at 0.1 mm. of mercury was recovered and redistilled over a boiling range of 141° to 146°C. at 0.9 to 1.0 mm. of mercury. The 3-iododiphenyl ether was recovered in the amount of 155 g.

The 3-iododiphenyl ether of the foregoing paragraph was combined with the product from a similar reaction and 250 g. (0.84 mol) together with 200 g. o-chloroiodobenzene (0.84 mol) and 450 g. copper-bronze (7.1 mols) introduced into the reaction vessel gradually in amounts of about 50 g. at about 15-minute intervals, i.e. was added over a period of about 2 hours and the reactants were refluxed for 4 hours at about 210°C. The reaction mixture was then taken up in benzene, the copper-bronze filtered therefrom, the filtrate dried over anhydrous magnesium sulfate, the benzene removed under vacuum, and the residual reaction mixture fractionally distilled under reduced pressure. The crude product having a boiling range of 160° to 200°C. at 1.0 mm. of mercury was recovered and redistilled through a Todd column to obtain the 2'-chloro-3-phenoxybiphenyl having a boiling range of 180° to 186°C. at 0.7 mm. of mercury.

Potassium phenate was formed in the same manner and with the same quantities of reactants as set out in Example 1. After the removal of the water, the reaction mixture was heated to 180°C. to evacuate the toluene. Then 5 g. of cupric chloride was added to the reaction vessel followed by the addition of 43.3 g. of 2'-chloro-3-phenoxybiphenyl (0.16 mol) over a time interval of about 15 minutes while maintaining the reaction mixture at about 180°C.

Thereafter, the reaction mixture was allowed to reflux for about 16 hours at about 190°C. while agitating the mixture. The reaction mixture was partially cooled but while still warm was taken up in benzene and the organic phase washed with an aqueous potassium hydroxide solution, filtered, and the aqueous phase separated therefrom. The organic phase was further washed with an aqueous solution of sodium hydroxide and dried over anhydrous magnesium sulfate. The benzene was removed from the reaction mixture under reduced pressure and the reaction mixture fractionally distilled. The crude product was observed to have a boiling range of 179° to 199°C. at 0.25 mm. of mercury. The aforesaid crude fraction was redistilled through a Todd column and the 2,3'-diphenoxybiphenyl was observed to have a boiling range of 220° to 222°C. at 0.65 mm. of mercury.

Calculated for $C_{24}H_{18}O_2$: C, 85.18; H, 5.36. Found: C, 85.20, 85.22; H, 5.36, 5.35. The 2,3'-diphenoxybiphenyl compound was observed to have a refractive index of 1.6385 at 25°C., a boiling point of about 824°F., and a thermal decomposition point of about 822°F.

Example 3: A charge of 163 g. of a mixed dichlorobiphenyl composition prepared by the chlorination of biphenyl in the presence of an iron catalyst to provide a composition containing about 32% chlorine, was introduced into the reaction vessel and the material warmed to 150°C. and saturated with dry hydrogen chloride. Then 8.6 g. of anhydrous aluminum trichloride (5% by weight) was added over a 10-minute period during which time the temperature of the reaction mixture rose to 160°C. Dry hydrogen chloride was swept through the hot reaction mixture for an additional time of about 0.5 hour, and then such flow of hydrogen chloride was terminated and the reaction mixture heated for an additional period of about 1.5 hours at a temperature between 160° and 170°C. and then allowed to cool.

The reaction mixture was taken up in carbon tetrachloride, filtered through a diatomaceous earth coated filter and the filter cake pressed dry after additional washing with carbon tetrachloride. The filtrate was washed with a saturated aqueous solution of sodium bicarbonate, the organic phase separated therefrom, dried over anhydrous magnesium sulfate, filtered, and the carbon tetrachloride removed therefrom under vacuum. The residual reaction mixture was then fractionated and the fraction having the boiling range of 80° to 118°C. at 0.07 mm. of mercury, consisting mainly of 3,3'-, 3,4'-, and 4,4'-dichlorobiphenyls. This fraction weighed 142 g. (87% recovery).

Potassium phenate was prepared in the same manner and with the same quantity of reactants as set out in Example 1. After all the water had been distilled off, the reaction mixture was heated to 170°C. to remove the final traces of toluene. Cupric chloride in the amount of 5 g. as catalyst was added thereto and 40 g. of the mixed dichlorobiphenyl (0.18 mol) prepared as shown in the above paragraph was added to the hot reaction mixture over a 30-minute period. The reaction mixture was then held at about 190°C. with stirring for a period of 10 hours, cooled, and while still warm taken up in excess benzene and an aqueous potassium hydroxide solution, filtered through Hy Flo Super Cel, the organic phase separated therefrom, washed with dilute aqueous potassium hydroxide, the organic phase dried over anhydrous magnesium sulfate, and the benzene removed under reduced pressure. The crude product boiling over the range 175° to 220°C. at 0.10 mm. of mercury was refractionated, and the product boiling over the range of 180° to 218°C. at 0.10 mm. of mercury recovered. This product was a fluid slurry which on standing set up to a nonfluid composition due to the precipitation of the 4,4'-dichlorobiphenyl product.

In similar manner to the procedure set out in the first paragraph of this example 170 g. of the dichlorobiphenyl was introduced into a reaction vessel, warmed to 150°C., and saturated with dry hydrogen chloride. Then 3.4 g. of anhydrous aluminum chloride (2 weight percent) was added over a 5-minute period and dry hydrogen chloride bubbled through the mixture for an additional 30 minutes. The temperature was observed to be about 160°C. The reaction mixture was heated for an additional 1.5 hours and the final temperature was 170°C. The reaction mixture was treated in a similar manner to that set out at the second paragraph of this example and the fraction distilling over the boiling range of 90° to 115°C. at 0.07 mm. of mercury recovered (155.7 g., i.e., 91.5% recovery). This isomerized mixture of dichlorobiphenyls, mainly consisting of the 3,3'-, 3,4'-, and 4,4'-isomers can be converted to a diphenoxybiphenyl mixed composition in similar manner to that set out in the third paragraph of this example.

The viscosities in centistokes at various temperatures ranging from 77° to 800°F. for various compositions are given in the table below. It should be noted that the manipulative examples were randomly selected as representative illustrative examples of particular procedures, whereby the viscosity determinations were not made at each listed temperature for all samples. Additionally, extrapolated values for the temperature in degrees F. for a viscosity of 20,000 centistokes are given in the table, which data provides a close approximation for the lower temperature at which the materials are effective fluid compositions.

	Viscosity in Centistokes at Temperature in °F.								°F. for 20,000 cs. Viscosity
Example	77	100	210	400	500	600	700	800	
1	859.1	196.7	8.532	1.541	0.965	0.662	0.490	0.380	37
2	3,001	394.1	8.808	1.397	0.863	0.596	0.444	0.342	60

In related work W.C. Hammann and R.M. Schisla; U.S. Patent 3,471,574; October 7, 1969; assigned to Monsanto Co. have found that functional fluids which have excellent physical properties and which are particularly suitable for use as heat transfer fluids are obtained through the use of phenoxyphenoxybiphenyl compounds selected from 3-(m-phenoxyphenoxy) biphenyl or 3-(p-phenoxyphenoxy) biphenyl.

Of the nine possible phenoxyphenoxybiphenyls, only two isomers have suitable physical properties to provide the desired wide liquid range necessary for use as a functional fluid. The suitable isomers of phenoxyphenoxybiphenyl having fluid properties at room temperature are 3-(m-phenoxyphenoxy)biphenyl and 3-(p-phenoxyphenoxybiphenyl). All of the other possible isomers of phenoxyphenoxybiphenyl are crystalline solids at room temperature having melting points ranging from about 124° to 271°F., i.e., the crystalline solid phenoxyphenoxybiphenyls have a melting point of the order of at least about 80°F. above that of the specific class of liquid phenoxyphenoxybiphenyls.

Mixtures of Diphenoxybenzenes and Diphenoxydiphenyls

F. Giolito and J. Philippe; U.S. Patent 3,429,816; February 25, 1969; assigned to Progil, France describe compositions which consist essentially of a mixture of diphenoxybenzenes, monophenoxydiphenyls or diphenoxydiphenyls and may also include ortho- or meta-terphenyl.

The respective ratios of the various compounds of the mixtures may vary between large limits. By indicating as (a) the whole of diphenoxybenzenes, (b) the mono- and/or diphenoxydiphenyls, and (c) the terphenyls, it is possible to say that especially interesting results are obtained with the following preferential ratios.

Binary mixtures: 1 to 60 parts of (a) and 99 to 40 parts of (b) for 100 parts by weight of (a) + (b).

Ternary mixtures: it is possible to add 1 to 40 parts of (c) to 100 parts of the above binary mixtures.

The following examples illustrate a number of compositions according to the process.

Example 1: Compositions which were true liquids at room temperature (25°C.) and thermally stable between 25° and 375°C. for more than 300 hours, without the addition of a stabilizing agent or antioxidant, were obtained by effecting the mixtures given in Table 1, in which the indicated figures correspond, on every horizontal line, to parts by weight for 100 parts of the whole composition. Table 1 is shown on the following page.

TABLE 1

Mixture No.	Phenoxy-diphenyls			Diphenoxybenzenes			Terphenyls	
	2-phenoxy	4-phenoxy	2,2'-diphenoxy	Ortho	Meta	Para	Ortho	Meta
1	34.2	18	7.8	-	40	-	-	-
2	50.4	18.9	20.7	-	-	10	-	-
3	44.8	16.8	18.4	-	-	20	-	-
4	30	12	12	-	27	9	10	-
5	35	25	-	-	30	10	-	-
6	35	25	-	5.2	24.8	14	-	-
7	28.5	15	6.5	-	40	-	7	3
8	37.24	13.96	15.3	9.12	5	19.38	-	-
9	31.36	11.76	12.88	7.68	20	16.32	-	-
10	30	20	10	5	35	-	-	-
11	48	17	20	-	-	15	-	-
12	42.9	16.0	17.6	-	10	13.5	-	-
13	38.0	14.3	15.6	-	20	12	-	-

Example 2: Table 2 below describes some other compositions, corresponding to the mixtures (a) + (b), which have the advantage of being liquid at temperatures lower than 0°C. The test has been made at -5°C. after recrystallizing the constituents after more or less long stays at between -40° and -5°C., in order to avoid superfusion phenomenon.

TABLE 2

Mixture No.	Phenoxy-diphenyls[1]			Diphenoxybenzenes[1]		
	2-phenoxy	4-phenoxy	2,2'-diphenoxy	Ortho	Meta	Para
14	35.9	18.9	8.2	9	10	18
15	31.9	16.8	7.3	8	20	16
16	27.9	14.7	6.4	7	30	14
17	39.2	14.7	16.1	-	30	-
18	33.6	12.6	13.8	-	40	-
19	27.44	10.29	11.27	6.72	30	14.28
20	33.3	12.5	13.7	-	30	10.5

[1] Parts by weight for 100 parts of the whole mixture.

It will be noted especially that the addition of the sole meta-isomer of diphenoxybenzene to mixtures of mono- and diphenoxydiphenyls yields very fluid compositions at a temperature lower than 0°C. Moreover it may be seen, in comparing the tests 12 and 13 with the test 20, that increasing additions of meta-diphenoxybenzene are especially interesting for decreasing the viscosity of mono- and diphenoxybenzene mixtures, already liquid at room temperature. The viscosities (expressed in centistokes) go from 103 for the composition of mixture 12 to 86.5 for mixture 13, and 70.1 for mixture 20.

Example 3: For studying the thermal stability of the compositions, pyrolysis tests in ampules have been made in the following way: The mixtures to be studied (liquid samples of 20 to 25 g.) were introduced in Pyrex glass ampules of a volume of about 50 ml. able to bear pressures of 10 atmospheres at 400°C. After degassing of the mixture, under vacuum, and closing of the ampules, they are placed in the pits of a multicellular electrical oven and brought to the desired temperature for about 500 hours. Moreover there has been introduced in some ampules, samples of various metals, such as especially, iron and copper, or still gaseous or liquid impurities such as oxygen, water, chlorinated products, etc. in order to determine the influence of these parameters on the pyrolytic decomposition of the studied fluids. Then after a heating time of 500 hours at a temperature of 380°C. for all the samples, the following characteristic measurements have been made:

(a) Measurement of the quantity of the formed gas, in ml. per gram of mixture (referred to normal conditions of temperature and pressure).

(b) Measurement of the quantity of the formed products, lighter and heavier than the starting products, through the microdistillation, under standard conditions, of 1 gram of product, that is:

Light — Products distilling in 30 minutes, under 0.1 mm. Hg at 100°C.

Heavy — Boiler residues after an ascent of temperature from 100° to 210°C. during 2 hours, then maintenance of this temperature for 30 minutes under 0.1 mm. Hg.

(c) Possibly, determinations, after pyrolysis, of phenol in a zone which is such that this viscosity is normally able to develop very much.

(d) Change of sample aspects: measurement of weight variation corrosion study.

(e) Possibly, determination, after pyrolysis, of phenol functions, metal traces, halogen contents, etc.

Testing was conducted under the above conditions, on the following binary composition (percent by weight):

Phenoxy-diphenyls	
2-phenoxy	35%
4-phenoxy	14%
2, 2'-diphenoxy	21%

(continued)

Diphenoxybenzene	
Meta-isomer	30%

The results given in Table 3 below were obtained:

TABLE 3

Analysis results	Atmosphere or impurities in sample ampules				
	Vacuum	Oxygen (100 mm. Hg)	Iron[1]	Copper[1]	Iron+copper +oxygen (100 mm. Hg)
Formed gases (ml./g.)	0.16	------	0.17	0.19	------
Light formed products (percent weight)	0.3	0.85	0.5	1	0.85
Heavy formed products (percent weight)	1.4	1.3	0.9	2.0	1.3
Corrosion			Null	Null	Null
Phenols, percent	0.7	0.66	0.67	0.65	0.66

[1] Dimensions of metal samples: 60 x 15 x 1.5 (mm.).

K.J. Sax, W.S. Saari and C.L. Mahoney; U.S. Patent 3,374,175; March 19, 1968; assigned to Shell Oil Company also describe blends of certain polyphenyl ethers which are especially suitable for use as synthetic lubricating oils and operating fluids capable of functioning over a wide temperature range and especially suitable for use at extremely high operating temperatures and in the presence of ionizing radiation.

Conversion of Solid Phenoxybiphenyls to Fluid Products

R.M. Schisla and H.I. Weingarten; U.S. Patent 3,565,960; February 23, 1971; assigned to Monsanto Company describe a process for the preparation of a fluid phenoxybiphenyl compound or phenoxybiphenyl composition enriched in fluid phenoxybiphenyl compounds. The process comprises the steps of contacting a solid phenoxybiphenyl compound or composition containing one or more solid phenoxybiphenyl compounds with hydrogen chloride and a class of catalysts representative of which are aluminum trichloride, aluminum tribromide, gallium trichloride, stannic chloride and antimony pentachloride. Typical examples of fluid phenoxybiphenyl compounds which can be prepared utilizing the process are

2,3'-diphenoxybiphenyl,
3,3'-di(o-phenoxyphenoxy) biphenyl, and
3-(p-phenoxyphenoxy)-4'-(m-phenoxyphenoxy) biphenyl.

The solid phenoxybiphenyls include, by way of example, the solid diphenoxybiphenyls, the solid tris- and tetraphenoxybiphenyls, solid phenoxy (phenoxyphenoxy) biphenyl and solid bis(phenoxyphenoxy) biphenyls. Examples of specific solid phenoxybiphenyls are

4,4'-diphenoxybiphenyl,
4-(m-phenoxyphenoxy-4'-(p-phenoxyphenoxy) biphenyl and
2-(m-phenoxyphenoxy-2'-(p-phenoxyphenoxy) biphenyl.

The above examples of solid phenoxybiphenyls can be converted singly or as a mixed composition to produce a fluid phenoxybiphenyl compound or composition enriched in fluid phenoxybiphenyls. The following examples are illustrative of the process.

Example 1: A 25 g. sample of 4,4'-diphenoxybiphenyl (0.074 mol) was heated to 210°C. to assure complete melting of the compound. The compound was saturated with dry hydrogen chloride and then 4 g. of anhydrous aluminum chloride catalyst was added over a 10 minute period and the reaction temperature was observed to rise to about 240°C. for about 0.5 hour. Dry hydrogen chloride gas was continuously introduced into the reaction mixture over the total reaction time of 1.5 hours to maintain the reaction mixture saturated with hydrogen chloride.

The final reaction temperature at the end of the 1.5 hour reaction time was observed to be 210°C. The reaction mixture was cooled to 100°C., taken up with carbon tetrachloride, and filtered through Hy Flo Super Cel. The organic filtrate was then washed with a saturated aqueous solution of sodium bicarbonate, filtered through Hy Flo Super Cel. The organic filtrate was then washed with a saturated aqueous solution of sodium bicarbonate, filtered through Hy Flo Super Cel, the organic layer separated and dried over anhydrous magnesium sulfate, filtered, and the carbon tetrachloride removed therefrom under reduced pressure.

The reaction mixture was then fractionally distilled and the crude converted diphenoxybiphenyl composition having a boiling range of 165° to 220°C. at 0.2 mm. of mercury recovered. This crude product was observed to sludge up or solidify at room temperature. The crude product was warmed in an oven until clear and homogenous, then treated with an equal volume of hexane with stirring and the small amount of unconverted residual 4,4'-diphenoxybiphenyl was observed to precipitate out of the composition within a few minutes. The precipitate was filtered out, the hexane removed from the diphenoxybiphenyl composition under vacuum and the product fractionally distilled, recovering a cut having a boiling range of 180° to 212°C. at 0.2 mm. of mercury. This clear diphenoxybiphenyl composition was observed to have a viscosity in centistokes at the temperatures indicated therewith of 20,000 at 44°F., 189.9 at 100°F., 8.314 at 210°F., 1.545 at 400°F., 0.965 at 500°F., 0.677 at 600°F., 0.507 at 700°F. and 0.397 at 800°F.

Example 2: A 118 g. sample of a solid mixed composition of diphenoxybiphenyls, prepared by the diphenoxylation of a mixed composition of dichlorobiphenyls (the 2,4'-, 2,2'- and 4,4'-isomers comprising the major components therein) by the Ullmann condensation reaction employing cuprous chloride catalyst, was introduced into a reaction vessel, heated to about 180°C., and saturated with anhydrous hydrogen chloride. Then 13.1 g. of anhydrous aluminum trichloride was added and the anhydrous hydrogen chloride continued to be bubbled through the reaction mixture during the total reaction time of 2 hours wherein the reaction temperature varied from about 180° to about 190°C.

The reaction mixture was then cooled and treated in a similar manner to that set out in Example 1. The crude converted diphenoxybiphenyl composition had a boiling range of from about 170° to about 210°C. at 0.04 mm. of mercury. This crude composition was then warmed on a steam bath and treated with hexane to precipitate and effect the removal of the small amount of unconverted 4,4'-diphenoxybiphenyl remaining in the mixed composition. The composition was then fractionally distilled to remove the hexane and the composition boiling over the range of 173° to 200°C. at 0.04 mm. of mercury was recovered as product. This product comprising a mixture of diphenoxybiphenyls, the 3,3'- and 3,4'-isomers comprising the major components, was found to be a stable fluid composition at normal temperatures.

As is seen from the examples, solid phenoxybiphenyls were converted to fluid phenoxybiphenyl compounds and compositions in excellent yield. In particular, Examples 1 and 2 clearly demonstrate the high yields which are obtained in carrying out the process. Thus, only small residual amounts of the phenoxybiphenyl remained after treatment of solid phenoxybiphenyl compounds.

Nonlinear Polyphenyl Ethers

In a process described by A.L. Williams and R.E. Kinney; U.S. Patent 3,476,815; November 4, 1969 and U.S. Patent 3,449,442; July 15, 1969; both assigned to Mobil Oil Corporation a class of nonlinear polyphenyl ether synthetic lubricants are prepared by reacting 2,4-diphenoxyphenyl bromide with phenolic compounds, by the Ullmann ether synthesis. These polyphenyl ethers have a branched configuration in which at least one internal phenyl group has more than two phenoxy group substituents. The nonlinear polyphenyl ethers display lubricant and stability characteristics comparable to those of the linear polyphenyl ethers. The following examples illustrate the process.

Example 1: A 500 ml. flask was fitted with a nitrogen inlet tube, a one foot Vigreux column, and a stirrer to prepare a solution of 55.8 g. (0.300 mol) of meta-phenoxyphenol in 200 ml. of pyridine. The solution was made at room temperature under a nitrogen atmosphere. The nitrogen covers the reaction mixture during all reactions in the flask. Stirring was continued while 15.6 (0.288 mol) of sodium methylate were added slowly. Then 135 ml. of methanol-pyridine mixture were distilled out through the Vigreux column over a period of one hour.

When the flask has cooled enough to stop boiling, the column was replaced by a reflux condenser. Then, 67.9 g. (0.199 mol) of 2,4-diphenoxybromobenzene were added. This was washed in with 25 ml. of pyridine, and followed at once by 2 g. of cuprous chloride. The mixture was refluxed with stirring for four hours. Pyridine was then distilled out until the pot temperature reached 160°C.

The reaction mixture was poured into a solution of 50 ml. of concentrated hydrochloric acid in 500 ml. of water and allowed to stand overnight. After decanting off the aqueous layer, the crude product was dissolved in 430 ml. of benzene. This benzene solution was then washed by 20 ml. of hydrochloric acid in 200 ml. of water, and again by 10 ml. of hydrochloric acid in 100 ml. of water.

Unreacted meta-phenoxyphenol was extracted into 12 g. of sodium hydroxide in 150 ml. of water, used in two equal portions. The benzene solution was then washed with 100 ml. of water. Benzene was removed by distillation at atmospheric pressure. Distillation was continued at 0.5 mm. of Hg to give a distilled product boiling at 262° to 268°C. The yield of 1-(meta-phenoxyphenoxy)-2,4-diphenoxy-benzene was 35.8 g. (40% based on 2,4-diphenoxy-bromobenzene). Gas chromatography indicated 99+ percent purity.

Example 2: The nonlinear polyphenyl ether of Example 1 was tested for lubricant properties in comparison with a linear polyphenyl ether having the same number of phenyl and oxygen moities, a bis(phenoxyphenoxy)-benzene, mixed isomers (5P4E). Tests were for Pour Point (ASTM D97-57) and Kinematic Viscosity (ASTM D445-61). Oxidation stability was measured as the time for absorbing 0.5 mol or 1.0 mol O_2 per kg. of oil at 600°F., using the method of Dornte, Ind. Eng. Chem., 28, 26 (1936). Test results are shown in the table on the following page.

	Example 1	5P4E
Pour Point, ° F	45	40
Kinematic Viscosity, cs.:		
At 130° F	154.7	93.20
At 210° F	15.54	13.11
At 400° F	2.00	2.19
Oxygen Absorption, hrs.:		
0.5 mole	11.3	10.5
1.0 mole	15.3	12.4

Phenol-Aryl Bromide Reaction Products

A process described by G.P. Brown; U.S. Patent 3,567,783; March 2, 1971; assigned to General Electric Company involves meta-polyphenyl ethers which are prepared through the Ullmann ether reaction of aryl bromide with the alkaline salt of a phenol such as the potassium or sodium salt in the presence of catalytic quantities of a metal, such as copper. In general, the reaction temperatures for preparing the materials range from about 175° to 250°C. and the reaction times from one to several hours to produce final products having a formula of the following general type:

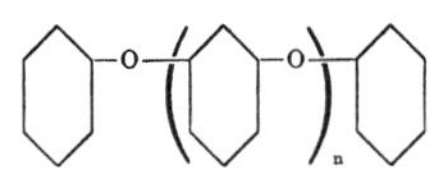

where n is a number from 1 to 5 and most preferably from 2 to 5. Generally speaking, the materials are prepared from the reaction of a monohydric phenol and a monobromo or dibromo compound. In the preparation of the products where the reaction is of a monohydric phenol and a dibromo compound, the reaction may be expressed as

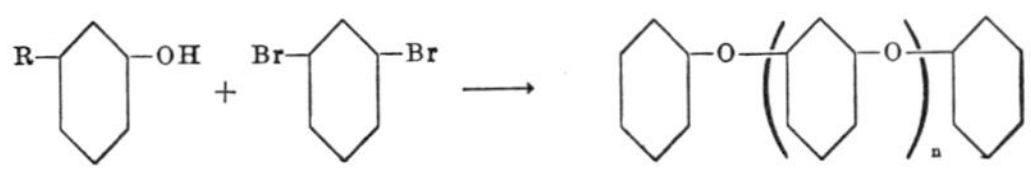

where R is hydrogen, C_6H_5O or $C_6H_5OC_6H_4O$, n being equal, respectively, to 1, 3 and 5. Mixtures of the various R's are also useful. In the preparation of the products from the reaction of a monohydric phenol and a monobromo compound, the reaction may be expressed simply as

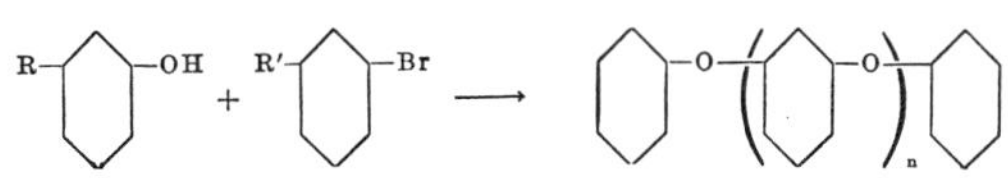

where R and R' are C_6H_5O, n is 2. Where R and R' are $C_6H_5OC_6H_4O$, n is 4. Where R is C_6H_5O and R' is $C_6H_5OC_6H_4O$, n is 3, and where R is $C_6H_5OC_6H_4O$ and R' is $C_6H_5OC_6H_4OC_6H_4O$, n is 5. Mixtures of the R's and R''s are also used. It has been found that only the compounds of the process or those in which n is a number from 1 to 5 and preferably from 2 to 5 exhibit flow points within the order of room temperature and at the same time have a desirably high temperature resistance and boiling point. The following examples illustrate the process.

Example 1: [Phenoxy(m-phenoxylene)] benzene was prepared by heating together in a flask fitted with a stirrer, nitrogen inlet, thermometer and steam heated condenser topped by a water-cooled take-off condenser, 169 grams (1.8 mols) of phenol, 58 grams (0.9 mol) of 86% KOH and 3 grams of copper powder. The reactants were blanketed with nitrogen. Upon heating, the reaction mass became fluid at about 40°C. and underwent exothermal heating to about 120°C. The reaction mixture was then heated to 180°C. and 70.8 grams (0.3 mol) of m-dibromobenzene added dropwise.

Next, about 45 cc of xylene was added to maintain reflux at from 180° to 185°C. During the dibromobenzene addition, no external heat was required to retain reflux, such reflux being continued for about 10 minutes after all of the dibromobenzene had been added. The product was washed with aqueous NaOH and water and dried over potassium carbonate. A total of 55.2 grams or 70% yield of [phenoxy(m-phenoxylene)] benzene was obtained by distillation and melted at about 136°F. and boiled at 698°F. atmospheric pressure. The formula of this material is

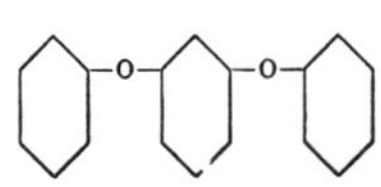

From its rather high melting point and relatively low boiling point, it will be gathered that this material has a relatively narrow useful temperature range.

Example 2: [Phenoxytri(m-phenoxylene)] benzene,

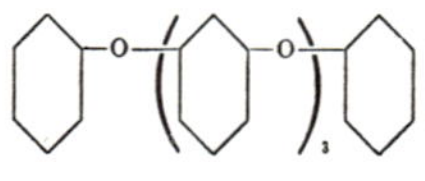

was prepared by charging in a flask fitted as in Example 1, 98 grams (0.525 mol) of 3-hydroxyphenyl phenyl ether, 59 grams meta-dibromobenzene, 5 grams of copper powder and 6 grams of xylene. The reaction mixture was heated to 200°C., and a 50% aqueous KOH solution consisting of 37.5 grams (0.575 mol) of 86% KOH and 26.9 grams of H_2O was added dropwise over a period of 42 minutes at 200° to 220°C., the reaction mixture then being heated additionally for one hour at 210°C.

The reaction mixture was then neutralized, extracted, washed and the solvent removed. The product was then redistilled three times under vacuum to produce a material having a refractive index n_D^{25} of 1.6300 and an average molecular weight of 408 against a theoretical molecular weight of 446 for the pure end product. This material had a flow point of 54°F. and an extrapolated atmospheric boiling point of 1030°F. The 3-hydroxyphenyl phenyl ether

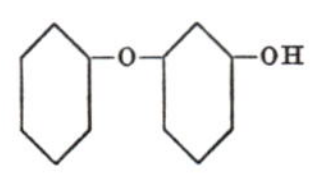

used in the preparation of this example was prepared by charging a flask with 524 grams (5.58 mols) of phenol, 114 grams (2.04 mols) 86% KOH and one gram of copper powder and 150 cc of diphenyl ether. The temperature was raised to 175°C. and 322 grams (1.86 mols) of m-bromophenol added. The reaction temperature rose to 210°C. and addition of bromophenol was stopped for about 30 minutes. The reaction mixture was then heated for a total of 3 hours between 175° and 210°C. Upon separation and distillation, a total weight of 171.6 grams of 3-hydroxyphenyl phenyl ether was obtained.

Example 3: [Phenoxypenta(m-phenoxylene)] benzene

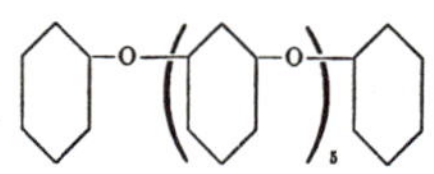

was prepared by charging into a 500 cc, 3-neck reaction flask equipped as in Example 1, 57.5 grams (0.207 mol) of 3-hydroxy-3'-phenoxydiphenyl ether, 21.9 grams (0.093 mol), meta-dibromobenzene, 13.5 grams (0.207 mol) of 86% KOH, 1 gram of copper powder and 5 grams of copper turnings. The reaction began at a temperature of 170°C. and was indicated by the precipitation of potassium bromide. After 1 1/2 hours at 220° to 230°C., the organic phase contained 7.2 ml. equivalents of bromine for a 96% reaction. After 3 hours' additional time at 240° to 250°C., it showed the same amount of reaction. The product was taken up with benzene and ether, washed with 5% NaOH, with water and then dried with potassium carbonate. The final product after distillation was shown to be 21.9 grams of [phenoxypenta(m-phenoxylene)] benzene. The 3-hydroxy-3'-phenoxy diphenyl ether

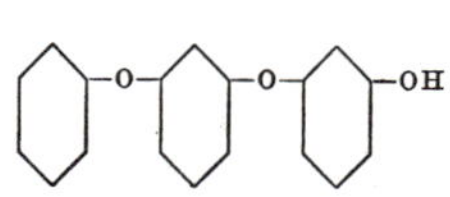

was prepared by charging to a flask with the usual fittings, 2 grams of copper powder, 384 grams (4.08 mols) of phenol and 131 grams (2.04 mols) of 86% KOH. The temperature was raised to 175°C. and 178 grams (0.68 mol) of 3-bromo-3'-hydroxy diphenyl ether added to allow the temperature to rise to 210°C. in 20 minutes.

The mixture was heated at a temperature of 208° to 210°C. for about 1 1/2 hours. A 68.7% yield or 130 grams of product were obtained after distillation and purification. The product had a flow point of 80°F. and an extrapolated atmospheric boiling point of 1202°F.

The 3-bromo-3'-hydroxy diphenyl ether

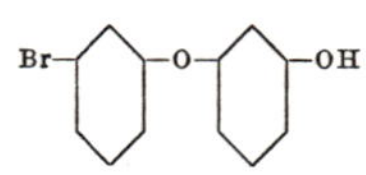

was prepared by reacting 75.7 grams (1.16 mols) 86% KOH, 1000 grams (5.8 mols) meta-bromophenol and 2 grams of copper powder. The reactants heated exothermically and the mixture was then further heated to 210°C. for 2 hours, cooled, purified and distilled to yield 114 grams of product.

Example 4: [Phenoxybi(m-phenoxylene)] benzene

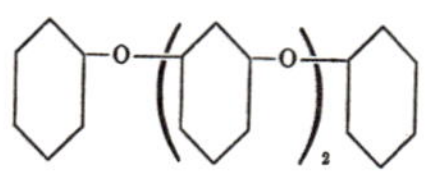

was prepared by charging a flask fitted as above with 150 cc of diphenyl ether, 68 grams (0.365 mol) of 3-hydroxyphenyl phenyl ether, 23.8 grams (0.365 mol) of 86% KOH and 1 gram of copper powder. The mixture was heated to 175°C. and then while heating was continued to about 225°C., 91 grams (0.365 mol) of 3-bromophenyl phenyl ether or m-bromophenyl phenyl ether was added dropwise over a period of 25 minutes.

The reaction mixture was stirred at from about 215° to 225°C., cooled and taken up in ether. After washing, purifying, drying and distilling, a 78.5% yield of the product [phenoxybi(m-phenoxylene)] benzene was produced, which had a flow point of 10°F. and a boiling point of 865°F. at atmospheric pressure. The m-bromophenyl phenyl ether

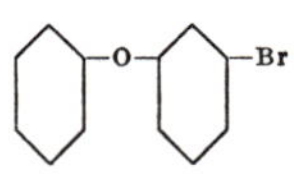

used in this preparation was prepared by reacting together in a flask fitted as above 354 grams (1.54 mols) of meta-dibromobenzene, 47 grams (0.50 mol) of phenol, 36 grams (0.55 mol) 86% KOH, 26 grams of water, 1 gram of copper powder and 5 grams of copper turnings, a reflux temperature of 200°C. being maintained over a time period of about 3 hours. The reaction product was diluted with benzene, filtered and distilled to obtain a 43% yield of m-bromophenyl phenyl ether.

Alkoxy Aromatic Ethers

A process described by E.P. Merica; U.S. Patent 3,441,615; April 29, 1969; assigned to The Dow Chemical Company provides a class of ethers characterized by the following general formula:

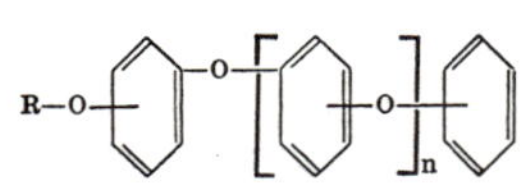

where R represents an alkyl radical having from about 8 to about 20 carbon atoms and n represents an integer from 0 to about 4. The compounds have been tested and found to be useful as high temperature lubricants characterized by their thermal stability at temperature up to 750°F. and excellent resistance to oxidative degradation at temperatures up to at least 500°F. and in many instances by having extremely low pour points, i.e., below zero to -80°F.

The compounds which exhibit the properties above set forth are those alkoxy aromatic ethers having from about 8 to 20 carbon atoms in the alkoxy moiety and from 2 to 6 aryl ether moieties. Thus, for example, one can prepare and employ: alkoxyphenoxy benzene, alkoxyphenyl phenoxyphenyl ether, alkoxyphenoxy phenoxyphenoxy benzene, alkoxyphenoxyphenyl phenoxyphenoxyphenyl ether and alkoxyphenoxyphenoxy phenoxyphenoxyphenoxy benzene.

The following general procedures are expedient methods for preparing the compounds:

(1) $\quad RBr + HO-ArH \xrightarrow[\text{HBR Acceptor}]{\text{reflux}} RO-Ar-H$

(2) [a] $RBr + HO-ArCl \xrightarrow[\text{HBr Acceptor}]{\text{reflux}} RO-ArCl$

[b] $RO-ArCl + HO-ArH \xrightarrow{\text{Cu}} RO-Ar-O-Ar-H$

(3) [a] $HO{-}ArH + HO-ArCl \xrightarrow{\text{Cu}} HO-Ar-O-Ar{-}H$

[b] $HO-Ar-O-Ar-H + RBr \xrightarrow[\text{HBR Acceptor}]{\text{reflux}} RO-Ar-O-Ar-H$

Ar represents a phenylene, phenyloxy phenylene, phenoxyphenoxyphenylene, and R represents an alkyl radical having from about 8 to 20 carbon atoms. The following examples are illustrative of the process.

Example 1: m-(n-Octyloxy) phenoxybenzene — A mixture of 193 grams (1 mol) of n-octyl bromide, 279 grams (1.5 mols) of phenoxyphenol (85% by weight meta isomer and 15% by weight para isomer) and 170 grams (1.25 mols) of potassium carbonate in 600 ml. of acetone was refluxed for 10 hours. The solids which formed were removed by filtration, the acetone distilled on a steam bath and 400 ml. of 5% sodium hydroxide added to the residue with stirring. The aqueous layer which formed was separated and the organic layer distilled to recover 223 grams of a fraction boiling at 178° to 180°C. which was identified as 3-(n-octoxy) phenoxybenzene. This represents an 82% yield based on n-octyl bromide consumed.

Example 2: In the manner of Example 1, employing the indicated materials according to the reactions:

(I) $\quad RBr + HO-\phi-Cl \longrightarrow RO-\phi-Cl$

(II) $\quad R{-}O-\phi-Cl + K-O-\phi-O-\phi H \longrightarrow R-O-\phi-O-\phi-O-\phi H$

ϕ represents a bivalent benzene nucleus, i.e. phenylene. The following materials were prepared.

	Reaction I					Reaction II			
Alkyl Bromide, mols	m-Chlorophenol, mols	K_2CO_3, mols	Acetone, ml.	Intermed. product	Intermed. product, mols	Hydroxy phenoxy benzene, mols	KOH, mols	Cu, gms.	Product
C_{12}, 0.5	0.58	0.55	200	$C_{12}O\ \phi Cl$	0.3	0.5	0.4	5	$C_{12}H_{25}O\ \phi O\ \phi O\ \phi H$
C_{12}, 2.0	3.0	2.7	1,000	$C_{12}O\ \phi Cl$	1.0	2.0	1.5	20	$C_{12}H_{25}O\ \phi O\ \phi O\ \phi H$
C_8, 2.0	3.0	2.7	1,000	$C_8O\ \phi Cl$	1.0	2.0	1.5	20	$C_8H_{17}O\ \phi O\ \phi O\ \phi H$
C_{10}, 0.6	0.9	0.81	400	$C_{10}O\ \phi Cl$	0.35	0.7	0.52	10	$C_{10}H_{21}O\ \phi O\ \phi O\ \phi H$
C_{10}, 2.0	3.0	2.7	1,000	$C_{10}O\ \phi Cl$	1.363	2.726	1.95	20+20 CuO	$C_{10}H_{21}O\ \phi O\ \phi O\ \phi H$
				$C_{10}O\ \phi Cl$	0.466	[1] 0.932	0.699	12	$C_{10}H_{21}O\ \phi O\ \phi O\ \phi O\ \phi H$
C_8, 1.0		1.0	500			1.0			$C_8H_{17}O\ \phi O\ \phi H$
C_8, 1.0		1.25	600			[2] 1.5			$C_8H_{17}O\ \phi O\ \phi H$
C_{16}		1.25	600			[2] 1.5			$C_{18}H_{37}O\ \phi O\ \phi H$
C_{12}		1.25	600			[2] 1.5			$C_{12}H_{25}O\ \phi O\ \phi H$

[1] Phenoxyphenoxyphenol. [2] 90% Metaisomer.

In a series of tests employing a Shell 4-Ball Tester, comparisons were made between the product 3-n-decoxy phenyl phenoxyphenyl ether, bis(phenoxyphenyl) ether, the unalkylated polyether of comparable molecular weight already accepted as a good lubricant, and also the alkyl polyether alpha-methylheptadecylphenoxybenzene. The pressure at seizure is set forth below:

	(Wt., kg.) at Seizure
Decoxy phenyl phenoxyphenyl ether	74
Bis(phenoxyphenyl) ether	75
Alpha-methylheptadecylphenoxybenzene	80

In another series of tests the oxidation stability of the compounds of this process and closely related compounds was determined by flowing 5 liters of air per hour through a 50-gram sample held at 500°F. for 24 hours. The viscosity of the sample was taken at 100°F. and 210°F. before and after the tests. The results of the viscosity measurements are set forth in percent increase in viscosity as shown on the following page.

	Viscosity Increase, percent	
	100° F.	210° F.
Decoxy phenyl phenoxyphenyl ether	83	41
Monododecyldiphenyl ether	142	48.5
Resorcinyl neoheptanoate (with inhibitors)	254	109

In another series of tests the volatility of the various compounds of the process was measured at 500°F. and 400°F. by passing a stream of 2 liters of air per minute for 6.5 hours at the indicated temperature (ASTM D-972-56). The results are set forth below.

	Evaporation Loss, at—	
	400° F.	500° F.
Bis(phenoxyphenyl)ether	3.2	41.7
Decoxy phenyl phenoxyphenyl ether	1.6	26.3
Resorcinyl neoheptanoate (with inhibitors)		99.7

ADDITIVES

Metal Acetylacetonates

K.L. McHugh, J.O. Smith and J.R. Stemniski; U.S. Patent 3,360,467; December 26, 1967; assigned to Monsanto Research Corporation describe the use of additives for the polyphenyl ether liquid fluids which are chelates of a heavy metal of Groups I-IV and VI-VIII of the Periodic Table and a carbonyl compound of the formula

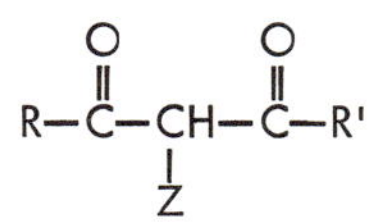

in which R and R' are selected from the class consisting of alkyl radicals from 1 to 6 carbon atoms and aryl, alkaryl and aralkyl radicals of from 6 to 10 carbon atoms, and Z is selected from the class consisting of hydrogen, R, R', and carbo-alkoxy radicals of from 2 to 6 carbon atoms.

Owing to their easy availability, a very useful class of chelates is that obtained from acetylacetone. They are readily available in known manner by reaction of acetylacetone with a salt of the appropriate metal, e.g., the acetate, chloride or sulfate. Examples of useful metal acetylacetonates include the copper, zinc, cadmium, aluminum, zirconium, ferrous, ferric, cobalt, manganous, nickel and indium acetylacetonates. The metal chelates are combined with the fluid polyphenyl ethers to the extent of 0.01 to 1.0% by weight, depending upon the nature of the individual chelate and of the ether fluid. The process is illustrated by the following examples.

Example 1: Manganous and lead acetylacetonates were compared with some commonly employed lubricant additives for antiwear effect when incorporated into a mixture of polyphenyl ethers consisting by weight of:

	Percent
m-Bis(m-phenoxyphenoxy) benzene	65
m-[(m-Phenoxyphenoxy)(p-phenoxyphenoxy)] benzene	30
m-Bis(p-phenoxyphenoxy) benzene	5

Testing was conducted by means of the "Shell 4-Ball Wear Machine." It consists of an equilateral tetrahedron formed by 4 stainless steel balls with the three lower balls immovably clamped in a ball holder. In the tests, the upper ball was rotated about the vertical axis in contact with the 3 lower stationary balls under a 40 kg. load for one hour at 400°F.

The contacting surfaces were immersed in the test fluid and the circular scars worn in the surface of the three stationary balls were measured by means of a low power microscope. A modified cup and heater assembly was used to evaluate lubricants at the 400°F. temperature; see "The Study of Lubrication Using the Four-Ball Type Machine," R.G. Larsen, Lubrication Engineering, 1, pages 35-43, 59, August 1945.

Employing the above procedure, the results shown in Table 1 on the following page, were obtained with the additives shown at the indicated concentrations.

TABLE 1

Additive	Concn., weight percent	Scar diameter, mm.
None	....	3.30
Manganous acetylacetonate	0.5	1.19
Lead acetylacetonate	0.5	1.18
Bis(triphenylphosphine)nickel dibromide	0.5	3.45
Diphenylmercury	1.0	3.32
Tropylidene molybdenum tricarbonyl	0.5	3.31
Boron trifluoride/Ni dimethylglyoxime complex	1.0	3.38
Bis(triphenylphosphine)nickel dithiocyanate	0.5	3.39
2,4,6-triphenylphenol	1.0	3.36
N-nitroso-N-phenylbenzylamine	1.0	3.04
2,2'-dipyridylamine	1.0	3.53

Example 2: The mixture of polyphenyl ethers described in Example 1 were incorporated with ferric or with manganous acetylacetonate and then submitted to the Falex test for the purpose of determining extreme-pressure resisting properties. This test is described in the articles by V.A. Ryan in "Lubrication Engineering," September 1946, and by S. Kyropoulos in "Refiner Natural Gasoline Mfr.," 18, 320-24 (1939). Briefly, the test was conducted as follows: There was employed a Faville-LeVally Falex lubricant testing machine with heating element, 4,500 lb. pressure gauge indicating bearing loads, calibrated, circular, toothed loader capable of providing wear estimates, and torque indicating gauge.

The Falex reading for the mixture of polyphenyl ethers containing a 1.0% by weight concentration of ferric acetylacetonate was 1750 and that for the mixture of polyphenyl ethers containing an 0.5% concentration of manganous acetylacetonate was 2250. That for the same mixture of polyphenyl ethers in absence of any additive was 500.

Example 3: A number of metal acetylacetonates were compared with some commonly employed lubricant additives for inhibition of viscosity increase at high temperature in the presence of oxygen. The base stock was the mixture of polyphenyl ethers described in Example 1. Testing was conducted as follows: Samples were prepared consisting of 20 ml. of the mixture of ethers and the concentration of additive shown below. The sample was heated to 100°F., and the viscosity of the sample was determined at this temperature.

The temperature was then increased to 600°F., and while holding the sample at this temperature, air was bubbled through it at a rate of 1 liter per hour, for either 24 or 48 hours as shown below. The sample was then allowed to cool, the viscosity was redetermined at 100°F., and the percent increase in viscosity was calculated. The results shown in Table 2 were obtained.

TABLE 2

Additive	Concn. percent weight	Time, hours	Viscosity Increase, percent
Co acetylacetonate	0.5	24	10.9
Mn^{II} acetylacetonate	0.5	24	10.1
Ni acetylacetonate	0.5	24	12.9
Fe^{II} acetylacetonate	0.5	24	16.8
Ti acetylacetonate	0.5	24	19.9
Al acetylacetonate	1.0	24	20.7
In acetylacetonate	0.5	48	20.2
Cu chelate of 2-phenyl-2-mercaptoethylphenyl ketone	1.0	48	45.9
Tin titanium malonate	0.5	48	40.2
Bis(N,N-dimethylamino)titanium dichloride	0.5	48	90.6
Diphenyltin bis(didecyldithiophosphate)	1.0	48	123.0
Reaction product of hydroquinone and dibutyltin oxide	0.5	48	56.2
Tropylidene molybdenum tricarbonyl	0.5	24	173.0
Bis(p-phenoxyphenyl)mercury	1.0	24	48.2
Ni bis(N-phenyl-5-nitrosalicylimine) and trichloroacetic acid (2:1)	3.0	24	64
Diphenylamine	1.0	24	73.3
2,4,6-triphenylphenol	1.0	24	59.0
Phenylbiphenylamine	1.0	48	116.9
Triphenylphosphine/triphenylboron complex	1.0	24	67.7
Tetraphenylsilane	1.0	24	52.1
Tetraphenylgermane	0.5	24	53.0
Triphenylarsine	1.0	24	50.0

Copper Oxides

G.R. Wilson; U.S. Patent 3,290,249; December 6, 1966; assigned to Monsanto Research Corporation describes the use of a copper oxide as antioxidant for the polyphenyl ether liquid fluids. The useful copper oxides include cupric oxide (CuO), cuprous oxide (Cu_2O), and copper suboxide (Cu_4O). Cuprous oxide (Cu_2O) is preferred. The copper oxides are combined with the fluid polyphenyl ethers to the extent of 0.01 to 1.0% by weight, depending upon the nature of the oxide and of the ether fluid. The copper oxides have an antioxidant effect on the polyphenyl ethers, generally. The process is illustrated by the examples shown on the following page.

Example 1: In this example, there was employed a mixture of polyphenyl ethers consisting by weight of: 65% of m-bis(m-phenoxyphenoxy) benzene, 30% of m-[(m-phenoxyphenoxy)(p-phenoxyphenoxy)] benzene, and 5% of m-bis(p-phenoxyphenoxy) benzene.

The rate of oxidation of the above mixture, in the presence or absence of cuprous oxide or cupric oxide was determined by means of an oxygen absorption test using a Dornte-type apparatus [J. Inst. Petrol., 41, 283 (1955)] equipped with an electrically heated deepwell unit in which the sample was heated by immersion of the test cell. Oxygen was passed, at a rate of 12 liters/hour into the test sample maintained at 600°F. and consisting of either 50 g. of the mixture of ethers or the same quantity of the mixture plus 0.2% by weight of one of the oxides.

The mixture of polyphenyl ethers, in the absence of any additive, was found to oxidize at the rate of 70 ml. of O_2/hour. In the presence of the cupric oxide the rate was reduced to 38 ml. of O_2/hour and in the presence of the cuprous oxide the oxidation rate was only 18 ml. of O_2/hour. That the decreased rate of oxidation is not due to the presence of metal was determined by operating in the presence of the same quantity of minutely suspended copper, zinc or silver. At the end of 4 hours at the 600°F. temperature and the same rate of oxygen introduction, the following results were noted:

Additive	O_2 Consumed, ml.
None	350
Zinc	312
Silver	302
Copper	225
Cuprous oxide	72

At the end of ca. 5.5 hours, the sample which contained the cuprous oxide had consumed only ca. 105 ml. of oxygen whereas that which contained the copper had consumed 350 ml. of oxygen. Observation of the "control," i.e., the sample of ether mixture which contained no additive, was discontinued at the end of about 5 hours at which point ca. 500 ml. of oxygen had been consumed.

Example 2: The antioxidant effect of cuprous oxide on a polyphenyl ether fluid was determined by bubbling air through a 20 ml. sample of the mixture of ethers described in Example 1, in the presence or absence of 0.2% weight of the oxide at 600°F. for 24 hours at 1 liter/hour, and determining the viscosity (at 100°F.) and percent loss in weight of the treated sample. The percent change in viscosity (before and after oxidation) was taken as an index of antioxidant activity.

Since some metals had been found to have an effect on the oxidation of polyphenyl ether fluids at high temperatures, the testing was also conducted in the presence of metals. In order to determine metal effect, one set of duplicate samples of ether fluid plus additive contained copper, iron, aluminum and silver wires, whereas another duplicate set contained only the ether fluid and additive. Duplicate sets of no-additive controls were also employed, one set of only the ether fluid and another set in which the metal wires were immersed in the ether fluids. The following results were obtained.

Additive	Viscosity Increase, Percent	
	No Wires	With Wires
None	54.0	39.0
Cuprous oxide	18.9	19.0

Organolead Compounds

R.E. Hatton and L.R. Stark; U.S. Patent 3,423,469; January 21, 1969; assigned to Monsanto Company have found that the oxidative stability, and thus the useful life of the polyphenyl ethers, can be greatly extended, even under the severe conditions encountered in jet engines and other devices operating at temperatures of the order of 600°F. and higher, by the addition to polyphenyl ethers of an organolead compound.

Examples of useful compounds are tetraphenyllead, triphenyllead chloride, triphenyllead fluoride, diphenyllead dichloride, diphenyllead diiodide, phenyllead trichloride, and diphenyllead dibromide, lead benzoate, lead vanillate, lead tetra-acetate, lead acetate, lead ethylcaproate, lead ethylhexoate, lead 2-ethylhexanoate, lead naphthenate and lead octoate.

Generally, the phenyllead compounds of this process can be prepared by the action of a Grignard reagent on lead chloride, and lead acid derivatives can be prepared by salt exchange, neutralization, etc., as from salts of sodium benzoate or acetate and lead chloride.

The major bench scale method used for evaluating the oxidative stability of a lubricant is the procedure given in Federal Specification 791, Method 5308.4, according to which the lubricant to be tested is heated at a specified temperature in the presence of certain metals and oxygen, and the viscosity increase of the lubricant is determined. Additionally, information as to the corrosivity of a lubricant to metals can be obtained.

Various polyphenyl ether compositions were tested according to the procedure of Federal Specification 791, Method 5308.4, except that the temperature was held at 600°F. instead of 250°F. as specified in that procedure, and the metal specimens used were steel, copper, silver, titanium, magnesium alloy and aluminum alloy. The compositions tested had essentially no effect on steel, titanium and aluminum alloy, and generally showed a slight increase in copper corrosion. In some tests, no metal specimens were included and only the viscosity increase was measured.

The results observed using the above described procedure are recorded in the table below. Viscosity measurements were made according to ASTM Method D-445-53T using a Cannon-Fenske modified Ostwald viscosimeter. The percentage of viscosity increase was determined by taking the difference in viscosity of a composition before and after it was heated, dividing that difference by the original viscosity, and multiplying the quotient by 100. The corrosivity to metals was determined by weighing the metal specimens before and after the test. The base stock used was composed, by weight, of about 65% m-bis(m-phenoxyphenoxy)benzene, about 30% m-[(m-phenoxyphenoxy)(p-phenoxyphenoxy)]benzene, and about 5% m-bis(p-phenoxyphenoxy)benzene.

Composition No.	Organolead compound		Viscosity Inc., Percent		Weight change, mg. per sq. cm.	
	Description	Percent by Wt.	100° F.	210 °F.	Mg.	Ag.
1	None-control		[1] 100	[1] 38	[1] +.38	[1] +.41
2	do		[2] 663	[2] 414	No metals	
3	Lead vanillate	1.0	44	18	0.14	0.34
4	Lead tetraacetate	1.0	45	17	0.47	0.39
5	Lead acetate	1.0	30	11	1.6	0.02
6	do	0.5	35	12	No metals	
7	Lead benzoate	0.2	44	16	0.10	0.16
8	do	0.5	33	12	No metals	
9	Tetraphenyllead	0.5	44	16	0.01	0.12
10	do	0.2	34	11	No metals	
11	Triphenyllead chloride	0.5	63	23	−0.78	−0.12
12	do	0.5	41	16	No metals	
13	Diphenyllead Dichloride	0.5	61	17	−2.02	−0.34
14	do	0.5	47	12	No metals	

[1] Average of 16 separate tests representing 3 lots of base fluid.
[2] Average of 7 separate tests representing 3 lots of base fluid.

From the above, it is evident that the addition of an organolead compound to the polyphenyl ethers provides polyphenyl ether compositions having a greatly increased oxidative stability and, therefore, a greatly extended useful life. In regard to the extension of useful life, it has been found that the test procedure described above correlates quite well with the results obtained under full-scale aircraft gas turbine-bearing tests and under conditions of actual use. It has been found that the magnitude of change in the viscosity at 100°F., as measured by the test procedure, is representative of the extent of increased service life obtainable under actual conditions.

It is also evident from the data presented above that the addition of an organolead compound to the polyphenyl ether provides polyphenyl ether compositions having significantly reduced rates of corrosion to certain metals and reduced tarnishing. It was also noted, by visual inspection, that the organolead stabilized polyphenyl ethers had significantly less sludge than did the unstabilized polyphenyl ethers.

Organotin Compounds

J.R. Stemniski; U.S. Patent 3,490,738; January 20, 1970; assigned to Monsanto Research Corporation has found that compositions consisting essentially of a polyphenyl ether base fluid and an additive amount of a disubstituted tin compound of the formula $(R_1—Sn—R_2)_n$ where each of R_1 and R_2 is an aromatic radical containing at least two aromatic rings, and n is an integer which has a value of at least one, have improved ability to resist oxidation degradation at high temperatures.

The tin compounds are disubstituted tin compounds, which may be either monomeric or polymeric. In general, the monomeric divalent tin compounds readily polymerize on standing: the polymers, which are compounds of the above stated formula $(R_1—Sn—R_2)_n$ where n is an integer greater than one, may have a cyclic structure, such as

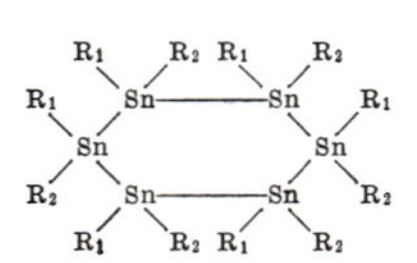

and in general appear to be low polymers, where n has a value of, say, up to 10. Illustrative of the useful diaryl tin compounds are dibiphenylytin, dinaphthyltin, dianthryltin, diphenanthryltin, diacenaphthyltin. Aryloxyaryltin compounds, which are particularly preferred in the process are exemplified by bis(phenoxyphenyl)tin, bis(biphenylyloxyphenyl)tin, bis(phenyloxybiphenylyl)tin and bis(naphthyloxyphenyl)tin. The process is illustrated by the following examples.

Example 1: This example illustrates preparation of a bis(aryloxyaryl)tin compound. p-Phenoxyphenylmagnesium bromide is prepared in the usual manner using 50 g. (0.20 mol) of p-bromophenyl phenyl ether and 4.86 g. (0.20 gram-atom) of magnesium using tetrahydrofuran as solvent. To the Grignard reagent is added 18.9 g. (0.1 mol) of stannous chloride (anhydrous) dissolved in 150 ml. of tetrahydrofuran. The mixture is refluxed for two hours and cooled. About 65 ml. of saturated ammonium hydroxide solution is added and the mixture is filtered. The filtrate is reduced to dryness. The yellow viscous oil obtained solidifies on standing: the compounds as formed initially is identified as monomeric bis(phenoxyphenyl) tin, and the solid formed on standing, as polymeric bis(phenoxyphenyl)tin.

Example 2: This example provides an illustration of the improved oxidation stability of compositions of this process as compared to the base fluid in the absence of additive. A lubricant composition is prepared by combining disubstituted tin compounds, as set forth below, with a polyphenyl ether of the following composition, by weight, 65% m-bis(m-phenoxyphenoxy)benzene, 30% m-[m-phenoxyphenoxy)-(p-phenoxyphenoxy)]benzene, 5% m-bis(p-phenoxyphenoxy)benzene.

For determination of the antioxidant effect of the additives, air is bubbled through duplicate samples at 600°F. for 24 or 48 hours at a rate of 1 l. per hour of air, in the presence and absence of Fe, Cu, Al and Ag wires. The percent change in viscosity (at 100°F.) from before to after oxidation is an index of antioxidant activity. Portions of the polyphenyl ether are combined, respectively, with the monomeric and with the polymeric, solid bis(phenoxyphenyl)tin product made as described in Example 1, each in a proportion of 1 g. of the tin compound to 100 g. of the base fluid.

A portion of the base fluid used to provide the above described composition is reserved, free of additive, and the stated compositions and the sample of the base fluid without additive are subjected to the above oxidation test, conducted with and without the presence of wires of Fe, Ag, Cu and Al, with the following results:

	Percent Viscosity Increase, 100°F.	
	In Presence of Wires	In Absence of Wires
Base Fluid	36 (24 hr.), 73 (48 hr.)	65 (24 hr.).
Plus monomer	13 (24 hr.), 18 (48 hr.)	10 (24 hr.).
Plus polymer	15 (48 hr.)	

In the same oxidation test, using a composition prepared by combining the stated polyphenyl ether base fluid with di-9-phenanthryltin in a proportion of 0.25 g. per 100 g. of base fluid, a reduction in viscosity increase is also obtained as compared to the base fluid.

Alkali Metal Compounds

R.W. Weiss; U.S. Patent 3,492,229; January 27, 1970; assigned to Monsanto Company has found that the oxidative stability and thus the useful life of functional fluids can be greatly extended, even under the severe conditions encountered in jet engines and other devices operating at temperatures of the order of 550°F. and higher, by the addition to functional fluids of a metal compound, the metal compound being selected from the group consisting of

(A) (1) an alkali metal compound, and (2) mixtures;

(B) (1) a metal compound where the metal is selected from the group consisting of antimony, bismuth and lanthanum, and (2) mixtures, and

(C) mixtures of (A) and (B).

Typical examples of metal compounds represented by (A) are alkali metal compounds represented by the structure M_aAnion where M is an alkali metal and a is a number having a value of at least one. Typical examples of metal compounds represented by (B) are metal compounds represented by the structure $(M_1)_b$Anion$_c$ where M_1 is selected from the group consisting of antimony, bismuth and lanthanum, b is a number having a value of at least one and c is a number having a value of from 1 to the product of b times the valence of M_1.

In order to demonstrate the properties of the compositions of this process, various metal salts were blended into base stocks and the resulting functional fluid compositions evaluated for oxidative stability and control of metal corrosion. One of the major bench scale methods used for evaluating the oxidative stability of a lubricant is the procedure given in Federal Test Method, Standard No. 791, Method No. 5308 according to which the lubricant to be tested is heated at a specific temperature (500°F.) in the presence of certain metals and oxygen and the viscosity increase of the lubricant is determined.

In addition, information as to the corrosivity of a lubricant to metals and the degree of sludge and deposit formation can also be obtained. Various functional fluid compositions containing a metal compound represented by (A), (B) and (C) were tested according to the above procedure except that the temperature was held at 650°F. with an air rate of 10 l. per hour in Examples 1 through 47 and at 500°F. and an air rate of 5 l. per hour in Examples 48 through 51. The metal specimens used were, as specified in the procedure, steel, copper, silver, titanium, magnesium alloy and aluminum alloy. The viscosity of the fluid before and after testing was determined at 100°F.

TABLE 1

Ex. No.	Metal salt	Concentration in milliequivalents per 100 grams of base stock	Degree of stabilization	Percent reduction in initial rate of oxidation	Rate of oxidation, cs./hour			Net change in rate of oxidation $R_{48}-R$ initial	Net change in metal mg./cm.[2]	
					R initial	R_{48}	R final		Cu	Ag
1	Neat base stock	0.00	0	0	16.5	[1] 42		[1] 29.5	−1.20	
2	Potassium cyclohexanoate	1.0	93	84	2.6	2.5		−0.1	+0.12	
3	Potassium acetate	1.0	91.5	77	4.0	3.6		−0.4	+0.02	
4	Potassium neodecanoate	1.0	95.5	82	3.0	1.6		−1.4	+0.24	
5	Potassium neotridecanoate	1.0	99	92	0.3	0.8		+0.5	+0.02	
6	Potassium oleate	1.0	95.5	88	2	1.7		−0.3	−0.60	
7	Potassium bitartrate	1.0	95.5	85	2.5	1.5		−1.0	+0.25	
8	Potassium thiocyanate	1.0	84.5							
9	Tetrapotassium ethylene diamine tetraacetate	1.0	92	73	4.5	2.7		−1.8	+0.56	
10	Dipotassium ethylene diamine diacetate	1.0	92	76	4.0	3.1		−0.9	+0.15	
11	Potassium ethylene diamine diacetate acid	1.0	90	73	4.5	4.7		+0.2	−0.12	
12	Potassium nitrilo triacetic acid	1.33	95.5	87	2.2	1.8		−0.4	+0.35	
13	Potassium benzoate	1.0	95	82	3.0	1.7		−1.3		
14	Potassium salicylate	1.0	92.5	79	3.5	3.1		−0.4	+0.58	
15	Potassium acetosalicylate	1.0	88	72	4.8	6.1		+1.3	+0.59	
16	Potassium m-phenoxybenzoate	1.0	95.5	86	2.3	1.7		−0.6		
17	Potassium o-phenyl azobenzoate	1.0	95.5	84	2.7	1.6		−1.1		
18	Potassium m-phenyl azobenzoate	0.94	95	85	2.4	2.4		0.00	[2] +0.57	
19	Potassium p-phenyl azobenzoate	2.0	3	76	4.0	2.7		−1.3		
20	Potassium phenyl acetate	1.0	94.5	84	2.6	2.2		−0.4	+0.37	
21	Potassium benzilate	0.74	95.5	85	2.5	1.4		−1.1	+0.40	
22	Lithium m-phenoxybenzoate	1.0	95	80	3.2	1.9		−1.3		
23	Lithium m-phenyl azobenzoate	2.0	94	82	3.0	2.2		−0.8	−0.46	
24	Lithium p-phenyl azobenzoate	0.47	91.5	82	3.0	3.7		+0.7		
25	Sodium o-phenoxybenzoate	1.0	96	90	1.6	2.0		+0.4		
26	Sodium 5-(m-nitrophenylazo)-salicylate	1.0	87						−0.75	
27	Rubidium m-phenyl azobenzoate	0.47	94.5	80	2.3	2.2		−0.1		
28	Rubidium p-phenyl azobenzoate	1.0	94	85	2.5	2.4		−0.1	[3] +0.22	
29	Potassium benzylate	0.3	97	91	1.4	1.4	1.5(75)	0.00		
30	do	2.0	97	90	1.6	1.4	2.5(307)	−0.2	[4] +0.30	
31	Potassium m-phenoxybenzylate	1.0	95.5	83	2.8	1.4		−1.4		
32	Lithium benzylate	1.0	93	73	4.5	2.4		−2.1		
33	Potassium diphenyl methoxide	1.0	96	88	2.0	1.6		−0.4	+0.55	
34	Potassium triphenyl methoxide	1.0	95	86	2.3	1.8		−0.5	+0.30	
35	Potassium t-butoxide	1.0	97	82	3.0	1.3	4.9(141)	−1.7	[5] +0.77	
36	do	2.0	96.5	82	3.0	1.3	2.3(307)	−1.7		
37	Potassium p-azobenzene sulfonate	3.0	48.5							
38	m,m′-phenoxyphenoxy potassium phenate	1.0	95.5	85	2.5	1.6		−0.9	+0.21	
39	Lanthanum m-phenoxybenzoate	1.0	91.5	76	4.0	4.4		+0.4		
40	Lanthanum m-phenylazobenzoate	1.0	91	74	4.3	4.4		+0.1		
41	Bismuth m-phenoxybenzoate	1.0	92	79	3.5	4.8		+1.3	−0.25	
42	Bismuth m-phenylazobenzoate	1.0	91						+0.12	
43	Potassium 3,5-dichloro anthranilate	2.0	57							
44	Potassium pentachloro phenate	2.0	86.5	48	8.5				[6] −2.20	
45	Potassium trichloro acetate	2.0	47							
46	Neat	0.0	00	00	5.3	7.8		+2.5	−14.4	−1.80
47	Potassium t-butoxide	3.0	69.5	36	3.4	2.5		−0.9	−0.4	−0.50
48	Neat	0.0	00	00					−2.13	−0.92
49	Potassium t-butoxide	0.1	65						−0.61	0.36
50	Potassium triphenyl methoxide	2.0	80						−0.80	−0.10
51	Potassium diphenyl methoxide	0.5	69						−0.85	−0.10

[1] Rate of oxidation determined at 25 hours test time.
[2] Copper weight change determined at 120 hours.
[3] Weight loss of copper determined at 65 hours.
[4] Copper weight determined at 300 hours.
[5] Copper weight change determined at 141 hours.
[6] Copper weight change determined at 29 hours.

In the table above, the percent stabilization is determined by the following formula:

$$\text{Degree of Stabilization (D)} = \frac{V_1 - V_2}{V_1} \times 100$$

and in addition

$$\frac{V_1}{V_2} = \frac{100}{100 - D}$$

where V_1 is the percent viscosity increase of the neat base stock after a certain test time and at a defined test temperature; V_2 is the percent viscosity increase of the base stock plus metal compound determined at the same test time and under the same test conditions as the neat base stock. V_1/V_2 relates to the extended usable life of the functional fluid as measured by the number of times that the percent viscosity increase is reduced over a given period of time.

In Examples 1 through 45 the degree of stabilization was determined after a test time of 24 hours at a test temperature of 650°F. In Examples 46 and 47 the degree of stabilization was determined after a test time of 48 hours at a test temperature of 650°F. and in Examples 48 through 51 the degree of stabilization was determined after a test time of 48 hours at a test temperature of 500°F. In addition, in Table 1 the percent reduction in the initial rate of oxidation is determined by the following formula:

$$\text{Percent reduction} = \frac{Rb - (Rb + a)}{Rb} \times 100$$

where Rb is the initial rate of oxidation of the base stock expressed as the change in viscosity in centistokes per hour determined at approximately 7 hours; Rb + a is the rate of oxidation in centistokes per hour of the base stock plus additive determined at 7 hours.

In Table 1, R Initial is the rate of oxidation of the fluid after approximately 7 hours, R_{48} is the rate of oxidation of the fluid in centistokes per hour determined after 48 hours, and R Final is the rate of oxidation of the fluid in centistokes per hour when the oxidation and corrosion test was continued beyond a 48 hour period. The test duration of R Final is given in parentheses next to the R Final rate of oxidation.

The base stocks that were used are as follows: in Examples 1 through 45 the base stock was a mixture of 5-ring polyphenyl ethers, in Examples 46 and 47 the base stock was bis(m-phenylmercaptophenyl)sulfide, and in Examples 48 through 51 the base stock was a mixture of m-bis(phenylmercapto)benzene, m-bis(phenylmercaptophenyl)sulfide, m-phenylmercaptophenyl-m-phenoxyphenyl sulfide and m-bis(phenoxyphenyl)sulfide.

A bearing test was run which was modified from MIL-L-27502. The test time was 100 hours using an Erdco bearing. The test conditions for Example 52 were a bearing temperature of 600°F., a bulk oil temperature of 600°F. and an oil in temperature of 550°F. The test conditions for Example 53 are the same as for Example 52 except the bearing temperature was increased to 650°F. In Table 2 the base stock that was used was a mixture of 5-ring polyphenyl ethers. The stabilizer that was used was potassium p-phenylazobenzoate.

TABLE 2

Example No.	Concentration by Weight	Overall Rating	Final Viscosity, cs. at 100°F.
52	0.00	79.3*	20,000*
53	0.124	61.8**	412**

* Test terminated and viscosity determined at 62 1/2 hours.
** Deposit rating and viscosity determined at 100 hours

A Ryder load test was run to determine the lubricating qualities of compositions comprising a polyphenyl ether, a metal compound represented by (A), (B) and (C) and a minor amount of a sulfur-containing base stock. The Ryder load test consists of subjecting a test lubricant (held at 165°F.) to a series of controlled tests at increasing gear tooth loads in a test machine which uses a set of spur gears (one of which has wider teeth than the other) as the test gears.

After each loading, the narrow gear of the test gears is examined to determine the scuffed area on each tooth. The load carrying ability of the scuff limit load of the lubricant is defined as the tooth load on the narrow test gear at which the average percent of tooth area scuffed is 22.5% for the narrow test gear. The test equipment and basic test procedure, which were first developed by E.A. Ryder, are described in ASTM Bulletin No. 184 entitled "A Test for Aircraft Gear Lubricants".

The Ryder load test utilizes the Ryder Gear Erdco universal tester which includes a Ryder Gear machine, a drive system, a support and load oil system, a test oil system and the necessary instrumentation and controls. The test gears are special spur gears made of AMS 6260 steel, case hardened and ground, having 28 teeth, 3 1/2 inches pitch diameter, 8 inch diametral pitch, 22.5° pressure angle with 0° tip relief.

In reporting the results from the Ryder load test, the scuffed area of each gear tooth is measured and is multiplied by 100 divided by the total working area of the gear tooth. The average percent of tooth area scuffed is the algebraic average of the percent of scuffed area of all 28 teeth of the test gear. The percent of tooth area scuffed is estimated visually to the nearest 5% of each individual tooth with the aid of a grid mounted in the left eye base of a microscope which is part of the test equipment.

As noted above the load carrying ability or the scuff limit load of a lubricant has been defined as the gear tooth load at which the average percent of tooth area scuffed is 22.5% for the narrow test gear by plotting the average percent of the area scuffed for the narrow test gear versus the load oil pressure. From the plot thereby obtained, the load oil pressure is read at which the average percent of tooth area scuffed is 22.5%. Then the load carrying ability of the lubricant is determined by the following formula:

$$P = \frac{kL}{W_2}$$

where P equals load carrying ability of the lubricant, pounds per inch; k equals Ryder gear machine constant, which is 18.55; L equals load oil pressure, pounds per square inch gauge; and W_2 equals the effective tooth width in inches and is the actual contact width of the gear tooth which is determined by multiplying the gross tooth width by the ratio of effective tooth width to gross tooth width as determined by visual examination. In Table 3 the base stock that was used in Example 54 was a 4-ring polyphenyl ether. In Examples 55 through 57 a mixture of 5-ring polyphenyl ethers was used as the base stock.

TABLE 3

Ex. No.	Sulfur-Containing Base Stock	Concentration, percent	Metal Salt	Concentration, percent	Ryder Load Test Result, lbs/in.
54	m-Bis(phenylmercapto)benzene	10			2,030
55			Potassium m,m'-phenoxy-phenoxy phenate	0.7	2,085
56	m-Bis(phenylmercapto)benzene	3	do	0.3	2,563
57	Mixture of: m-bis(phenylmercapto)-diphenyl sulfide; m-(phenylmercapto)phenoxy diphenyl sulfide; m-bis(phenoxy)diphenyl sulfide.	3	do	0.3	2,505

As is demonstrated by Tables 1 and 2, it is evident that the incorporation of the metal compounds represented by (A), (B) and (C) into a base stock provides a functional fluid composition which has a high degree of resistance towards oxidative degradation and therefore, a greatly extended useful life.

Polyalkylaminotriazines

J.Dazzi, K. Schwarzenbach and E. Keller; U.S. Patent 3,424,683; January 28, 1969; assigned to Geigy Chemical Corporation describe functional fluids comprising a synergistic mixture of (a) a polyalkylaminodiazine or polyalkylaminotriazine, and (b) a polyphenyl ether in which the ratio of (a) to (b) is from 1:9 to about 9:1. The viscosity of the composition is substantially less than the viscosity of either component. As polyalkylamino di- or triazines which are suitable as component (a) in the mixtures according to the process, principally those of Formula 1 are used.

(1)

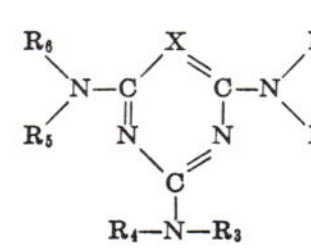

In this formula, X represents N or CH, R_1 represents hydrogen or an alkyl radical having 1 to 16 carbon atoms, R_2 represents an alkyl radical having 1 to 16 atoms, and R_3, R_4, R_5 and R_6 each represent an alkyl radical having 1 to 6 carbon atoms, the sum of all carbon atoms in all nitrogen substituents being at least 12.

Advantageously each of R_1, R_2, R_3, R_4, R_5 and R_6 represents an n-alkyl radical. X is preferably nitrogen; i.e., the compounds are polyalkylamino-1,3,5-triazines. They are described in French Patent 1,328,168. They can be obtained, for example, by reacting the 3 chlorine atoms of the cyanuric chloride, in steps, with the desired alkylamines by known methods. Examples of such polyalkylamino-1,3,5-triazines are:

2-dimethylamino-4,6-bis-dibutylamino-1,3,5-triazine,
2-diethylamino-4,6-bis-dibutylamino-1,3,5-triazine,
2,4,6-tris-diethylamino-1,3,5-triazine,
2,4,6-tris-dibutylamino-1,3,5-triazine and
2,4,6-tris-dihexylamino-1,3,5-triazine.

Polyalkylamino compounds of Formula (1) where X is CH, i.e., polyalkylaminopyrimidines, are obtained, e.g., by reacting in steps by methods known per se, the three chlorine atoms of the 2,4,6-trichloropyrimidine with the desired alkylamines. Examples of such polyalkylaminopyrimidines are:

2-diethylamino-4,6-bis-dibutylaminopyrimidine, or
2,4,6-tris-dibutylaminopyrimidine.

These synthetic functional fluids can also be mixed with usual functional fluids of other classes to attain special effects. The following examples show the low viscosity of the mixtures according to the process compared with that of the individual components. The units of measure of the metric system are used therein. The viscosities are measured in centistokes.

Viscosity of Mixtures of Polyalkylamino Di- or Triazines with Polyphenyl Ethers

Polyalkylamino di- or triazines of the general formula

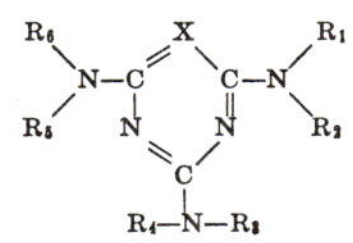

Polyphenyl ethers of the general formula

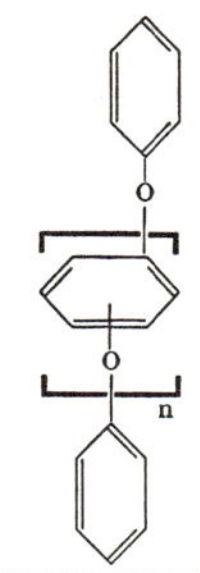

Mixtures of polyalkylamino di or triazines and polyphenyl ethers

No. Ex.	X	No. of carbon atoms R_1	R_2	R_3	R_4	R_5	R_6	Viscosity at 100° F. (37.5° C.) in centistokes	n	Viscosity at 100° F. (37.5° C.) in centistokes	Viscosity at 100° F. (37.5° C.) in centistokes	Content of polyphenyl ethers in percent by weight
1	N	2	2	4	4	4	4	103.5	2	69.6	50.0	60
2	N	1	1	4	4	4	4	68.8	2	69.6	48.5	50
3	N	2	2	2	2	2	2	73.0	[1] 2	69.6	46.1	50
4	N	10	1	4	1	1	1	37.1	[1] 2	69.6	35.2	30
5	N	4	4	4	4	4	4	131.6	[1] 2	69.6	57.6	70
6	N	6	6	6	6	6	6	105.0	[1] 2	69.6	56.3	50
7	N	[2] 0	4	4	4	2	2	136.6	[1] 2	69.6	54.0	70
8	N	1	1	4	1	2	4	[3] 7.62	[4] 1	[3] 4.31	[3] 4.75	40
9	N	4	4	4	4	4	4	131.6	[5] 3	341.0	113.8	30
10	N	6	6	6	6	6	6	105.0	[5] 3	341.0	97.0	10
11	N	2	2	4	4	4	4	103.5	[5] 3	341.0	100.0	30
12	CH	4	4	4	4	4	4	38.8	[1] 2	69.6	36.3	50
13	CH	4	4	4	4	6	6	112.0	[1] 2	69.6	55.7	30
14	CH	10	10	4	4	4	4	57.9	[1] 2	69.6	45.3	55
15	CH	12	12	4	4	4	4	79.9	[1] 2	69.6	51.3	45
16	CH	2	2	4	4	6	6	101.0	[1] 2	69.6	54.7	30
17	CH	2	2	4	4	6	6	101.0	[5] 3	341.0	96.8	80
18	CH	4	4	4	4	6	6	112.0	[5] 3	341.0	103.0	80

[1] Mixture of about 85% of bis(m-phenoxyphenyl)-ether, 13% of m-phenoxyphenyl-o-phenoxyphenyl ether, and 2% of p-phenoxyphenyl-o-phenoxyphenyl ether.

[2] $R_1 = H$.

[3] Measured at 170° F. (76.7° C.)

[4] p-Bisphenoxybenzene.

[5] Mixture of about 65% of m-bis(m-phenoxyphenoxy)-benzene, 30% of m-[(m-phenoxyphenoxy)-(p-phenoxyphenoxy)]-benzene and 5% of m-bis(p-phenoxyphenoxy)-benzene.

The mixtures numbering 1 through 18 given above are suitable for greasing metal and glass surfaces.

Example 19: Production of a Functional Fluid Containing Antioxidant — 98 g. of a mixture of polyalkylamino triazine and polyphenyl ether according to No. 1 of the above table are stirred with 2 g. of di-tert-octyl-diphenylamine as antioxidant until the latter has completely dissolved. Compared with a mixture not containing antioxidant, the mixture obtained has increased stability to oxidation. The other mixtures given in the above table can be made more resistant to oxidative influences in the same way.

If, instead of the 2 g. of di-tert-octyl-diphenylamine, 2 g. of one of the antioxidants given below are used and otherwise the procedure given is followed, then the stability to oxidation of the polyalkylamino di- and tri-azine/polyphenyl ether mixture is likewise increased:

Didodecyl selenide,
1,1,3-tri(2'-methyl-4'-hydroxy-5'-tert-butylphenyl)-butane,
2,6-di-tert-butyl-4-methyl phenol,
2-tert-octylimino-dibenzyl, or
5-ethyl-10,10-diphenyl-phenosilazine.

Example 20: Production of a Functional Fluid Containing Anticorrosive — 99.8 g. of a mixture of polyalkylaminotriazine and polyphenyl ether according to No. 1 of the above table are mixed with 0.2 g. of benztriazole. Compared with mixture No. 1 not containing this active substance, the mixture obtained has a less corrosive action particularly on copper and alloys thereof.

HALOGEN-CONTAINING ETHERS

Fluorinated Polyphenyl Ethers

A process described by E.S. Blake and G.A. Richardson; U.S. Patent 3,391,195; July 2, 1968; assigned to Monsanto Research Corporation provides a class of partially fluorinated polyphenyl ethers. According to the process, uncrystallizable liquids having very good thermal stability are provided when the terminal benzene ring of m-polyphenyl ether having from 3 to 7 phenyl groups is selected from the class consisting of monofluorophenyl, pentafluorophenyl and bis(m-phenoxyphenoxy) trifluorophenyl. The terminal benzene ring is either monosubstituted by fluorine or, it is entirely substituted, either by 5 fluorine atoms or by three fluorine atoms and two m-phenoxyphenoxy groups. Such compounds are represented by the general formula:

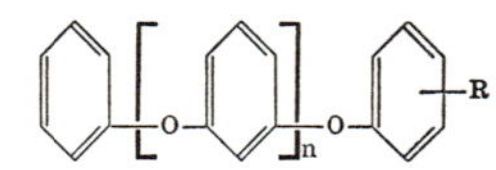

where R designates substitution from the class consisting of (1) one fluorine atom, (2) five fluorine atoms, and (3) three fluorine atoms and two m-phenoxyphenoxy radicals, and n is a number of 1 to 5. One class of compounds of the above formula comprises the monofluoro polyphenyl ethers.

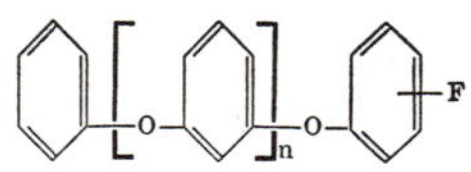

where n is a number of from 1 to 5. Another group of compounds are the pentafluorophenyl polyphenyl ethers:

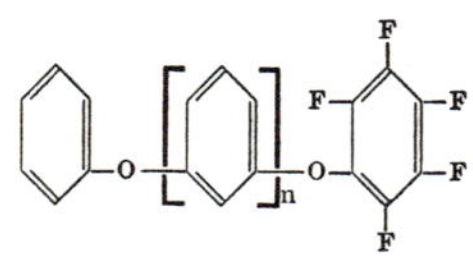

where n is defined above. Still another class of compounds which is provided by the process comprises the bis(phenoxyphenoxy)trifluorophenyl polyphenyl ethers. They have the formula:

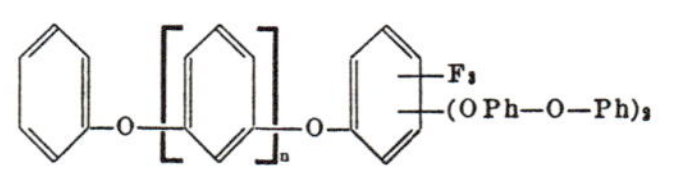

where OPhOPh designates phenoxyphenoxy and n is as above defined.

Examples of normally liquid, thermally stable compounds provided by the process are the monofluoropolyphenyl ethers such as o-, m- or p-fluorophenyl m-phenoxyphenyl ether, m-(o-, m- or p-fluorophenoxy)phenyl m-phenoxyphenyl ether, 1-[m-(o-, m- or p-fluorophenoxy)phenoxy]-3-(m-phenoxyphenoxy)-benzene, m-[m-(o-, m- or p-fluorophenoxy)phenoxy] phenyl m-(m-phenoxyphenoxy)phenyl ether, and 1-{m-[m-(o-, m- or p-fluorophenoxy)phenoxy]phenoxy}-3-[m-(m-phenoxyphenoxy)phenoxy]-benzene.

Examples of normally liquid, heat resistant bis(m-phenoxyphenoxy)-trifluoro compounds are m-[3,4-bis(m-phenoxyphenoxy)-2,5,6-trifluorophenoxy]phenyl m-phenoxyphenyl ether and m-[2,4-bis(m-phenoxyphenoxy)-3,5,6-trifluorophenoxy]phenyl m-(m-phenoxyphenoxy)phenyl ether. The process is illustrated by the following examples.

Example 1: Potassium m-(m-phenoxyphenoxy)phenoxide was prepared by azeotropic distillation of a mixture of 58.4 g. (0.2 mol) of m-(m-phenoxyphenoxy)phenol, 12.4 g. (0.2 mol, 90% pure) of potassium hydroxide and 35 ml. of toluene. The above obtained phenoxide was dissolved in 100 ml. of diglyme and added to a solution of m-bromofluorobenzene (35.0 g., 0.2 mol) and copper powder (0.5 g.) in 100 ml. of diglyme over a period of 2.25 hours at 155°C. The whole was stirred for 25.5 hours at 165°C.

The solid which formed upon cooling was filtered off, and the diglyme was removed from the filtrate by distillation in vacuo. The residue was taken up in ether and washed with acid, dilute alkali and water to neutrality. Distillation of the washed product gave 44.5 g. (59.5% theoretical yield) of the substantially pure m-(m-fluorophenoxy)phenyl m-phenoxyphenyl ether, BP 212°C./0.15 mm., n_D^{25} 1.6094. Vapor phase chromatography confirmed the homogeneity of the sample. In infrared absorption, the frequencies correspond to those for m-phenyl ether structure, and in nuclear magnetic resonance analyses the chemical shift was 32.7 ppm with a peak width of 30 cps, which is characteristic of the m-fluoro-m-phenyl ether moiety. The m-(m-fluorophenoxy)phenyl m-phenoxyphenyl ether, a noncrystallizable liquid, was found to have a decomposition point of 850°F., as determined on the isoteniscope and employing substantially the procedure described in E.S. Blake et al, J. Chem. Eng. Data, 6, 87 (1961). The following vapor pressure data were thus determined:

Vapor Pressure, mm. Hg	Temperature, °C.
10	271
100	352
600	436
760	448

The analogous tetrafluoro compound, m-(tetrafluorophenoxy)phenyl m-phenoxyphenyl ether, prepared by reaction of pentafluorobenzene with potassium m-(m-phenoxyphenoxy)phenoxide in diglyme at 165°C. for 17.5 hours, was a solid, MP 58.5° to 59.5°C.

Example 2: Operating substantially as in Example 1, but using p-bromofluorobenzene instead of the m-bromofluorobenzene of Example 1, there was obtained the substantially pure, normally liquid m-(p-fluorophenoxy)phenyl m-phenoxyphenyl ether, having the BP 198°C./0.07 mm., a decomposition point of 837°F. as determined on the isoteniscope by the method referred to in Example 1, and the following vapor pressure characteristics.

Vapor Pressure, mm. Hg	Temperature, °C.
10	276
100	355
600	435
760	447

The autogenous ignition temperature of the compound, determined by ASTM D60-7 procedure, was 1155°F. for 0.07 ml. with 13 seconds lag.

Example 3: Potassium m-[m-(m-phenoxyphenoxy)phenoxy]phenoxide was prepared by azeotropic distillation of a mixture of m-[m-(m-phenoxyphenoxy)phenoxy]phenol (27.8 g., 0.073 mol), potassium hydroxide (4.6 g., 0.069 mol) and toluene (50 ml.).

The above prepared phenoxide was dissolved in 100 ml. of diglyme, the solution was added over a 2 hour period to a solution of 27.2 g. (0.146 mol) of hexafluorobenzene in 200 ml. of diglyme at 140°C., and the whole was stirred for 16 hours at 150°C. The precipitate which formed upon cooling was filtered off and diglyme was removed from the filtrate by distillation in vacuo.

The residue was taken up in ether and washed with acid, alkali, and finally with water to neutrality. It was then distilled to give a fraction BP 250°C./0.18 mm., which did not crystallize when cooled to -75°C. It was redistilled to give 20.9 g. (53% theory) of the substantially pure 1-[m-(pentafluorophenoxy)phenoxy]-3-(m-phenoxyphenoxy)benzene, boiling point 240°C./0.18 mm., an uncrystallizable liquid. Upon heating it for 18 hours at 675°F. with an absolute pressure of 760 millimeter Hg, there was no evidence of decomposition, and the compound was found to have a decomposition point of 783°F. as determined by the procedure described in Example 1.

Chlorine Derivatives of m-Phenoxybenzene

A process described by A.L. Williams; U.S. Patent 3,499,840; March 10, 1970; assigned to Mobil Oil Corporation provides lubricant compositions which comprise a mixture of a monochloro compound and a dichloro compound, each of which is represented by the general formula

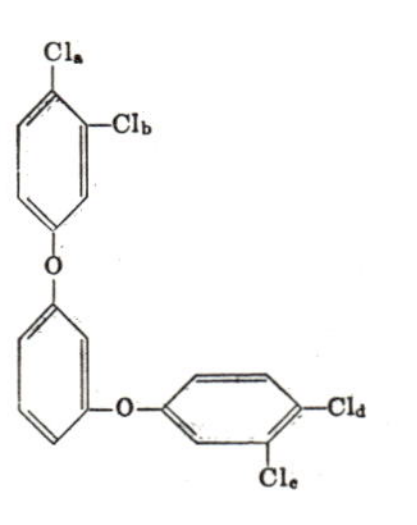

where each of a to d is zero or 1, the sum of a to d is 1 or 2, and a ring contains a maximum of one chlorine atom. The process comprises reacting, in the absence of molecular oxygen, about one molar proportion of a dialkali metal salt of resorcinol, with from about 1 to 3 molar proportions of monohalobenzene where the halogen is bromo or iodo, with from about 3 to 1 molar proportions of a meta- or para-halochlorobenzene, where the halogen is bromo or iodo, in the presence of a copper salt catalyst and a solvent that forms coordinate covalent bonds with copper ions. Preferably, approximately one molar proportion of each of the monohalobenzene and of the halochlorobenzene are used, to produce a monochloro derivative in high yield and to minimize production of diphenoxybenzene. Compounds of the general formula given above include: 1-(p-chlorophenoxy)-3-phenoxybenzene; 1-(m-chlorophenoxy)-3-phenoxybenzene; 1-(3,4-dichlorophenoxy)-3-phenoxybenzene; m-bis(m-chlorophenoxy)benzene, and m-bis(p-chlorophenoxybenzene).

The chloro derivatives of m-diphenoxybenzenes can be used in liquid or solid phase lubricating compositions, where they can constitute part or all of a lubricant vehicle. To take greater advantage of the high temperature stability of the chloro derivatives, it is preferred to use them in combination with other lubricants also characterized by relatively high temperature stability. They can also be used with other stable lubricants to prevent or deter crystal formation.

When used in a liquid lubricant composition together with a mineral or synthetic oil, the chloro derivatives can be present in amounts ranging from about 5 to 95% by volume (weight). As a preferred range, the chloro derivatives can be so used in amounts ranging from about 25 to 75% by volume (weight). The following examples illustrate the process. Examples 1 and 2 illustrate compounds which are useful as lubricants. Example 3 illustrates a dichloro m-diphenoxybenzene suitable as a lubricant. Examples 4 and 5 illustrate the process and the products that are also suitable as excellent high temperature lubricants.

Example 1: 1-(p-Chlorophenoxy)-3-Phenoxybenzene — A 1 l. 4 neck flask was fitted with stirrer, a 10" vacuum jacketed Vigreux column and a nitrogen inlet tube. The flask was flushed with nitrogen and an atmosphere of nitrogen was maintained therein during all subsequent operations in the flask. Pyridine (400 ml.) was introduced. In this was dissolved 111.5 g. (0.600 mol) of m-phenoxyphenol, by stirring in the flask. Then 31.0 g. (0.575 mol) of sodium methoxide powder was added slowly with stirring. Stirring was continued and 300 ml. of a mixture of methanol with pyridine was distilled out. The Vigreux column was then replaced by a reflux condenser and a thermometer was introduced into the liquid in the 1 l. flask.

Next 172 g. (0.900 mol) of melted 4-bromochlorobenzene at 80°C. was poured into the still warm contents of the flask. This was followed at once by 5.0 g. of cuprous chloride with stirring continued. The resulting mixture was then brought to slow reflux at a pot temperature of 125°C. The mixture was held at this temperature with stirring under nitrogen for 18 hours. The reaction mixture was then poured into a solution of 400 ml. of concentrated hydrochloric acid in 2 l. of water and stirred for 5 minutes.

The organic layer which formed was allowed to settle and the aqueous layer which had formed was decanted off the top. The organic layer was dissolved in 500 ml. of benzene and transferred to a separatory funnel. There it was washed by 200 ml. of concentrated hydrochloric acid in 200 ml. of water, followed by 100 ml. of concentrated hydrochloric acid in 200 ml. of water.

The benzene solution was next washed by 250 ml. of aqueous 10% sodium hydroxide solution, and finally by 300 ml. of water. Benzene was then distilled off through a 10" vacuum jacketed Vigreux column. The residue was distilled under vacuum to give 56 g. of recovered 4-bromochlorobenzene and 143 g. of 1-(p-chlorophenoxy)-3-phenoxybenzene, distilling at 168° to 170°C. at 0.4 mm. Hg. Analysis: Calculated for $C_{18}H_{13}ClO_2$: C, 72.85; H, 4.43; Cl, 11.95. Found: C, 72.51; H, 4.42; Cl, 11.8.

Example 2: 1-(m-Chlorophenoxy)-3-Phenoxybenzene — This preparation was carried out just as in Example 1, except that 110 g. (0.575 mol) of 3-bromochlorobenzene were substituted for the 172 g. (0.900 mol) of 4-bromochlorobenzene. Amounts of all other materials were unchanged. The final distillation under vacuum gave 137 g. of 1-(m-chlorophenoxy)-3-phenoxybenzene, distilling at 163° to 167°C. at 0.3 mm. Hg. Analysis: Calculated for $C_{18}H_{13}ClO_2$: C, 72.85; H, 4.43. Found: C, 72.33; H, 4.36.

Example 3: m-Bis(m-Chlorophenoxy)Benzene — A 2 l., 4 neck flask was fitted with stirrer, a Dean-Stark trap under a condenser and a nitrogen inlet tube. All reactions in the flask were carried out under nitrogen. A suspension of 55.0 g. (0.500 mol) of resorcinol was made in 800 ml. of benzene in the flask. Then, 51.8 g. (0.960 mol) of sodium methoxide was added slowly with stirring. Methanol and benzene were then distilled out to leave a dry resorcinol disalt in the flask. The Dean-Stark trap was removed from under the condenser, and a thermometer was introduced into the pyridine.

The contents of the flask were stirred and heated to 90°C. Then 287 g. (1.50 mol) of 3-bromochlorobenzene was poured in. This was followed at once by 15.0 g. of cuprous chloride. The mixture was brought up to a pot temperature of 115°C. and held at this temperature with stirring under nitrogen for 18 hours. The Dean-Stark trap was then reinserted and 500 ml. of pyridine was distilled out over a period of 1 1/2 hours.

The reaction mixture was then poured into 400 ml. of concentrated hydrochloric acid in 2 l. of water. The mixture was stirred for 5 minutes and the upper aqueous layer which formed was then decanted off. A pasty organic layer remained; it was washed by decantation with 110 ml. of concentrated hydrochloric acid in 100 ml. of water. The crude product was taken up in 700 ml. of benzene and filtered through diatomaceous earth. The residue was washed on the filter with 200 ml. of benzene, and the washings added to the filtrate.

The benzene solution was then washed in a separatory funnel by 100 ml. of concentrated hydrochloric acid in 100 ml. of water. Next the benzene solution was washed by two 200 ml. portions of aqueous 10% sodium hydroxide, and finally by 400 ml. of water. Benzene was then distilled off through a 12" vacuum jacketed Vigreux column. The residue was distilled under vacuum to give 61 g. of recovered 3-bromochlorobenzene and 91 g. of m-bis(m-chlorophenoxy)benzene distilling at 185° to 187°C. at 0.2 mm. Hg.

Example 4: m-Bis(Mixed-Chlorophenoxy)Benzene of High m-Chlorophenoxy Content — A 5 l. 4 neck flask was fitted with stirrer, a Dean-Stark trap under a condenser and a nitrogen inlet tube. All reactions in the flask were carried out under nitrogen. A suspension of 137.6 g. (1.250 mol) of resorcinol was made in 2 l. of benzene in the flask. Then 132.4 g. (2.460 mol) of sodium methoxide was added slowly with stirring. Methanol and benzene were then distilled out to leave a dry resorcinol disalt in the flask. Next, 2 l. of pyridine was added to the flask. The Dean-Stark trap was removed from under the condenser, and a thermometer was introduced into the pyridine. The contents of the flask were stirred and heated to 100°C. Then, 945 g. (4.95 mol) of mixed bromochlorobenzene isomers were poured in.

These isomers were obtained by isomerizing 4-bromochlorobenzene at 70°C. by aluminum bromide and concentrating the desired isomers by distillation at atmospheric pressure through a 12" vacuum jacketed Vigreux column. The isomerization procedure employed is that of G.A. Olah, W.S. Tolgyesi and R.E.A. Dear in J. Org. Chem., 27, 3455 to 3464 (1962). Analysis of the isomers used, by means of 12' UCON 550X gas chromatography column, showed the distribution: o-bromochlorobenzene, 6.4%; p-bromochlorobenzene, 26.8% and m-bromochlorobenzene, 66.8%.

Immediately after adding the bromochlorobenzene isomers, 38.0 g. of cuprous chloride was added. The resulting mixture was brought to a pot temperature of 117°C. and held at this temperature with stirring under nitrogen for 15 hours. The reflux condenser was then replaced by a 12" vacuum jacketed Vigreux column and 1600 ml. of pyridine was distilled off over a period of 6 hours, with a rise in pot temperature to 134°C. The reaction mixture was then poured into 800 ml. of concentrated hydrochloric acid in 4 l. of water. The aqueous layer which formed was decanted off and discarded.

A pasty organic layer remained; it was then washed by decantation by 2 l. of water. The crude product was taken up in 1,200 ml. of benzene and filtered through diatomaceous earth. This residue was washed on the filter by 300 ml. of benzene, and the washings added to the filtrate. The benzene solution was transferred to a separatory funnel and washed twice by 100 ml. portions of concentrated hydrochloric acid in 500 ml. of water. The benzene solution was then washed with two 200 ml. portions of aqueous 10% sodium hydroxide, and finally by 1,200 ml. of water.

Benzene was then distilled off through a 12" vacuum jacketed Vigreux column. Distillation was continued at atmospheric pressure to give 450 g. of recovered bromochlorobenzene isomers distilling a 185° to 196°C. The residue was distilled through the column under vacuum to give 258 g. of m-bis(mixed chlorophenoxy)benzene, distilling at 175° to 178°C. at 0.3 mm. Hg. Analysis: Calculated for $C_{18}H_{12}Cl_2O_2$ (percent): C, 65.28; H, 3.65. Found (percent): C, 64.88; H, 3.73.

Example 5: The preparation of the following compounds is described; (A) Mixed m- and p-chloro derivatives in 1-(mixed-chlorophenoxy)-3-phenoxybenzene and (B) mixed m- and p-chloro derivatives in m-bis(mixed-chlorophenoxy)benzene. A 5 l., 4 neck flask was fitted with stirrer, a 20" vacuum jacketed Vigreux column and a nitrogen inlet tube. All reactions in the flask were carried out under nitrogen. A suspension of 147.0 g. (1.335 mol) of resorcinol was made in 1,600 milliliters of benzene in the flask. The suspension was warmed to 60°C. with rapid stirring. Then 141.5 g. (2.62 mols) of sodium methoxide was added as a stream of powder with rapid stirring under nitrogen continued.

Methanol plus 1,200 ml. of benzene was distilled out over a period of 3 hours to leave a slurry of resorcinol disalt in benzene. Next, 1,500 ml. of pyridine was added and 700 ml. of a mixture of benzene with pyridine was distilled out. Then, an additional 900 ml. of pyridine was added to the flask. A mixture of 420 g. (2.67 mols) of bromobenzene with 512 g. (2.67 mols) of a mixture of meta- and para-bromochlorobenzenes was then poured into the flask.

The mixture of bromochlorobenzene isomers was obtained earlier by isomerization of 4-bromochlorobenzene by aluminum bromide, using the procedure identified in Example 4. The ortho-isomer content was reduced to 3% by fractional distillation through a 20" Vigreux column. 4-bromochlorobenzene was then blended with this distillate to reduce the ortho-isomer impurity to 2%. Gas chromatography on the blend of isomers showed the 3-bromochlorobenzene content to be 48% and the 4-bromochlorobenzene content to be 50%. After the bromides had been added to the reaction mixture, 40.0 g. of cuprous chloride was introduced. The Vigreux column was replaced by a reflux condenser and a thermometer was immersed in the reaction mixture.

The pot temperature was brought to 115°C. and held at that temperature with stirring under nitrogen for 16 hours. Then, the Vigreux column was put in place of the reflux condenser and 1,400 ml. of pyridine was distilled off over a 5 hour period. The resulting reaction mixture was then poured into 1,200 ml. of concentrated hydrochloric acid in 4 l. of water and stirred for 5 minutes. The aqueous layer which formed was decanted off and discarded. Then, the remaining crude product was dissolved in 1 l. of benzene and filtered through diatomaceous earth.

The residue was washed on the filter by 800 ml. of benzene, and the washings added to the filtrate. In a separatory funnel, the benzene solution was washed by 200 ml. of concentrated hydrochloric acid in 200 ml. of water, and then by 1 l. of water. The benzene solution was then washed by two 400 ml. portions of aqueous 5% sodium hydroxide, followed by 800 ml. of water. Benzene and bromobenzene were distilled off through a 12" Vigreux column. Distillation was then continued at reduced pressure through a 20" vacuum jacketed Vigreux column.

Bromochlorobenzenes recovered were 175 g. and bromobenzene 184.5 g. Product A, 122 g. distilled at 162° to 172°C. at 0.2 mm. Hg. and was mainly 1=(mixed-chlorophenoxy)-3-phenoxybenzene with minor amounts of m-diphenoxybenzene and m-bis-(mixed-chlorophenoxy)benzene. Product B, 124 g., distilled at 172° to 185°C. at 0.2 mm. Hg was m-bis(mixed-chlorophenoxy)benzene, approximately 68% by weight, with a lesser amount, about 31 weight percent, of 1-(mixed-chlorophenoxy)-3-phenoxybenzene. The composition and properties of the entire reaction product and of products A and B are given below:

TABLE 1

	Reaction Product	Product A	Product B
Weight, grams	246	122	124
m-Diphenoxybenzene (m-DPB), wt.%	10	19	1
Chloro-DPB, wt.%	45	59	31
Dichloro-DPM, wt.%	45	22	68
KV at 0°F., cs	10,400	5,400	24,500
KV at 100°F., cs	18.01	15.96	20.89
KV at 210°F., cs	3.12	2.96	3.33
Pour point, °F.	-30	-50	-55
Hours at 585°F. for absorption of 1 mol of O_2 per kg.		43.3	42.6

Viscosities and pour points of monochloro and dichloro derivatives of m-diphenoxybenzene are provided in Table 2 below. Advantages over a 4 ring polyphenyl ether in both viscosity at 0°F. and pour point are shown. The mixed-dichloro product of Example 4 contains about 6% by weight of the chlorine atoms ortho to the ether linkage, and has a higher viscosity than that of the product of Example 3 where all chlorine atoms are meta to the ether linkage. The two products, A and B, of Example 5 are both oils of low 0°F. viscosity and contain only 2% of the chlorine ortho to ether linkages. Mixing of monochloro and dichloro products provides the advantages of lower pour points, -50°F. and -55°F. as compared with -15° to -30°F.

TABLE 2: VISCOSITIES AND POUR POINTS OF MONOCHLORO AND DICHLORO DIPHENOXYBENZENES

Example	KV at 210°F., cs	KV at 100°F., cs	KV at 0°F., cs	Pour Point °F.
1	2.99	16.18	6,280	-25
2	2.92	15.68	4,860	-25
3	3.43	22.54	30,040	-30
4	3.59	24.55	53,540	-25
5A	2.96	15.96	5,400	-50
5B	3.33	20.89	24,500	-55
*4P3E comparison	6.3	69.6	>99,000	+20

*This is a commercial polyphenyl ether having four benzene rings.

Oxidation stability of both monochloro and dichloro derivatives of m-diphenoxybenzene is approximately twice as great as that of a commercial polyphenyl ether lubricant. This is shown in Table 3, on the following page. Rates of oxidation at 585°F. in the Dornte-type oxidation test [Ind. Eng. Chem. 28, 26 (1936)], are approximately 1/2 of that for the commercial polyphenyl ether. Since this oxidation stability is inherent in the chloro derivatives, such derivatives are advantageous in blends with less oxidation resistant oils with or without antioxidants present.

Table 3 lists noncatalytic oxidation tests on monochloro and dichloro m-diphenoxybenzenes at 585°F. in oxygen.

TABLE 3

Example	Time, hours*
1	38.9
2	62.0
3	61.5
4	55.0
5A	43.3
5B	42.6
5P4E comparison**	19.2

*These are the times required for the sample to absorb 1 mol of O_2 per kg. of oil.

**This is a commercial polyphenyl ether having five rings.

Oxidation corrosion test data is provided in Table 4. The chloro derivatives of m-diphenoxybenzene exhibit very little acid formation or viscosity change at 500°F. They are substantially superior to a commercial type 2 jet oil (full formulation) in these respects as well as in sludge and lacquer formation. There is significant corrosion, but almost none for a bronze.

TABLE 4: OXIDATION CORROSION TESTS AT 500°F. FOR 48 HOURS*

Oil or Metal After Oxidation	Example 1	Example 4	Commercial Type 2 Jet Oil (inhibited)
Percent viscosity increase at 100°F.	-4	+4	+34,500
Total acid number	0.02	0.13	7.3
Sludge in oil	None	None	Heavy
Lacquer	None	None	Very heavy
Percent lost as vapor	1.0	1.0	6.0
Weight loss of metal, mg./sq.cm.:			
Ag	0.1	0.4	0.0
Steel	0.1	0.2	0.0
Al	0.1	0.0	
Mg	0.0	0.0	Disintegrated
Ti	0.1	0.0	
General Electric bronze	0.1	0.1	
Copper	2.1**	3.8**	0.0

*This is Federal Test Method 791, Method 5308, modified. Metals are present as standard washers. Tests are on 100 g. samples of oil with an air flow of 5 l. per hours.

**These tests on copper were made separately and prolonged to 72 hours.

Lubrication test data for the 4 ball machine test are given in Table 5 below. The chloro derivatives of m-diphenoxybenzene exhibit significantly superior lubrication, to that obtained by use of commercial polyphenyl ethers. For steel-on-steel lubrication, wear values are in the range of values obtained with esterlubes.

TABLE 5: 4-BALL MACHINE TESTS WITH STEEL ON STEEL[1]

Sample	Wear Rate cc./cm. x 10″ 350° F.	Wear Rate cc./cm. x 10″ 500° F.	Average f^2 at 500° F.	Wear Scar diameter, mm. 350° F.	Wear Scar diameter, mm. 500° F
Commercial ester lubricant	0.20	7.60	0.103	0.43	0.95
5P4E comparison	37.0	13.9	0.205	1.42	1.11
4P3E comparison	44.4	14.1	0.171	1.48	1.12
Example 1	5.32	15.5	0.167	0.88	1.14
Example 4				1.087	
Example 5B	6.98	8.96	0.145	0.94	1.00

[1] Tests were conducted for 60 minutes at 1200 r.p.m. on 52100 Steel with 40 kg. load.

[2] f is coefficient of friction.

Data set forth in Table 5 reveal that the chloro derivatives have properties making them suitable for use in lubricants for use in lubricating high temperature jet engines. The derivatives are even more resistant to oxidation than polyphenyl ethers, and lubricate better in the 4-ball machine. Low temperature properties are substanticlly superior to those of unsubstituted polyphenyl ethers; pour points are as low as -55°F. as opposed to +20°F. for 4P3E. The derivatives have low corrosivity for metals, including magnesium, under oxidizing conditions. The chloro derivatives are also desirable for use as lubricants in view of their relatively high fire points. For example, the fire point of the 1-(p-chlorophenoxy)-3-phenoxybenzene described in Example 1 is 500°F. and of the mixed dichloro product of Example 4 is 600°F.

G. Irick, Jr., and G.F. Hawkins; U.S. Patent 3,379,771; April 23, 1968; assigned to Eastman Kodak Company have found that polyphenyl ethers having halogen substituents selected from the group consisting of fluorine, chlorine and bromine are in and of themselves effective antioxidants for polyphenyl ethers. The antioxidant effect of such halogenated polyphenyl ethers is not significant, however, unless the halogen concentration is at least about 0.01% by weight of the polyphenyl ethers.

Halogenated Bicyclic Ethers

In a process described by P.R. Stapp; U.S. Patent 3,532,720; October 6, 1970; assigned to Phillips Petroleum Company halogenated bicyclic ether compounds are obtained by reacting a cycloalkene with formaldehyde and a hydrogen halide. Halogenated bicyclic ether compounds produced in the process are useful as metalworking or cutting oils. The production of halogenated bicyclic ethers from cycloalkenes is illustrated by the following reaction:

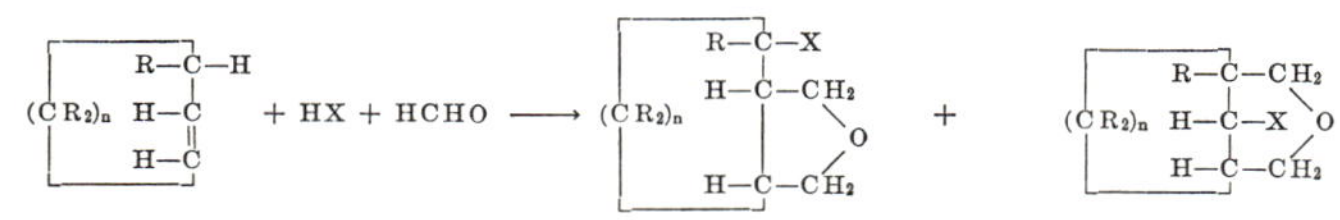

where X is a halogen (chlorine, bromine, iodine or fluorine), n is an integer of from 1 to 9 and any one R group is hydrogen or an alkyl group having one to six carbon atoms, the total number of carbon atoms in all of the R groups not to exceed 12.

Example 1: A stirred reactor was charged with 135 g. (4.5 mols) of paraformaldehyde, 174 g. (3.0 mols) of cyclopentene, and 500 ml. of methylene chloride. At a temperature of about -65°C. hydrogen chloride was introduced into the mixture for about three hours. Isolation of the product by distillation gave 108.9 g. of product comprised largely of a mixture of 8-chloro-3-oxabicyclo[3.2.1]octane and 8-chloro-3-oxabicyclo[3.3.0]octane having a boiling point of 97° to 101°C. (35 to 1.5 mm. of mercury) and having infrared characteristics indicative of a cyclic compound.

This product was refractionated to give a center fraction having a boiling point of 84° to 86°C. (15 mm. of mercury) which was found to have the following elemental composition: C, 56.9; H, 7.5; Cl, 24.1. Elemental composition calculated for 8-chloro-3-oxabicyclo[3.2.1]octane and 8-chloro-3-oxabicyclo[3.3.0]octane ($C_7H_{11}ClO$) is: C, 57.4; H, 7.5; Cl, 24.3.

A total of 50 g. of the above product of halogenated bicyclic ethers having a boiling point in the range of 97° to 101°C. was charged to a stirred reactor. Also charged were 300 ml. of methanol. Then a total of 46 g. (2.0 gram atoms) of sodium was added in small pieces. The reaction mixture was then poured into water and extracted into ether. The ether was dried with magnesium sulfate, filtered, and the ether was stripped.

Distillation of the residue gave 23.1 g. of a colorless liquid having a boiling point of 62° to 64°C. (47 mm. of mercury) which was determined by gas liquid chromatography to be comprised of two compounds in a proportion of 1:2. The respective isomers of the composition were separated by preparatory gas chromatography, characterized, and identified by nuclear magnetic resonance. The major component, a liquid, was identified as cis-3-oxabicyclo[3.3.0]octane.

Example 2: The following procedure is used in the synthesis of 9-chloro-3-oxabicyclo[3.3.1]nonane and 2-chloro-8-oxabicyclo[4.3.0]nonane. A stirred reactor was charged with 135 g. (4.5 mols) of paraformaldehyde and 246 g. (3.0 mols) of cyclohexene. At a temperature of -65°C. hydrogen chloride was passed through the reactor for 4 hours. A total of 201.7 g. (65.5% crude yield) of the product comprised largely of 9-chloro-3-oxabicyclo[3.3.1]nonane and 2-chloro-8-oxabicyclo[4.3.0]nonane was recovered by distillation, boiling point 88° to 135°C. (17 mm. of mercury).

Refractionation of this product gave 121.5 g. of a product having a boiling point of 90° to 96°C. (11 mm. of mercury), which partly solidified. A portion of this semisolid material was recrystallized from methanol to give a crystalline solid, melting point 112° to 116°C., which was comprised of two isomeric compounds as characterized by gas liquid chromatography.

This product was found to have the following elemental composition: C, 59.8; H, 8.3. Elemental composition calculated for 9-chloro-3-oxabicyclo[3.3.1]nonane and 2-chloro-8-oxabicyclo[4.3.0]nonane, $C_8H_{12}OCl$ is: C, 59.8; H, 8.1. The mixture of halogenated bicyclic ethers was comprised of about 65 mol percent of 9-chloro-3-oxabicyclo[3.3.1]nonane and 35 mol percent 2-chloro-8-oxabicyclo[4.3.0]nonane.

A stirred reactor was charged with 16.1 g. (0.1 mol) of the above solid isomer mixture and 100 ml. of methanol. To this mixture was added 11.5 g. (0.5 gram atoms) of sodium as in Example 1. After dilution with water and extraction into ether, the resultant ether solution was dried with magnesium sulfate and the solvent was stripped. The crystalline residue (10.6 g., melting point 120° to 122°C.) was determined by gas liquid chromatography to be of purity greater than 95%. The nuclear magnetic resonance spectrum was in agreement with the assigned structure of 3-oxabicyclo[3.3.1]nonane. A high resolution mass spectrum confirmed the composition $C_8H_{14}O$ (theoretical mass, 126.1448; measured mass 126.1466).

Example 3: A mixture of halogenated bicyclic ethers comprised of about 65 mol percent of 9-chloro-3-oxabicyclo[3.3.1]nonane and about 35 mol percent of 2-chloro-8-oxabicyclo[4.3.0]nonane were applied to a die used to cut threads on stainless steel pipe and functioned as a cutting oil.

4-Chlorotetrahydropyran

M.M. Johnson and P.R. Stapp; U.S. Patent 3,546,118; December 8, 1970; assigned to Phillips Petroleum Company have found that a cyclic ether having the formula

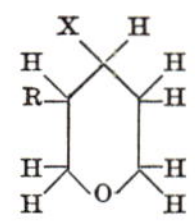

where R is hydrogen or an alkyl or cycloalkyl radical or combinations such as alkylcycloalkyl having up to and including 17 carbon atoms and where X is a halogen can be employed as a lubricant in metal working operations. Specific examples of such cyclic ethers are 4-chlorotetrahydropyran, 3-methyl-4-chlorotetrahydropyran, 3-propyl-4-bromotetrahydropyran, 3-amyl-4-chlorotetrahydropyran, 3-heptyl-4-iodotetrahydropyran, 3-nonyl-4-chlorotetrahydropyran, 3-cyclohexyl-4-fluorotetrahydropyran and 3-(4-butylcyclohexyl)-4-chlorotetrahydropyran. The cyclic ether lubricants can be prepared by reacting 1-olefins having the formula

$$R{-}CH_2{-}CH{=}CH_2$$

where R is as described above with formaldehyde and a halogenated compound selected from the hydrogen halides, thionyl halides, and carbonyl halides. The following examples illustrate the process.

Example 1: Threads on new stainless steel connectors using no lubricant, tightened up and backed out very easily with no pulling and galling. Threads on used stainless steel connectors using no lubricant, tightened up and backed out with considerable pulling and galling. Threads on used stainless steel connectors using a lubricant 4-chlorotetrahydropyran, tightened up and backed out with no pulling and galling. The used connectors were smoother threading after lubricating with 4-chlorotetrahydropyran.

Example 2: Using an AE coning tool, 1/4 inch o.d. x 0.083 inch i.d. 304 stainless steel tubing was cut using 4-chlorotetrahydropyran as a cutting lubricant. The cutting was every fast leaving a smooth surface on the tubing. Threads were cut on the abovementioned tubing using 4-chlorotetrahydropyran as a cutting lubricant. Threads were cut with little effort leaving smooth threads and collars, which tightened up and backed out with no dragging or binding.

MISCELLANEOUS

Phenylene Sulfides

A process described by L.W. Bannister; U.S. Patent 3,455,846; July 15, 1969; assigned to Monsanto Company provides lubricant compositions comprising mixtures of aryl sulfides represented by the following formulas:

(a)

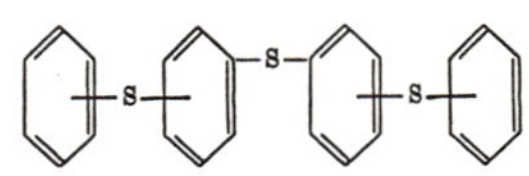

(b) 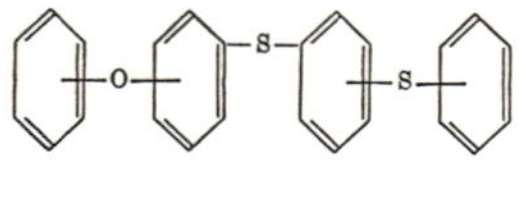

(c) 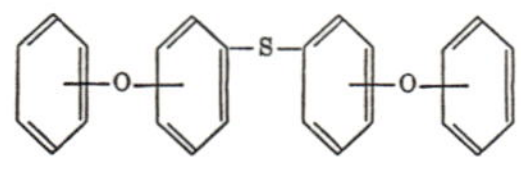

where the majority of the phenylene radicals of each compound are meta orientated. It has been found that compositions comprising aryl sulfides represented by the above formulas (a), (b) and (c) have properties which make them well suited for the use disclosed above and particularly those applications such as jet engine lubricants requiring thermal and oxidative stability and wide liquid range. Preferred compositions comprise mixtures of (1) bis(m-phenylmercaptophenyl) sulfide, (2) 3-phenoxy-3'-phenylmercaptodiphenyl sulfide and (3) bis(m-phenoxyphenyl) sulfide.

Compositions of this process, particularly useful as jet engine lubricants, are prepared by combining the components designated (1), (2) and (3) above in the weight percent ranges of from 1 to 40% of (1), from 40 to 80% of (2) and from 1% to 40% of (3). More particularly, compositions comprising components (1), (2) and (3) in the weight percent range of from 10 to 30% of (1), from 40 to 80% of (2) and from 10 to 30% of (3) are preferred.

Compositions of this process can be produced by blending the individual components together with adequate mixing to produce a homogeneous mixture or they can be prepared by reacting chlorophenyl phenyl sulfide and chlorophenyl phenyl ether wherein the majority of the halogen atoms are in the meta-position with sodium sulfide in mol ratios of about 1 to 2:1 to 2:1.5 to 2, respectively, as exemplified in Example 1 below where parts are by weight.

Example 1: Into a suitable reaction vessel there were charged a mixture of 90 parts of m-chlorophenyl phenyl ether, 79.4 parts of m-chlorophenyl phenyl sulfide, 61.4 parts of 61% sodium sulfide hydrate and 260 parts of N-methylpyrrolidone. The mixture was heated to 210°C., water and low boilers being collected as the temperature rose. The reaction mixture was held at 210° to 215°C. for 16 hours, the solvent was then distilled and water was added.

The organic layer was separated, washed with water and distilled, yielding 100 parts of product which boiled at 200° to 250°C. (0.1 mm.). The product was a mixture comprised by weight of 28% bis(m-phenylmercaptophenyl) sulfide, 50% of 3-phenoxy-3'-phenylmercaptodiphenyl sulfide and 22% of bis(m-phenoxyphenyl) sulfide.

The individual components of the compositions of this process can be prepared by reacting an alkali metal salt of a thiophenol or phenol with a suitable halide in a suitable carboxamide solvent such as N, N'-dimethylacetamide or N-methyl-2-pyrrolidone at temperatures of the order of 100° to 250°C., but preferably 160° to 230° C. The individual components of the compositions of this process can also be prepared by the Ullmann synthesis by reacting an alkali metal phenate or thiophenate and a suitable halide in the presence of copper at temperatures in the order of 180° to 260°C. The preparation of the individual components of the compositions is illustrated by the following examples.

Example 2: The following procedure is used in the preparation of bis(m-phenylmercaptophenyl) sulfide. Into a suitable reaction vessel fitted with agitation means, means for measuring reaction mass temperatures and vapor temperatures, heating means, reactant and product inlets and outlets and also fitted with a reflux condenser, there were charged a mixture of 44 g. of thiophenol and 26 g. of 86% potassium hydroxide in 150 ml. of dimethylacetamide. The mixture was heated to obtain the potassium salt of thiophenol and to drive off water of reaction.

The mixture was cooled to about 120°C. and 59.2 g. of bis(m-bromophenyl) sulfide were charged and the resulting mixture heated at 165° to 170°C. for about 14 hours. The reaction mass was then allowed to cocl to room temperature and then washed with 200 ml. of 12.5% sodium hydroxide. The organic layer was separated, diluted with benzene, then washed with water and dried under vacuum. The residue solidified and was crystallized from an isopropanol benzene mixture to give bis(m-phenylmercaptophenyl) sulfide.

Example 3: The following procedure is used in the preparation of 3-phenoxy-3'-phenylmercaptodiphenyl sulfide. Into a suitable reaction vessel there were charged 14 ml. of water and 70 g. of 85% potassium hydroxide and the resulting mixture was heated to 110°C., after which there was added 125 g. of phenol. Then, 15 ml. of toluene was added and the water removed by azeotropic distillation until a pot temperature of 265°C. was reached. The reaction mass was then cooled to 240°C. and 2 g. of cupric chloride were added. The reaction mass was then heated slightly as 329 g. of m-(m-phenylmercaptophenylmercapto)phenyl chloride was slowly added. The reaction mass was then heated at 255° to 260°C. for about 6 hours and quenched.

An aqueous layer was then separated, extracted with toluene and the extract combined with the organic layer. The organic material was extracted with water, filtered, and the toluene stripped to give 233 g. of 3-phenoxy-3'-phenylmercaptodiphenyl sulfide, which had a boiling range of 236° to 240°C. at 0.1 mm. of mercury.

Example 4: The following procedure is used in the preparation of bis(m-phenoxyphenyl) sulfide. Following the procedure set forth in Example 1, 163.5 parts of m-chlorophenyl phenyl ether was reacted with 61.4 parts of 61% sodium sulfide hydrate in 260 parts of N-methylpyrrolidone to produce bis(m-phenoxyphenyl) sulfide.

Some of the properties often measured in determining the suitability of a fluid as a jet engine lubricant and the method used for such determination include the following.

Property	Method
Viscosity	ASTM D-445-61.
Pour point	ASTM D-97
Evaporation loss	ASTM D-972 (modified as to temperature).
Oxidation and corrosion (O and C)	Federal Test Method Standard 791, Method 5208.4 (modified as to temperature).

Solution points were determined by placing a test composition in a jacketed tube provided with an agitator and with the agitator running, circulating through the jacket, a cooling solution held a temperature of 16° to 23°F. This temperature range was considered high enough to prevent a glass from forming and low enough to speed up potential crystallization. After a test composition had been agitated for about a day, seeds of one of the components were added.

Those mixtures which did not crystallize after one week were warmed to room temperature to make the fluids pourable and were transferred to small bottles with lids. The bottles were then placed in cold storage at 0° and -30°F. In either case, upon the inducement of crystallization, the container was heated gradually with agitation and the temperature noted at which the last crystal dissolved. Using the aforedescribed tests and procedures, properties of various compounds utilized here were determined and such properties are set forth below.

Property	Bis(m-phenylmercaptophenyl) sulfide	3-phenoxy-3'-phenylmercapto diphenyl sulfide	Bis(m-phenoxyphenyl) sulfide
Viscosity cs.:			
210° F.	6.19	6.24	6.03
100° F.	60.37	62.5	62.23
Melting point, ° C.	56	−4	27–29
Evaporation loss 6½ hrs. at 500° F., percent	8	16.4	32.4
Lubricity, shell 4-ball wear at 15 minutes, 400° F., 1,200 r.p.m. steel on steel:			
40 kg. load	1.39	1.14	.930
50 kg. load	1.35	.86	.967

Some properties of a typical composition of this process are listed below. A mixture produced according to Example 1 above was used and the various properties were determined as described above.

Property	Result
Viscosity, cs.:	
210°F.	6.1
100°F.	5.9
Solution point, °C.	28
Pour point, °F.	5
Evaporation loss (6 1/2 hrs. at 500°F.; 760 mm.) percent	16.9
Lubricity, shell 4-ball wear @ 15 minutes, 400°F., 1,200 rpm, steel on steel:	
40 kg. load	0.913
50 kg. load	0.992

(continued)

Property	Result
O and C test 500°F.:	
Viscosity change, 100°F., %	2.2
Copper corrosion, mg./cm.^{2}	-1.94

Octadecane Ethers

A process described by L.I. Osipow, F.J. Brashear and W.Rosenblatt; U.S. Patent 3,344,195; September 26, 1967; assigned to Fats and Protein Research Foundation, Inc., involves isomeric mixtures of a product having a formula selected from the group consisting of

(1) $$CH_3-(CH_2)_x-\underset{\displaystyle R}{\underset{|}{CH}}-(CH_2)_y-R_1$$

(2) $$CH_3-(CH_2)_x-\underset{\displaystyle R}{\underset{|}{CH}}-(CH_2)_{2y}-\underset{\displaystyle R}{\underset{|}{CH}}-(CH_2)_x-CH_3$$

(3) $$[CH_3-(CH_2)_x-\underset{\displaystyle R}{\underset{|}{CH}}-(CH_2)_y-]_2O$$

where R is selected from the group consisting of alkoxy radicals having 1 to 18 carbon atoms, phenyl and alkylphenyl radicals having 1 to 8 alkyl carbon atoms, R_1 is selected from the group consisting of hydrogen, alkoxy radicals having 1 to 8 carbon atoms, phenyl and alkylphenyl radicals having 1 to 8 alkyl carbon atoms, x is an integer from 6 to 8 and y is an integer from 8 to 10.

The products are derived from monounsaturated alcohols, such as oleyl alcohol by adding substituents across the double bond. Using oleyl alcohol as an example, the substituents add mainly at the ninth and tenth carbon atoms but since the double bond is labile, some shifting of the double bond occurs and addition occurs in other places in the molecule.

The products show negligible thermal decomposition at temperatures up to 700°F., have low pour points and a good viscosity index and are excellent base stocks to which various known additives may be added for detergency, lubricity, pour point depression, corrosion inhibition, oxidation inhibition, etc. to form high temperature lubricants and hydraulic fluids. The following examples illustrate the process.

Example 1: The following procedure shows the preparation of isomeric mixtures of 10-phenyl-octadecane. Step A is the preparation of 10-phenyloctadecyl bromide. Commercial grade oleyl alcohol (iodine value 91.1, refractive index n_D^{30}=1.4571) was reacted with benzene and $AlCl_3$ under Friedel-Crafts reaction conditions to form 10-phenyl-octadecanol BP 198° to 204°C. at 1 mm. Hg, n_D^{20}=1.4910. Anhydrous HBr was slowly added to the 10-phenyl-octadecanol at 100° to 110°C. for 4 hr. The mixture was washed successively with cold H_2SO_4, 50% methyl alcohol, 15% NH_3 and 50% methyl alcohol to obtain 10-phenyloctadecyl bromide as a clear yellow liquid BP 185° to 190°C. at 0.2 mm. Hg and n_D^{20}=1.4950. Analysis - Calculated for $C_{24}H_{41}Br$: Br, 19.5%. Found, 19.3%. Step B is the preparation of 10-phenyl-octadecane. 110 g. (0.27 mol) of the product of Step A were added dropwise over a period of thirty minutes to 0.3 mol of magnesium turnings on 500 cc of anhydrous ether. The mixture was warmed slightly to initiate the reaction and after all the bromide had been added, the reaction mixture was refluxed for 4 hours.

The reaction mixture was cooled in an ice bath and the reaction mixture was hydrolyzed by the careful addition of first 100 cc of a cooled 14% ammonium chloride solution and then 100 cc of 5% hydrochloric acid. The ether solution was then washed until neutral and dried over sodium sulfate. Upon distillation of the ether solution of 0.5 mm. Hg, 44 g. of a colorless liquid fraction at 166° to 172°C. was obtained which was an isomeric mixture of 10-phenyl-octadecane having a refractive index of n_D^{20} = 1.4850. The yield was 49% based on the 10-phenyl-octadecyl bromide. Analysis: $C_{24}H_{42}$ Calculated: C, 87.2; H, 12.8. Found: C, 87.0; H, 13.2.

Example 2: The following procedure shows the preparation of isomeric mixtures of 10-(tert-butylphenyl)-octadecane. Step A is the preparation of 10-(tert-butylphenyl)-octadecyl bromide. Commercial grade oleyl alcohol (Novol) was reacted with tert-butyl benzene having a boiling point of 167.5° to 170.5°C. and aluminum chloride under Friedel-Crafts reaction conditions to form 10-(tert-butylphenyl)-octadecanol as an amber fluorescent oil having a boiling point of 211° to 215°C. at 0.8 mm. Hg and a refractive index n_D^{20} = 1.4975. Anhydrous hydrogen bromide was slowly added to the 10-(tert-butylphenyl)-octadecanol at 100° to 110°C. for four hours and the reaction mixture was then washed successively with cold sulfuric acid, 50% methyl alcohol, 15% ammonia and 50% methyl alcohol to obtain 10-(tert-butylphenyl) octadecyl bromide having a boiling point of 215° to 220°C. at 0.2 mm. Hg. Analysis: $C_{28}H_{49}Br$, percent Br, Calculated: 17.2. Found: 17.3. Step B is the preparation of 10-(tert-butylphenyl)-octadecane. 0.27 mol of 10-(tert-butylphenyl)-octadecyl bromide produced in Step A was added to 0.3 mols of magnesium turnings in 500 cc of anhydrous ether over a period of thirty minutes while slightly warming the reaction mixture. After all of the bromide had been added, the reaction mixture was refluxed for four hours.

The reaction mixture was cooled in an ice bath and the reaction mixture was hydrolyzed by the careful addition of a cooled 14% ammonium chloride solution and then 5% hydrochloric acid. The ether solution was washed until neutral and dried over sodium sulfate. Upon distillation of the ether solution at 0.5 mm. Hg, 10-(tert-butylphenyl)-octadecane was obtained in 35% yield as a light yellow oil having a refractive index of $n_D^{20} = 1.4900$. Analysis: $C_{28}H_{51}$ Calculated: C, 86.7; H, 13.3. Found: C, 86.2; H, 13.1.

Example 3: The following procedure shows the preparation of isomeric mixture of 9,28-dimethoxyhexatriacontane. Step A is the preparation of 9,28-dibromohexatriacontane. Oleyl alcohol was reacted with phosphorus tribromide in carbon tetrachloride at -10° to 0°C. according to the method of Deathenage et al (JACS, volume 61, 1939, page 630). After the removal of the carbon tetrachloride, the residue was distilled through a 30 cm. Vigreux column and oleyl bromide having a trace odor of phosphine was obtained as the fraction boiling at 175° to 180°C. at 0.5 mm. Hg. Analysis: $C_{18}H_{35}Br$, percent Br, Calculated, 24.1. Found: 23.5.

A fine suspension of sodium was obtained by refluxing a mixture of 17.5 g. (0.76 mol) of sodium and 200 cc of dry toluene. 165 g. (0.5 mol) of the oleyl bromide were added in small increments over a period of one hour to the suspension and then the reaction mixture was refluxed for 24 hours. The excess sodium was decomposed with cold dilute hydrochloric acid and the toluene layer was washed with saturated solutions of sodium chloride, sodium bicarbonate and water.

The toluene layer was dried over sodium sulfate and the solvent was removed by distillation. The crude residue of 111.5 g. was recrystallized from acetone at -25°C. to yield 26.1 g. of 9,27-hexatriacontadiene having a wax-like consistency melting at 44° to 45°C. and 58.0 g. of an acetone insoluble oil. Fractionation of the oil did not yield any 9,27-hexatriacontadiene.

9,27-hexatriacontadiene was hydrobrominated by the free radical hydrobromination method of Jungermann et al (J. American Oil Chemists Society, vol. 35, 1958, p. 393). 9,27-hexatriacontadiene in benzene was reacted with hydrogen bromide in the presence of benzoyl peroxide as the catalyst at room temperature to obtain an isomeric mixture of 9,28-dibromohexatriacontane as a dark amber oil. Analysis: $C_{36}H_{72}Br_2$, percent Br, Calculated, 24.1. Found: 22.9.

Step B is the preparation of isomeric mixtures of 9,28-dimethoxyhexatriacontane. 40.0 g. (0.06 mol) of the isomeric mixture of dibromohexatriacontane and 6.5 g. of sodium methoxide in 15 cc of anhydrous methanol were added to a 300 ml. round bottom flask equipped with a mechanical stirrer and a reflux condenser fitted with a calcium chloride drying tube. After the initial reaction had subsided, the reaction mixture was refluxed on a water bath for four hours.

The reaction mixture was cooled and 20 ml. of ice water were added. The crude product was extracted with ether and the ether extract was washed with water until a test for bromide ion was negative. After drying the ether extract over calcium chloride, the ether was removed to obtain 26.0 g. (77% yield) of an isomeric mixture of 9,28-dimethoxyhexatriacontane having a refractive index $n_D^{30} = 1.4654$. Analysis: $C_{38}H_{78}O_2$, Calculated: C, 80.5; H, 13.9. Found: C, 79.6; H, 13.5.

Example 4: The following shows the preparation of isomeric mixture of 9,28-di-(2-ethylhexoxy)-hexatriacontane. Using the process described in Example 3, 9,28-dibromohexatriacontane was reacted with the sodium derivative of 2-ethylhexyl alcohol to form an isomeric mixture of 9,28-di-(2-ethylhexoxy)-hexatriacontane which after recrystallization from ethanol and then acetone had a melting point of 18°C. Analysis: $C_{52}H_{106}O_2$, Calculated: C, 81.8; H, 14.0. Found: C, 82.6; H, 13.3.

Example 5: The following procedure shows the preparation of isomeric mixtures of 1,10-diphenyl-octadecane. 191.9 g. (0.58 mol) of oleyl bromide prepared as in Step A of Example 3 were dissolved in 250 cc of anhydrous benzene and the resulting solution was added slowly over a period of four hours to a stirred solution of 170 g. (1.28 mols) of anhydrous aluminum chloride in 750 cc of benzene at 35°C. The mixture was then stirred at 65° to 70°C. for two hours and was then cooled and hydrolyzed with cold dilute hydrochloric acid. After removing the benzene by steam distillation, 66 g. (28% yield) of an isomeric mixture of 1,10-diphenyl-octadecane were obtained as a red oil distilling at 178° to 180°C. at 0.2 mm. Hg and having a refractive index $n_D^{20} = 1.5472$. Analysis: $C_{30}H_{46}$, Calculated: C, 88.6; H, 11.4. Found: C, 88.9; H, 11.0.

Example 6: The following procedure shows the preparation of isomeric mixture of 1,10-di(tert-butylphenyl)-octadecane. Using the procedure of Example 5, tert-butyl benzene was reacted with oleyl bromide to obtain a 22% yield of 1,10-di-(tert-butylphenyl)-octadecane as a highly fluorescent red oil boiling at 185° to 202°C. and having a refractive index

n_D^{20} = 1.5880. Analysis: $C_{38}H_{64}$, Calculated: C, 87.6; H, 12.4. Found: C, 87.9; H, 12.2.

Example 7: The following procedure shows the preparation of isomeric mixture of 10,10'-dimethoxy-dioctadecyl ether. Step A is the preparation of dioleyl ether. 537 g. (2.0 mol) of oleyl alcohol and 72 g. (0.51 mol) of boron trifluoride ethyl ether were placed in a 2 l. round bottom flask equipped with a reflux condenser, a mechanical stirrer and a calcium chloride drying tube. The solution was refluxed for four hours at 180°C. and then cooled to room temperature. The crude product was extracted with ethyl ether and the ether extract was washed with water until neutral and then was dried over sodium sulfate.

After removing the solvent by distillation, dioleyl ether was collected as a viscous oil boiling at 278° to 290°C. at 1 mm. Hg. Upon recrystallization of the product from acetone, 122.5 g. (47% yield) of dioleyl ether were obtained as a white solid having a melting point of 34° to 35°C., a refractive index n_D^{38} = 1.4550 and a hydroxyl value of 3.9.

Step B is the preparation of isomeric mixture of 10,10'dibromodioctadecyl ether. Dioleyl ether was reacted with hydrogen bromide in benzene at room temperature in the presence of a benzoyl peroxide catalyst for six hours to form 10,10'-dibromodioctadecyl ether. The reaction was exothermic. The product was washed with water until neutral and then was dried over calcium chloride. Analysis: $C_{36}H_{72}OBr_2$, percent Br, Calculated, 23.5. Found: 22.0.

Step C is the preparation of isomeric mixture of 10,10'-dimethoxy dioctadecyl ether. 102.0 g. (0.15 mol) of 10,10'-dibromo di-n-octadecyl ether and 16.2 g. (0.3 mol) of sodium methoxide were added with stirring to a three neck flask equipped with a mechanical stirrer and a reflux condenser fitted with a drying tube. After the initial reaction had subsided, 20 cc of dry methanol were added to the flask and the mixture was refluxed on a water bath for 9 hours. The reaction mixture was cooled and cold water was cautiously added.

The product was washed successively with water, sodium bicarbonate solution and water and was then dried over sodium sulfate. After removing the ether by evaporation, an amber oil was obtained which was treated with decolorizing charcoal and purified by absorption on an alumina column from a hexane solution. 34.9 g. (40% yield) of 10,10'-dimethoxy-di-n-octadecyl ether were obtained as a light amber oil having a refractive index of n_D^{30} = 1.4634. Analysis: $C_{38}H_{78}O_3$, Calculated: C, 78.3; H, 13.5. Found: C, 78.2; H, 13.1.

Example 8: The following procedure shows the preparation of isomeric mixture of 10,10'-di-(tert-butoxy)-di-n-octadecyl ether. Using the procedure of Example 7, 10-10'-dibromo-di-n-octadecyl ether was reacted with sodium tert-butoxide to obtain a 26% yield of 10,10'-di-(tert-butoxy)-di-n-octadecyl ether having a refractive index n_D^{30} = 1.4645. Analysis: $C_{44}H_{90}O_3$, Calculated: C, 79.2; H, 13.6. Found: C, 79.0; H, 13.3.

Example 9: The following procedure shows the preparation of isomeric mixtures of 10,10'-di-(2-ethylhexoxy)-di-n-octadecyl ether. Using the procedure of Example 7, 10-10'-dibromo-di-n-octadecyl ether was reacted with sodium 2-ethylhexoxide to obtain a 39% yield of 10,10'-di-(2'-ethylhexoxy)-di-n-octadecyl ether having a refractive index n_D^{30} = 1.4635. Analysis: $C_{52}H_{106}O_3$, Calculated: C, 80.1; H, 13.7. Found: C, 79.8; H, 13.3.

Example 10: The following procedure shows the preparation of isomeric mixtures of 10,10'-di-(tert-butylphenyl)-di-n-octadecyl ether. 145 g. (0.28 mol) of oleyl ether prepared as in Example 7, Step A, dissolved in 250 cc of tert-butyl benzene were added to a stirred solution of 82.6 g. (0.62 mol) of anhydrous aluminum chloride in 250 cc of tert-butyl benzene at a temperature not greater than 35°C. The mixture was then heated to 80°C. and stirred for 10 hours and then was cooled and hydrolyzed with dilute hydrochloric acid.

The excess tert-butyl benzene was removed by steam distillation and the product was taken up in ethyl ether which was then dried over sodium sulfate. 10,10'-di-(tert-butylphenyl)-di-n-octadecyl ether was obtained as a red oil distilling at 215° to 233°C. at 0.5 mm. Hg. Analysis: $C_{56}H_{90}O$, Calculated: C, 86.3; H, 11.6. Found: C, 86.1; H, 11.4.

Example 11: The following procedure shows the preparation of isomeric mixtures of 9,28-diphenylhexatriacontane. 81.8 g. (0.2 mol) of 9-phenyl-octadecyl bromide as prepared in Example 1 were added to a stirred suspension of 1.15 g. (0.05 mol) of clean sodium in dry toluene at reflux temperature and the reaction mixture was refluxed for 24 hours during which a characteristic blue color developed. After cooling the reaction mixture, cold dilute hydrochloric acid was added to decompose the excess sodium and the solution was then washed with water until neutral and was dried over calcium chloride.

After removing the toluene by vacuum distillation, the product was dissolved in hexane and passed several times through a chromatographic column of alumina. 23.0 g. (35% yield) of 9,28-diphenylhexatriacontane were obtained as a clear yellow oil having a refractive index n_D^{30} = 1.4871 and an iodine value of 0.

Example 12: The following procedure shows the preparation of isomeric mixture of 9,28-di-(tert-butylphenyl)-hexatriacontane. Using the procedure of Example 11, tert-butylphenyl-octadecyl bromide prepared as in Example 2 was reacted with sodium to obtain a 26% yield of 9,28-di-(tert-butylphenyl)-hexatriacontane having a refractive index n_D^{20} = 1.4910.

Example 13: The following procedure shows the preparation of isomeric mixtures of 10-ethoxy-octadecane. Oleyl magnesium bromide prepared from oleyl bromide was hydrolyzed to obtain 9-octadecene having a boiling point of 125° to 137°C. at 1.0 mm. Hg. 9-octadecene was then reacted with hydrogen bromide in the presence of a benzoyl peroxide catalyst to form 10-bromo-octadecane. Sodium ethoxide and 10-bromo-octadecane were reacted to form 10-ethoxy-octadecane.

The residual products were purified by passing a hexane solution through several chromatographic columns of alumina. After removal of the solvent, 10-ethoxy-octadecane was obtained as a clear yellow oil having a refractive index n_D^{31} = 1.4477. Analysis: $C_{20}H_{42}O$, Calculated: C, 80.2; H, 14.2. Found: C, 79.7; H, 13.9.

Example 14: The following procedure shows the preparation of isomeric mixtures of 1,10-di-(tert-butoxy)-octadecane. Oleyl bromide was reacted with sodium tert-butoxide to form 1-tert-butoxy-octadecene which was then reacted with hydrogen bromide in the presence of a benzoyl peroxide catalyst to form 1-tert-butoxy-10-bromo-octadecane. The latter was then reacted with sodium tert-butoxide to obtain 1,10-di-(tert-butoxy)-octadecane as a colorless oil distilling at 170° to 182°C. at 0.2 mm. Hg. Analysis: $C_{26}H_{54}O_2$, Calculated: C, 78.3; H, 13.7. Found: C, 78.0; H, 13.5.

Example 15: The following procedure shows the preparation of isomeric mixtures of 1-tert-butoxy-10-phenyl-octadecane. 10-phenyloctadecyl bromide as produced in Example 1 was reacted with sodium tert-butoxide to obtain a 65% yield of 1-tert-butoxy-10-phenyl-octadecane distilling at 153° to 168°C. at 0.2 mm. Hg having a refractive index n_D^{20} = 1.4590. Analysis: $C_{28}H_{50}O$, Calculated: C, 83.5; H, 12.5. Found: C, 82.9; H, 12.9.

Example 16: The following procedure shows the preparation of isomeric mixture of 1-methoxy-10-(tert-butoxy)-octadecane. Oleyl bromide was reacted with sodium methoxide to obtain a 69% yield of 1-methoxy-9-octadecene having a boiling point of 136° to 140°C. at 0.5 mm. Hg and a refractive index n_D^{20} = 1.4510. The latter product was reacted with hydrogen bromide in the presence of a benzoyl peroxide catalyst to obtain a 57% yield of 1-methoxy-10-bromo-octadecane having a boiling point of 178° to 190°C. at 0.2 mm. Hg and a refractive index n_D^{20} = 1.4715.

The latter was reacted with tert-butanol and sodium to obtain a 30% yield of 1-methoxy-10-(tert-butoxy)-octadecane having a boiling point of 164° to 170°C. at 0.2 mm. Hg and a refractive index n_D^{20} = 1.4650. Analysis: $C_{23}H_{48}O_2$, Calculated: C, 77.9; H, 13.6. Found: C, 77.2; H, 12.9.

The viscosity index, pour points and the thermal stability of the compounds of Example 1 through 16 were determined as follows. The viscosity indices were determined according to ASTM D567-53 and the pour points were determined by ASTM D97-57.

The thermal stability test was effected by placing a weighed sample and sample tube into the inner chamber of an Abderhalden drying apparatus. The air in the system was evacuated and nitrogen was admitted thereto. A small positive pressure of nitrogen was maintained on the system during the entire test. The Arochlor 1254 in the apparatus was brought to reflux at about 690°F. and was maintained there for ten hours after which the samples and contents of the inner chamber were collected, reweighed and the percent loss calculated. The test results are summarized below and on the following page.

Octadecanes Wherein—		Isomeric Mixture of Example	Pour Point, °F.	Viscosity, cs:		ASTM Viscosity Index	After 10 hours at 690° F.	
R	R_1			100° F.	210° F.		Change In 100° F. Viscosity, Percent	Weight Loss, Percent
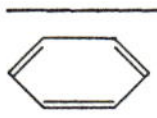	H	1	+15	11.0	2.7	84	13.6	5.8
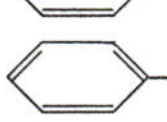—$C(CH_3)_3$	H	2	+14	25.4	10.5	187	6.3	11.7
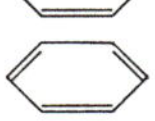	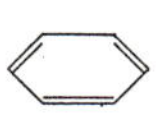	5	−25	32.3	4.1	−63	8.7	6.2
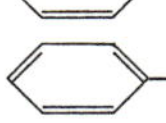—$C(CH_3)_3$	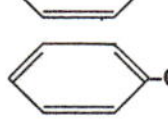—$C(CH_3)_3$	6	−35	33.0	4.2	−50	2.7	1.0
OC_2H_5	H	13	−12	4.4	2.7	310	------------	7.0
$OC(CH_3)_3$	$OC(CH_3)_3$	14	−25	5.5	1.9	142	−3.6	14.4
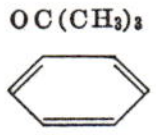	$OC(CH_3)_3$	15	−15	5.5	1.9	142	−1.8	3.4
$OC(CH_3)_3$	OCH_3	16	+45	11.3	2.5	29	0	2.4

Hexatriacontanes Wherein R—	Isomeric Mixture of Example	Pour Point, °F.	Viscosity, cs.		ASTM Viscosity Index	After 10 hours at 690° F.	
			100° F.	210° F.		Change in 100° F. Viscosity, Percent	Weight Loss, Percent
OCH_3	3	>+32	14.6	4.3	214	----------	----------
OC_8H_{17}*	4	+45	16.7	3.3	50	13.8	7.0
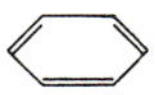	11	<+32	22.4	----------	----------	----------	5.6
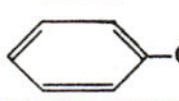—$C(CH_3)_3$	12	0	17.7	3.7	109	37.8	14.4
Dioctadecyl Ethers Wherein R—							
OCH_3	7	>+32	17.0	5.1	205	----------	14.4
$OC(CH_3)_3$	8	+50	19.6	4.6	167	−2.0	2.2
OC_8H_{17}	9	+60	22.7	5.3	169	45.0	3.0
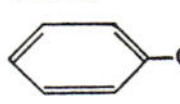—$C(CH_3)_3$	10	+25	40.0	4.7	−28	4.2	0.6

*2-ethylhexyl ether.

The compounds of Formula 1 are generally preferred since they have lower pour points. As can be seen from the table, the best compounds tested were 1,10-di-(tert-butoxy)-octadecane of Example 14 and 1-tert-butoxy-10-phenyl-octadecane of Example 15 when considering pour point, viscosity index and thermal stability.

FLUOROCARBONS

GREASE FORMULATIONS

Tetrafluoroethylene Thickened Grease

J.T. Skehan; U.S. Patent 3,505,229; April 7, 1970; assigned to E.I. du Pont de Nemours and Company describes a grease composition which comprises

(A) a grease containing —

(1) as base oil, a perfluorinated polyether having a structure selected from the group consisting of

$$R_fO[CF(CF_3)CF_2O]_nCF(CF_3)F$$

$$R_fO[CF(CF_3)CF_2O]_nCF(CF_3)H$$

$$R_fO[CF(CF_3)CF_2O]_nCF(CF_3)CF(CF_3)[OCF_2CF(CF_3)]_nOR_f$$

where R_f is perfluoroalkyl group and n indicates the degree of polymerization and having a maximum volatility of 50% at 400°F. according to Federal Test Method Standard FTMS-791, Method 351, and a maximum pour point of 50°F., and

(2) as a thickener, about 5 to 50%, based on the weight of grease component (A), of a polymer of tetrafluoroethylene containing at least 85% by weight $-CF_2CF_2-$ units, and having a molecular weight of at least 2,000, and

(B) about 0 to 70%, based on the weight of the total composition, of a volatile liquid solvent for the perfluorinated polyether.

Preferred classes are those containing perfluorinated polyethers of molecular weights within the range 3,000 to 13,000 and tetrafluoroethylene polymers of molecular weights within the range of 35,000 to 50,000. Preferred polymers are those containing 90% or more tetrafluoroethylene units.

The greases are useful at relatively low temperatures as lubricants and by proper choice of base oil and thickener can be devised to be useful at well below 0°F. All of the greases are useful at temperatures in excess of 400°F. (204°C.) which sets them out from other greases. Greases within the scope of this process have been found to operate for over 1,000 hours at 400°F. The preferred types operate for over 2,000 hours at 500°F. (260°C.) and operate for considerable periods of time at temperatures as high as 600°F. (316°C.)

Example 1: A dispersion (7.5% by weight in 1,1,2-trichloro-1,2,2-trifluoroethylene) of a telomer of tetrafluoroethylene and 1,1,2-trichloro-1,2,2-trifluoroethane (1 kilogram) was added to 500 g. of a hexafluoropropylene oxide polymer. The trichlorotrifluoroethane was evaporated at 117°F. (47.2° C.) with rapid mechanical agitation. The mixture was then passed through an 8-inch (20.3 cm.) 3-roll ink mill with a roller clearance of 0.0015 inch (38 microns). After three passes, the last traces of trichlorotrifluoroethane were removed by spreading the mixture in a thin layer on trays in a vacuum oven at 120° F. (54.4° C.) and 10-inch (254 mm.) Hg vacuum. Finally, the mixture was passed through the ink mill an additional 10 times. The resultant product was a smooth, buttery NLGI grade 2 grease containing 13% telomer. The grease had the following properties: consistency mm./10 at 77° F. unworked 276, worked 292, mechanical stability after six hours Shell Roll Test 325, water loss 2.2% at 100° F. (FTMS-791-3252), high temperature bearing test (FTMS-791-333) greater than 2,000 hours at 500°F. and 10,000 rpm, greater than 650 hours at 550°F. and greater than 60 hours at 600°F., no oxygen takeup in the oxidation stability test ASTM-D-942 at 210°F., no copper corrosion in FTMS-791-5309, passed the liquid oxygen and nitrogen tetraoxide impact tests described in USAF Spec. Bulletin 527.

The tetrafluoroethylene/trichlorotrifluoroethane telomer was prepared using the procedure of Example 2 of U.S. Patent 3,067,262, omitting the methylcyclohexane, at 150°C. and 600 psig. The following mol ratios were maintained: 0.081 mol tetrafluoroethylene/mol trichlorotrifluoroethane, 0.0014 mol di-tert-butyl peroxide/mol trichlorotrifluoroethane. The tetrafluoroethylene telomer had a crystalline melting point of 323° to 327°C. and an average molecular weight of 35,000 to 50,000.

The hexafluoropropylene oxide polymer was prepared by polymerizing hexafluoropropylene oxide according to the procedure of Canadian Patent 725,740 to a product of structure

$$C_3F_7O[CF(CF_3)CF_2O]_nCF(CF_3)COF$$

which was then coupled according to the procedure of U.S. Patent 3,214,748 with ultraviolet light to product predominantly of structure

$$C_3F_7O[CF(CF_3)CF_2O]_nCF(CF_3)CF(CF_3)[OCF_2CF(CF_3)]_nOC_3F_7$$

having an average molecular weight of 6,230, a pour point of -30°F. (-34.4°C.), a 100°F. viscosity of 309 cs., a 400°F. viscosity of 3.86 cs. and a volatility of 1.9% after six and one-half hours at 500°F. (ASTM-D-972, FTMS-Method 351.2).

Example 2: A grease was prepared using the procedure of Example 1 from 257 g. of a 7.0% (weight) dispersion of the tetrafluoroethylene-1,1,2-trichloro-1,2,2-trifluoroethane telomer of Example 1 in trichlorotrifluoroethane and 102 g. of a hexafluoropropylene oxide polymer. After milling and removing the trichlorotrifluoroethane as in Example 1, the resulting grease, containing 15% telomer, was a smooth, buttery product of NLGI grade 2 with a worked ASTM consistency of 271 mm./10 (ASTM-D-1403). The grease operated for greater than 1,000 hours at 400° F. and 10,000 rpm in the bearing performance test FTMS-791-333 and had a -30°F. torque of 1,840 g. cm. starting at 800 g. cm. running.

The hexafluoropropylene oxide polymer was prepared by polymerizing hexafluoropropylene oxide as in Example 1, then replacing the terminal acid fluoride group with fluorine according to the procedure of U.S. Patent 3,242,218. The resulting polymer of structure

$$C_3F_7O[CF(CF_3)CF_2O]_nCF(CF_3){-}F$$

had an average molecular weight of 3,130, a pour point of -50°F. (-45.6°C.), a 100°F. viscosity of 72 cs., a 400°F. viscosity of 1.58 cs., a volatility of 14.7% at 400°F. and 53.4% at 500°F., both after six and one-half hours (ASTM-D-972, FTMS-Method 351.2).

Example 3: A grease was prepared using the procedure of Example 1 from 185 g. of the tetrafluoroethylene telomer dispersion of Example 2 and 85 g. of hexafluoropropylene oxide polymer. After milling and removing the trichlorotrifluoroethane as in Example 1, a smooth, buttery NLGI grade 2 grease was obtained containing 12% telomer with a worked ASTM consistency of 290 mm./10 (ASTM-D-1403). The grease would perform in an equivalent manner to that of Example 1 in the Ball Bearing Performance Test at 10,000 rpm and 500°F. (FTMS-791-333).

The hexafluoropropylene oxide polymer, prepared in the same manner as that of Example 1, had an average molecular weight of 13,000, a pour point of -5°F. (-20.6°C.), a 100°F. viscosity of 1,307 cs., a 400°F. viscosity of 12.33 cs., a volatility of 500°F. of 0.07% after six and one-half hours (ASTM-D-972, FTMS-Method 351.2).

Metal-Free Phthalocyanine Thickener

J.L. Dreher; U.S. Patent 3,432,432; March 11, 1969; assigned to Chevron Research Corporation has found that superior high-temperature greases may be compounded from a major portion of a perfluoropolyether and a minor portion, sufficient to produce proper grease consistency, of metal-free phthalocyanine. Additionally, it has been found that the further addition of a minor portion of sodium nitrite to the greases results in further improvement in the high-temperature stability of the grease.

The base materials of the greases are, perfluoropolyethers. Reference to the polymers and their preparation may be found in U.S. Patents 3,214,478, 3,242,218, 3,257,466, and 3,274,239. Although the preferred polymers consist entirely of perfluorocarbonether monomer units, materials which are capped by other groups are suitable. A description of the preferred base materials, which are Du Pont PR-143 fluids, may be found in ASLE Paper 65LC-3, Lubrication Conference, Oct. 18 to 20, 1965, Academic Press, entitled, "PR-143-A New Class of High-Temperature Fluids," by W.H. Gumpricht. These fluids are colorless and odorless, have high densities and low indices of refraction. The fluids are basically polymers of hexafluoropropylene oxide.

The phthalocyanine, which is employed as a thickener in the grease compositions, is the metal-free material and is used as a dyestuff and is commercially available. It is employed in the grease in amounts sufficient to produce proper grease consistencies; amounts of from 5 to 50% by weight are usually sufficient. Amounts of from 10 to 25% are preferred.

Sodium nitrite, which is used to extend the high temperature performance of the greases, is optionally employed. It is used in amounts of from 0.10 to 5.00% by weight; amounts of from 0.2 to 2.0% by weight are preferred. The following examples illustrate the preparation of the grease compositions.

Example 1: Preparation of Phthalocyanine Thickened Perfluoropolyether Grease — 400.5 g. (89% by weight) of perfluorocarbon (Du Pont PR-143-AC) and 49.5 g. (11% by weight) of phthalocyanine were placed in a 400 ml. beaker. The mixture was heated with stirring to 390°F. and maintained with stirring at that temperature for 10 minutes. The material was allowed to cool. It was then milled three times on a three-roll lab mill. The mixture was then allowed to stand overnight; it was stirred slowly with a spatula for five minutes and allowed to stand overnight again. The yield was smooth, blue-colored grease having an ASTM Worked Penetration of 270.

Example 2: Preparation of Perfluoropolyether Grease Thickened with Copper Phthalocyanine — The procedure of Example 1 was repeated employing copper phthalocyanine thickener (Copper Phthalocyanine Blue Toner, Holland-Suco Color Company) instead of metal-free phthalocyanine. A blue grease resulted having an ASTM Worked Penetration of 325.

Example 3: Preparation of Perfluoropolyether Grease with Sodium Nitrite Rust Inhibitor — The general procedure of Example 1 was followed except that 0.5% by weight of sodium nitrite replaced a like amount of the base fluid. The nitrite addition was accomplished by dissolving it in several milliliters of water, adding the solution to the fluid-phthalocyanine mixture, from which the water was driven off at 370°F. The procedure yielded a blue-colored grease material having an ASTM Worked Penetration of 277.

Example 4: Preparation of Silicone Grease Thickened with Phthalocyanine — For purposes of comparison, a grease produced from a silicone fluid containing the metal-free phthalocyanine thickener was also employed. The composition of this grease is as follows:

	Percent by Weight
Dow-Corning, DC510 (50 cs.)	50
Dow-Corning, DC510 (100 cs.)	26.5
Phthalocyanine	23.0
Sodium nitrite	0.5

The grease had an ASTM Worked Penetration of 290

Example 5: Commercial Perfluoropolyether Grease — Also for purposes of comparison, samples of a commercially available grease containing the same perfluoropolyether base fluid as in Examples 1 to 3 were tested. The grease was thickened with a fluorocarbon telomer.

In order to show the excellent high-temperature performance of the metal-free phthalocyanine thickened material, the samples were subjected to a high-speed bearing test, Federal Test Method Standard 791A-331. A brief description of the test is as follows: A ball bearing lubricated by the subject grease and containing heat-treated ball bearings is operated at 10,000 rpm continuously 22 hours at 500°F. The apparatus is then cooled to room temperature during a period of two hours. This procedure of operating and cooling is repeated until there is bearing failure. "Bearing Life" is the number of hours to bearing failure. Data from the high-speed Bearing Test using the greases of Examples 1 to 5 are included in the following table.

Grease Type	Bearing Life (hrs.) 500° F., 10,000 rpm
Metal-free phthalocyanine in perfluoropolyether (Example 1)	1,543
Metal-free phthalocyanine in perfluoropolyether 0.5% $NaNO_2$ (Example 2)	2,498
Copper phthalocyanine in perfluoropolyether (Example 3)	378
Metal-free phthalocyanine in silicone base (Example 4)	295
Fluorocarbon telomer in perfluoropolyether (Example 5) *	
Sample 1	628
Sample 2 (different lot)	748

* Upon disassembly of the apparatus at the end of the tests run with the greases containing fluorocarbon telomer thickeners, excessive corrosion was noted and extensive overhaul of the test equipment was required. This corrosion was not noted in the cases where the metal-free phthalocyanine was employed as a thickener.

It may be seen from these data that the perfluoropolyether base grease thickened with metal-free phthalocyanine gives longer bearing life than the grease employing the copper phthalocyanine thickener. In addition, it may be seen that the silicone base grease does not achieve an extraordinary bearing life using the metal-free phthalocyanine thickener.

Fluorocarbon Resin and Boron Nitride Particles

J.B. Christian; U.S. Patent 3,525,690; August 25, 1970; assigned to the U.S. Secretary of the Air Force describes a grease formulation which comprises (1) a thickener selected from the group consisting of 19 to 24% by weight of ammeline, 7 to 10% by weight of spherical boron nitride particles ranging from 0.1 to 0.5 micron in diameter and 20 to 35% in weight of a fluorinated ethylene-propylene polymer, and (2) the balance a base fluid of hexafluorcpropylene epoxide.

The triazine may be the technical grade powder form sold as Ammeline, the fluorocarbon resin may be any of the tetrafluoroethylene (TEF) or fluorinated ethylene-propylene (FEP) polymers or resins such as Teflon, and the boron nitride is in spherical particles of a size ranging from 0.1 to 0.5 micron in diameter. Ammeline is known in the art to be 2,4-diamino-6-hydroxy-1,3,5-triazine.

A grease formulation was made by utilizing 78.7 parts by weight of Du Pont's P.R. 143 hexcfluoropropylene epoxide liquid and 21.3 parts by weight of ammeline. After these ingredients were thoroughly mixed, first by spatula blending and then by milling several passes through a 3-roll mill, the mixture was subjected to conventional testing with the following results. On the ASTM D217-60T unworked penetration cone drop test, a value of 260 was achieved; and, after working the cone sixty strokes in the test device, a worked penetration reading of 275 was achieved. In standard Pope Spindle bearing test where the bearings were loaded with five pounds and operated at 10,000 rpm, the grease performed and allowed the test to continue demonstrating satisfactory lubricity for 1,760 hours at temperatures of 500°F., for 240 hours at 550°F. and for 160 hours at 600°F.

In another example where 76 parts by weight of the hexafluoropropylene epoxide liquid and 24% by weight of the ammeline thickener were combined in a grease formulation, the same unworked penetration test gave a value of 319 while the worked penetration value was 348. In the same five pound loaded bearing test at 10,000 rpm, the grease performed as a lubricant for more than 1,000 hours at 500°F., for 300 hours at 550°F. and for 128 hours at 600°F. When this same formulation was applied to the same bearing test under a three pound load at 10,000 rpm, but in a vacuum of 10^{-7} mm. of mercury at 400°F., the apparatus functioned properly for more than 1,400 hours and when the load was increased to five pounds under the same revolutions and vacuum consitions, the grease functioned properly for 595 hours indicating its satisfactory performance under the vacuum condition existing in outer space.

In a related formulation based upon the spherical boron nitride particles where they represent 7% by weight of the total grease and the hexafluoropropylene epoxide liquid represents 93 weight percent of the total formulation, a grease-like consistency is achieved capable of performing substantially as does the mixture composed of the fluid and thickener in the ratio of 9 to 1. Tests conducted on the formulation containing only 7% of spherical boron nitride particles, however, according to U.S. Air Force Specification Bulletin No. 527, produced a reading of more than eighty foot-pounds indicating that this formulation possesses excellent explosion resistance upon impact in the presence of liquid oxygen. In view of this particular property, these grease formulations, of course, offer considerable promise for use in aerospace launch vehicles for example where liquid oxygen is employed in the combustion system. Beyond this, the availability of boron nitride thickened hexafluoropropylene epoxide greases for use at temperatures on the order of from 400° to 500°F. is demonstrated by the fact that, on the Pope Spindle Bearing Test, when the bearings were lubricated by this formulation, loaded at five pounds and operated at 10,000 rpm, grease-like consistency and lubricity were maintained for over 1,900 hours at 400°F. and for over 200 hours at 500°F.

The superiority of spherical particles of boron nitride over platelet particles is illustrated by the following table:

	Grease 1 [1]	Grease 2 [2]
Unworked penetration	261.5	303.4
Worked penetration	279.7	339.0
Pope spindle test, 10,000 r.p.m., 400° F. (hrs.)	81.3	2,432.0
Pope spindle test, 10,000 r.p.m., 450° F. (hrs.)		1,265.0
Pope spindle test, 10,000 r.p.m., 500° F. (hrs.)		320.0

[1] 83.0% hexafluoropropylene epoxide, 17.0% boron nitride, commercial grade, platelet form.
[2] 93.0% hexafluoropropylene epoxide, 7.0% boron nitride, spherical form.

From the above table it will be apparent that the use of less than half as much boron nitride but in the spherical form, is much more effective than the boron nitride in platelet form. The following table illustrates that the hexafluoropropylene epoxide as a base fluid is more effective than a well known silicone oil, particularly at high temperatures:

	Grease 1 [1]	Grease 2 [2]
Unworked penetration	260	272
Worked penetration	275	275
Pope spindle performance, 10,000 r.p.m., 600° F. (hrs.)	160	74

[1] 78.7% hexafluoroprepylene epoxide, 21.3% ammeline.
[2] 65.0% Dow Corning QF 6-7039 silicone, 35.0% ammeline.

FLUORINATED ESTERS

Cyanoalkyl Perfluoromethylphenyl Phosphates

A process described by E.S. Blake, R.E. DeBrunner and J.A. Webster; U.S. Patent 3,359,349; December 19, 1967;

assigned to Monsanto Research Corporation provides mixed esters of phosphoric acid with an alkylene cyanohydrin and a perfluoromethylphenol by the reaction of certain phosphorohalidates or phosphorodihalidates with certain cyanohydrins, substantially according to the scheme:

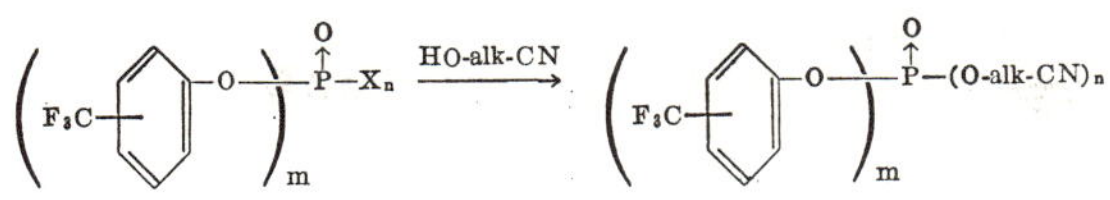

where each of m and n is 1 or 2 and the sum of m + n is 3, X is halogen having an atomic weight greater than 34, and alk denotes a bivalent alkylene radical of from 2 to 4 carbon atoms. Depending upon whether there is employed a phosphomonohalidate or a phosphorodihalidate the products are either bis(perfluoromethylphenyl) cyanoalkyl phosphates

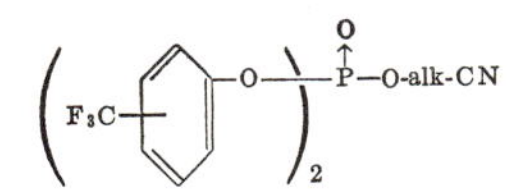

or bis(cyanoalkyl) perfluoromethylphenyl phosphates.

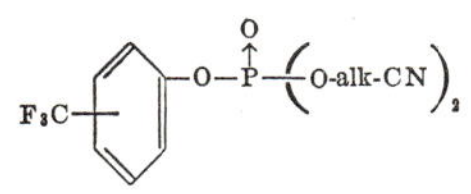

Reaction of the phosphorohalidate or the phosphorodihalidate with the alkylene cyanohydrin takes place by simply contacting the halogen compound with the cyanohydrin at ambient temperature or with heating. In some instances, reaction rate is accelerated by heating; and, particularly when working with the higher molecular weight cyanohydrins, temperatures of from, 50°C. to below the decomposition temperature of either the reactants or the ester product are employed. Generally, depending upon the nature of the individual reactants, temperatures of from 0° to 150°C. and preferably of from about 0° to 75°C. are useful.

The cyanoalkyl perfluoromethylphenyl phosphates are stable, well-defined compounds which are generally clear, high-boiling liquids. They usually have kinematic viscosities which are well below 5,000 cs. at 25°F. and are particularly useful as functional fluids. The phosphates are generally liquid over wide temperature ranges, possess high flash points and high ignition points and are characterized by good thermal stability. They generally remain liquid at low temperatures. Hence they are suited for use as hydraulic fluids, especially in hydraulic systems which are subjected to widely varying temperature conditions. They possess good viscosity/temperature relationships and are also useful, e.g., as heat-exchange media, gyro fluids, and lubricants. The process is illustrated by the following examples.

Example 1: m-Trifluoromethylphenyl phosphorodichloridate and bis(m-trifluoromethylphenyl) phosphorochloridate were prepared as follows: A mixture consisting of 180 g. of crude m-trifluoromethylphenyl phosphorodichloridate boiling below 185°C./17 mm., which had been obtained by refluxing phosphoryl chloride with m-trifluoromethylphenol and distilling the reaction product, 539 g. (3.5 mols) of phosphoryl chloride and 1,000 g. (6.18 mols) of m-trifluoromethylphenol, was heated to reflux within a period of 20 hours, at the end of which time the temperature was 230°C. Distillation of the resulting reaction mixture gave 389.5 g. of the substantially pure m-trifluoromethylphenyl phosphorodichloridate, BP 125° to 130°C./20 mm. and 814.2 g. of bis(m-trifluoromethylphenyl) phosphorochloridate, BP 195°C./20 mm.

To a mixture consisting of 17.8 g. (0.25 mol) of hydracrylonitrile in 22.9 g. (0.29 mol) of pyridine there was added, dropwise over a period of 75 minutes, 101.1 g. (0.25 mol) of the bis(m-perfluoromethylphenyl) phosphorochloridate prepared above. The whole was then stirred for 2 hours at 65° to 75°C. and then allowed to cool to ambient temperature. After washing with dilute hydrochloric acid and with water to neutral, the organic phase was dried over magnesium sulfate and filtered. The filtrate was evacuated to remove low-boiling volatiles and then distilled to remove material boiling up to 110°C./0.25 mm. The residue was filtered several times through clay and carbon black to give 83.5 g. (79.4% theoretical yield) of the substantially pure bis(m-perfluoromethylphenyl) 2-cyanoethyl phosphate, n_D^{25} 1.4705. The compound was found to have a pour point of minus 10°F., and the following kinematic viscosities were determined at the temperatures shown below:

Temperature, °F.	Centistokes
25	4291
100	74.27
210	6.49

The autogenous ignition temperature was found to be 1150°F. for 0.04 ml. with a lag of 4 seconds. In the molten metal test, the compound idi not burn without applying a spark.

Example 2: To a mixture of 43.7 g. (0.66 mol) of hydracrylonitrile in 54.5 g. (0.69 mol) of dry pyridine there was added, dropwise over a 70 minute period, 82.6 g. (0.33 mol) of m-perfluoromethylphenyl phosphorocichloridate, BP 125° to 130°C./20 mm. which was prepared in Example 1. The whole was heated for 2 hours at 70° to 80°C. After cooling to ambient, it was washed several times with dilute hydrochloric acid and several times with water, and the resulting organic phase was dried over magnesium sulfate and filtered. Vacuum treatment of the filtrate to a pot temperature of 120°C. removed any remaining by-product hydrochloride and unreacted material residue was successively filtered through clay, carbon black, and alumina to give the substantially pure bis(2-cyanoethyl) m-perfluoromethylphenyl phosphate, n_D^{25} 1.4640, which analyzed as follows. Calculated for $C_{13}H_{12}N_2O_4PF_3$: C, 44.83%; H, 3.47%; F, 16.36%. Found: C, 44.58%; H, 3.36%; F, 16.13%.

The phosphate was found to have a pour point of minus 15°F. and the following kinematic viscosities were determined at the temperatures shown below:

Temperature, °F.	Centistokes
25	3367
100	76.01
210	7.1

The autogenous ignition temperature was 1160°F. for 0.04 ml. with a lag of 2 seconds.

Polyfluorinated Phosphonate Esters

A process described by C.A. Seil, R.H. Boschan and J.P. Holder; U.S. Patent 3,337,655; August 22, 1967; assigned to McDonnell Douglas Corporation involves bis-(polyfluoroalkyl) benzenephosphonates. The preferred bis-(polyfluoroalkyl) benzenephosphonates have the formula

$$[CX_3(CX_2)_nCH_2O]_2\overset{O}{\overset{\|}{P}}C_6H_5$$

where X is a member selected from the group consisting of hydrogen and fluorine, and n is an integer of from 0 to 10. Also, in preferred practice, there are at least 2 fluorine atoms in each alkyl group and at least one of the X substituents on at least one of the carbon atoms commencing with the carbon atom in the 2-position is hydrogen. Preferably, the major portion of the X substituents are fluorine. A particularly desirable class of compounds, according to the process are those which have the formula

$$[HCF_2(CF_2)_nCH_2O]_2\overset{O}{\overset{\|}{P}}C_6H_5$$

where n has the values noted above. It has been found that the above defined bis-(polyfluoroalkyl) benzenephosphonates are particularly valuable as functional fluids, e.g., as base stocks for hydraulic fluids, and are also useful as heat transfer media and as lubricants due to the advantageous physical properties of such compounds. These properties include high fire resistance, increased temperature stability, and liquidity over a relatively wide temperature range. Thus, for example, generally these fluorinated phosphonate esters have a thermal stability in the range of about 400° to 700°F., autoignition temperatures up to and in excess of 1000°F., and remain liquid down to a temperature of the order of -20° to -65°F. Further, the viscosity of such fluoroalkyl benzenephosphonates at low temperatures of the order of -20° to -65°F. permit the operative use of such liquids as hydraulic fluids, cooling media and lubricants in these low temperature ranges. These compounds are also relatively nonvolatile at elevated temperatures due to the high boiling point of these materials. The phosphonate esters have relatively low pour points, e.g., in the range of about -20° to -75°F., or lower, which, together with the relatively low viscosity of these materials at low temperatures permits the liquid compounds to be pumped without high expenditure of energy at low temperatures.

Preferred compounds of the process are, bis-(3,3,4,4,5,5,5-heptafluoropentyl) benzenephosphonate; bis-(1,1,5-trihydroperfluoropentyl) benzenephosphonate; and bis-(4,4,5,5,6,6,6-heptafluorohexyl) benzenephosphonate. The following are examples of the process.

Example 1: The preparation of bis-(3,3,4,4,5,5,5-heptafluoropentyl) benzenephosphonate is as follows. To a mixture of 90.0 g. (0.421 mol) of 3,3,4,4,5,5,5-hepafluoro-1-pentanol, 36.5 g. (0.461 mol) of pyridine and 100 ml. benzene, is added dropwise with stirring 41 g. (0.21 mol) of redistilled benzene phosphonyl dichloride. The temperature is maintained at 15° to 18°C. during the addition, which is complete in 1 hour and 25 minutes. The mixture is warmed to approximately 25°C. and stirred at this temperature for 1 hour, then heated to reflux temperature and refluxed overnight. The mixture is then cooled and poured into 600 ml. of water. The bottom organic layer is withdrawn and the aqueous layer is extracted with two 100 ml. portions of ether. The ether extracts are then added to the organic layer, which is then washed with two 100 ml. portions of 5% hydrochloric acid, 100 ml. of 5% sodium bicarbonate, and 100 ml. of water. After drying over anhydrous magnesium sulfate, the solvents are removed by lowering the pressure with a water aspirator while heating.

The residue is distilled at lower pressure; the main product fraction 100.5 g. (87.0% yield) distills at 102.5° to 103°C. (0.10 to 0.12 mm. mercury pressure), which analyzed as follows. Calculated for $C_{16}H_{13}F_{14}O_3P$: C, 34.92%; H, 2.38%; P, 5.63%. Found: C, 35.08%; H, 2.86%; P, 5.55%. This bis(polyfluoropentyl) benzenephosphonate ester thus produced, has a viscosity at 210°F. of 2.27 cs., a pour point of the order of -60°F., a density of 1.54 g./ml. at 77°F., and a thermal stability in the range of about 400° to 500°F. Such phosphonate ester has a high autoignition temperature, good hydrolytic stability and high fire resistance. Such a compound can be employed as a hydraulic fluid in the hydraulic systems of aircraft.

Example 2: The preparation of bis(1,1,5-trihydroperfluoropentyl) benzenephosphonate is as follows. To a mixture of 300 g. (1.29 mols) of 1,1,5-trihydroperfluoropentanol, 114 g. (1.44 mols) of pyridine and 260 ml. of benzene, are added dropwise with stirring 128 g. (0.656 mol) of redistilled benzene phosphonyl dichloride. The temperature is maintained below 45°C. during the addition, which is complete in 2 1/4 hours. The mixture is heated slowly to reflux and refluxed overnight.

The mixture is then cooled and washed with four times its volume of water. The botton organic layer is withdrawn and washed with successive portions of 5% hydrochloric acid, water, and 5% sodium bicarbonate solution. The organic layer is then dried over anhydrous potassium carbonate, and the carbonate appeared to have some solubility in the organic material. Overnight a clear bottom layer separates out. This layer is extracted with ether and the extracts are added to the top organic layer, which is then dried over anhydrous magnesium sulfate. The solvents are evaporated by lowering the pressure with a water aspirator while heating. The residue is distilled at lower pressure; the product, 337.9 g. (88% yield) distills at 123° to 125°C. (0.2 mm. mercury), with a yellow color, and 284.2 g. of the product are redistilled to remove the color. The main distillate fraction, 237.4 g. (61.7% yield) distills without color at 118° to 121.2°C. (0.12 to 0.18 millimeter mercury).

The bis(polyfluoropentyl) benzenephosphonate ester thus produced, has a kinematic viscosity at 210°F. of 3.71 cs. and 7,240 cs. at -20°F. Such compound has a pour point of about -62°F. and a density at 77°F. of 1.61 g./ml. This compound has a thermal stability of at least 600°F., a high autoignition temperature, good hydrolytic stability and high fire resistance. The bis(polyfluoroalkyl) benzenephosphonate ester of this example is a good functional fluid, particularly useful as a base stock for hydraulic fluids of an aircraft system operating over a wide temperature range of from -30° to 500°F.

Polyfluorinated Glutarate Esters

A process described by C.A. Seil, R.H. Boschan and J.P. Holder; U.S. Patent 3,412,410; November 19, 1968; assigned to McDonnell Douglas Corporation involves bis(polyfluoroalkyl)-3-methyl glutarate esters in which the fluoroalkyl groups have the formula

$$CF_3(CF_2)_nCX_2CH_2CH_2-$$

where n is an integer of from 0 to 5, preferably 1 to 3, and X is a member selected from the group consisting of hydrogen and fluorine, and is preferably fluorine. The glutarates have properties of high fire resistcnce, increased temperature stability, liquidity over a relatively wide temperature range, and improved hydrolytic stability. The combined fluorine content of the bis(polyfluoroalkyl)-3-methyl glutarates can range from about 31 to 61%, preferably about 40 to 57%, by weight.

The bis(polyfluoroalkyl)-3-methyl glutarates are produced according to one convenient mode of procedure by reacting the corresponding polyfluorinated alcohol with 3-methyl glutaric acid in a proportion of about 2 mols of the alcohol to 1 mol of the acid in a solvent, e.g., toluene. The mixture is heated at reflux for a period of hours, e.g., about 6 to 20 hours, while collecting water generated in the esterification reaction. The solvent, e.g., toluene, is removed by distilling at reduced pressure, and the remaining mixture is distilled at still lower pressure to recover the desired bis(polyfluoroalkyl)-3-methyl glutarate compounds.

Specific examples of bis(polyfluoroalkyl)-3-methyl glutarates of the process are as follows, the symbol R' representing the

$$-CH_2\overset{\displaystyle CH_3}{\overset{|}{C}H}-CH_2-$$

portion of the 3-methyl glutarate radical in the compounds shown below.

(1) $CF_3(CF_2)_2CH_2CH_2OOCR'COOCH_2CH_2(CF_2)_2CF_3$

(2) $CF_3(CF_2)_2CH_2CH_2CH_2OOCR'COOCH_2CH_2CH_2(CF_2)_2CF_3$

(3) $CF_3(CF_2)_3CH_2CH_2OOCR'COOCH_2CH_2(CF_2)_3CF_3$

(4) $CF_3(CF_2)_4CH_2CH_2OOCR'COOCH_2CH_2(CF_2)_4CF_3$

(5) $CF_3(CF_2)_3CH_2CH_2CH_2OOCR'COOCH_2CH_2CH_2(CF_2)_3CF_3$

Example 1: The preparation of bis(3,3,4,4,5,5,5-heptafluoropentyl)-3-methyl glutarate is as follows. To a mixture of 76.2 g. (0.356 mol) of 3,3,4,4,5,5,5-heptafluoropentanol, 24.4 g. (0.167 mol) of 3-methylglutaric acid, and 1 g. of p-toluenesulfonic acid monohydrate, is added 300 cc of toluene and the mixture is refluxed with an attached Stark-Dean trap to collect the water generated in the esterification.

The toluene layer is shaken twice with saturated sodium bicarbonate solution and dried over anhydrous potassium carbonate, then vacuum stripped at about 4 mm. to remove toluene. The residue is distilled to yield 73.5 g. (75%) of product, BP 104.5°C. (0.35 mm. pressure).

The bis(polyfluoropentyl)-3-methyl glutarate ester thus produced, Compound 1, above, has a viscosity at 210°F. of 1.78 cs. and a viscosity at -30°F. of 2,885 cs. The compound has a low pour point, a density of 1.47 g./ml. at 77°F., and a thermal stability of about 473°F. Such glutarate ester has a high autoignition temperature of about 750°F., good hydrolytic stability and high fire resistance. Such compound can be employed as a hydraulic fluid in the hydraulic systems of aircraft, and is essentially noncorrosive to metals such as iron, steel or copper. The hydrolytic stability of the glutarate ester of Example 1, designated Compound A, is compared with the hydrolytic stability of a related prior art compound

$$CHF_2CF_2CF_2CF_2CH_2OOCCH_2\underset{}{\overset{CH_3}{\overset{|}{C}}}H{-}CH_2COOCH_2CF_2CF_2CF_2CHF_2$$

designated Compound B, by shaking each compound with water and titrating according to ASTM procedure D 947-58T to obtain the acid number. A comparision of these results for Compounds A and B, together with a comparision of certain other important properties of these compounds, are set forth in the table below. The acid numbers noted in the table below are obtained after exposure of both compounds to water and titrating according to the ASTM procedure indicated above.

	Viscosity (centistokes)			Acid Number	Density at 77° F., g./ml.
	-30° F.	100° F.	210° F.		
Compound A	2,885	7.2	1.78	0.47	1.47
Compound B	7,820	23.2	3.49	54.50	1.55

From the table above it is seen that at temperatures of the order of -30°F. the viscosity of fluid Compound A of the process is substantially lower than the viscosity of prior art Compound B. Further, Compound A has a lower density than Compound B, requiring less weight per unit volume of fluid in a hydraulic system of an aircraft than in the case of Compound B, and such reduced density together with the lower viscosity of Compound A at low temperatures, permit pumping of fluid A at low temperatures with a smaller expenditure of pump energy than for fluid B. Of particular significance, it is seen that the acid number representing the extent of hydrolysis of Compound A is negligible as compared to the high acid number of prior art Compound B, and hence the hydrolytic stability of Compound A is highly superior to that of Compound B. This latter characteristic of resistance to hydrolysis for the process compounds is of particular importance in the operation of hydraulic systems, since the elimination of moisture in such systems is extremely difficult and requires the use of special and complicated equipment and techniques.

Example 2: The preparation of bis(3,3,4,4,5,5,6,6,6-nonafluorohexyl)-3-methyl glutarate is as follows. The procedure of Example 1 is carried out except employing as the fluoroalcohol 3,3,4,4,5,5,6,6,6-nonafluorohexanol in the same molar proportion as the fluoropentanol of Example 1.

The bis(polyfluoroalkyl)-3-methyl glutarate ester thus produced, Compound 3 above, has a low viscosity at low temperatures, good thermal stability, a high autoignition temperature, good hydrolytic stability and high fire resistance. The bis(polyfluoroalkyl)-3-methyl glutarate ester of this example has utility as a functional fluid, and is particularly useful as a base stock for hydraulic fluids in an aircraft system operating over a wide temperature range.

Polyfluorinated Orthosilicates

C.A. Seil, R.H. Boschan, and J.P. Holder; U.S. Patent 3,491,134; January 20, 1970; assigned to McDonnell Douglas Corporation describe a class of tetrakis(polyfluoroalkyl) orthosilicates having the formula shown on the following page.

(1) 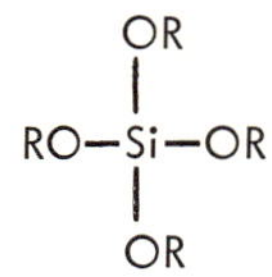

where R has the formula

(2) $CF_3(CF_2)_nCX_2CH_2CH_2-$

where n is an integer of 0 to 4, preferably 1 to 3, and X is a member selected from the group consisting of hydrogen and fluorine, and is preferably fluorine. The combined fluorine content of the polyfluoroalkyl orthosilicates can range from 42 to 67%, preferably from 51 to 66%, by weight.

It has been found that the above-defined tetrakis(polyfluoroalkyl) orthosilicates are particularly valuable as functional fluids, e.g., as base stocks for hydraulic fluids, and are also useful as heat transfer medic and as lubricants due to the advantageous physical properties of such compounds. These properties include high fire resistance, increased temperature stability, and liquidity over a relatively wide temperature range.

The tetrakis(polyfluoroalkyl) orthosilicates are produced according to one convenient mode of procedure by a transesterification reaction which involves reacting the corresponding polyfluorinated, alcohol with a tetraalkyl orthosilicate such as tetraethyl orthosilicate, in a proportion of about 4 mols of the alcohol to one mole of the tetraethyl orthosilicate, preferably in the presence of a minor amount of sodium or sodium alcoholate. The mixture is heated to reflux, distilling off unfluorinated alcohol, e.g., ethanol, which is split off in the reaction, and unreacted polyfluorinated alcohol is removed by distilling at reduced pressure. The remaining mixture is distilled at still lower pressure to recover the desired tetrakis-(polyfluoroalkyl) orthosilicates.

Specific examples of tetrakis(polyfluoroalkyl) orthosilicates are as follows:

(1) $(CF_3CF_2CH_2CH_2O)_4Si$
(2) $[CF_3(CF_2)_2CH_2CH_2O]_4Si$
(3) $[CF_3(CF_2)_2CH_2CH_2CH_2O]_4Si$
(4) $[CF_3(CF_2)_3CH_2CH_2O]_4Si$
(5) $[CF_3(CF_2)_4CH_2CH_2O]_4Si$
(6) $[CF_3(CF_2)_3CH_2CH_2CH_2O]_4Si$
(7) $[CF_3(CF_2)_4CH_2CH_2CH_2O]_4Si$

Preferred compounds are compounds 2 to 6 above. The following are examples of preparation of the compounds of the process.

Example 1: The preparation of tetrakis(3,3,4,4,5,5,5-heptafluoropentyl) orthosilicate is as follows. In a flask is placed 20.8 g. (0.1 mol) of redistilled tetraethyl orthosilicate, 107 g. (0.5 mol) of 3,3,4,4,5,5,5-heptafluoro-1-pentanol and 0.5 g. of sodium.

The mixture is heated to reflux; and ethanol is distilled off through a vacuum-jacketed Vigreaux column. The unreacted alcohol is removed by distillation at between 15 and 5 mm. mercury pressure. The orthosilicate residue is distilled at 0.2 mm.; the main fraction of orthosilicate, 69.0 g. (78.3% yield) distills at 98° to 113°C. (0.2 mm. mercury).

The tetrakis(polyfluoroalkyl) orthosilicate thus produced, compound (2) above, has a viscosity at 210°F. of 1.28 cs. and a viscosity at -65°F. of 687 cs. The compound has a pour point below -150°F., a density of 1.60 g./ml. at 77°F., and a thermal stability in the range of about 500° to 600°F. The orthosilicate has a high autoignition temperature, good hydrolytic stability and high fire resistance. The compound can be employed as a hydraulic fluid or as a heat transfer medium, e.g., as a coolant, in the hydraulic systems of aircraft, for operation at very low temperatures and also at high temperature, and is essentially noncorrosive to metals such as iron, steel and copper.

A comparison of certain important properties of the orthosilicates of Example 1, designated compound A, with a related prior art compound $(CHF_2CF_2CF_2CF_2CH_2O)_4Si$ designated compound B, is set forth in the table below:

	Viscosity (centistokes)			Pour Point, °F.	Autoignition Temperature, °F.	Density at 77° F., g./ml.
	−65° F.	−60° F.	210° F.			
Compound A	687	----------	1.28	<−150	800	1.60
Compound B		6,725	2.44	<−110	735	1.74

From the table on the preceding page it is seen that at temperatures as low as —65°F. the viscosity of compound A of the process is relatively low and is substantially lower, by about ten-fold, than the viscosity of the prior art compound B at even the higher temperature of —60°F., and that compound A has a substantially lower pour point than compound B. Hence, it is clearly seen that compound A can be employed as a hydraulic fluid or heat transfer medium in hydraulic systems, particularly of aircraft, operating at temperatures of the order of —150°F. or below, and substantially lower than the temperature of operability of compound B. Further, compound A has a lower density than compound B, thus requiring less weight per unit volume of fluid in a hydraulic system of an aircraft than in the case of compound B, and such reduced density together with the lower viscosity of compound A at lower temperatures and also the lower pour point of compound A, permit the pumping of fluid A at low temperatures with a smaller expenditure of pump energy than fluid B. Further, it is seen that compound A has a higher autoignition temperature and hence a higher thermal stability than prior art compound B.

Example 2: The preparation of tetrakis(3,3,4,4,5,5,6,6,6-nonafluorohexyl) orthosilicate is as follows. A mixture of 1.50 mols tetraethyl orthosilicate, 0.35 mol of 3,3,4,4,5,5,6,6,6-nonafluorohexanol and 1 g. of sodium is heated to reflux; ethanol is distilled from the mixture. The unreacted alcohol is removed by distillation at 15 to 5 mm. mercury pressure. The orthosilicate residue is distilled off at lower pressure between about 0.2 and 0.3 mm. mercury.

The tetrakis(polyfluoroalkyl) orthosilicate thus produced, compound 4 above, has a low viscosity at low temperature, a low pour point, high thermal stability, a high autoignition temperature, good hydrolytic stability and high fire resistance. The tetrakis(polyfluoroalkyl) orthosilicate of this example can be employed as a functional fluid, particularly useful as a base stock for hydraulic fluids and as a heat transfer fluid in an aircraft system operating over a wide temperature range of from -75° to 500°F.

GENERAL

Fluorocarbon Diethers

H.R. Nychka and L.G. Anello; U.S. Patent 3,435,078; March 25, 1969; assigned to Allied Chemical Corporation describe the preparation of α,ω-bis(fluoroperhaloisopropoxy)-perfluoroalkanes which are useful as thermally stable dielectric coolants.

The compounds can be prepared by the condensation of two mols of a 1-(fluoroheptahaloisopropoxy)-ω-iodoperfluoroalkane, by reaction with zinc in nonreactive solvent media, most favorably in methylene chloride-acetic anhydride medium, with elimination of iodine as ZnI_2. Starting with a hexahaloacetone, the iodo compound, which is the starting material in the process, can be prepared by a series of two known reactions where hexahaloacetone is reacted with potassium fluoride, and the resulting KF adduct is reacted with tetrafluoroethylene and iodine to produce 1-(heptahaloisopropoxy)-2-iodotetrafluoroethane. Two mols of this iodo compound are then condensed to form the first member of the series of diethers. For preparation of the longer chain compounds, the 1-(heptahaloisopropoxy)-2-iodotetrafluoroethane is reacted with additional tetrafluoroethylene in the presence of a catalyst such as t-butyl peroxide whereby the chain may be lengthened by increments of 2 carbon atoms, forming a 1-(fluoroperhaloisopropoxy)-omega-iodo-perfluoroalkane of the desired even number of alkane carbons, for use in the final condensation reaction. The following examples illustrate the process.

Example 1:, Into a 500 ml. flask was placed 653 g. (1.58 mols) of $(CF_3)_2CFOCF_2CF_2I$ (BP 86°C.). Twelve grams (0.18 mol) of zinc dust was added. Then about 45 ml. of absolute methanol were slowly added to the stirred mixture at 25°C. The reaction was quite exothermic and the temperature rose rapidly to 53°C. The addition was stopped and then continued when temperature dropped to 40°C. This temperature was maintained overnight. The mixture was washed with water, dried and filtered. The 598 g. of crude oil was distilled. There was recovered: 15 g. (0.052 mol) of $(CF_3)_2CFOCF_2CF_2H$, BP 46°C.; 449 g. (1.09 mol) of $(CF_3)_2CFOCF_2CF_2I$, BP 86°C.; and 20 g. (0.035 mol) of $[(CF_3)_2CFOCF_2CF_2]_2$, BP 128 to 132°C. equivalent to an 19.5% yield of $[(CF_3)_2CFOCF_2CF_2]_2$ based on zinc added.

Example 2: Into a 500 ml. flask was placed 357 g. (0.87 mol) of $(CF_3)_2CFOCF_2CF_2I$. Eight grams (0.12 mol) of zinc dust was added. About 28 g. of absolute methanol was slowly added while maintaining the reaction temperature at 20°C. The mixture was stirred for 4 hours at this temperature. The mixture was washed, dried and filtered. The 339 g. of crude oil was distilled. There was recovered: 17 g. (0.060 mol) of $(CF_3)_2CFOCF_2CF_2H$, BP 46°C.; 295 g. (0.71 mol) of $(CF_3)_2CFOCF_2CF_2I$, BP 86°C.; and 1.35 g. (0.24 mol) of $[(CF_3)_2CFOCF_2CF_2]_2$, BP 132° to 134°C. equivalent to a 20% yield of $[(CF_3)_2CFOCF_2CF_2]_2$ based on zinc added, which analyzed as follows. Calculated for $[(CF_3)_2CFOCF_2CF_2]_2$: C, 21.5%; F, 73.33%. Found: C, 20.93%; F, 72.69%. The analysis above indicated the product is the diether.

Example 3: Into a reaction flask were placed 60 g. of acetic anhydride, 125 ml. (167 g.) of methylene chloride and 20 g. (0.31 mol) of zinc. The resulting slurry was stirred and into the stirred slurry was charged 103.6 g. (0.25 mol) of 1-heptafluoroisopropoxy)-2-iodoethane (BP 86°C.). The mixture became mildly exothermic causing its temperature to rise from 22° to 43°C. over a two hour period. Thereafter the mixture was gently refluxed (43°C.) for 20 hours. The reaction mixture was then cooled and 100 ml. of 5% sulfuric acid was added, whereupon three liquid layers formed. The bottom layer, containing the product, was washed twice with water, dried over Na_2SO_4 and distilled through a spinning band column. The product fraction was 39 g. of a water white liquid of boiling point 133° to 134°C. It was 99.5% pure according to a vapor phase chromatographic analysis. The percent yield was 63%, based on the 1-(heptafluoroisopropoxy)-2-iodoethane, after adjusting for a small amount of intermediate cut and column holdup. A fluorine nuclear magnetic resonance analysis (performed on a Varian DP-56 NMR Spectrometer using $CFCl_3$ as standard) of the resulting product confirmed the composition as being 1,4-bis(heptafluoroisopropoxy)-octafluorobutane, and supports the structure of

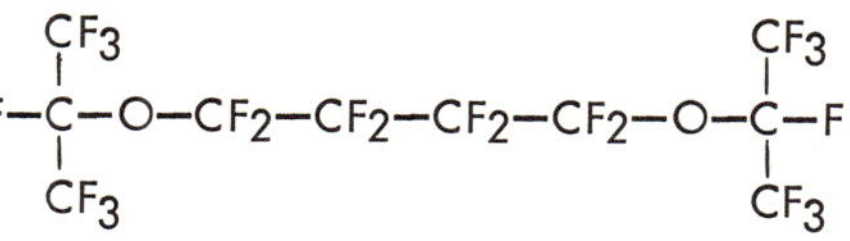

as shown by the NMR results tabluated below:

Types of fluorine	Chemical shift (p.p.m.)	Fluorine area ratio	
		Theory	Found
$4 \times CF_3$ } $2 \times CF_2O$ }	82.0	16	16
$2 \times CF_2$	127.1	4	4
$2 \times CF$	146.2	2	1.9

A mass spectrographic analysis showed the characteristic ionic fragments for the products including the fragment of the parent compound minus fluorine at mass number 551. The infrared analysis was consistent for a perfluoroether, and exhibited characteristic absorption peaks at 7.25 microns, 8.00 microns, 8.22 microns, 8.65 microns, 10.08 microns and 11.18 microns. The product, calculated for $C_{10}F_{22}O_2$, analyzed as follows. Theory: C, 21.1%; F, 73.7%. Found: C, 20.8%; F, 70.7%. Molecular weight —Theory: 570. Found: 585 (NMR).

Example 4: The 1,4-bis(heptafluoroisopropoxy)-octafluorobutane prepared in Example 3 was tested to determine various physical and electrical properties in comparison with two perfluorocarbon ethers (B and C) available commercially and recommended for use as dielectric vapor coolants, with results shown below:

Property	A	B	C
Boiling point, ° C	135	99-107	152
Freezing point, ° C	−85	−113	[1] −115
Density, g./cc. at 25° C	1.75	1.76	1.72
Viscosity centiposies at—			
25° C	1.2	0.82	1.2
54° C	27.0	7.40	21.6
Thermal conductivity, 25° C. B.t.u./hr. ft.²°F/ft.	0.092	0.091	0.041
Dielectric strength, kv./0.1" ASTM-D-877	35	31	29
Dielectric constant, 1 kc. ASTM-D-150	1.86	1.89	2.57
Dissipation factor, 1 kc. ASTM-D-150	<.0005	<.0005	<.0005
Volume Resistivity, ohm-cm. ASTM-D-257	8.1×10^{11}	7.6×10^{15}	1.4×10^{13}

[1] Pour point.

Note: A is 1,4-bis(heptafluoroisopropoxy)-octafluorobutane of this process; B is a six membered cyclic fluorocarbon ether having a three carbon side chain and formula $C_8F_{16}O$ commercially available as FC-75; C is commercially available as Freon E-3.

It will be noted from the figures in the table that the 1,4-bis(heptafluoroisopropoxy)-octafluorobutane is substantially equal to or superior to the two commercial fluoroethers in all of the critical properties listed in the table. The fluorocarbon diether is superior to the commercial products in dielectric strength with a value of 35 kv./0.1" as against 31 and 29 respectively in the two commercial materials, and has a desirably higher thermal conductivity than either of the two commercial compounds. A thermal stability test of the 1,4-bis(heptafluoroisopropoxy)-octafluorobutane and the two commercial ethers, E—3 and FC—75, was performed by storing the three ether samples in sealed Pyrex glass tubes at 400°C. for seven days. After the first day the E—3 sample showed signes of decomposition. It had become turbid and had severely etched the glass tube. After seven days the FC—75 sample had coated the Pyrex tube with a heavy opaque carbon deposit and had become strongly acidic. In contrast, the diether sample was still clear and water white after the seven day test. There was also no sign of attack on the Pyrex tube. Comparison vapor phase chromatographic analysis of the diether before and after the heat test showed virtually no change, thus indicating that the diether is significantly more heat resistant than the prior art compounds tested.

Fluoroalkyl Ethers of Alicyclic Fluoroolefins

A process described by V.W. Gash; U.S. Patent 3,497,563; Feburary 24, 1970; assigned to Monsanto Company involves fluoroalkyl ethers of alicyclic fluoroolefins. The process provides ethers of the formula shown on the next page in which A is selected from the group consisting of

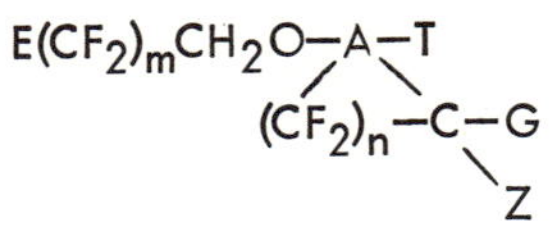

E is selected from the group consisting of hydrogen, fluorine and CH_2OH and may be CH_2OH only when T, G and Z are halogen;

G is selected from the group consisting of fluorine and $OCH_2(CF_2)_mD$ and is fluorine when T is $OCH_2(CF_2)_mD$.

T is selected from the group consisting of bromine, chlorine and $OCH_2(CF_2)_mD$ and is a halogen when G and Z are $OCH_2(CF_2)_mD$.

X is selected from the group consisting of hydrogen, bromine, chlorine and fluorine;

Z is selected from the group consisting of fluorine and $OCH_2(CF_2)_mD$ and is fluorine when T is $OCH_2(CF_2)_mD$.

D is selected from the group consisting of hydrogen and fluorine; and

n is a number from 1 to 7 and each m expression is a number from 1 to 12.

The products are advantageously prepared by reacting a perfluoroalkanol with a perhalocycloolefin in the presence of a base and water. Particularly advantageous yields and desirable purities are obtained when the reaction system is an aqueous reaction system although nonaqueous systems may also be used. The following examples are illustrative of the process.

Example 1: To a solution of 2 g. (0.05 mol) of sodium hydroxide, 16.6 g. (0.05 mol) dodecafluoroheptanol, 20 ml. water, and 20 ml. tetrahydrofuran was added 4.1 g. (0.017 mol) of 1,2-dichloro-3,3,4,4,5,5-hexafluoro-1-cyclopentene. After an initial exothermic reaction, the stirred mixture was refluxed 24 hours. The cooled mixture was poured into water and extracted with two 100 ml. portions of carbon tetrachloride. The organic layer was washed with water, dried over magnesium sulfate and distilled yielding 14.05 g. (54.9% yield) of nearly colorless oil boiling point 168° to 178°C. at 0.08 mm. of mercury vapor pressure, with a refractive index of n_D^{25} 1.3421. The product was 1-chloro-3,3,4,4-tetrafluoro-2,5,5-tri(2,2,3,3,4,4,5,5,6,6,7,7-dodecafluoroheptyloxy)-1-cyclopentene which for convenience will be referred to as compound C.

Other triethers were prepared using the procedure of Example 1. For convenience these compounds are identified as follows:

(A) 1-chloro-3,3,4,4-tetrafluoro-2,5,5-tri(2,2,3,3,4,4,4-heptylfluorobutyloxy)-1-cyclopentene
(B) 1-chloro-3,3,4,4-tetrafluoro-2,5,5-tri(2,2,3,3,4,4,5,5-octafluoropentyloxy)-1-cyclopentene
(D) 1-chloro-3,3,4,4-tetrafluoro-2,5,5-tri(2,2,3,3,4,4,5,5,6,6,7,7,8,8,9,9-hexadecafluorononyloxy)-1-cyclopentene
(E) 1-chloro-3,3-difluoro-2,4,4-tri(2,2,3,3,4,4,5,5,6,6,7,7-dodecafluoroheptyloxy)-1-cyclopentene
(F) 1-chloro-3,3,4,4,5,5-hexafluoro-2,6,6-tri(2,2,3,3,4,4,5,5,6,6,7,7-dodecafluoroheptyloxy)-1-cyclohexene

The pertinent data is tabulated along with compound C in the tables on the following page.

Example 2: A previously prepared solution of 116 g. (0.35 mol) of dodecafluoroheptanol, 14.4 g. (0.36 mol) of sodium hydroxide, 30 ml. of water, and 100 ml. of tetrahydrofuran was added slowly to a solution of 25 g. (0.154 mol) of perfluorocyclobutene in 40 ml. tetrahydrofuran at −10°C. A strong exothermic reaction was noted during the addition. Stirring was continued overnight at room temperature, then the mixture was heated slowly to reflux conditions for approximately 3 hours. The reaction product was then recovered as in Example 1. The product was 1,2-di(2,2,3,3,4,4,5,5,6,6,7,7-dodecafluoroheptyloxy)-3,3,4,4-tetrafluoro-1-cyclobutene which for convenience will be referred to as Compound G.

The physical properties of Compound G and other diethers, 1,2-di(2,2,3,3,4,4,5,5,6,6,7,7-dodecafluoroheptyloxy)-3,3,4,4,5,5-hexafluoro-1-cyclopentene (Compound H), 1,2-di(2,2,3,3,4,4,5,5,6,6,7,7-dodecafluoroheptyloxy)-3-3,4,4,5,5,6,6-octafluoro-1-cyclohexene (Compound I), prepared in the same manner, are tabulated in the tables on the following page. Similar properties are obtainable when other fluoroalkanols and perfluoro cyclic olefins are employed.

Example 3: Chlorine was bubbled through 10.6 parts of Compound G on a steam bath for a period of 4 hours and then distilled to obtain 10 parts of a clear oil having a boiling point of 110°C. at 0.05 mm. mercury vapor pressure and a refraction index of $n_D^{25.2}$ 1.3437. Infrared analysis showed no absorption in the double bond region confirming the product as 1,2-dichloro-1,2-di(2,2,3,3,4,4,5,5,6,6,7,7-dodecafluoroheptyloxy)-3,3,4,4,-tetrafluorocyclobutane. For convenience this compound will be identified as J.

The ethers possess a combination of fire resistance and wide usable temperature range which makes them quite unique and well suited for use as a force transmission fluid (hydraulic fluid). The excellent fire resistance of these ethers can be demonstrated by the results obtained in the Molten-metal Test in which the fluid under evaluation is dropped from a medicine dropper or poured from a test tube onto the surface of molten aluminum alloy which has been heated to about 1250°F. If spontaneous ignition does not occur, an ignition source, such as a spark or flame is placed in the vapors to determine if they can be ignited. In this test these ethers do not flash without a spark and with a spark are nonburning or are self-extinguishing.

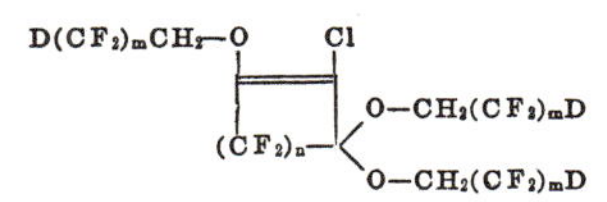

	D	m	n	Percent, yield	B.P., °C. at mm. mercury pressure	B.P., °F.[1]	Ml. wt. calcd./found [2]	n_D^{25}	Infrared C=C absorption (cm.$^{-1}$)
Compound:									
A	F	3	2	82.9	75/0.05	474	768/736	1.3325	1,673
B	H	4	2	73.6	135/0.1	586	865/820	1.3503	1,670
C	H	6	2	72.2	170/.08	618	1,165/1,234	1.3421	1,660
D	H	8	2	74.4	200/0.1	676	1,465/1,472	1.3386	1,665
E	H	6	1	54.9	170/0.1	561	1,115/1,077	1.3429	1,697
F	H	6	3	68.3	175/0.1	629	1,215/1,200	1.3440	1,649

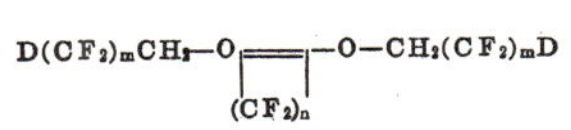

	D	m	n	Percent, yield	B.P., °C. at mm. mercury pressure	B.P., °F.[1]	Ml. wt. calcd. found [2]	n_D^{25}	Infrared C=C absorption (cm.$^{-1}$)
Compound:									
G	H	6	2	53.1	105/0.1	543	786/823	1.3322	1,750
H	H	6	3	40.4	110/0.1	536	836/837	1.3331	1,728, 1,707
I	H	6	4	52.6	115/0.1	561	886/863	1.3317	1,711, 1.679

[1] Extrapolated boiling point from vapor pressure data.
[2] Molecular weights determined by osmometer ($CHCl_3$ solvent).

	Viscosity (cs., °F.)					ASTM slope (100-210° F.)	Density (° F.)		T_D(° F.)
	-40	0	100	210	400		100	210	
Compound:									
A	7,800	392	8.0	1.47	0.38	1.21	1.6996	1.5790	518
B	79,000	3,700	42.4	4.11	0.73	0.98	1.7460	1.6366	523
C	40,500(−20)	8,800	68.6	5.87	0.93	0.91			489
D		4,700(32)	162.0	9.69	1.26	0.86	1.8423	1.7289	504
E		13,181	87.9	6.47	0.96	0.91	1.7891	1.6761	390
F		33,915	132.9	8.05	1.05	0.89	1.8231	1.7132	480
G [1]	5,770	461	12.6	2.22	0.54	1.02	1.7676	1.6533	>550
H	5,390	429	12.0	2.13	0.53	1.02	1.7891	1.6759	482
I	18,094	1,053	18.5	2.63	0.58	1.01	1.8010	1.6897	>550
J	33,808	1,818	28.3	3.42	0.67	0.97	1.7874	1.6785	433

[1] A four-ball lubrication test was run on this fluid; conditions, 40 kg., 1,260 r.p.m., one hour, 400° F., tool steel on tool steel. Result, 0.88 mm. scar.

Polychlorotrifluoroethylene Oils for Corrosion Inhibition

E.T. Fogg and W.H. Gumprecht; U.S. Patent 3,306,853; Feburary 28, 1967; assigned to E.I. du Pont de Nemours and Company have found that reduced corrosion of metals in the presence of poly(hexafluoropropylene oxide) oils can be obtained by using a composition which comprises poly(hexafluoropropylene oxide) oil of the structure

$$R_fO{+}CF(CF_3)CF_2O{+}_nCF_2CF_3$$

or

$$R_fO{+}CF(CF_3)CF_2O{+}_pCF(CF_3)CF(CF_3){+}OCF_2CF(CF_3){+}_rOR_f$$

where n, p and r are integers indicating the degree of polymerization and Rf is a perfluoroalkyl group havine 1 to 6 carbon atoms. The poly(hexafluoropropylene oxide) oil has an average molecular weight of at least 3,000 and a pour point not in excess of 50°F. and; as corrosion inhibitor, an effective amount of a high boiling polychlorotrifluoroethylene oil of the formula $Cl{+}CF_2CFCl{+}_sCl$, where s indicates the degree of polymerization, the polychlorotrifluoroethylene oil having an average molecular weight of 500 to 1,000.

In general the amount of polychlorotrifluoroethylene inhibitor in the oil composition will be in the range of 0.1 to 2% by weight. So long as at least about 0.1% of inhibitor is present, a significant amount of corrosion inhibition is obtained. As the amount of inhibitor is increased, a quantitative reduction in corrosion is obtained up to a point. At about 1% by weight concentration the effect reaches a maximum, higher concentration seldom producing any significant increase in effect. For a general purpose, high temperature oil, it is usually preferable to add about 0.5 to 1.5% by weight of inhibitor. This amount is sufficient to meet the requirement of the vast majority of needs for the oils. The following examples illustrate the process.

Examples 1 through 11: The compositions of this process are tested using a modified form of the apparatus specified in the WADD Microoxidation-Corrosion Test of High Temperature Fluids, Fluids and Greases Section, Aeronautical Systems Division, Wright-Patterson Air Force Base, replacing the specified Pyrex tube with an Inconel tube. Basically the apparatus consists of a 1/8 in. nickel tube adapted for mounting three washers outside the lower end. The 1/8 in. tube is inserted inside a 13/16 in. vertical nickel tube so that the washers are immersed in a body of oil contained in the larger tube. The larger tube, which is adapted with a condenser for recovering any oil which may be stripped from the tube, is inserted in an aluminum heating block. Various steel washers having the following compositions are used.

Steel	Composition
QQ-S-636 Steel	3% Cr. 0.9% Mn, balance Fe.
M-50 Steel	4.3% Mo, 4.1% Cr, 1% V, 0.4% Ni, 0.3% Mn, balance Fe
52100 Steel	1.6% Cr, 0.5% Mn, 0.4% Si, balance Fe
301 Stainless	18% Cr, 8% Ni, 2% Mn, 1% Si, balance Fe
304 Stainless	20% Cr, 12% Ni, 2% Mn, 1% Si, balance Fe
310 Stainless	20% Cr, 22% Ni, 2% Mn, 1.5% Si, balance Fe
316 Stainless	18% Cr, 14% Ni, 3% Mo, 2% Mn, 1% Si, balance Fe
321 Stainless	19% Cr, 12% Ni, 2% Mn, 1% Si, balance Fe
440C Stainless	18% Cr, 1% Mn, 1% Si, 0.8% Mo, balance Fe

The test consists of placing weighed and measured test washers of the metals to be tested in the apparatus, adding sufficient oil composition to insure the washers are covered, assembling the remaining apparatus and inserting it in the heating block. The fluid is then heated to the test temperature and air is passed down through the smaller nickel tube into the body of oil containing the washers for 72 hours. Test temperatures of 650° and 700°F. are used. Corrosion is determined by weighing the test washers before and after the test. The corrosion rate is calculated from the weight change and the known surface area of the washer. Before weighing, both before and after the test, the test washers are metallurgically cleaned, i.e., scrubbed with scouring powder and water, rinsed with water, degreased and dried with acetone.

The oil is a hexafluoropropylene oxide polymer of the structure

$$CF_3CF_2CF_2O[CF(CF_3)CF_2O]_pCF(CF_3)CF(CF_3)[OCF_2CF(CF_3)]CF_2CF_2CF_3$$

having an average molecular weight of 6,200. The inhibited oil compositions contain 1% of a polychlorotrifluoroethylene oil of the structure $Cl[CF_2CFCl]_sCl$ having an average molecular weight of about 940 and known as Kel-F-40.

For comparison, control experiments in which no inhibitor is added to the polymer oil are carried out for each example. The following table summarizes the test conditions and results for these examples.

Example	Temperature °F.	Metal	Weight Change, mg./cm.²/day: Control	Weight Change, mg./cm.²/day: With Inhibitor
1	650	QQ-S-636 Steel	-1.66	-0.71
2	650	52100 Steel	-1.53	-0.83
3	650	M-50 Steel	-5.47	-1.06
4	650	301 Stainless	-0.81	-0.47
5	650	304 Stainless	-2.87	-1.55
6	650	310 Stainless	-0.63	-0.64
7	-	321 Stainless	-2.62	-1.62
8	-	440C Stainless	-2.61	-0.87
9	650	Titanium (8% Mn)	-1.35	-0.50
10	700	QQ-S-636 Steel	-3.50	-1.54
11	700	316 Stainless	-1.95	-1.66

Examples 12 through 22: The procedure of Examples 1 through 11 above is repeated using a poly(hexafluoropropylene oxide) oil of the formula $n\text{-}C_3F_7O[CF(CF_3)CF_2O]_nCF_2CF_3$ having an average molecular weight of 6,000. Results essentially identical with those shown above are obtained.

Nonfoaming Systems

W.H. Gumprecht and R.E. Johnson, Jr.; U.S. Patent 3,445,392; May 20, 1969; assigned to E.I. du Pont de Nemours and Company have found that the tendency of perfluorinated polyethers to foam can be overcome by providing a mixture of perfluorinated polyethers of the formula $C_3F_7O[CF(CF_3)CF_2O]_nCF_2CF_3$ where n indicates the degree of polymerization, having a number average molecular weight of 2,000 to 8,500 and a volatility range no broader than 140 Centigrade degrees. It has been found that the wide molecular weight distribution normally associated with perfluorinated polyethers is the cause of their foaming. Thus a narrowing of the molecular weight distribution of these polyethers, as indicated by their limited volatility range, has been found to result in an elimination of their foaming tendency.

Perfluorinated Phenyl Phosphine Additives

In a process discribed by R.E. Dolle, Jr., F.J. Harsakey and C. Tamborski; U.S. Patent 3,483,129; December 9, 1969; U.S. Patent 3,393,151; July 16, 1968; U.S. Patent 3,481,872; December 2, 1969; all assigned to U.S. Secretary of the Air Force perfluorinated substituted phosphines; for example, tris[4-(pentafluorophenoxy)tetrafluorophenyl] phosphine and tris(4-heptafluorotolyl) phosphine are used as additives to perfluorinated aliphatic polyethers. A minor percentage, up to 5% but preferably 0.01 to 0.2%, of the additives are used. The resulting compositions are useful as lubricants and hydraulic fluids at temperatures as high as 650°F.

It is known that perfluorinated aliphatic polyethers are useful as lubricants at high temperatures. Among the compounds which can be used are, for example, compounds of the general formula:

$$X{-}CF_2{-}CF_2{-}O\left(\begin{matrix}X\\|\\CF{-}CF_2{-}O\end{matrix}\right)_n CF_2{-}X$$

where n is a positive integer including zero and X is a member of the class consisting of fluorine and the perfluoromethyl radical. The preparation of these compounds (referred to as base fluid I) is described in U.S. Patent 3,242,218. Also among the compounds which are particularly useful are compounds of the type (referred to as base fluid II):

$$CF_3{-}CF_2{-}CF_2{-}O\left(\begin{matrix}CF_3\\|\\CF{-}CF_2{-}O\end{matrix}\right)_{n'}\begin{matrix}CF_3\ CF_3\\|\quad\ |\\CF{-}CF\end{matrix}\left(\begin{matrix}CF_3\\|\\O{-}CF_2{-}CF\end{matrix}\right)_m O{-}CF_2{-}CF_2{-}CF_3$$

where n' and m are integers greater than 2. The preparation of these compounds is described in U.S. Patent 3,214,478.

U.S. Patent 3,393,151 suggested the use of lubricant compositions comprising a perfluorinated aliphatic polyether and a compound such as tris(pentafluorophenyl)phosphine or tris(pentafluorophenyl)phosphine oxide. It was found that the latter compositions were highly effective in reducing corrosion and fluid degradation at temperatures of 550° to 600°F. but their usefulness is somewhat limited by their volatility at these and, particularly, higher temperatures. These formulations can be improved by using a composition comprising a perfluorinated aliphatic ether and a minor proportion of a perfluorinated phenoxyphenyl phosphine having the structure:

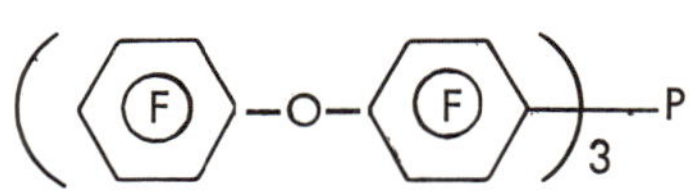

Particularly useful compositions comprise, as the phosphine, tris[4-(pentafluorophenoxy)tetrafluorophenyl]phosphine; and comprise as the ether, base fluid I or base fluid II.

Another example of a substituted perfluorinated phenyl phosphine that can be used is, tris(4-heptafluorotolyl)phosphine. Tris(4-heptafluorotolyl)phosphine, having the structure

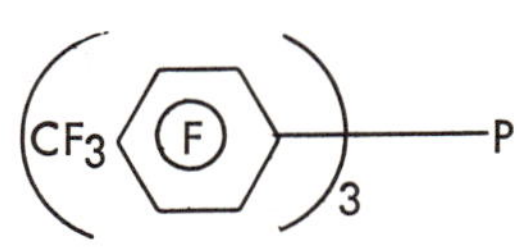

is highly effective in reducing corrosion of ferrous and titanium alloys in the perfluorinated aliphatic polyether fluids and inhibits the degradation of the fluids themselves. This additive is effective in concentrations as low as 0.10 wt. percent with optimum concentrations near 0.25 wt. percent. The optimum concentrations of previous perfluoroarylphosphine additives, set forth in U.S. Patent 3,481,872, were as high as 0.5 to 1.0 wt. percent. Thus optimum effectiveness of tris(4-heptafluorotolyl)phosphine is obtained at lower concentrations than with former additives as evidenced in 600°F. oxidation-corrosion tests. Tris(4-heptafluorotolyl)phosphine is soluble in the perfluorinated aliphatic polyether fluids in 0.25 wt. percent for several days at room temperature. Concentrations of the additive less than 0.25 wt. percent remain soluble for longer periods of time below room temperature. Concentrations of the additive as high as 5 wt. percent are not believed to be seriously detrimental and can be used if additive solubility is not required at room temperature.

Example 1: Compositions consisting of base fluid I with 0.05 and 0.20% tris[4-(pentafluorophenoxy)tetrafluorophenyl]-phosphine were formulated and compared with base fluid I with no additives in an oxidation-corrosion test in contact with various metal washers. The test procedure involves the total immersion of the metal specimen in the test composition and, for a period of 24 hours, the bubbling of one liter per hour of dey air through the composition and over the metal specimens at a temperature of 600°F. The weight change (mg. per cm.^{2}) of the metal washers is used as the measure of corrosion. The metal specimens chosen represent a variety of ferrous alloys likely to be encountered in the commercial use of the compositions. Results are shown in Table 1, weight changes of less than 0.10 mg. per cm.^{2} being considered negligible and weight changes less than 0.20 mg. per cm.^{2} being minor.

TABLE 1

Wt. percent $(C_6F_5OC_6F_4)_3P$ additive	Weight change in mg. per cm.^{2} 4140 Steel	52100 Bearing steel	M-2 Tool steel	M-50 Tool steel	410 Stainless steel	440C Stainless steel
None	+0.73	+0.24	+1.07	+0.66	+0.23	+0.35
0.05	+0.10	+0.06	+0.13	+0.08	+0.05	+0.02
0.2	+0.02	+0.07	+0.30	+0.08	+0.05	+0.04

Example 2: The test procedure of Example 1 was carried out using 0.05% of the additive with base fluid II. Results are shown in Table 2.

TABLE 2

Wt. percent $(C_6F_5OC_6F_4)_3P$ additive	Weight change in mg. per cm.^{2} 4140 Steel	52100 Bearing steel	M-2 Tool steel	M-50 Tool steel	410 Stainless steel	440C Stainless steel
None	+0.88	+0.29	+0.86	+0.98	+0.21	+0.57
0.05	+0.14	+0.09	+0.08	+0.10	+0.04	+0.05

Example 3: The test procedure of Example 1 was carried out using 0.01, 0.02, 0.05 and 0.1% of the additive with base fluid I. The metal specimens used included commercially pure titanium and two titanium alloys likely to be encountered in the commercial use of the compositions. Fluid evaporation losses and percent viscosity changes during the experiment were also measured. Results are shown in Table 3.

TABLE 3

Wt. percent $(C_6F_5OC_6F_4)_3P$ additive	Fluid evap. loss, percent	Kinematic viscosity change at 100° F., percent	Weight change in mg. per cm.^{2} Titanium (commercially pure)	Titanium alloy (4% Al, 4% Mn)	Titanium alloy (6% Al, 4% V)
None	7.8	−7.4	−0.09	−0.32	+0.16
0.01	3.3	−0.4	+0.06	+0.04	+0.08
0.02	0.7	+1.1	−0.01	+0.03	−0.01
0.05	0.2	+1.4	+0.02	+0.02	0.00
0.1	0.3	+1.9	0.00	+0.01	0.00

Example 4: Compositions consisting of base fluid I and 0.10, 0.25 and 0.50 wt. percent tris(4-heptafluorotolyl)phosphine were formulated and compared with base fluid I with no additives at 600°F. according to the test procedure of Example 1 using various ferrous alloys. Results are shown in Table 4.

TABLE 4

Wt. percent $(CF_3C_6F_4)_3P$ additive	Weight change in mg. per cm.^{2} 4140 Steel	52100 Bearing steel	M-2 Tool steel	M-50 Tool steel	410 Stainless steel	440C Stainless steel
None	+0.73	+0.24	+1.07	+0.66	+0.23	+0.35
0.1	+0.15	+0.11	+0.11	+0.04	−0.01	0.00
0.25	+0.10	+0.09	0.00	+0.01	+0.01	0.00
0.5	+0.09	+0.05	+0.21	+0.10	+0.01	+0.03

Example 5: Composition of base fluid II and 0.25 wt. percent tris(4-heptafluorotolyl)phosphine was formulated and compared with base fluid II without an additive at 600°F. according to the test procedure of Example 1 using various ferrous alloys. Results are shown in Table 5.

TABLE 5

Wt. percent $(CF_3C_6F_4)_3P$ additive	Weight change in mg. per cm.^{2} 4140 Steel	52100 Bearing steel	M-2 Tool steel	M-50 Tool steel	410 Stainless steel	440C Stainless steel
None	+0.88	+0.29	+0.86	+0.98	+0.21	+0.57
0.25	+0.16	+0.08	+0.24	+0.08	+0.04	+0.03

Cyclic Phosphonitrilate Esters for Corrosion Inhibition

W.H. Gumprecht; U.S. Patent 3,306,854; Feburary 28, 1967; assigned to E.I. du Pont de Nemours and Company describes poly(hexafluoropropylene oxide) oil compositions containing cyclic phosphonitrilate esters as corrosion inhibitors. The following examples illustrate the process.

Examples 1 through 8: The compositions are tested using a modified form of the apparatus specified in the WADD Micro-oxidation-Corrosion Test of High Temperature Fluids, Fluids and Greases Section, Aeronautical Systems Division, Wright-Patterson Air Force Base, replacing the specified Pyrex tube with an Inconel tube. Various steel washers having the following compositions are used:

Steel	Composition
QQ-S-636 Steel	3% Cr, 0.9% Mn, balance Fe.
52100 Steel	1.6% Cr, 0.5% Mn, 0.4% Si, balance Fe.
301 Stainless	18% Cr, 8% Ni, 2% Mn, 1% Si, balance Fe.
316 Stainless	18% Cr, 14% Ni, 3% Mo, 2% Mn, 1% Si, balance Fe.

Two different oils are used in these examples. Oil A is a hexafluoropropylene oxide polymer of the structure

$$CF_3CF_2CF_2O+CF(CF_3)CF_2O+_pCF(CF_3)CF(CF_3)+OCF_2CF(CF_3)+_rCF_2CF_2CF_3$$

having an average molecular weight of 6,200. Oil B is a hexafluoropropylene oxide polymer of the same structure which has been heat treated to remove less stable components. It has an average molecular weight of 6,100. The inhibited oil compositions contain 1% of a cyclic phosphonitrilate ester of the structure

$$\{CF_3CF_2CF_2+CF(CF_3)CF_2O+_4CF(CF_3)CH_2O\}_8(PN)_4$$

For comparison, control experiments in which no inhibitor is added to the polymer oil are carried out for each example. The following table summarizes the test conditions and results for these examples.

Example	Oil	Temp., °F.	Metal	Weight change, mg./cm.²/day	
				Control	With Inhibitor
1	A	650	QQ-S-636 Steel	−2.24	−1.60
2	A	650	52100 Steel	−1.53	−0.96
3	A	650	301 Stainless	−0.81	−0.05
4	A	650	Titanium (8% Mn)	−1.35	−1.09
5	A	650	Silver	−0.28	−0.23
6	A	700	QQ-S-636 Steel	−2.15	−0.66
7	B	700	301 Stainless	−0.76	−0.48
8	A	700	316 Stainless	−1.95	−0.066

Examples 9 through 15: When Examples 1 through 6 and 8 above are repeated using Oil C, a poly(hexafluoropropylene oxide) oil of the structure $n\text{-}C_3H_7O+CF(CF_3)CF_2O+_nCF_2CF_3$ having an average molecular weight of 6,000, results essentially identical with those shown for Oil A above are obtained.

THREAD SEALANTS

Solid Polytetrafluoroethylene

C. Hinds; U.S. Patent 3,445,393; May 20, 1969; assigned to Oil Center Research, Incorporated describes a packing and sealing composition which comprises comminuted particles of a polymer selected from the group consisting of polytetrafluoroethylene, polytrifluorochloroethylene, polyvinylidene fluoride and the copolymer of hexafluoropropylene and tetrafluoroethylene, where the polymer is sintered and incorporated in a lubricant carrier characterized by an easily extrudable fluid or paste consistency.

Example 1: 55 parts sintered polytetrafluoroethylene were comminuted to granulated particles which will pass through a mesh size of approximately 20 (Tyler Standard). The particles are combined with 45 parts high pressure, high film strength grease type lubricant having the following physical properties: specific gravity, 1.1; worked penetration, 315; melting point greater than 500°F. and waterproof. Such a grease lubricant is identical to that available commercially under the name Lubriplate 930AA. The ingredients are thoroughly mixed to produce a composition having a specific gravity of approximately 1.4 and a paste-like consistency which can be easily extruded into the packing gland or stuffing box through an opening in the exterior of such compartment. As the particles become worn to such a size that they are no longer retained in the packing gland or stuffing box due to the motion of the reciprocating or revolving shaft, it is only necessary to extrude into the box or gland additional composition having the requisite particle size through the external opening in order to replenish and refill the box or gland. This operation can be performed without interrupting the operation.

This composition and method for packing glands and stuffing boxes affords continuous maintenance for the life of the shaft for temperatures between -450° and +500°F. and pressures less than 20,000 psi. A composition including the sintered copolymer of tetrafluoroethylene and hexafluoropropylene instead of sintered polytetrafluoroethylene is operable over the temperature range between -450° and +400°F. Utilization of sintered polytrifluorochloroethylene or sintered polyvinylidene fluoride instead of sintered polytetrafluoroethylene in the composition affords operable temperature ranges of between -328° and +390°F. and between -328° and +300°F., respectively. While the above identified particle size is most applicable it has been found that the particle size of the sintered fluorocarbon resin can vary between 400 mesh and 2 mesh (Tyler Standard) depending upon the maximum clearance between the shaft and the end of the housing of the particular stuffing box or packing gland without substantially detracting from the process.

Example 2: 45 parts sintered polytetrafluoroethylene were comminuted into flaked particles having a thickness no greater than 0.01 inch and a surface area no greater than 0.1 square inch. The flakes were then combined with 55 parts high pressure, high film strength grease lubricant, similar to that in Example 1. The ingredients were admixed to form a material having a specific gravity of approximately 1.2 and a paste-like consistency which can be easily extruded into the packing gland or stuffing box. The composition is introduced and maintained in the gland or box in a manner similar to Example 1. While the above identified flaked particle size has been found most desirable due to the acceptable tolerances between the shaft and the end of the housing of a packing gland or stuffing box for conventional pumps and the like, where the clearance is exceptionally high, it has been found satisfactory in connection with the process to increase the flake size up to 1/4 of an inch in thickness and up to 1 aquare inch in surface area.

J.E. Oliver, Jr.; U.S. Patent 3,526,593; September 1, 1970; assigned to Chevron Research Corporation has found that an effective high pressure pipe thread dope is provided in the composition consisting essentially of an oil of lubricating viscosity, from about 0.5 to 10% by weight of a calcium salt of a fatty acid containing from 10 to 22 carbon atoms, from about 1 to 10% by weight of calcium acetate, from about 0.1 to 5% by weight of base expressed as calcium hydroxide, and from 33 to 80% by weight of ground polytetrafluoroethylene resin having an average particle diameter of 0.0005 to 0.05 inch.

Talc-Containing Formulations

D.W. Van Winkle and F.M. Carlisle; U.S. Patent 3,437,630; April 8, 1969; assigned to Baker Oil Tools Incorporated describe a thread sealing composition consisting essentially of a substantial proportion of a grease, a minor proportion of a solid, deformable plastic substance, preferably tetrafluoroethylene resin, and a minor proportion of a pulverulent mineral substance, preferably talc. The composition improved pressure sealing and lubricating characteristics and can be employed in both oil well applications and food processing.

The following are examples of representative grease-like vehicles which can be employed in the thread sealing compositions of the process.

Composition 1

Components:	
Aluminum stearate base grease containing mineral oil and aluminum stearate, percent by weight	35.9
Graphite, percent by weight	18.0
Lead powder, percent by weight	30.5
Copper flake, percent by weight	3.3
Zinc dust, percent by weight	12.3
Worked penetration at 77°F.	310-340
Dropping point, °F. minimum	190

Composition 2

Components:	
Mineral oil, percent by weight	63
Calcium acetate soap stock, percent by weight	26
Graphite, percent by weight	8
Basic lead carbonate (white lead) percent by weight	3
Viscosity of the mineral oil at 100°F., SSU	1500
Worked penetration at 77°F.	450-480
Dropping point, °F.	500

Composition 3

Mineral oil containing microgel thickener, %	100
Worked penetration at 77°F.	280
Viscosity of the mineral oil at 100°F., SSU	540

Composition 4

Components:	
Mineral oil, percent by weight	78
Calcium acetate soap stock, percent by weight	22
Viscosity of the mineral oil at 100°F., SSU	350
Worked penetration	410-425
Dropping point, °F.	500

Composition 1 is known as API modified thread compound; Composition 2 is marketed as Chevron Drill Pipe Lubricant; Composition 3 is marketed as Shell Darina; and Composition 4 is marketed as Chevron Duraplex W-1, and is a white grease. Note that Compositions 1 and 2 contain graphite, and the Compositions 3 and 4 are graphite-free greases.

The plastic filler substances incorporated in the abovedescribed grease-like vehicles for producing the thread compositions are solid, deformable plastic materials, preferably thermoplastic materials, and which are preferably employed in granular or flake form. The addition of such plastic granules or particles to the grease-like vehicle improves the sealing and lubricating properties of the vehicle. The plastic granular material should be capable of remaining in suspension in the grease-like vehicle, and preferably should have a compression strength of about 1,000 to 20,000 psi, and preferably a modulus of elasticity in tension of about 50,000 to 500,000 psi. Illustrative types of such plastic fillers which may be employed include fluorocarbon resins such as the tetrafluoroethylene resins represented by Teflon, and trifluorochloroethylene resins represented by Kel-F, polyamide resins such as nylon, rubbers, e.g., butadiene-acrylonitrile copolymers as represented by Hycar, polyethylene and polypropylene resins, styrene polymers and vinyl polymers, such as polyvinyl chloride. Preferred additives of this type include Teflon, rubber, particularly Hycar, and nylon.

The particle size of the deformable plastic granules, as well as the distribution of the various sized particles in the filler mix, affects the proper sealing ability of the thread sealing composition. Those particles in the granular plastic filler mix which are of smaller mesh size, when present in sufficient quantity and distribution in the vehicle, guarantee the sufficiency of enough plastic material so that it will act not only as a filler, but also as a lubricating agent. Thus, for example, the plastic granule particles can be of a mesh size not substantially smaller than about 200 mesh, and ranging from 20 to 200 mesh, the major portion of the plastic granules being of a size through 20 mesh and on 80 mesh (-20 to +80), at least about 75% of the granules having the last-mentioned mesh size range, and a substantial portion of the granular plastic particles, for example, in the range of about 15 to 50% by weight of the plastic particles, having a mesh size through 20 and on 40 mesh (-20 to +40).

It has been found that the granular plastic material, e.g., Teflon, having a particle gradation or mix of the proportions listed below is suitable.

	Percent by Weight Passing	Size Range	Percent by Weight
Mesh Size:			
#20	99.6	Under #20	0.4
#40	69.1	#20-#40	30.5
#80	7.8	#40-#80	61.3
#100	4.9	#80-#100	2.9
#140	2.0	#100-#140	2.9
#200	0.9	Over #200	0.9

The following are examples of the process.

Example 1: A series of 1/2" -14 thread standard type nipples and couplings are tested employing the thread sealing compositions noted below. The crests of the threads of these nipples are flattened to simulate worn threads. The respective nipples and couplings are screwed together and pressure tested at a make-up torque of 200 pound-inches, with water at ambient temperature as the testing fluid, using Compositions A and B below as thread sealing compositions on various test nipple-coupling sets.

Composition A

Chevron Duraplex graphite-free grease (Composition 4), percent by weight	83.5
Teflon granules (screen analysis substantially as shown in the table above) percent by weight	16.5
	100.0

Composition B

Chevron Duraplex graphite-free grease (composition 4), percent by weight	84
Teflon, percent by weight	8
Talc (passing a 325 mesh screen) percent by weight	8
	100

The nipples treated with Composition A above leak at about 4,900 psi average pressure and the nipples treated with Composition B commence to leak at about 5,400 pounds average pressure. Thus, it will be seen that according to the process, the substitution of talc in Composition B for a substantial portion of the Teflon in Composition A, and forming a mixture of the talc and Teflon in suspension in the grease as in Composition B, results in improved sealing of the threads as indicated by the higher leakage pressure for the nipples treated with Composition B as compared to Composition A. Note that also the substitution of talc for a substantial proportion of the Teflon in Composition A substantially reduces the cost of the thread sealing composition.

Example 2: Pressure testing of another set of the same type of nipples and couplings noted in Example 1 is carried out, testing these nipples and couplings for pressure leak after make-up at 200 pound-inches of torque, using Composition C below as a thread sealant.

Composition C

Chevron Duraplex graphite-free grease (Composition 4), percent by weight	77
Teflon, percent by weight	8
Talc (particle size passing a 325 mesh screen), percent by weight	15
	100

Pressure testing of the coupling-nipple combinations made up with thread sealing Composition C, using water as in Example 1, results in a commencing of leakage at the joint at about 6,500 psi average pressure. Thus, comparing the results employing Composition C as compared to process Composition B in Example 1, it is seen that by increasing the amount of talc to 15% in Composition C as compared to about 8% in Composition B, significantly improved pressure sealing is obtained.

Example 3: Pressure testing of the same types of nipple-coupling combinations as in Example 1 and under the same conditions is carried out using Compositions D and E below on the threads of the respective nipple-coupling sets.

	Percent by Weight	
	Composition D	Composition E
Chevron Duraplex W-1	67	54
Teflon	13	13
Talc (passing a 325 mesh screen)	20	33
	100	100

In these tests, the nipples treated with thread sealing Composition D commence to leak at about 7,900 psi average pressure, and the nipple threads treated with Composition E commenced to leak at about 8,800 psi average pressure. Thus, it is seen, comparing the results using Compositions D and E of the process containing 20 and 33% talc, respectively, as compared to Composition C of Example 2, containing 15% talc, that by increasing the percentage of talc in the composition including the grease and Teflon, pressure sealing results are substantially improved. Also, comparing the results using Composition E as compared to Composition D, it is seen that the use of 33% talc in E as compared to only 20% in Composition D, the amount of Teflon being substantially the same in both, results in substantially increased pressure sealing.

Compositions B, C, D and E of Examples 1 through 3, employing a white graphite-free grease and white talc accordingly are all "white" thread sealing compositions which have utility not only in oil well applications but also in the food processing industry.

PETROLEUM

VISCOSITY INDEX IMPROVERS AND DISPERSANTS

Syndiotactic Alkyl Polymethacrylates

A process described by J.E. Fields and E.H. Mottus; U.S. Patent 3,304,260; February 14, 1967; assigned to Monsanto Company involves highly syndiotactic oil-soluble methacrylate polymers having superior viscosity index improving properties when incorporated in mineral and synthetic oils (functional fluids). These polymers can also have dispersancy properties by incorporating a minor amount of an appropriate comonomer during polymerization to give the polymer dispersancy properties or by treatment after the formation of the polymer.

Conventional methacrylate polymers useful as viscosity index improvers in functional fluids are described in U.S. Patent 2,091,627. These polymers are made by free radical polymerization at about 70°F. or above. It has been found that the higher the degree of syndiotacticity in methacrylate polymers the better will be the viscosity index improving characteristics of the polymer. Improved oil-soluble methacrylates having relative syndiotacticity in excess of 87 have been made, which polymers have improved viscosity index improving qualities as compared to the conventional methacrylate polymers. It is preferred that the relative syndiotacticity of the polymers be at least 90.

The polymers are made by polymerizing alkyl methacrylate having from about 6 to 20 carbon atoms in the alkyl group by free radical polymerization at low temperatures of about 30°C. or lower, preferably 0°C. or lower. Alternatively, alkyl methacrylates having from about 8 to 20 carbon atoms in the alkyl group can be copolymerized with C_1 to C_4 alkyl methacrylates with the amount of the lower alkyl methacrylate in the copolymer being limited to insure that the copolymer is oil soluble, e.g., when a 50/50 molar mixture of methyl methacrylate and lauryl methacrylate is polymerized the resulting polymer is insoluble in mineral lubricating oil; whereas, if the mixture is 35/65 molar of methyl methacrylate and lauryl methacrylate the resulting polymer is soluble in mineral lubricating oil.

In general, it can be said that the average number of carbon atoms in alkyl groups of the methacrylate monomers for homopolymers as well as polymers of mixtures of monomers should be at least about 7.5 for the polymer to be soluble in mineral lubricating oil to the extent of at least 1% by weight; whereas, if the oil is a mineral or synthetic hydraulic oil having an appreciably lower viscosity than mineral lubricating oil, the average number of carbon atoms in the alkyl groups can be as low as about 6.0 to 6.5 depending on the particular oil being used. Preferred polymers have an average number of from 6.0 to 18 carbon atoms in the alkyl groups.

The polymers can also be provided with dispersancy characteristics by polymerizing with the methacrylate monomers a minor amount of polar or hydrophilic monomers. Polar groups are groups such as OH, COOH, NH_2, NHR, NRR where the R's are hydrocarbon radicals which can be either substituted or unsubstituted, preferably alkyl group having up to 4 carbon atoms or cyclohexyl groups. An illustrative listing of such polar monomers is as follows: vinyloxyethanol, dimethylaminoethyl methacrylate, diethylaminopropyl methacrylamide, N-vinylpyridine, N-vinyl-2-pyrrolidone and methacrylic acid.

The polymers are normally incorporated in mineral or synthetic oils to the extent of at least 0.1% by weight based on the composition of the polymer in oil and normally in an amount not in excess of about 15%, except where concentrates are prepared to be blended back with oil, in which case the polymer in the concentrate may be from 20 to 50% or higher. For use as viscosity index improvers usually the polymers will normally be blended in the oils to the extent of from about 1 to 7% by weight based on the oil composition, but in case the polymer is also a dispersant it can be desirable to use higher than 7%. The following examples illustrate the process.

Example 1: This example describes the making of a 90/10 lauryl methacrylate/methyl methacrylate polymer. To a glass reaction flask was added 22.86 g. (0.09 mol) of lauryl methacrylate, 1.0 g. (0.01 mol) of methyl methacrylate and 20 ml. of dry toluene. The monomers and toluene were mixed and cooled in an ice bath to 0°C. The reaction mixture was blanketed with nitrogen and 0.20 ml. of triethylboron was added. The nitrogen was then displaced with air and the reaction mixture stirred occasionally with the reaction being allowed to run at 0°C. overnight. The product polymer was precipitated in methanol, dissolved in benzene, reprecipitated from methanol, and this sequence of purification steps was repeated once again. The purified polymer was then dried at 50°C. in a vacuum oven overnight. Dry weight of the purified polymer was 18.0 g. An elemental analysis of a sample of this polymer yielded the following results. Found: C, 74.1%; H, 11.4%. Calculated: C, 74.9%; H, 11.7%.

Example 2: This example describes the making of an 80/20 lauryl methacrylate/methyl methacrylate polymer. To the reaction vessel was charged 20.32 g. (0.08 mol) of lauryl methacrylate, 2.0 g. (0.02 mol) of methyl methacrylate and 20 ml. of dry toluene. This experiment was run by exactly the same procedure as the experiment in Example 1 with the same amount of triethylboron catalyst being added. The yield of purified dry polymer was 11.1 g. An elemental analysis of a sample of this polymer yielded the following results. Found: C, 73.4%; H, 11.4%. Calculated: C, 74.1%; H, 11.5%.

Example 3: This examples describes the making of a 65/35 lauryl methacrylate/methyl methacrylate copolymer. This experiment was run in the same manner as was Example 1 except that the charge of monomers to the reactor was 16.51 g. (0.065 mol) of lauryl methacrylate and 3.5 g. (0.035 mol) of methyl methacrylate. A sample of 20 ml. of toluene was charged along with the monomer and the same amount of catalyst was added later as in Example 1. Yield of purified dried polymer was 11.5 g. An elemental analysis of a sample of this polymer yielded the following results. Found: C, 72.2%; H, 11.0%. Calculated: C, 72.8%; H, 11.2%.

Example 4: This example describes the preparation of an isotactic polymer from tallow methacrylate, which is a mixture of about 33% by weight of C_{16} and about 67% by weight of C_{18} straight chain alkyl methacrylates. To the reaction vessel was charged 200 ml. of toluene at 0°C. and nitrogen blanketing was applied. To this toluene in the reaction vessel was added the Grignard catalyst which was 3 ml. of 3 molar phenyl magnesium bromide in ethyl ether. The tallow methacrylate monomer was purified by percolation over alumina and a sample of this purified material 31.2 g. was added to the reaction vessel over a 24 minute period at a rate to keep the temperature between 0° and 5°C. After four hours, during which time the reaction mixture was maintained at about 0°C., the reaction mixture was poured into methanol. The crude polymer was recovered from the methanol with this purification procedure being repeated two more times. The purified polymer was then dried in a vacuum oven for 58 hours. Yield of dried polymer was 22.8 g. This polymer is a solid at room temperature.

Example 5: This example describes the preparation of a sample of conventional tallow methacrylate polymer. To a Coke bottle was charged 65.8 g. of tallow methacrylate, 44 g. of benzene and 0.082 g. of azobisisobutyronitrile catalyst. The bottle was then flushed with nitrogen and sealed. This bottle was placed in a rotating air oven to facilitate agitating reactants and maintained at 70°C. for a period of 65 hours. The polymer product was purified in a manner similar to that described in Example 4. Yield of dried polymer was 65.5 g.

Different mineral base oils were used in making up and testing the additive compositions of the process. These base oils are as follows.

Base Oil No. 1

This is a solvent refined Mid-Continent petroleum lubricating oil having the following properties:

Viscosity at 210°F., centistokes	3.94
Viscosity at 100°F., centistokes	21.58
Viscosity index	75.3
Specific gravity 25°/25°C.	0.886
Flash point, Cleveland open cup, °F.	375

Base Oil No. 2

This is a solvent refined Mid-Continent petroleum lubricating oil having the following properties:

Viscosity at 210°F., centistokes	10.39
Viscosity at 100°F., centistokes	91.73
Viscosity index	103.4
Flash point, Cleveland open cup, °F.	450

The table on the following page contains a summary of Examples 1 through 5 and the viscosity index improving characteristics of the polymers of these examples.

Ex. No.	Monomers [1]	Monomer, Mole Ratio	Type of Polymer	Sp. Visc. in Benzene at 1% Conc. and 25° C.	Sp. Visc.[2] at 100° F.	Sp. Visc.[2] at 210° F.	Efficiency [2]
1	LM/MM	90/10	Syndiotactic		0.299	0.415	1.39
2	LM/MM	80/20	do		0.540	0.777	1.44
3	LM/MM	65/35	do		0.172	0.308	1.79
4	TM		Isotactic	0.315	0.483	0.519	1.08
5	TM		Conventional		1.793	2.052	1.14

[1] LM—Lauryl methacrylate. MM—Methyl methacrylate. TM—Tallow methacrylate.

[2] These specific viscosities and efficiencies were determined from a 30% concentrate of the polymer in Base Oil No. 1 diluted to 3% polymer in Base Oil No. 2.

In general, the "Type of Polymer" has been determined qualitatively by the method of making the polymer, but also a number of the polymers have been subjected to infrared analysis. Isotactic or syndiotactic polymers have their own characteristic infrared patterns. Efficiency is a measure of the viscosity-temperature coefficient of a polymer in the solvent in which the measurements are made.

$$\text{Efficiency} = \frac{\text{specific viscosity at 210°F.}}{\text{specific viscosity at 100°F.}}$$

In general, the higher the efficiency the higher is the viscosity index.

Ethylene-Lauryl Methacrylate-Hydroxyethyl Methacrylate Copolymers

R.J.A. Eckert; U.S. Patent 3,518,197; June 30, 1970; assigned to Shell Oil Company describes lubricant compositions containing one or more lubricants and polymeric compounds which have been obtained by copolymerization of one or more olefins containing 2 to 6 carbon atoms per molecule with one or more alkyl esters of unsaturated carboxylic acids and with one or more esters of unsaturated carboxylic acids containing at least one free hydroxyl group located in the part of the molecule derived from the alcohol.

By an appropriate choice of the molar ratio of the olefins to the esters of the unsaturated carboxylic acids in the mixture to be polymerized, and by the use of alkyl esters with a suitable alkyl chain length, one may obtain thermally stable copolymers which are not only active as dispersants and viscosity index improvers and/or thickeners but are also capable of depressing the pour point of the lubricating oils into which they have been incorporated.

The preparation of these polymers may be accomplished in two ways. The polymers may be prepared directly by copolymerization of a monomer mixture consisting of one or more olefins with 2 to 6 carbon atoms per molecule, one or more alkyl esters of unsaturated carboxylic acids and one or more esters of unsaturated carboxylic acids which contain at least one free hydroxyl group located in the part of the molecule derived from the alcohol. However, the polymers are preferably prepared by an indirect route comprising copolymerization of a monomer mixture consisting of one or more olefins having 2 to 6 carbon atoms per molecule, one or more unsaturated carboxylic acids and one or more alkyl esters of unsaturated carboxylic acids, followed by aftertreatment of the acid copolymers so obtained in which the carboxyl groups are converted into —COOR groups, in which R represents a monovalent alkyl radical containing at least one free hydroxyl group.

Examples of suitable monoolefins are ethylene, propylene, butene and pentene. Examples of suitable diolefins are, among others, butadiene, isoprene and dimethylbutadiene. Examples of suitable alkyl esters of unsaturated monobasic carboxylic acids which may be mentioned are methyl methacrylate, isobutyl acrylate, lauryl methacrylate and stearyl methacrylate. Some examples of suitable alkyl esters of unsaturated polybasic carboxylic acids are diethyl maleate, dioctyl fumarate, and di(2-ethylhexyl)itaconate. Preference is given to polymers prepared with the aid of alkyl esters of monobasic carboxylic acids where the alkyl group contains 8 or more carbon atoms, in particular lauryl methacrylate or a mixture of lauryl and stearyl methacrylate.

The esters of unsaturated carboxylic acids, where the part of the molecule derived from the alcohol contains at least one free hydroxyl group, may be esters of both monobasic carboxylic acids and polybasic carboxylic acids. Examples of such esters are partial esters of polyvalent alcohols, such as glycol, glycerol, trimethylol propane, pentaerythritol and p-dimethylol benzene, with monobasic carboxylic acids, such as acrylic acid and methacrylic acid, or with polybasic carboxylic acids, such as maleic acid, fumaric acid and itaconic acid, in which partial esters part of the hydroxyl groups originally present in the polyvalent alcohols remain as such. Preference is given to copolymers which have been derived from one or more hydroxyalkyl esters of unsaturated monocarboxylic acids, in particular from β-hydroxyethyl methacrylate.

The aftertreatment of the copolymers containing free carboxyl groups may be carried out in different ways. What is important is that the carboxyl groups are converted into —COOR groups, in which R represents a monovalent alkyl radical containing at least one free hydroxyl group. For example, the copolymers containing free carboxyl groups may be caused to react with polyvalent alcohols such as glycol, glycerol, trimethylol propane, pentaerythritol or p-dimethylol benzene. The aftertreatment is preferably carried out by reacting the copolymers containing free carboxyl groups with cyclic ethers where the ring contains an oxygen atom and two or three carbon atoms, i.e., ethylene oxide and trimethylene oxide. Ethylene oxide is especially preferred.

The process relates in particular to lubricant compositions containing polymeric compounds which have been obtained by the copolymerization of ethylene with one or more alkyl esters of C_3 to C_4 unsaturated monocarboxylic acids, in which the alkyl group contains 8 to 20 carbon atoms, and with one or more hydroxyalkyl esters of C_3 to C_4 unsaturated monocarboxylic acids in which the hydroxyalkyl group contains 2 to 3 carbon atoms. Favorable results were obtained by application of the following copolymers of this type in a mineral lubricating oil: (1) ethylene-lauryl methacrylate-β-hydroxyethyl methacrylate copolymers; (2) ethylene-lauryl methacrylate-stearyl methacrylate-β-hydroxyethyl methacrylate copolymers. The following examples illustrate the process.

Example 1: Five copolymers of ethylene, methacrylic acid stearyl methacrylate and lauryl methacrylate and four copolymers of ethylene, methacrylic acid and lauryl methacrylate (designated as polymers 1 through 9) were prepared as follows.

A specified mixture of methacrylic esters, methacrylic acid, azodiisobutyronitrile, 300 g. of tert-butanol and 100 g. of benzene was passed into a 2 liter autoclave. After the air had been expelled from the autoclave with the aid of pure nitrogen, ethylene was introduced into the autoclave at room temperature under a pressure of up to 170 to 200 atm. The mixture was heated for a specified period, during which the pressure increased to 380 to 480 atm. After the pressure had been released, the mixture was discharged and concentrated by partial evaporation of the solvent. The polymer was purified by precipitating three times in methanol. Finally the polymer was dissolved in benzene, the solution filtered and the polymer separated by freeze drying.

Further details on the copolymerizations described above and the resulting acid copolymers are shown in the tables below.

Polymer No.	1	2	3	4	5	6	7	8	9
Quantity of lauryl methacrylate in the mixture to be polymerized, grams	22.8	22.8	22.8	22.8	22.8	22.8	45.3	24.4	15.4
Quantity of stearyl methacrylate in the mixture to be polymerized, grams	30.4	30.4	30.4	30.4	30.4	---	---	---	---
Quantity of methacrylic acid in the mixture to be polymerized, grams	4.7	4.7	4.7	6.3	6.3	3.0	6.1	2.5	1.6
Quantity of azodiisobutyronitrile in the mixture to be polymerized, grams	0.2	0.2	0.2	0.2	0.2	0.12	0.2	0.2	0.15
Duration of copolymerization, hours	2	4.5		6	4	4	3.5	4	6
Temperature during copolymerization, °C	75	65	65	65	65	65	65	65	65
Yield of copolymer, grams	19.9	19.3	29.4	21 8	13.2	12.9	12.6	8.6	9.1
Average molecular weight of the copolymer ($\overline{M}_n$)			57,000	87,000	---	55,000	---	50,000	45,500

Polymer	Molar ratio of the monomers in the acid copolymer: Lauryl methacrylate	Stearyl methacrylate	Methacrylic acid	Ethylene
1	5	5	3.1	22
2	5	5	2.8	20
3	5	5	3.0	21
4	5	5	3.7	19
5	5	5	3.9	12
6	10	---	2.6	30
7	10	---	3.7	19
8	10	---	2.9	27
9	10	---	2.7	44

Example 2: The acid copolymers (1 through 9) from Example 1 were converted into polymers I through IX as follows. To a solution composed of 20 g. of the acid copolymer and 48 mg. of lithium hydroxide in a solvent mixture comprising 9 g. of toluene and 71 g. of methanol, ethylene oxide was added at room temperature. The solution was kept at 95°C. for 24 hours and subsequently cooled down to room temperature. The ethoxylated polymer was purified by precipitating two times in methanol. Further details on the ethoxylation described above are noted below.

Polymer No.	I	II	III	IV	V	VI	VII	VIII	IX
Quantity of ethylene oxide added, grams	2.1	2.0	2.1	2.7	2.7	2.5	2.9	2.2	1.8
Yield of ethoxylated copolymer, grams	11.4	18.0	17.7	19.2	15.2	18.5	18.2	19.4	17.4

Example 3: In order to assess the viscometric behavior of the polymers I through IX, they were dissolved in a concentration of 1.5% by weight in a paraffinic base oil. The results of the viscometric determinations are listed below.

	V_k 100° F., cs.	V_k 210° F., cs.	VI	VT
Base oil	57.82	7.38	95	------
Base oil plus 1.5% polymer I	71.03	9.41	116	132
Base oil plus 1.5% polymer II	73.02	9.82	120	137
Base oil plus 1.5% polymer III	80.51	10.68	121	116
Base oil plus 1.5% polymer IV	76.90	10.93	128	148
Base oil plus 1.5% polymer V	71.65	10.43	129	177
Base oil plus 1.5% polymer VI	72.93	11.08	134	196
Base oil plus 1.5% polymer VII	64.33	9.31	125	221
Base oil plus 1.5% polymer VIII	67.20	8.96	115	136
Base oil plus 1.5% polymer IX	76.05	9.96	117	113

These results definitely show the advantages of the polymers of the process with regards to improvement in viscosity index (VI) of a lubricating oil composition.

Example 4: The thermal stabilities of the polymers according to the process are recorded below. They were determined in a vacuum at a rate of heating of 3°C./minute. The values T_{10}, T_{20} and T_{50} represent the temperatures at which 10, 20 and 50% weight loss, respectively, of the polymer was attained. To test the polymers as dispersants they were subjected to a peptization test. In this test the lowest concentration of additive is determined that is capable of keeping 0.015% by weight of carbon in mineral oil in suspension for 15 minutes at 250°C. Concentrations below 0.008 are considered favorable. The results of these peptization tests are also listed below. The results of the peptization tests (in all cases <0.005% by weight) may be considered very favorable.

Polymer	T_{10}	T_{20}	T_{50}	Peptization test, percent by weight
I	330	355	390	0.001-0.003
II	325	350	380	0.001-0.003
III	350	375	405	0.004
IV	345	360	390	0.001-0.003
V	335	360	390	0.003
VI	360	385	420	0.001-0.003
VII	340	365	395	0.001-0.003
VIII	340	365	395	0.005
IX	340	365	395	0.005

From the above it is evident that the polymers impart dispersancy and viscosity index characteristics to lubricating oil as well as possessing excellent thermal stability.

Tetrahydrofurfuryl Methacrylate Copolymers

E.H. Mottus and J.E. Fields; U.S. Patent 3,321,405; May 23, 1967; assigned to Monsanto Company describe a process for providing oil-soluble tetrahydrofurfuryl methacrylate polymers which are useful as oil additives and which have improved viscosity-stability in the presence of heavy duty barium detergent additives.

In this process the alkyl methacrylate monomers are polymerized for a sufficient length of time prior to the addition of the tetrahydrofurfuryl methacrylate monomer that the final product is stable in the presence of heavy duty detergents, such as barium detergents. The time for polymerizing the alkyl methacrylate monomers before adding the tetrahydrofurfuryl methacrylate monomer will vary somewhat depending on the particular alkyl methacrylate monomers being used and the amount and method of addition of the tetrahydrofurfuryl methacrylate monomer; however, normally the alkyl methacrylate monomers are polymerized to incorporate not more than about 90 mol percent of the alkyl methacrylate monomers in the polymer before adding the tetrahydrofurfuryl methacrylate monomer, and usually from about 40 to 75 mol percent of the alkyl methacrylates will be incorporated into the polymer before the addition of the tetrahydrofurfuryl methacrylate monomer. Then, the polymerization is continued to completion causing substantially all of the monomers to be incorporated into the polymer.

A free radical type catalyst is used as is customarily used for alkyl methacrylate polymerization, and reaction conditions of temperature and pressure are those customarily used. The monomer 5-methyl tetrahydrofurfuryl methacrylate can be used instead of tetrahydrofurfuryl methacrylate or mixtures can be used. The polymer additives made by the process are the tetrahydrofurfuryl methacrylate and alkyl methacrylate polymers containing a minor amount of tetrahydrofurfuryl methacrylate and a major amount of alkyl methacrylates. The following examples illustrate the process.

Example 1: The apparatus used for making this run was a 4 necked glass flask fitted with a condenser, a stirrer, a thermometer and a Y-fitting to take two dropping funnels. In the stirrer motor circuit are ammeters to determine the power requirements. Ammeter readings are observed to follow the increase in viscosity in the polymerization mixture. The apparatus is flushed with nitrogen both before and during polymerization.

To the flask was charged 64.6 g. of tridecyl methacrylate, 46.8 g. of tallow methacrylate, 8.5 g. of methyl methacrylate and 6.2 g. of tetrahydrofurfuryl methacrylate. A sample of 25 ml. of the monomer mixture was removed from the flask, and to this was added 0.25 g. of benzoyl peroxide catalyst. The catalyst was dissolved in the monomers and added to a dropping funnel. The polymerization was carried out at about 95°C. At the beginning of the run, 10 ml. of the monomer catalyst mixture was added to the flask. Periodically during the run, 10 ml. amounts of base oil No. 1 was added to the reaction mixture to reduce the viscosity. At the end of one hour of polymerization, 3 ml. of catalyst monomer mixture was added to the reaction flask, 3 ml. more at the end of 1 3/4 hours, 3 ml. more at the end of 2 1/2 hours, 3 ml. more at the end of 3 1/4 hours, and the balance of the catalyst monomer mixture was added to the reaction mixture at the end of 4 hours.

During the polymerization run a total of about 80 ml. of base oil No. 1 was added to the reaction mixture in 10 ml. amounts. At the end of the polymerization run, an additional 120 ml. of base oil No. 1 was added to the reaction mixture. Additional base oil No. 1 was added to the reaction mixture to provide a 37% concentrate of polymer in oil. To the concentrate was added 1.25 g. of an antioxidant which was 4,4'-methylene-bis(2,6-di-tert-butylphenol), and the antioxidant was incorporated into the concentrate by mixing and heating to 100° to 110°C./100 mm. A sample of the 37% concentrate was diluted with base oil No. 2 to 3% polymer concentration and viscosity determinations were made of the 3% polymer in oil. Specific viscosity at 100°F. was 0.858 and at 210°F. was 1.089. Efficiency was 1,269. Efficiency is a measure of the viscosity-temperature coefficient of a polymer in the solvent in which the measurements are made.

$$\text{Efficiency} = \frac{\text{specific viscosity at 210°F.}}{\text{specific viscosity at 100°F.}}$$

In general, the higher the efficiency the higher is the viscosity index.

Two different mineral base oils were used in making up and testing the additive compositions of Example 1, and these base oils were as follows.

Base Oil No. 1

This oil is a highly paraffinic refined petroleum lubricating oil having the following properties:

Viscosity at 100°F., SUS	113
Viscosity at 210°F., SUS	40
Flash point, °F.	380
Specific gravity at 60°F.	0.8639

Base Oil No. 2

This oil is a solvent refined Mid-Continent petroleum lubricating oil having the following properties:

Viscosity at 210°F., centistokes	10.39
Viscosity at 100°F., centistokes	91.73
Viscosity index	103.4
Flash point, Cleveland open cup, °F.	450

Example 2: This example was carried out in a similar manner to Example 1, except that the tetrahydrofurfuryl methacrylate monomer was not added to the flask as were the other monomers at the beginning of the polymerization run. After 1 3/4 hours of polymerization, one-half of the tetrahydrofurfuryl methacrylate monomer was added in base oil No. 1 to the reaction mixture, and after 2 1/2 hours of polymerization the balance of the tetrahydrofurfuryl methacrylate monomer was added in base oil No. 1 to the polymerization mixture. Viscosity and efficiency determinations were made on the product of Example 2 in a similar manner as described for the product of Example 1.

Examples 3 and 4: Except for the tetrahydrofurfuryl methacrylate monomer, all of the monomers were charged to the flask at the beginning of the polymerization. In addition to the other monomers of Example 1, a small amount of hydroxypropyl methacrylate was used. After 1 3/4 hours of polymerization, one-half of the tetrahydrofurfuryl methacrylate monomer was added to the polymerization mixture in base oil No. 1, and after 2 1/2 hours of polymerization the balance of the tetrahydrofurfuryl methacrylate monomer was added to the polymerization mixture in base oil No. 1. At the end of the polymerization run sufficient base oil No. 1 was added to make a 37% concentrate of the polymer. Other than the exceptions noted above, this run was carried out in a similar manner to Example 1. The final concentrate product was divided into two parts and to one part of the concentrate was added an ashless detergent in minor amount, and this sample with the ashless detergent in it was labeled Example 4.

Examples 5 and 6: In this example the hydroxypropyl methacrylate rather than the tetrahydrofurfuryl methacrylate was added late in the polymerization reaction. All of the monomers except the hydroxypropyl methacrylate were charged to the polymerization vessel at the beginning of the polymerization period. After 1 3/4 hours of polymerization, one-half of the hydroxypropyl methacrylate dissolved in base oil No. 1 was added to the polymerization mixture, and after 2 1/2 hours of polymerization, the balance of the hydroxypropyl methacrylate was added to the polymerization mixture. At the end of the polymerization a 37% concentrate was made by adding additional base oil No. 1. Other than the exceptions noted above, Example 5 was conducted in the usual fashion of Example 1. The concentrate was divided in two portions, and to one portion was added a minor amount of the anitoxidant 4,4'-methylene-bis(2,6-di-tert-butylphenol). The antioxidant was incorporated into the concentrate in the usual fashion, and the sample with the antioxidant in it was labeled Example 6.

Examples 7 and 8: Example 7 was a run where all of the monomers except the tetrahydrofurfuryl methacrylate and methyl methacrylate were charged to the reaction flask at the beginning of the polymerization run. One-half of the tetrahydrofurfuryl methacrylate monomer and one-half of the methyl methacrylate monomer were added to the polymerization mixture in base oil No. 1 after one hour of polymerization time and the other half of these two monomers was added to the polymerization mixture after 1 3/4 hours of polymerization time. Other than this, Example 7 was carried out in the usual batch fashion. At the end of the polymerization run a 37% concentrate was made of the product by adding additional base oil No. 1. The concentrate product was divided in two parts, and to one part was added a minor amount of ashless detergent. The portion containing the ashless detergent was designated Example 8.

Examples 9 and 10: This example was a typical polymerization run where all of the monomers were charged to the reaction flask at the beginning of the polymerization run with the exception of methyl methacrylate and hydroxypropyl methacrylate. The methyl methacrylate and hydroxypropyl methacrylate monomers were dissolved in base oil No. 1, and one-half of each of these monomers was added to the polymerization reaction at the end of one hour of polymerization and the other half at the end of 1 3/4 hours of polymerization. At the end of the polymerization run the 37% concentrate of the additive was made by adding base oil No. 1. The concentrate product was divided into two portions and to one portion was added a minor amount of ashless detergent. This portion containing the ashless detergent was labeled Example 10.

In the following table are summarized the ten experimental examples.

Ex. No.	Monomers	Monomers, Mole Ratio	Specific Viscosity, at—		Efficiency	Time, Hrs.	Process Details		
			100° F.	210° F.			Temp., ° C.	Solvent	Catalyst
1	TrM/TaM/MM/THFM	48.2/27.6/16.9/7.3	0.858	1.089	1.269	6	95	Base Oil #1	Bz_2O_2
2	TrM/TaM/MM/THFM	48.2/27.6/16.9/7.3	0.968	1.119	1.156	6	95	do	Bz_2O_2
3, 4	TrM/TaM/MM/THFM/HPM	46.4/26.5/16.3/6.9/3.9	0.814	1.053	1.294	6½	95	do	Bz_2O_2
5, 6	TrM/TaM/MM/THFM/HPM	46.4/26.5/16.3/6.9/3.9	1.038	1.347	1.298	6	95	do	Bz_2O_2
7	TrM/TaM/MM/THFM/HPM	46.4/26.5/16.3/6.9/3.9	1.109	1.372	1.237	6¼	95	do	Bz_2O_2
8			1.123	1.345	1.198				
9	TrM/TaM/MM/THFM/HPM	46.4/26.5/16.3/6.9/3.9	1.056	1.281	1.213	6¼	95	do	Bz_2O_2
10			1.081	1.260	1.166				

TrM=Tridecyl methacrylate.
TaM=Tallow methacrylate.
MM=Methyl methacrylate.
THFM=Tetrahydrofurfuryl methacrylate.
HPM=Hydroxypropyl methacrylate.
Bz_2O_2=Benzoyl peroxide.

To portions of a number of additive concentrates of the process were added minor amounts of a heavy duty barium detergent comprising sulfurized alkylated phenols reacted with barium and carbonated. A summary of the effect of varying the time of addition of certain monomers in producing polymer additives is given below. Additives which are more viscosity stable in the presence of barium heavy duty detergents are produced in certain instances.

Example	Results
1	Gel.
2	Clear compatible solution.
3	Clear compatible solution.
4	Superior to Example 3.
5	Gel in 20 to 25 seconds; stiff gel at 1 minute.
6	Same as Example 5.
7	More stringy than Example 3. Hazy in 20 minutes.
8	Similar to Example 3.
9	Not as bad as Example 5. No flow at 1 minute.
10	Slow flow at 1 minute. Starts to gel at 2 minutes. Gelled at 3 minutes.

An examination of the above data indicates that the tetrahydrofurfuryl methacrylate monomer should not be added with the other monomers at the beginning of the run, but rather this particular monomer should not be added until a substantial amount of polymerization of the other monomers has taken place in order to make a polymer having improved stability in the presence of a heavy duty barium detergent.

Thus, in runs 1, 5, 6, 9 and 10, where the tetrahydrofurfuryl methacrylate monomer was added with the other monomers at the start in the polymerization, gels were formed when heavy duty barium detergent was added to the concentrates containing the polymer, and of course this is very undesirable for lubricating oil additive use. In the other runs 2, 3, 4, 7 and 8, where the tetrahydrofurfuryl monomer was added late in the polymerization, gels were not formed.

It seemed from the results that the procedure of Example 7 is not as desirable as that of Examples 2 and 3; however, the product of Example 7 was considerably more stable in the presence of the barium detergent than the examples where the tetrahydrofurfuryl methacrylate was added at the beginning of the polymerization run.

Methyl Vinyl Ether Copolymers

H.S. Schultz, W. Katzenstein and E.P. Williams; U.S. Patent 3,541,015; November 17, 1970; assigned to GAF Corp. have found that when methyl vinyl ether is copolymerized in certain weight percent proportions with certain weight percent proportions of an aryl or alkyl vinyl ether in which the alkyl contains from 3 to 30 carbon atoms, a class of copolymers are obtained which are soluble and which exhibit a combination of properties in hydrocarbon lubricating oils. In view of this solubility and combination of properties, the copolymers are ideally suited as multifunctional additives to hydrocarbon lubricating oils. When used in lubricating oils in amounts in the concentration range of 1 to 4 weight percent, the multifunctional additive copolymers impart high viscosity index, satisfactory thickening, excellent dispersancy, good shear stability, and good 0°F. properties. The multifunctional additive copolymers may be added advantageously in the same concentration range to fuel oils or jet fuels to disperse sludge.

The multifunctional additive copolymers can be made up as concentrates in a hydrocarbon lubricating oil stock in the order of 30 to 40% for use as a convenient way of incorporating or dissolving them in a base stock. The copolymers having the desired combination of properties are prepared by copolymerizing the following comonomers in the stated weight percent:

	Weight percent	Methyl vinyl ether	Weight percent
Alkyl vinyl ether:			
From 3 to 4 carbon atoms	55-80	20-45	
From 8 to 10 carbon atoms	30-65	35-70	
From 16 to 18 carbon atoms	40-70	30-60	
From 20 to 30 carbon atoms	15-70	30-85	
Phenyl vinyl ether	40-70	30-60	

The following examples illustrate the process. In evaluating the copolymeric multifunctional additives the kinematic viscosities at 100°F. and 210°F. were determined according to ASTM Procedure D 445-61. In the examples, the viscosity indices were calculated according to ASTM Procedure D 567-53. Shear stability of the copolymer containing oils by means of a sonic shear test was determined according to "Proposed Method of Test for Shear Stability of Polymer-Containing Oils," Appendix XII, page 1160, vol. 1, 38 edition, October 1961, ASTM Committee D-2 on Petroleum Products and Lubricants. The shear time used is one-half hour.

The sludge dispersancy test is that described in paper 23, pages 308-318 of Section 5 Proceedings of the Fifth World Petroleum Congress, 1959, New York, by R.M. Jolie. All of the copolymers that tested satisfactorily gave opaque suspensions or dispersions, hence all "passed" the dispersancy test. Extrapolated oil viscosities, the official procedure generally used in the trade, were determined according to the method described on pages 275-276 of the 1961 SAE Handbook.

Examples 1 through 6: The following reactions were carried out in carefully cleaned, dried and capped citrate of magnesia bottles equipped with perforated crown cap and butadiene-acrylonitrile copolymer rubber self-sealing lines. Methyl vinyl ether (purified and dried over sodium ribbon) was transferred through a closed initially evacuated system from the storage vessel into the cooled reactor bottle by way of a hypodermic needle. The other reagents were then charged into the reactors by the use of hypodermic syringe techniques.

The comonomer vinyl ether in Examples 1 and 2 is a distillation fraction of isooctyl vinyl ether made from oxo isooctyl alcohol. In Examples 3 and 4 it is 2-ethylhexyl vinyl ether, and in Examples 5 and 6 it is a fraction from the distillation of the vinyl ether made from Conoco Alfol 810 alcohol containing about 44% alkyl C_8 and 55% alkyl C_{10} vinyl ethers. Aliquots of a stock solution containing 0.0929 g. BF_3-diethyl etherate per cc 1,4-dioxane total solution was finally injected into the temperature equilibrated bottle in a 0°C. ice water bath and hand mixed immediately. The table below summarizes the reagents used in the examples. The reaction took place immediately after catalyst injection. The reactors were allowed to stand in the ice water bath for 22 hours. The reactions were quenched with a small amount of methanolic ammonia, diluted with reaction solvent, the solution filtered through a sintered glass funnel to remove the catalyst complex, and the solution pumped under vacuum several days, first at ambient temperature and then at 40° to 50°C. to remove solvent.

Example No.	Gms. MVE	Gms. comonomer	Cc. solvent	Reaction solvent	Cc. BF_3 etherate in 1,4 dioxane	Wt. percent comonomer in monomer mixture	Yield, percent	η_{inh}, 0.1% toluene
1	13.9	16	18.4	Toluene	0.259	53.5	90.5	0.925
2	14.3	16.8	19.1	Heptane	0.264	54.0	90.6	0.889
3	14.1	16.8	18.9	Toluene	0.262	54.4	100	0.506
4	15.1	17.8	20.2	Heptane	0.283	54.1	87.6	1.095
5	16.4	19.6	22.0	Toluene	0.309	54.5	100	0.506
6	19.6	23.2	26.2	Heptane	0.369	54.2	89.6	1.08

The following table summarizes the results of testing 2 weight percent solution of the copolymers in 100 second solvent refined naphthenic oil.

Example No.	Before shear: Kv.-210° F./kv.-100° F., centistokes	VI	Kv.-0° F. in centistokes extrapolated	Dispersancy	After shear: VI/VI decrease	Kv.-100° F., percent visc. decrease
1	8.32/46.47	143	1,400	Pass	136/7	
2	7.39/41.06	144	1,200	do	135/9	17.0
3	5.75/32.8	130				9.0
4	8.07/45.0	142	1,300	Pass	135/7	16.0
5	5.75/32.74	127				
6	7.70/43.47	140	1,200	Pass	136/4	11.0
Acryloid 966*	7.44/37.22	150	850	do	127/23	28.5
Base oil	4.02/22.03	80				

* Rohm & Haas.

It is to be noted that the copolymers of Examples 3 and 5, which have an inherent viscosity of 0.5, are deficient in viscosity index (VI) as compared with related copolymers of Examples 4 and 6. It is to be further noted that the shear stability of the commercially available lube oil additive Acryloid 966 is significantly poorer than the copolymers of Examples 2, 3, 4 and 6 as shown by the kv.-100°F. percent viscosity decrease.

Example 7: This polymerization was carried out in a 500 cc Morton flask appropriately equipped to efficiently stir the reactor mixture, record the temperature of reaction continuously and efficiently keep volatile reaction components condensed in the reaction vessel. A great effort was made to make the reaction system and the polymerization manipulation as anhydrous as possible. The reactor was kept under slight nitrogen pressure and all reagents were injected using specially prepared hypodermic syringes.

55 grams of purified and dried methyl vinyl ether were vaporized from a storage vessel and condensed into the cooled reactor. 65.4 grams purified and dried isooctyl vinyl ether (same as used in Examples 1 and 2) were then injected, followed by 78 cc purified and dried heptane. The reaction mixture was equilibrated with a 0°C. ice water bath. 0.946 cc solution of BF_3-etherate in 1,4-dioxane (0.102 g./cc titer) was injected to the rapidly stirring solution. Temperature rose quickly to 18.5°C. in a few minutes and slowly fell back to 6°C. in 18 minutes. The reaction was allowed to proceed overnight at mainly 0° to 2°C. The next day the viscous reaction mixture was warmed to 40°C. and then quenched with methanolic ammonia. The yield after workup was 90.6% and η_{inh} (0.1% toluene) is 0.94. The table below summarizes testing in the same lube oil stock as Example 1 at 2 weight percent concentration.

Example 8: A stirred run was carried out as in Example 7. 65 grams methyl vinyl ether were charged as before. 75.4 grams of an isooctyl vinyl ether fraction were charged followed by 98 cc heptane. 1.15 cc BF_3-diethyl etherate in 1,4-dioxane (titer is 0.0995 g. BF_3-diethyl etherate/cc solution) were charged into the rapidly stirring 0°C. solution. Temperature rose immediately to 22°C. Temperature was quickly brought back to below 5°C. A second charge of catalyst of the same size as before was added after 3 hours and 50 minutes. The reaction ran overnight and was worked up the same way as before. The yield was 92% and η_{inh} (0.1% toluene) was 0.83. The lube oil test data is noted below.

Example No.	Kv.-210° F./kv.-100° F., centistokes	VI	Kv.-0° F. in centistokes extrapolated	Dispersancy
7	7.98/44.23	144	1,500	Pass.
8	7.51/42.41	140	1,250	Do.

Examples 9, 10 and 11: The copolymerizations were carried out in the same fashion as in Examples 1 through 6. The comonomer was 2-ethyl hexyl vinyl ether. The reaction solvent was 20 volume percent toluene in heptane. The BF_3-diethyl etherate in 1,4-dioxane titer was 0.995 g./cc. The reaction bottles were equilibrated with a 0°C. ice water bath before catalyst injection. The following table summarizes the reagents in the examples.

Ex. No.	Gms. MVE	Gms. 2-EHVE	Cc. solvent	Cc. BF_3-diethyl etherate in 1,4-dioxane	Wt. percent comonomer in monomer mixture	Yield, percent	η_{inh}, 0.1% toluene
9	10.7	7.6	14.7	0.167	41.5	95.0	0.57
10	10.7	9.9	16.5	0.178	48	94.8	0.60
11	8.0	9.2	13.8	0.141	53.5	95.3	0.69

The reactions were carried out overnight in a 0°C. bath followed by a usual workup. The following table summarizes testing data in the standard 100 second solvent refined stock.

Example No.	Conc., wt. percent	Before shear: Kv.-210° F./Kv.-100° F.	VI	Kv.-0° F. in centistokes extrapolated	After shear: Dispersancy	Kv.-100° F., percent visc. decrease
9	1	5.44/28.33	139	800		
9	2	5.67/30.05	138	1,100	Passed	5.7
9	3	7.15/38.45	144	900		
10	1	4.84/27.02	112	1,000		
10	2	6.18/34.68	134	1,200	Passed	8.5
10	3	7.66/43.88	139	2,200		
11	1	5.10/28.57	119	1,100		
11	2	6.66/38.21	134	1,200	Passed	9.8
11	3	8.49/48.11	142	2,100		

Example 12: This polymerization was carried out in a fashion similar to Example 7 except that a five neck, appropriately equipped resin pot was used and the catalyst was added in two increments. 1,300 cc purified anhydrous toluene was added to the reactor by use of a special transfer vessel. The reactor was cooled to 0°C. and 508 g. purified dried 2-ethyl hexyl vinyl ether were added in a similar fashion. The 603 g. purified dried methyl vinyl ether was added as a liquid by the use of a special transfer cylinder to the reactor at 0°C. 1.30 cc BF_3-diethyl etherate in 1,4-dioxane (titer is 0.2 g./cc of BF_3-diethyl etherate in 1,4-dioxane) was charged into the rapidly stirring 0°C. solution. The temperature rose to 24°C. in 25 minutes and fell slowly to 10°C. in another hour. After a total of three hours, 2 cc more catalyst solution was injected and the temperature rose to 13°C. in 15 minutes. After a total of 22 hours with the final temperature at 4°C. the reaction was quenched with methanolic ammonia and worked up as before. The yield was 100% and η_{inh} (0.1% toluene) is 1.0.

The resulting polymer was evaluated with two commercially available lube oil additives in engine tests as shown in the table below. The viscosity index values in this table were determined by ASTM Procedure D 2270. A viscosity index value while employing ASTM Procedure D 567-53 is also given for comparison. From the data it is clearly evident that the polymer of Example 12 is not only more shear stable, but actually improves in VI performance while the two commercial products lose performance on shearing.

	Polymer of Example 12	Rohm & Haas	
		Acryloid 763	Acryloid 622
Viscosity at 210° F., centistokes [1]	10.48	10.77	10.50
Viscosity at 100° F.	61.1	60.4	55.7
Concentration, Vol. percent (33.1 wt. percent in oil for Ex.12	5.75	5.90	6.30
Viscosity index, new soln. before testing (142 by ASTM D 567-53 for Ex.12)	173	177	204
Viscosity index, after testing in dynamometer engine	178	165	184
Viscosity at 0° F., centistokes	2,090	1,810	1,784
Shear losses: Kv.-210° F., percent viscosity decrease:			
Car, high mileage [2]	12.9	17.1	20.1
Car, low mileage [3]	12.8	16.0	19.5
Dynamometer engine [4]	14.1	20.8	21.7
Shear losses: Kv.-0° F., percent viscosity decrease:			
Car, high mileage [2]	6.2	17	15
Dynamometer engine [4]	7.1	7.9	24.6

[1] Base oil viscosity at 210° F.=5.54 centistokes; 100° F.=35.7 centistokes.
[2] 1963 Ford Galaxie 289 CID V-8, 60,000 engine miles, 500 mi./test.
[3] 1966 Buick Le Sabre 340 CID V-8, 10,000 engine miles, 500 mi./test.
[4] 1960 Olds Seq. I Engine, 2,500 r.p.m. no load, 5 hrs./test.

Alkyl Aryl Ether Polymers

D.R. Wilgus; U.S. Patent 3,409,552; November 5, 1968; assigned to Chevron Research Company has found that improved viscosity indexes are provided in a lubricant composition comprising a major proportion of an oil of lubricating viscosity and a minor proportion sufficient to improve the viscosity index of the oil of the polymer of 2,6-dialkyl-1,4-phenylene ether in which one alkyl group is methyl or ethyl and the other is a straight chain alkyl radical containing from 8 to 12 carbon atoms, the polymer having a molecular weight in the range of 10,000 to 100,000. The polymers are ordinarily used in lubricating oils at concentrations from about 1 to 20% by weight. The following examples illustrate the preparation of the dialkyl phenylene ether polymers.

Example 1: o-Cresyl Pelargonate — A 250 ml. Erlenmeyer flask was charged with 79 g. (0.5 mol) pelargonic acid and 24.7 g. (0.18 mol) phosphorus trichloride. This is a 10 mol percent excess of the phosphorus trichloride. The mixture was heated and swirled intermittently on the steam plate for 45 minutes. The acid chloride was decanted from the phosphorus acid into a separatory funnel. 54 g. (0.5 mol) o-cresol was charged to a 500 ml., 3 neck flask and the acid chloride was added slowly. The temperature of the reaction was allowed to rise at will during the addition, and the mixture was stirred and heated at 300° to 330°F. for another three hours. The mixture was then purged for 1/2 hour with nitrogen and distilled under a vacuum of 0.6 mm. of mercury. 95 g. of product was obtained boiling from 122° to 129°C.

Example 2: 2-Methyl-6-Pelargonylphenol — A 1 liter flask was charged with 86 g. (0.346 mol) of o-cresyl pelargonate and 56 g. (0.42 mol) of pulverized aluminum trichloride. On intimate mixing, the temperature rose to 75°C. The flask was placed in an oil bath which was slowly heated and maintained at 160° to 170°C. for 1/2 hour. After cooling, the contents were poured into 300 cc of 20% aqueous hydrochloric acid. 500 cc of benzene were added to dissolve the product and the aqueous mixture was separated and discarded. The benzene solution was washed twice with water, dried over anhydrous sodium sulfate, and freed of solvent. The dry product was distilled and yielded 28.9 g. of the ortho isomer and 40.7 g. of the para isomer. The ortho isomer boiled at 132° to 143°C. at a pressure of 0.1 mm. of mercury. The ratio of ortho to para isomer was 2:3.

Example 3: 2-Methyl-6-Nonylphenol — 200 g. mossy zinc, 300 cc water, 5.0 g. mercuric chloride and 5 cc concentrated hydrochloric acid were stirred for 1/2 hour at room temperature. The aqueous solution was decanted and the remaining zinc was washed with water. 150 cc each of water and concentrated hydrochloric acid were then added along with a solution containing 300 cc of ethanol and 28.5 g. (0.115 mol) of 2-methyl-6-pelargonylphenol. The mixture was heated at reflux (188° to 190°F.) for a 24 hour period. The liquid phases were separated from the zinc and the aqueous portion separated from the organic phase. The aqueous portion was extracted with benzene and the benzene extract added to the organic phase. This benzene solution was water washed until neutral and the benzene was removed by distillation. The crude product was distilled and 23.8 g. recovered boiling at 124° to 126°C. at a pressure of 0.1 mm. of mercury. An infrared spectrum indicated disappearance of the carbonyl band of the original ketophenol.

Example 4: Poly-2-Methyl-6-Nonyl-1,4-Phenylene Ether — A 1 liter, 3 neck flask was equipped with a stirrer, thermometer and gas inlet tube. The flask was charged with 250 cc of benzene, 45 cc of pyridine, 0.4 g. of cuprous chloride and 11.5 g. of 2-methyl-6-nonylphenol. Oxygen was added into the gas inlet tube and the mixture was stirred at ambient temperature. The amount of oxygen used was measured and amounted to 98% of theory at the end of two hours. The mixture was filtered to remove inorganic salt and the filtrate was concentrated on the steam plate. The polymer was precipitated with methanol, dissolved in 100 cc of benzene, and reprecipitated. This procedure was repeated. No benzene-insoluble polymer was observed, and the supernatant solution of the last precipitation was clear. The polymer was dried in a vacuum oven at 100°C. for 3 hours. 9.7 g. was recovered.

The polymer of 2-methyl-6-nonyl-1,4-phenylene ether was blended with a typical solvent refined mineral lubricating oil, namely, 150 Neutral Oil, and with an aromatic synthetic oil, namely, 1-n-nonyl-4-"iso"-octylbenzene. It has a marked thickening action in the oils, and the viscosity indexes of the blended oils were markedly improved. The polymer also has excellent thermal stability as determined by subjecting the thickened oils to a temperature of 700°F. for 6 hours. The properties of the blended oils are listed in the following table.

Base oil	Percent	Pour Point, °F.	Visc. at 100° F., cs.	Visc. at 210° F., cs.	V.I.
1-n-nonyl-4-"iso"-octylbenzene	Nil	<−107	9.25	2.65	138
Do	4.0	<−90	27.6	7.03	176
Do	10.0	−80	89.23	19.64	153
Do	10.0	−92	28.30	6.69	170
Do	13.0	--------	39.7	8.77	161
Do	10.0	--------	16.9	4.15	172
Do	5.0	--------	17.5	4.50	187
Do	10.0	--------	30.2	7.36	169
150 Neutral Oil	Nil	--------	30.7	4.92	89
Do	2.8	--------	65.0	10.12	134
Do	3.5	--------	70.37	10.83	144
Do	5.0	--------	102.0	15.23	137

Alkyl Hydrocarbonthiophosphonates

A process described by J.M. Petersen and J.F. Lyons; U.S. Patent 3,296,132; January 3, 1967; assigned to Texaco Inc. involves alkyl hydrocarbonthiophosphonates of the formula

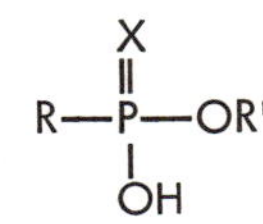

where R is a monovalent hydrocarbon derived radical (hydrocarbyl), R' is alkyl and X is a chalcogen selected from the group consisting of sulfur and a mixture of oxygen and sulfur. The alkyl hydrocarbonthiophosphonates are useful as detergent and dispersant additives in lubricant fuels, such as mineral oil, synthetic ester oil and jet fuel. They are prepared by reacting trialkyl phosphite with hydrocarbonthiophosphonic acid at an elevated temperature.

Specific examples of the alkyl hydrocarbonthiophosphonates are ethyl polybutene(940 MW)thiophosphonate, butyl polypropylene(2,500 MW)thiophosphonate, butyl polyisobutylene(1,500 MW)thiophosphonate, methyl propylene-isobutylene copolymer (2,000MW)thiophosphonate. In concentrates used in the formulation of the finished lubricants, the concentration of the alkyl hydrocarbonthiophosphonate can be as high as 50%. In finished lubricants, the concentration of additives falls between 0.2 and 10 weight percent with a concentration between 1 and 5 weight percent normally preferred. The following examples illustrate the process.

Example 1: To a 2 liter, 3 necked flask fitted with a glass stirrer, condenser and thermocouple, there were added 700 g. of a naphthene oil (100 SUS at 100°F.) containing 0.25 mol of polybutene(940 MW)thiophosphonic acid of the formula

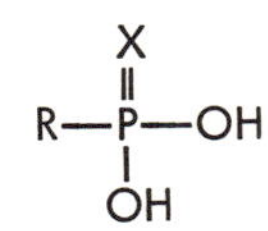

where R is a polybutene radical having an average molecular weight of 940 and X is a mixture of sulfur and oxygen. To the thiophosphonic acid there were added 42 g. (0.25 mol) triethyl phosphite and the stirred reaction mixture was heated for a total of 7 hours at 150°C. The reaction mixture was then stripped at 150°C. under reduced pressure for 2 hours utilizing a water aspirator.

Analysis determined the product to be an oil solution of ethyl polybutene(940 MW)thiophosphonate of the formula

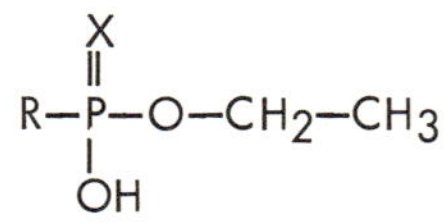

where R is a polybutene radical having an average molecular weight of 940 and X is a mixture of sulfur and oxygen. Analysis of the thiophosphonic acid reactant and thiophosphonate product found the following:

Description	Neut. No.		P, wt. percent		S, wt. percent	
	Calc.	Found	Calc.	Found	Calc.	Found
Polybutene (940 M.W.) thiophosphonic acid	-------	20	1.11	0.97	*1.15	0.64
Ethyl polybutene (940 M.W.) thiophosphonate	0	4.8	1.10	0.9	*1.13	0.60

*Based on X=100% sulfur.

Example 2: To a 3 liter flask fitted with a glass stirrer, condenser and thermocouple, there were placed 1,400 g. of a naphthene oil (100 SUS at 100°F.) solution containing 0.5 mol of polybutene(940 MW)thiophosphonic acid of the formula

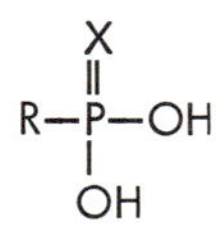

where R is a polybutene radical of an average molecular weight of 940 and X is a mixture of sulfur and oxygen. To the thiophosphonate reactant 70 g. (0.56 mol) of trimethyl phosphite were added. The reaction mixture was heated to 100°C. for 6 hours with stirring and then stripped at 100°C. utilizing a water aspirator. The product was found to be an oil solution of methyl polybutene(940 MW)thiophosphonate of the formula where R and X are as previously defined.

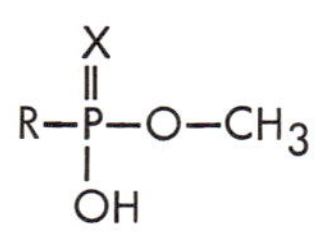

Analysis of the product showed the following:

Description	Calculated	Found
Neut. No.	0	6.6
Percent Phosphorus	1.10	1.17

Example 3: This examples illustrates the lubricant composition of the process and the detergent-dispersant properties of the thiophosphonates utilizing the well-known CLR engine sludge test. A description of the SAE 10W-30 motor oil containing the thiophosphonates of the process and a comparative SAE 10W-30 control sample are reported below. The CLR engine test results are also reported. It can be seen that the control after 54 hours of operation had a poorer sludge rating than the thiophosphonate containing oils after 70 hours.

Description	A	B	Control
Refined paraffinic distillate oil (SUS-100 at 100° F	90.72	88.0	93.73
Methyl polybutene (940 M.W.) thio phosphonate (Ex. 2)	2	----------	----------
Ethyl polybutene (940 M.W.) thiophosphonate (Ex. 1)	----------	4.70	----------
Barium C_{12} alkyl phenolate	1.83	1.83	1.83
Zinc isopropyl 1,3-dimethylbutyl dithiophosphate	0.88	0.87	0.87
CO_2 neutralized basic barium sulfonate	0.57	0.57	0.57
Mineral oil concentrate containing 25 wt. percent of a copolymer of mixed methacrylate alkyl esters in which the alkyl groups range from butyl to octadecyl	4.00	4.00	4.00
Dimethyl silicon anti-foam concentrate, p.p.m.	150	150	150

CLR* LOW TEMPERATURE SLUDGE ENGINE TEST RESULTS

Hours	Oil A	Oil B	Control
38	----------	8	6.9
54	6.3	6.5	5.4
70	5.6	5.9	----------
94	----------	5.2	----------

*Clean oil rating=10.

W.T. Shepherd; U.S. Patent 3,459,660; August 5, 1969; assigned to Texaco Inc. has found that a combination of additives comprising a hydrocarbyl phosphite, a hydrocarbyl tin ester compound and an alkyl substituted phenol when incorporated in a suitable base oil provided a compounded lubricant for refrigeration systems that is characterized by being remarkably stable against deterioration and sludge formation.

Phosphorus-Containing Olefin Polymers

J.F. Ford and J.M. Wood; U.S. Patent 3,397,219; August 13, 1968; and U.S. Patent 3,484,373; December 16, 1969; both assigned to The British Petroleum Company Limited, England describe the preparation of phosphorus-containing olefin polymers which are suitable for use as additives for improving the viscosity index and detergent-dispersant properties of lubricating oils, and to lubricating compositions containing such polymers.

According to the process, a liquid phosphorus-containing olefin polymer is prepared by reacting an olefin, preferably an alpha olefin having up to 20 carbon atoms, with a phosphorus halide, preferably chloride or bromide, the molar ratio of olefin to phosphorus halide being 100:0.1 to 25, preferably 100:2 to 10, in the presence of at least 1.5 mols of a Friedel-Crafts catalyst per mol of phosphorus halide and at a temperature below 0°C. so as to produce a phosphohalogenated polymer, and hydrolyzing the latter compound to convert halogen groups present in the polymer to oxygen or hydroxyl groups. In order to produce products that are liquid at normal ambient temperatures, it will be necessary in most cases to carry out the phosphorylation-polymerization reaction at a temperature above -60°C. Suitable Friedel-Crafts catalysts include aluminum chloride, zinc chloride, antimony chloride, ferric chloride, stannic chloride, titanium chloride, boron trifluoride and hydrogen fluoride. Aluminum chloride is preferred.

According to the process, the hydrolyzed phosphorus-containing polymer produced as described above may be reacted with a basic compound, e.g., a hydroxide, oxide or alkaline-reacting salt of an alkaline earth metal, alkali metal or aluminum, to produce a neutral product which is more suitable for use as a lubricating oil additive. This neutralization step may be carried out by adding excess of the basic compound, e.g., 2 to 4 times the required amount based on the neutralization value of the polymer, and refluxing the reactants in an inert solvent, e.g., methylene chloride, ethyl chloride, ethyl bromide or carbon tetrachloride, or in this case a hydrocarbon solvent such as n-heptane. Alternatively, where there is used in the phosphorylation-polymerization reaction a Friedel-Crafts catalyst that can easily be converted into a basic compound, the neutralization step may be carried out in situ. For example, where aluminum chloride is used as the catalyst, after the hydrolysis reaction the aqueous layer may be neutralized with ammonia to convert the aluminum chloride to aluminum hydroxide and the mixture then refluxed to neutralize the polymer.

The preparation of three polymers according to the process, termed Polymer MW7(a), MW7(b) and MW8(a), is shown below.

Polymer MW7(a): The following reactants were used.

4-Methylpentene-1 (4-MP-1) (100 mol percent)	40.93 g.
$AlCl_3$ (10 mol percent)	6.497 g.
PCl_3 (5 mol percent)	2.15 cc

A mixture of 200 cc methylene chloride, 6.497 g. $AlCl_3$ and 2.15 cc PCl_3 was cooled to -40°C., under a nitrogen atmosphere. 4-MP-1 was added slowly with stirring and some reaction occurred. After addition of about 15 g. of 4-MP-1 in this way, the rate of addition was speeded up by adding it in portions of about 5 g. A temperature gradient of 30°C. was obtained but the flask contents were not allowed to rise above -20°C. Polymer separated out in the vessel. After complete addition of the 4-MP-1, stirring was continued for a further two hours, allowing the contents of the flask to warm up to 0°C. The solution was then left overnight and by the following morning was found to have turned from a slight yellow color to brown.

About 50 cc water and 100 cc CCl_4 were added and the mixture refluxed for 1 1/2 hours to complete the hydrolysis. The organic phase was separated, washed three times with water, dried over anhydrous Na_2SO_4 and evaporated on a water bath under reduced pressure. The residual polymer was then heated to 200°C. at a pressure of 0.05 to 0.02 mm. Hg until no further light distillate was obtained. A nitrogen bleed was used. The product was a very viscous polymer, slightly cloudy and of a pale brown color.

Polymer MW7(b): The following reactants were used.

4-MP-1 (100 mol percent)	42.8 g.
$AlCl_3$ (10 mol percent)	6.796 g.
PCl_3 (5 mol percent)	2.20 cc

The reaction was carried out substantially as for MW7(a) except that the initial mixture was cooled to -50°C. Addition of 4-MP-1 produced no temperature increase and eventually all the monomer had been added. There was then a very rapid reaction after a further delay of 15 minutes and the temperature rose but was quickly reduced to -30°C. Polymer was formed. In this case there was no delay between the end of the reaction and the addition of water. The hydrolysis reaction and working up of the product was carried out as for MW7(a) and the final product was a viscous polymer faintly cloudy and of a pale yellow color.

Polymer MW8(a): 3.2 g. of Polymer MW7(a) were dissolved in n-heptane (100 ml.). $Ba(OH)_2 \cdot 8H_2O$ (0.325 g., i.e., three times the amount required for neutralization) was added and the mixture refluxed for one hour, dried with anhydrous Na_2SO_4, filtered and evaporated. The product was a pale brown viscous liquid, similar in appearance to the parent polymer. It had a barium:phosphorus ratio of 1:4 which is in agreement with the neutralization value of the parent (unneutralized) polymer MW7(a) which had a hydroxyl:phosphorus ratio of 1:2.

The properties of Polymers MW7(a), MW7(b) and MW8(a) are given in the table below. The base oil used for the viscosity and viscosity index tests was a fully refined lubricating oil obtained from a Middle East crude oil. The effectiveness of the polymers in regard to VI improvement and detergent-dispersant properties is clearly demonstrated by the results shown.

The carbon sedimentation test gives an indication of the detergent-dispersant properties of the polymers by measuring their ability to retard the settling of a suspension of carbon black in n-heptane. In this test, 0.05 g. of the material to be tested is weighed in a test tube 6" x 1/2" and 10 milliliters of n-heptane are added. The material is dissolved by heating if necessary. 0.1 g. American Carbon Black Kosmos BB is added to the solution which is shaken vigorously and allowed to stand overnight. The next day the test tube is inverted ten times and the rate at which the carbon black falls to the bottom of the test tube assesses the dispersant properties of the material. Comparative assessments are done by measuring the time taken for carbon black layer to fall below 1.2 cm. in height. The height of the solution at the beginning is 7.5 cm. Without an additive, the time for settling of the carbon black to 1.2 cm. was approximately one minute, and with 0.05 g. poly-4-methylpentene-1 it was approximately 10 minutes.

	MW7 (a)		MW7 (b)		MW8 (a)	
Phosphorus content, percent wt.	1.35		1.34		1.2	
Barium content, percent wt.	Nil		Nil		1.34	
Neutralization value, mg. KOH/g. polymer	11.9		12.0		----------	
	38° C.	99° C.	38° C.	99° C.	38° C.	99° C
Kinematic Viscosity (centistokes):						
No polymer	45.64	6.2	45.64	6.2	45.64	6.2
2% wt. polymer	53.33	7.46	54.75	7.64	52.76	7.46
4% wt. polymer	65.15	8.80	68.77	9.16	64.23	8.73
Viscosity Index:						
No polymer	103		103		103	
2% wt. polymer	110		112		112	
4% wt. polymer	116		116		116	
Carbon sedimentation test (minutes):						
No polymer	1		1		1	
0.05 g. polymer	530		300		1,800	

Carbonylated Polymers

A process described by A.F. Preuss and R.W. White; U.S. Patent 3,417,020; December 17, 1968; assigned to Rohm and Haas Company involves oil-soluble aliphatic hydrocarbon polymers having a carbonyl group. As olefinic polymers there may be used any of the polymers formed from olefins containing 3 to 12 carbon atoms. The process for preparing the oil-soluble saturated aliphatic hydrocarbon polymers comprises: (1) reacting an olefinic polymer of a molecular weight of about 300 to 5,000 with ozone, the polymer being in a liquid state, whereby an ozonolysis product is formed; and (2) reducing the resulting ozonolysis product until a carbonyl group is formed.

Example 1: A solution is prepared from 200 parts of hexane and 42 parts of a polyolefin having a molecular weight of 800 and prepared from a mixture of olefins which are chiefly 1-butene and isobutylene. A stream of oxygen containing ozone is passed into the solution at about room temperature (25°C.) for 76 minutes. The stream supplied 0.69 millimol of ozone per minute. There results a solution of ozonolyzed polyolefin. When a sample of this solution is heated, solvent is vaporized to give an oily residue. This is heated at 65° to 70°C. to decompose peroxides which may be present. The resulting product has an acidity of 0.118 milliequivalent per gram.

The bulk of the above solution of ozonolyzed product is mixed with one part of platinum dioxide and cooled to about -45°C. Hydrogen is admitted and the reaction mixture changes in temperature to about 0°C. Hydrogen is vented, catalyst is filtered off, and solvent is evaporated under reduced pressure to give a residue. Examination of this by infrared absorption shows the presence of a carbonyl group. The residue is soluble in lubricating oils and in synthetic lubricants such as di-2-ethylhexyl sebacate. The residue at room temperature is a viscous liquid.

Example 2: There are mixed 150 parts (0.125 mol) of a commercial polybutene of a molecular weight of about 1,200, 300 parts of n-butanol and 300 parts of n-hexane. The mixture is cooled to about 10°C. and a stream of oxygen-containing ozone is passes therethrough at the rate of 1.45 millimols of ozone per minute for 86 minutes. The resulting solution is treated with about one part of platinum and cooled to -50°C. Hydrogen is passed into the mixture while the temperature thereof rises to 0°C. About 0.12 mol of hydrogen is absorbed. Catalyst is filtered from the treated solution, which is then heated to 75°C. under reduced pressure, which is finally lowered to 3 mm. absolute pressure. There is obtained 132.5 parts of product which is found by infrared analysis to contain a carbonyl group. This product is soluble in oils and synthetic lubricants. Study of the product by nuclear magnetic resonance shows that the carbonyl is present chiefly as an aldehyde. There also appear to be some keto groups present.

Example 3: A mixture is prepared from 100 parts of a commercial polybutene of a molecular weight of about 1,200, 400 parts of n-heptane and 30 parts of methanol. Temperature of the mixture is adjusted to about 3°C. and ozone in oxygen is passed in at the rate of 1.46 millimols per minute for 50 minutes. This solution is treated with two parts of a platinum hydrogenation catalyst held at 0° to 10°C., and reacted with 0.08 mol of hydrogen. The mixture is then filtered and solvent evaporated to give a residue of 98 parts which contains 86% of the theoretical carbonyl groups which may be introduced into this polymer.

For purposes of comparison a mixture of 60 parts of the polybutene used in Example 3 and 160 parts of heptane is treated with ozone in air at the rate of 1.45 millimols of ozone per minute for 30 minutes. The solution is heated under reflux for 16 hours to decompose peroxides. Solvent is evaporated to give the product as a residue. By infrared analysis it is found to contain carbonyl groups, but only 53% of the theoretical carbonyl groups are present.

Example 4: Into a solution of 50 parts of polyisobutene having a molecular weight of about 2,500 in 300 ml. of 50% octanol-50% decanol is passed a stream of ozone in oxygen (1.45 millimols of ozone per minute) for 13 minutes. The solution is then treated with hydrogen in the presence of two parts of 1% palladium on alumina. The catalyst and solvents are removed to leave a viscous product whose infrared spectra shows a strong absorption at 1,718 $cm.^{-1}$ and which has a carbonyl value of 0.18 millimol per gram as determined by hydroxylamine titration.

The carbonyl-containing aliphatic polymers prepared as described are useful as lubricants themselves and as additives for lubricating compositions. They may be mixed with petroleum oils or synthetic lubricants and the resulting compositions may contain other additives such as stabilizers, antioxidants, pour-point depressors, corrosion inhibitors, dispersing agents, oiliness agents, antifoam agents, viscosity modifiers or antiwear agents. In these lubricating compositions the carbonyl-containing polymers serve to increase viscosity. They also supply dispersing activity. The polymers are employed in petroleum oils or synthetic lubricants in any amounts being completely compatible therewith. Particularly advantageous compositions are prepared by using about 0.1 to 50%, preferably 0.5 to 40%, by weight of a carbonyl-containing polymer of this process.

A synthetic lubricant was prepared from 18 parts of a carbonyl-containing polymer of a molecular weight of about 620, 0.5 part of phenothiazine and 81.5 parts of bis(isodecyl) adipate. This composition has viscosities of 5.28 cs. at 210°F. and 25.70 cs. at 100°F., corresponding to a viscosity index of 149.

Another composition was prepared from 15 parts of the same carbonyl-containing polymer used above, 0.5 part of phenothiazine, 0.75 part of phenyl-α-naphthylamine, 1.3 parts of a copolymer from a mixture of 40% butyl methacrylate and 60% lauryl-myristyl methacrylate, and 82.55 parts of bis(isodecyl) adipate. This composition has viscosities of 5.27 cs. at 210°F. and 25.36 cs. at 100°F., giving a viscosity index of 151. These compositions are useful in turbojet and turboprop engines.

Lubricating compositions are prepared by blending (1) 20 parts of a polybutene of a molecular weight of 1,200 with 80 parts of di-2-ethylhexyl sebacate containing 0.025% of free sebacic acid; (2) 20 parts of the saturated polymer obtained from the same polybutene by hydrogenation and 80 parts of the same preparation of di-2-ethylhexyl sebacate; and (3) 20 parts of the carbonyl-containing polymer derived from the above polybutene by the method defined and 80 parts of the same preparation of di-2-ethylhexyl sebacate. Properties of these blends are summarized below.

Properties of blends	1	2	3
Viscosities in cs. at —			
210° F	7.23	7.10	7.32
100° F	36.36	35.58	36.55
−40° F	9,916	9,582	9,811
Pour point (° F.)	−50	−55	−70

The lower pour point (determined by the ASTM method) for blend 3 demonstrates the greater compatibility of the carbonyl-containing aliphatic hydrocarbon polymers and one of the substantial advantages.

Epoxide Treated Sorbitol-Polyisobutenyl Succinic Acid Diesters

W.M. Lesuer; U.S. Patent 3,579,450; May 18, 1971; assigned to The Lubrizol Corporation describes a process for post-treating oil-soluble esters of mono- or polycarboxylic acids and polyhydric alcohols with organic epoxides. The acyl moeity of the esters is derived from mono- or polycarboxylic acids containing at least about 50 aliphatic carbon atoms exclusive of the carboxyl carbon atoms. The products are useful as lubricant and fuel additives. A typical example of the process is the post-treatment of a diester of polyisobutenyl substituted succinic acid and sorbitol with propylene oxide.

The process comprises contacting (1) at least one oil-soluble ester of a mono- or polycarboxylic acid and a polyhydric alcohol where the carboxylic acid moiety of the ester is characterized by at least about 50 aliphatic carbon atoms exclusive of the carboxyl carbon atoms and the alcohol moiety contains up to about 40 aliphatic carbon atoms with (2) at least one organic epoxide, the amount of (1) to (2) in the reaction mixture being such that the ratio of equivalents alcohol present in the ester to equivalents of epoxide is about 1:0.05 to 1:5. The following examples illustrate the process.

Example 1: An ester is prepared by reacting 600 parts of polyisobutenyl substituted succinic anhydride (average molecular weight 1,100) with 230 parts of polypropylene glycol (average molecular weight 425) in the presence of 547 parts of a mineral oil for about 17 hours at 150° to 160°C. while blowing the reaction mixture with nitrogen. Then 32.8 parts of an acidified clay (commercially available as Super Filtrol) is added and the mixture heated to 200°C. for an additional 11 hours with hydrogen blowing and subsequently filtered. The filtrate is an oil solution of the desired ester. To the filtrate there are added 43.2 parts of propylene oxide and this mixture is heated at 85° to 90°C. for 17 hours. The reaction mixture is then stripped to 85°C. at a pressure of 80 mm. Hg. The resulting material is an oil solution of the desired propylene oxide treated ester.

Example 2: A carboxylic acid ester is prepared by slowly adding 3,240 parts of a high molecular weight carboxylic acid (prepared by reacting chlorinated polyisobutylene and acrylic acid in a 1:1 equivalent ratio and having an average molecular weight of 982) to a mixture of 200 parts of sorbitol and 1,000 parts of diluent oil over a 1.5 hour period while maintaining a temperature of 115° to 125°C. Then 400 parts of additional diluent oil are added and the mixture is maintained at about 195° to 205°C. for 16 hours while blowing the mixture with nitrogen. An additional 755 parts of oil are then added, the mixture cooled to 140°C., and filtered. The filtrate is an oil solution of the desired ester.

The above ester is post-treated with propylene oxide by adding 108 parts of propylene oxide to 5,105 parts of the filtrate and 25 parts of pyridine while maintaining a temperature of 80° to 90°C. Then the mixture is heated to 110° to 120°C. for 2 to 3 hours and stripped to 170°C. at a pressure of 15 mm. Hg. The stripped product is the desired epoxide treated product.

Example 3: An ester is prepared by heating 658 parts of a carboxylic acid having an average molecular weight of 1,018 (prepared by reacting chlorinated polyisobutylene with acrylic acid) with 22 parts of pentaerythritol while maintaining a temperature of 180° to 205°C. for about 18 hours, during which time nitrogen is blown through the mixture. The mixture is then cooled to about 90°C. and maintained at 90° to 100°C. for 2 hours while 14 parts of ethylene oxide are slowly added. The mixture is subsequently heated to 190° to 200°C. for about 4.5 hours while slowly adding an additional 16 parts of ethylene oxide to the mixture. This mixture is then maintained at this latter temperature for an additional 2 hours, then stripped to 150°C. at a pressure of 18 mm. Hg, and filtered. The filtrate is an oil solution of the desired ethylene oxide treated ester. This ester is post-treated with epichlorohydrin following the same general procedure by substituting an equivalent amount of epichlorohydrin for the ethylene oxide.

The post-treated esters produced by this process are useful as additives in lubricants and fuels in the same manner as the ester starting materials. They function effectively as sludge dispersants in both lubricants and fuels. When employed as lubricating oil additives they are usually present in amounts of from about 0.01 to 30% by weight in the final lubricating composition. Ordinarily, when used as additives for lubricating oil compositions, the post-treated esters will be present in amounts of from 0.5 to 10% by weight although under unusually adverse conditions, such as in the operation of certain diesels, they may comprise up to about 30% by weight of the lubricant. The products are particularly useful as dispersants in lubricating oil compositions used in the crankcase of various internal combustion engines.

EXTREME PRESSURE ADDITIVES

Di(Organo) Phosphonates and Sulfurized Hydrocarbons

H. Myers; U.S. Patent 3,583,915; June 8, 1971; assigned to Mobil Oil Corporation has found that the combination of a di(organo) hydrogen phosphonate, in which at least one organo group is an aliphatic group containing at least 14 carbon atoms, in a mixture with an active sulfur compound evidences synergistic load-carrying properties in organic base media. The di(organo) phosphonates of this process have the structure

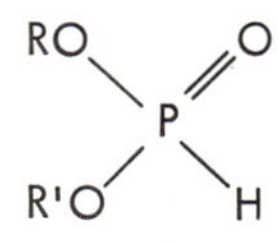

where R and R' are individually alkyl or alkenyl from 1 to 30 carbon atoms and at least one of which is an aliphatic group of at least 14 carbon atoms, and preferably over 16 carbon atoms. Dioctadecyl and dioleyl phosphonates are of particular interest.

The preferred example of the sulfur additive used in this process are the polysulfurized alkylene hydrocarbons, particularly those derived from olefins having from 2 to 6 carbon atoms or their polymers. Suitable examples of such organic polysulfides are the polysulfides of ethylene, propylene, butylene and amylene. Sulfurized alkylene compounds, such as sulfurized isobutylene containing from 30 to 65% sulfur, and preferably from 35 to 50% sulfur, are used. The additive concentration may vary from about 0.05 to 10% for each additive; preferably the range is from 0.5 to 5% by weight. Optimum performance characteristics are evidenced when the fluid composition contains from 0.005 to 0.25% and especially 0.1% phosphorus and from 0.5 to 1.5% and especially 0.9% sulfur. The following example illustrates the process.

Example: To a solvent refined mineral lubricating oil were added 2% by weight of dioctadecyl phosphonate and 2% by weight of a sulfurized isobutylene containing from about 45 to 47% of sulfur. Both components were prepared by known means. The composition was tested in the Timken Load Test. The result for this oil composition was over 60 pounds pass.

The results for two compositions, one containing 5% by weight of dioctadecyl phosphonate and one containing 5% of the sulfurized isobutylene, were pass loads of 25 and 30 pounds, respectively. Other oil compositions were tested in the Timken Load Test, and the results along with those of the above compositions are tabulated in the following table. The results for components used alone are also included at the beginning of the table.

Lubricant additive	Conc., percent by wt.	Percent P	Percent S	Timken load, lbs. Pass	Timken load, lbs. Fail
Dioctadecyl phosphonate	5	0.22		25	30
Dioleyl phosphonate	5	0.22		20	25
Sulfurized isobutylene	5		2.2	30	35
Tert-butyl trisulfide	4.5		2	30	35
Oleyl-2-chloroethyl phosphonate.	2.5	.2		15	20
{Dioctadecyl phosphonate	2	0.1		>60	
{Sulfurized isobutylene	2		0.9		
{Dioleyl phosphonate	2	0.1		>70	
{Sulfurized isobutylene	2		0.9		
{Dioleyl phosphonate	2	0.1		>60	65
{Tert-butyl trisulfide	2		0.9		
{Oleyl-2-chloroethyl phosphonate.	1.25	0.1		55	60
{Sulfurized isobutylene	2		0.9		
{Dioleyl phosphonate	2	0.1		>70	
{Sulfurized fat [1]	2		0.9		
{Dioleyl phosphonate	2	0.1		50	55
{Di-tert butyl disulfide	2.5		0.9		
{Dioleyl phosphonate	2	0.1		60	65
{Di-tert butyl trisulfide	2.0		0.9		

[1] Sulfurized sperm oil containing 9% to 10% sulfur.

As evidence of the unusual effect of the compositions of this process, it has been found that phosphonates of lower carbon content do not respond as do the higher phosphonates. A mixture of 1.4% by weight of di-n-dodecyl phosphonate (0.1% phosphorus) and 2% sulfurized isobutylene (0.9% sulfur) in the same mineral oil used previously had a Timken Load result of 35 and 40. This result is almost as poor as the sulfurized isobutylene alone.

In a process described by A.G. Papay; U.S. Patent 3,510,426; May 5, 1970; assigned to Mobil Oil Corporation lubricant compositions are provided which contain a small amount, sufficient to impart extreme pressure properties, of a mixture comprising in synergistic proportions an alkyl phosphite and at least one sulfurized olefin selected from the group consisting of sulfurized butylenes and sulfurized dipentene.

Dihalomethylene Diphosphonates

R.E. Wann, D.A. Nicholson and T.J. Logan; U.S. Patent 3,579,449; May 18, 1971; assigned to Procter & Gamble Company describe lubricant compositions comprising a major amount of a lubricating base fluid and a minor amount of extreme pressure additives which contain either a methylene, monohalomethylene or dihalomethylene group between two moieties selected from the group consisting of phosphono, dihydrocarbyl phosphoryl, phenyl, substituted phenyl, keto, cyano and carboxylate ester moieties. The following example illustrates the process.

Example: The following compositions were tested in a standard four ball tester for weld point (extreme pressure test) and wear scar (wear test). The weld point is defined as that pressure where the balls weld together in two successive runs of one minute under the indicated pressure. The wear scar is the average, based on three balls, of the average of both length and width of the scars when the balls are run under a 40 kg. load for one hour at 1,200 rpm and at a temperature of 130°F. All of the indicated additives in the table shown on the following page were added to a Kendall SAE 10, paraffin base, additive free oil (lubricating base fluid) at the indicated level.

Compositions 2 through 12 and 32 demonstrate the effectiveness of certain representative prior art extreme pressure additives. It should be noted that, with the exception of the trichloroacetic acid and the diethyl trichloromethyl phosphonate, the weld points are well under 400 kg. at an additive level of about 5%. The trichloroacetic acid and trichloromethyl phosphonate, however, are quite corrosive. Compositions 13 through 19, 23, 24, 25 and 40 through 51 show the effect of varying the level of six representative extreme pressure additives of this process. At a level of only about 1% they are still better than most conventional additives at a level of 5%.

Compositions 20, 21 and 22 show the effect of changing the length of the alkyl chain. The longer alkyl chains merely dilute the extreme pressure additive effect, since on a molar basis the improvements of the additives of Compositions 20, 21 and 22 are about the same. There is a major difference however in antiwear properties since the tetrakis(decyl) and tetrakis(stearyl) additives provide a major improvement as demonstrated by the wear scar. The tetramethyl additive does not help the antiwear properties of the base fluid.

A comparison of Compositions 17 and 23, 14 and 26, and 37 and 38 shows that, in general, the dibromo additives are more effective than the dichloro additives. This is especially true with the malonates. Similarly, a comparison of Compositions 14, 26 and 58 shows that for the particular tetraisopropyl methylene diphosphonate additives the diiodo derivative is better than the dibromo derivative which is better than the dichloro derivative. However, the ranking of the diiodo compound is undoubtedly affected by the fact it contained a substantial amount of monoiodomethylene diphosphonate.

Composition	Additive	Percent by weight	Weld point (kg.)	Wear scar (cm.)
1	(1)		130	0.82
2	Sulfur, chlorine and lead naphthenate in fish oil (SCL).	5	240-260	0.76
3	SCL	2	216	0.68
4	SCL	1	180	0.70
5	Tricresyl phosphate (TCP)	5	140	0.63
6	Benzyl disulfide	5	340	0.64
7	Hexachloroethane	5	210	0.64
8	Tributyl phosphite	5	200	0.70
9	Trichloroacetic acid [2]	5	>700	1.01
10	Randomly chlorinated paraffin	5	210	0.62
11	Lead naphthenates	5	230	0.72
12	Molybdenum disulfide	5	140	0.80
13	Tetraisopropyl dichloromethylenediphosphonate (iPr_4Cl_2MDP).	10	468	-----
14	iPr_4Cl_2MDP	5	390	1.18
15	iPr_4Cl_2MDP	2.5	360	-----
16	iPr_4Cl_2MDP	1	270	0.94
17	Tetraethyl dichloromethylenediphosphonate (Et_4Cl_2MDP).	5	500	1.16
18	Et_4Cl_2MDP	2.5	460	1.06
19	Et_4Cl_2MDP	1	270	0.94
20	Tetramethyl dichloromethylenediphosphonate (Me_4Cl_2MDP).	5	400	-----
21	Tetrakis(decyl)dichloromethylenediphosphonate	5	210	0.49
22	Tetrakis(stearyl)dichloromethylenediphosphonate	5	160	0.345
23	Tetraethyl dibromomethylenediphosphonate (Et_4Br_2MDP).	5	>800	1.19
24	Et_4Br_2MDP	2.5	660	1.12
25	Et_4Br_2MDP	1	250	0.60
26	Tetraisopropyl dibromomethylenediphosphonate (iPr_4Br_2MDP).	5	400	1.12
27	Tetraisopropyl monobromomethylenediphosphonate	5	270	1.14
28	do	5	280	1.07
29	Tetraisopropyl methylenediphosphonate	5	260	1.02
30	Tetraethyl methylenediphosphonate	5	190	-----
31	Tetrakis(stearyl)methylenediphosphonate ($C_{18}MDP$).	5	140	0.375
32	Diethyl trichloromethylphosphonate. [2]	5	>800	0.96
33	$iPr_4Cl_2MDP+C_{18}MDP$ (1:1)	10	360	0.78
34	Triethyl dichlorophosphonoacetate	5	660	0.95
35	Triethyl dibromophosphonoacetate	5	600	2.41
36	Diphenylmethylene dichloride	5	400	1.42
37	Diethyl dibromomalonate [2]	5	>800	0.68
38	Diethyl dichloromalonate	5	250	0.71
39	Diethyl malonate	5	130	0.76

Composition	Additive	Percent by weight	Weld point (kg.)
40	Di(n-pentyl) dibromomalonate (nC_5Br_2Mal)	5	>800
41	do	3	500
42	do	2	380
43	do	1	280
44	Di(isopentyl) dibromomalonate (iC_5Br_2Mal)	5	>800
45	do	3	380
46	do	2	320
47	do	1	280
48	Diethyl dibromomalonate (Et_2 Br_2Mal) [2]	5	>800
49	do [2]	3	>800
50	do [2]	2	780
51	do [2]	1	500
52	Tris(decyl) dibromophosphonoacetate ($C_{10}Br_2PA$)	5	240
53	Triethyl chlorophosphonoacetate (Et_3 ClPA)	5	260
54	Triethyl phosphonoacetate (Et_3 PA)	5	120
55	Tetrakis(decyl) methylenediphosphonate ($C_{10}MDP$)	5	140
56	Tetrabenzyl methylenediphosphonate (benzyl MDP)	5	120
57	Tetrabenzyl dichloromethylenediphosphonate (Benzyl Cl_2MDP).	5	220
58	Tetraisopropyl diiodomethylenediphosphonate	5	500
59	Dimethyl 3-keto-2,2,4,4-tetrabromoglutarate	5	>800
60	Dibromomalononitrile	5	>800
61	Dibenzoyl dibromomethane	5	560

[1] Base oil.
[2] Very corrosive.

A comparison of Compositions 26, 27 and 29; 14, 28 and 29; 17 and 30; 21 and 55; 23 and 30; 37 and 39; 38 and 39; and 34, 53 and 54 shows the effect of going from a nonhalogenated to a monohalogenated to a dihalogenated extreme pressure additive having the same basic structure. In general the nonhalogenated compounds are not effective extreme pressure additives. However, surprisingly, the methylene diphosphonate esters have substantial extreme pressure activity. The monohalogenated compounds in general are not much better than the nonhalogenated compounds as extreme pressure additives and it is only with the dihalogenated compounds that the best extreme pressure activity is achieved. It is significant that the change from monohalogenated to dihalogenated compounds would give such a tremendous increase in extreme pressure activity.

As Compositions 22 and 31 show, the antiwear properties are affected only slightly by the presence or absence of the halogen atoms on a compound. Also, in Composition 33 a mixture of an extreme pressure additive (iPr_4Cl_2MDP) and an antiwear additive ($C_{18}MDP$) gives both good extreme pressure activity and antiwear activity.

A comparison of the compositions shows that the dichloro- and dibromomethylene diphosphonates, the dichloro- and dibromophosphonoacetates, the dibromomalonates, the dibromomalononitrile and the tetrabromoglutarates give the best extreme pressure activity. The dichloromalononitrile and the tetrachloroglutarates are also good. The diiodomethylene diphosphonates, dibenzoyldibromomethane, and the diphenylmethylene dichloride are also very good extreme pressure additives. These compounds, in general, will give a weld point in the four ball test of at least about 400 kg. when used at the 5% level in this base fluid.

The dimethyl-3-keto-2,2,4,4-tetrabromoglutarate (Composition 59) is a difunctional dibromomethylene compound which shows that more than one YCX_2Y group can be present in a molecule of the extreme pressure additives. Tetrabenzyl methylene diphosphonates are not very soluble in the base fluid of this example, and accordingly are not as effective in this fluid as some of the other extreme pressure additives. The diethyl dibromomalonate, e.g., Compositions 37, 48, 49, 50 and 51, are extremely corrosive. The dipentyl and diisopentyl esters of the dichloro- and dibromomalonates, e.g., Compositions 40 through 47, are not corrosive.

Phosphorous Acid-Lanolin Reaction Product

C.E. Johnson; U.S. Patent 3,368,973; February 13, 1968; assigned to Nalco Chemical Company describes the use of the reaction product of phosphorous acid and lanolin as a primary component for an extreme pressure lubricant. This material can be readily prepared by merely reacting the two ingredients at room temperature or under slightly elevated temperatures to promote a quicker reaction. The respective ingredients may be employed in a wide range of weight percentages based on total weight of the final product. However, it is preferred that 5 to 15% by weight of phosphorous acid be reacted with about 85 to 95% by weight of lanolin.

In order to demonstrate the high load-bearing capacity imparted to lubricants by addition of the phosphorous acid-lanolin composition, the following test was run. Lubricating compositions containing extreme pressure (EP) agents were subjected to the Timken load-bearing test in the Timken Lubricant Tester in which a hardened steel ring rotating at 800 rpm was flooded by the test lubricating composition, while a hardened steel block was pressed against it by a system of weights and levers.

Two specific compositions were prepared. The first, Composition A, was comprised of 20% by weight of fatty material including fatty acid and triglycerides, 77.5% by weight of a waxy hydrocarbon oil, and 2.5% of an extreme pressure lubricant additive derived from reaction of 10% phosphorous acid and 90% lanolin. This lubricant composition as run in the above test did not produce scoring of the block at a 30 pound load after a run of approximately 10 minutes. In another test run a lubricating composition, Composition B, comprising 20% by weight of fatty material, 78% by weight of waxy hydrocarbon oil and 2% by weight of a reaction product derived from 7% by weight of phosphorous acid and 93% by weight of lanolin, also passed the 30 pound load test. A blank run involving the above materials, but without benefit of phosphorous acid-lanolin EP additive could only be run at about 12 to 13 pounds. Higher loads resulted in severe scoring.

In an actual plant run a typical lubricating composition of the process was tested for its ability to prevent scoring or scratching of a 0.09 heavy gauge steel strip being rolled into an approximate 15 ton coil. Specifically, the lubricant composition made up contained 21.0% fatty material, 2.0% of a reaction product of 7% phosphorous acid and 93% lanolin, 1.0% emulsifier, 74.0% of waxy hydrocarbon and 2.0% of a reaction product derived from 8% phosphorous acid and 92% lanolin. The above was employed as the lubricating media during the actual rolling operation. Excellent lubricant protection was afforded during the whole of the operation and substantially no scratching or gouging occurred on the surface of the rolled strips.

p-Polyphenyl

M.A. McMahon, Jr., H. Chafetz and W.J. Coppoc; U.S. Patent 3,384,588; May 21, 1968; assigned to Texaco Inc. have discovered that when poly-p-phenylene of the general formula

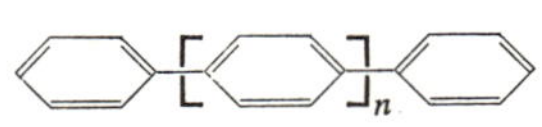

where n is an average integer of at least 38 and up to 200 and higher, is incorporated in a lubricating oil or grease in an amount to impart extreme pressure properties, e.g., between about 0.5 and 30 weight percent, a significant improvement is imparted to the lubricating oil and grease formulation in respect to extreme pressure (film strength) properties as measured by the standard film strength tests.

The poly-p-phenylene additive of the process is a solid polymer and essentially insoluble in lubricating oil and grease composition. It is incorporated into these compositions or employed as a dry lubricant as a finely divided solid advantageously in a particle size between 2 and 100 microns, preferably between 2 and 10 microns.

Example 1: This example illustrates the preparation of the poly-p-phenylene extreme pressure agent utilized in the compositions of the process. A slurry of 396 grams (3.0 mols) of aluminum chloride powder and 399 grams (3.0 mols) of anhydrous cupric chloride powder were refluxed with 1,500 ml. of dry benzene at 80°C. for a period of two hours. The slurry was filtered and the solid was washed repeatedly with an aqueous hydrochloric acid solution until the filtrate was colorless and then washed with water until neutral. The brownish solid was dried in vacuo to give about 80 grams of poly-p-phenylene solid. Poly-p-phenylene solid of the type thus prepared gave the following analysis.

Infrared bands, $cm.^{-1}$	805 (strong), 765 (weak), 695 (weak)
X-ray d-spacing, A.	450, 3.90, 2.10, 202
Elemental analysis, weight percent:	
Carbon	91.4
Hydrogen	5.2
Impurity	3.4
Carbon/hydrogen atomic ratio	1.46
Color	Brown-black
Solubility	Insoluble all solvents
Electrical resistance, ohm-cm.	3.5×10^{10}
Average molecular weight	>3,000

The impurity in the above table is composed primarily of inorganic copper and aluminum compounds entrapped by the polymer molecules as well as a minor amount (e.g., 0.1 weight percent) of chlorine substituent in the polymer itself.

Example 2: This example illustrates the extreme pressure properties imparted to mineral lubricating oil compositions by the poly-p-phenylene solid polymer of the type prepared in Example 1. Poly-p-phenylene of the type prepared above was ground to a particle size averaging about 17 microns and incorporated in varying amounts in the following described base oil composition at 25° to 30°C. under mixing conditions. For comparative purposes two additional lubricating oil compositions were prepared, one containing no extreme pressure additive and the other containing the well-known molybdenum sulfide extreme pressure additive. The molybdenum sulfide employed was of an average two micron particle size. The base lubricating oil composition employed was designated as Base Oil A and consists of 99 weight percent of a mineral lubricating oil having a viscosity at 100°F. of about 310 SUS, and containing 1 weight percent of an interpolymer of vinyl pyrrolidone, stearyl methacrylate, lauryl methacrylate and butyl methacrylate in a respective weight ratio of 8:30:50:12.

The prepared lubricating oil compositions were subjected to the Mean Hertz Load Test and the 4-Ball Wear Test. The Mean Hertz Load Test is described in U.S. Patent 2,600,058 and the 4-Ball Wear Test is described in U.S. Patent 3,050,466. These two tests measure the relative extreme pressure properties of a lubricating oil composition. In the Mean Hertz Load Test the greater the weight load that can be applied the better the extreme pressure of the lubricant material tested. In the 4-Ball Wear Test the smaller the average scar diameter on the ball the better the extreme properties of the tested lubricant. The test results are reported below.

Extreme Pressure Agent Dispersed in Base Oil A	E.P. Agent Content, Weight Percent	Mean Hertz Load, kg.	4-Ball Wear Test Average Scar Diameter, mm.
None	0	20	2.329
Poly-p-phenylene	1.0	------	1.542
Do	4.7	25	0.958
Do	9.1	29	0.914
Do	20.0	38	------
Molybdenum disulfide	5.0	25	------

As can be seen from the above, the poly-p-phenylene solid polymer imparts extreme pressure properties to lubricating oils, comparable to the well-known molybdenum disulfide extreme pressure agent.

Example 3: This example illustrates the effectiveness of the poly-p-phenylene as extreme pressure additive in a synthetic lubricating oil. The synthetic lubricating oil employed was polydimethylsiloxane having a viscosity of 1,000 cs. at 77°F. The poly-p-phenylene containing siloxane lubricating oil composition was formed by grinding the solid poly-p-phenylene of the type prepared in Example 1 to an average 17 micron particle size and adding the polyphenylene polymer with mixing to the siloxane fluid at 25° to 30°C. For comparative purposes two additional compositions were tested, namely, the polydimethylsiloxane fluid per se and the siloxane fluid containing molybdenum disulfide of about 2 to 7 microns average particle size. The three compositions, namely, the pure silicone fluid, the poly-p-phenylene and molybdenum disulfide containing silicone fluids were subjected to the 4-Ball Wear Test. The test results are reported below.

Material Dispersed in Silicone Fluid	Amount Dispersed Material, wt. Percent	4-Ball Wear Test Average Scar Diameter, mm.
Poly-p-phenylene	25	0.846
Molybdenum disulfide	25	0.932
None	0	1.359

As can be seen from the table, the poly-p-phenylene containing synthetic fluid has a better load-carrying capacity than the fluid without any extreme pressure additive and the silicone fluid containing the well-known molybdenum disulfide extreme pressure additive.

OTHER ADDITIVES

Phosphosulfurized Terpene and Aralkyl Phenol Compounds

N.M. Jones and R.J. Stephenson; U.S. Patent 3,310,491; March 21, 1967; assigned to Monsanto Chemicals Limited, England describe a lubricating oil composition comprising a mixture of a phosphosulfurized terpene and an aralkyl phenol having the formula

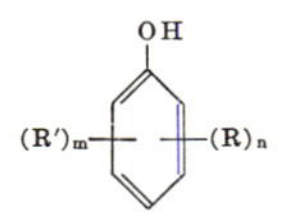

where R can be aralkyl or substituted aralkyl, R' can be alkyl or substituted alkyl, n is a whole number having a value of from 1 to 5, m is a whole number having a value of from 0 to 4, provided that the sum of n + m is not greater than 5 and provided that where R' is ortho to the phenolic hydroxyl group, R' is primary or secondary with respect to the bond between the aromatic phenolic ring and the alkyl radical.

An aralkyl phenol when used alone imparts a measure of oxidation resistance to lubricating oil, but like the phosphosulfurized terpenes, it may exhibit an adverse effect on bearing corrosion at high temperatures. The effect which is observed, whereby bearing corrosion is reduced significantly below the level which occurs when either the phosphosulfurized terpene or the aralkyl phenol is used alone, is thus synergistic. The phosphosulfurized terpene in the composition is preferably a phosphosulfurized dicyclic terpene, preferably one obtained by the phosphosulfurization of α-pinene or a commercial terpene product containing a major proportion of α-pinene.

Preferred aralkyl phenol compounds generally contain from 1 to 3 aralkyl groups and diaralkyl phenols, especially 2,4- and 2,6-diaralkyl phenols or mixtures of aralkly phenols consisting predominantly of aralkyl phenols, are particularly effective. Specific examples of aralkyl phenols useful according to the process are as follows.

- 2-α-methylbenzylphenol
- 2,4-bis(α-methylbenzyl)phenol
- 2,6-bis(α-methylbenzyl)phenol*
- 4,6-bis(α-methylbenzyl)-2-methylphenol
- 2-methyl-4,6-dibenzylphenol
- 2,6-bis(α-methylbenzyl)-4-methylphenol

*Product consisting largely of an isomeric mixture of bis(α-methylbenzyl)phenols obtained by the reaction of approximately two molar proportions of styrene with one molar proportion of phenol.

Example 1: A series of concentrates containing a range of proportions of a phosphosulfurized terpene and an aralkyl phenol in mineral oil was prepared. The amount of oil varied from 14.5 to 36.5% by weight of the concentrates. The phosphosulfurized terpene was obtained by heating at about 160°C. a mixture of one molar proportion of phosphorus pentasulfide and four molar proportions of α-pinene in the presence of a mineral oil as diluent. The aralkylated phenol was obtained by the reaction of about 2 molar proportions of styrene with one molar proportion of phenol in the presence of an acid catalyst, and contained a mixture of 2,4- and 2,6-bis(α-methylbenzyl)phenols with some mono- and tris(α-methylbenzyl)phenols. An appropriate amount of each concentrate was then added to a separate portion of a base stock having the following composition:

	Parts by Weight
Mixed detergent containing barium and calcium sulfonates with a polymeric ashless dispersant	6.5
Phosphate ester	0.2
Mineral oil, viscosity 11 cs. at 210°F.	Balance

to give the percentages of the phosphosulfurized terpene and aralkylphenol shown in the table on the following page.

Each lubricating oil composition was then evaluated using the Petter W1 test procedure, bearing weight loss being determined after a period of 36 hours. Some formulations associated with low bearing loss in addition were subjected to test for a further 36 hour period. The results obtained are shown in the same table.

Run No.	Antioxidant and percent by weight of lubricating oil compositions	Bearing weight loss (mgs.) after 36 hours	Bearing weight loss (mgs.) after 72 hours
1	Phosphosulfurized terpene, 0.56	19	
2	Aralkyl phenol, 1.0, 0.25	64 38	
3	Phosphosulfurized terpene, 0.5 Aralkyl phenol, 0.2	6	
4	Phosphosulfurized terpene, 0.4 Aralkyl phenol, 0.25	1	
5	Phosphosulfurized terpene, 0.28 Aralkyl phenol, 0.5	1	
6	Phosphosulfurized terpene, 0.28 Aralkyl phenol, 0.75	1	30
7	Phosphosulfurized terpene, 0.42 Aralkyl phenol, 0.75	1	8

Example 2: Aralkyl phenols were obtained by the reaction of 2 molar proportions of styrene with one molar proportion of o-cresol to give a product (A) consisting largely of 4,6-bis(α-methylbenzyl)-2-methylphenol, and by the reaction of two molar proportions of styrene with one molar proportion of p-cresol to give a product (B) consisting largely of 2,6-bis(α-methylbenzyl)-4-methylphenol.

Samples of these aralkyl phenols and samples of the phosphosulfurized terpene described in Example 1 were added to the base oil stock described in Example 1 to give one lubricating oil composition containing 0.25% by weight of product (A) and 0.4% by weight of the phosphosulfurized terpene (Composition 1), and a second lubricating oil composition containing 0.25% by weight of product (B) and 0.4% by weight of the phosphosulfurized terpene (Composition 2). The compositions were tested by the Petter W1 procedure and gave the following results which are listed below.

Run No.		Bearing Weight Loss (mgs.) after 36 hours
8	Composition 1	4
9	Composition 2	8

Example 3: Mineral oil concentrates were prepared containing different proportions of the phosphosulfurized terpene described in Example 1 and an aralkyl phenol obtained by the reaction of two molar proportions of styrene with one molar proportion of 4-nonylphenol and consisting essentially of 2,6-bis(α-methylbenzyl)-4-nonylphenol. An appropriate amount of each concentrate was then added to a separate portion of the base oil stock described in Example 1 to give the percentages of the phosphosulfurized terpene and aralkyl phenol shown in the table below. Each lubricating oil composition was then evaluated using the Petter W1 test procedure; the results obtained are listed below.

Run No.	Antioxidant and its percent by weight in lubricating oil composition	Bearing Weight Loss (mgs.) after 36 hours
10	Phosphosulfurized terpene, 0.4 Aralkyl phenol, 0.25	7
11	Phosphosulfurized terpene, 0.28 Aralkyl phenol, 0.5	6
12	Aralkyl phenol, 0.25	20

The above data effectively demonstrate the reduction in corrosion rates which are obtained with lubricating oil compositions containing the additive combination of an aralkyl phenol compound and a phosphosulfurized terpene.

Quinizarin Dilaurate and Amino Silanes

H.G. Burkard and D.S. Bosniack; U.S. Patent 3,470,099; September 30, 1969; assigned to Esso Research and Engineering Company have found that esters of certain hydroxylated fused ring aromatic compounds are extremely effective in anticorrosion and antioxidation agents in lubricating oils in general and lubricating oils of high performance jet aircraft in particular. Broadly, the ester additive is represented by the following structural formula

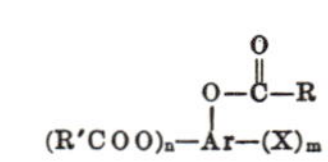

where R and R' can be the same or different and contain from 2 to 20 carbon atoms, preferably 6 to 11 carbon atoms, and are each selected from the group consisting of normal alkyl, branched chain alkyl, aryl, aralkyl, cycloaliphatic, normal alkenyl, branched chain alkenyl, and aralkenyl, the preferred group being normal alkyl and branched chain alkyl groups; n is 0 to 3, preferably 1; X is a carbonyl or hydroxy group or both; m is 1 to 4; and Ar is a hydrocarbon radical of naphthalene, a hydrocarbon radical of anthracene, or a hydrocarbon radical of phenanthrene.

Most preferably the additive is a diester of quinizarin which is represented by the structural formula

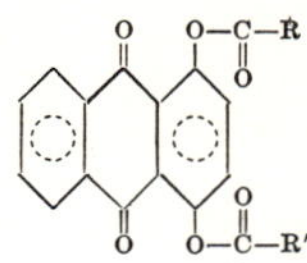

where R and R' are C_6 to C_{11} normal alkyl groups. Examples of R include hexyl, heptyl, octyl, nonyl, decyl and undecyl. The lubricating compositions will comprise a major amount of lubricating stock and an amount sufficient to inhibit oil oxidation and corrosion, usually 0.1 to 10 weight percent, preferably 0.5 to 3 weight percent, based on the total amount of lubricating composition, of the ester additive. The process is illustrated by the following examples.

Example 1: To a 2 liter flask fitted with a thermometer, stirrer and condenser were added 240 g. (1.0 mol) 1,4-dihydroxy-9,10-anthraquinone (quinizarin), 480 g. (2.2 mols) dodecanoyl chloride (lauroyl chloride), 600 ml. xylene and 5 grams of H_2SO_4. The reaction mixture was stirred at reflux for 8 hours and thereafter cooled to 80°C., whereupon 150 grams of pyridine were added. The mixture was stirred for an additional two hours at a temperature between 80° and 100°C., then cooled and filtered. The filtrate was distilled under a pressure of 10 mm. Hg until a final temperature of 100°C. was obtained. 40 g. of the residue were dissolved in 1,200 ml. of boiling methanol. Bone charcoal (1 g.) was added, the mixture filtered, and the filtrate cooled to 0°C. This cooled material was again filtered leaving 36 g. of 1,4-bis(dodecanoyloxy)-9,10-anthraquinone (quinizarin dilaurate) having a melting range of 83.5° to 85°C.

Example 2: To the apparatus of Example 1 was charged a mixture of 480 g. (2 mols) of purified 1,4-dihydroxy-9,10-anthraquinone (quinizarin), 850 cc of xylene and 1 g. of concentrated sulfuric acid. This mixture was then brought to reflux. To this refluxing mixture 300 g. (2 mols) of 2,2-dimethylpentanoyl chloride (neoheptanoyl chloride) were added dropwise with stirring over a 12 hour period. After this addition was completed the reaction mixture was refluxed for a further 40 hour period. The mixture was then cooled to 120°C. and 480 g. (2.2 mols) of dodecanoyl chloride (lauroyl chloride) were added in one portion. Reflux was resumed for a 24 hour period. At the end of this period the reaction mixture was vacuum stripped with the pot temperature maintained at 140°C. The residue was dissolved in 2 liters of boiling hexane containing 30 g. of bone charcoal. This solution was stirred for 30 minutes and filtered. The filtrate was cooled to 0°C., affording 780 g. of crude product. Recrystallization from hexane produced 590 g. of pale yellow colored product having a melting point range of 41° to 43°C. Infrared analysis confirmed the product to be 1-(2,2-dimethylpentanoyloxy)-4-dodecyloxy-9,10-anthraquinone (quinizarin mononeoheptanoate monolaurate).

Example 3: In the manner of the preceding examples, 240 g. (1 mol) of 1,4-dihydroxy-9,10-anthraquinone (quinizarin), 475 ml. of xylene, 1 g. of concentrated H_2SO_4 and 330 g. (2.2 mols) of 2,2-dimethylpentanoyl chloride (neoheptanoyl chloride) were reacted at the reflux temperatures of the mixtures for 48 hours. At the end of this period, the reaction mixture was filtered and the filtrate stripped of the excess acid chloride and xylene under 10 mm. Hg vacuum to a pot temperature of 100°C. The residue was dissolved in 500 ml. of boiling hexane and chilled to 0°C. This material was then filtered by suction filtration leaving 395 g. of 1,4-di(2,2-dimethylpentanoyloxy)-9,10-anthraquinone (quinizarin dineoheptanoate) having a melting point of 137° to 140°C.

The anticorrosive properties of the 1,4-bis(dodecanoyloxy)-9,10-anthraquinone which, for convenience, has been designated as quinizarin dilaurate, were evaluated according to known oxidation-corrosion stability tests, the results of such tests being shown below.

Oxidation, corrosion stability tests	Metals corroded, mg./cm.² Cu	Mg	Fe	Al	Ag	Percent Δ Visc. at 100°F.	Δ T.A.N.
400° F., 72 hours:							
Oil A[1]	+0.02	0.00	0.00	+0.01	+0.01	22.3	1.54
Oil A+1 pt. by wt. quinizarin dilaurate	0.12	0.00	0.00	0.00	0.00	18.4	1.55
425° F., 72 hours:							
Oil A	−0.04	−4.80	0.00	−0.01	0.00	41.6	4.80
Oil A+1 pt. by wt. quinizarin dilaurate	−0.17	−0.05	0.00	0.00	0.00	31.4	3.73
450° F., 48 hours:							
Oil A	−0.18	−12.63	+0.02	−0.01	−0.04	58.5	5.94
Oil A+1 pt. by wt. quinizarin dilaurate	−0.17	−0.01	0.00	0.00	−0.02	31.5	2.93

[1] Oil A is a lubricant composition containing 75 parts by weight of the triester prepared by fully esterifying trimethylolpropane with a $C_{8½}$ average n-alkyl carboxylic acid, i.e., esterified with octanoic and nonanoic acids, 25 parts by weight of the ester prepared by fully esterifying dipentaerythritol with a C_6 average n-alkyl carboxylic acid, i.e., esterified with C_5 to C_9 n-alkyl carboxylic acids, 1.3 parts by weight p,p'-dioctyldiphenylamine, 1.1 parts by weight of phenyl-B-naphthylamine, 1.0 part by weight of tricresyl phosphate, 0.1 part by weight of quinizarin, 0.025 part by weight of sebacic acid, and 0.001 part by weight of silicone.

The oxidation corrosion stability tests referred to above were carried out by blowing air at the rate of 100 volumes per hour through one volume of ester maintained at the indicated temperature for the indicated periods of time. Thus, in the 425°F., 72 hour oxidation-corrosion stability test, the ester was maintained at a temperature of 425°F. for a period of 72 hours. These tests simulate to a degree the operating conditions of jet engines. The percent increase in centistoke viscosity, metals corroded, and the increase in total acid number (T.A.N.) were then determined. The effectiveness of the quinizarin ester was evaluated in combination with other additives conventionally employed in a fully finished lubricant. As seen by the data, the additives of this process were very effective in inhibiting oxidation and corrosion at higher temperatures. The lubricant containing the fused ring aromatic ester was substantially less corrosive than the lubricant containing no such compound. The percent change in viscosity and change in total acid number are smaller when the additive of this process is employed, thus indicating the antioxidant properties of the substance.

In a process described by M.J. Napoli; U.S. Patent 3,350,311; October 31, 1967; assigned to Dow Corning Corporation lubricating fluids are stabilized by the addition of 4,4'-bis(triorganosilyl)diphenylalkylamines.

Diester Lubricity Additives

In a process described by M.J. Furey and A.F. Turbak; U.S. Patent 3,429,817; February 25, 1969; assigned to Esso Research and Engineering Company the lubricity and load-carrying ability of a lubricating oil is improved by addition of an ester formed by reacting about two mols of C_2 to C_5 glycol with about one mol of C_{36} dicarboxylic acid dimer of a C_{18} unsaturated fatty acid. The following example illustrates the process.

Example 1: In order to illustrate the process, a number of tests were carried out using the additives in base jet fuels and the load-carrying capacity of the fuels determined. One mol of C_{36} dimer acid (Empol 1014, consisting of 95% dimer acid, a trace of monobasic acids and the remainder essentially trimer acid) was reacted with two mols each of either of two glycols (ethylene and neopentyl) by refluxing in benzene in the presence of p-toluene sulfonic acid monohydrate as a catalyst. The water evolved was measured and the benzene solutions were water washed. On evaporation of the benzene under a vacuum at 35° to 40°C., the products were found to be dark amber, clear, viscous fluids quite soluble in hydrocarbons.

Effect on Scuffing — As shown by the following data, the addition of 0.1% of the C_{36} dimer acid/ethylene glycol diester to a jet fuel greatly increased the load-carrying capacity as measured by the Ryder Gear Test. The dimer acid alone has little effect.

Fluid	Ryder Rating (lb./in.)
Base jet fuel*	400
Base + 0.1% diester of C_{36} dimer acid and ethylene glycol	1,100**
Base + 0.1% C_{36} dimer acid	480

*Highly isoparaffinic fuel of 375° to 500°F. boiling range, high thermal stability, low freezing point and low sulfur content.

**Run under more severe condition of 5# load steps rather than 2#. Under this condition (5# step) the base fails on the first step.

As shown by the above data, the addition of 0.1% of the diester increases the antiscuffing properties of jet fuel as measured by the Ryder Gear Test [Ryder, E.A., ASTM Bulletin 184, 41 (1952)]. The ratings represent the load in pounds per inch of tooth width to produce a given amount (22.5%) of gear scuffing. It can also be seen that the C_{36} dimer acid itself has substantially no effect.

Example 2: Another test was carried out with the following results. Thus, it is apparent that the diester is very effective.

	Additive, wt. percent	Ryder scuff test, lb./in.
Neutral mineral Oil, viscosity 43 SUS at 210° F.	0	1,170
Mineral oil with diester of ethylene glycol and dimer of linoleic acid	0.5	3,200

Example 3: The preparations and tests of Example 1 are repeated except that the C_{36} dimer acid is obtained from oleic acid. The ethylene glycol diester of the dioleic acid is observed to have the effect of improving the antiscuffing properties of jet fuel as measured by the Ryder Gear Test.

Example 4: The preparations and tests of Example 1 are repeated except that the C_{36} dimer acid is obtained from the mixed dimer of oleic and linoleic acids. The ethylene glycol diester of the dioleic acid is observed to have the effect of improving the antiscuffing properties of jet fuel as measured by the Ryder Gear Test.

In the use of partial esters of dicarboxylic acids having at least 9 carbon atoms between the carboxylic acid groups in the molecule, the partial esters of the acids generally consist of a mixture containing a major portion of the monoester with minor proportions of the diester and the unreacted acid. The preparation of these partial esters can best be illustrated by reference to the following equation:

$$HOOC{-}R{-}COOH + HO{-}R'{-}OH \longrightarrow HO{-}R'OOC{-}R{-}COOH + H_2O$$

where R is the hydrocarbon skeleton of the dicarboxylic acid having more than 9 carbon atoms between the acid groups and R' is either the hydrocarbon skeleton of a C_2 to C_8 glycol or the oxa-alkyl sketeton of an oxa-alkane diol.

The procedures used in preparing the partial ester consisted of weighing out the dimer acid and glycol in equimolar quantities and carrying out the esterification in benzene under reflux conditions (80°C.) with a small amount of p-toluene sulfonic acid as a catalyst. A condenser and trap were used, the refluxing being stopped when the theoretical amount of water was collected. (The water azeotropes off with the benzene.) Then the benzene solution was cooled and water washed to remove the catalyst. The benzene was then stripped off under vacuum at 35° to 40°C., leaving the product usually a clear, dark, amber, viscous fluid.

Example 5: The procedure outlined immediately above was used to make six different monoesters. In each case, 0.1 mol of C_{36} dimer acid (Empol 1014, essentially linoleic dimer acid), 0.1 mol of glycol and 1.5 grams p-toluene sulfonic acid were mixed with 1 pint of benzene and then the reaction was carried out. Monomeric esters of the following glycols were made: ethylene; triethylene; neopentyl-(2,2-dimethyl-1,3-propane diol); 1,4-butane diol; 1,6-hexane diol; 1,12- dihydroxy octadecane. Each of these monoesters was quite soluble in hydrocarbons.

As shown by the data below, the addition of 0.1% of several of the monoesters increases the antiscuffing properties of jet fuel as measured by the Ryder Gear Test. The ratings represent the load in pounds per inch of tooth width to produce a given amount (22.5%) of gear scuffing. It can be seen that the ethylene glycol derivative is by far the most effective, while the neopentyl compound is the least effective. It can also be seen that the C_{36} dimer acid itself has substantially no effect.

Run	Additive in base jet fuel [1]	Ryder rating (lb./in.)
A	None	400
B	0.1% C_{36}/ethylene glycol monoester	1,530
C	0.1% C_{36}/triethylene glycol monoester	1,100
D	0.1% C_{36}/neopentyl glycol monoester	770
E	0.1% C_{36}/1,6 hexane-diol monoester	1,800
F	0.1% C_{36} dimer acid	480
G	Run B except dimer acid hydrogenated	2,290

[1] Highly isoparaffinic fuel of 375 to 500° F. boiling range, high thermal stability, low freezing point and low sulfur content.

Furthermore, it has also been found that the ethylene glycol monoester blend (at 0.1% concentration) passes a critical thermal stability test for jet fuels. This is a 5 hour test in which the fuel, held at 300°F. in bulk, is pumped through a heat exchanger to reach 500°F. and then through a filter. The requirements are that no deposits should form on the metal heat exchanger tube and that there should be no (or little) pressure drop across the filter. Many conventional load-carrying additives fail the test.

GREASE FORMULATIONS

Phosphorodithioate and Boroxaryl Compounds

C.N. Rowe; U.S. Patent 3,532,626; October 6, 1970; assigned to Mobil Oil Corporation has found that organic compositions, normally susceptible to deterioration, particularly in the form of lubricating oils and greases, can exhibit reduced deterioration and improved antiwear properties by incorporating into such compositions a small amount, sufficient to inhibit the deterioration and wear of a mixture comprising (a) a metal salt of an O,O-diorganophosphorodithioate, and (b) a boroxaryl compound. Particularly effective are the zinc and mercuric O,O-dialkylphosphorodithioates, particularly representative of which are zinc and mercuric O,O-diisopropylphosphorodithioates. Some examples of the 9,10-boroxarophenanthrene compounds used in the process are as follows:

- 9-hydroxy-9,10-boroxarophenanthrene
- dioctadecyl-9-hydroxy-9,10-boroxarophenanthrene
- anhydride of 9-hydroxy-9,10-boroxarophenanthrene
- 9-tert-butoxy-9,10-boroxarophenanthrene
- 9-methoxy-9,10-boroxarophenanthrene
- 9-chloro-9,10-boroxarophenanthrene
- 5,12-di-n-hexadecoxy-5,12-diboro-6,13-dioxarodibenz[a,h]anthracene

The following examples illustrate the synergistic effect of the components (a) and (b) in the additive mixtures in inhibiting wear, particularly with respect to lubricants in the form of lubricating oils. Component (a) in the examples in the table on the following page is either zinc O,O-diisopropylphosphorodithioate or mercuric O,O-diisopropylphosphorodithioate, and component (b) is either 9-hydroxy-9,10-boroxarophenanthrene or 5,12-di-n-hexadecoxy-5,12-dibora-6,13-dioxarodibenz-[a,h]anthracene. The base fluid is cetane (n-hexadecane). As shown in the table, the uninhibited base oil containing the additive compounds, individually as well as in admixture, were subjected to an antiwear test in order to ascertain synergistic effects. This test is carried out as follows. A stationary pin is held upright against an axially mounted rotatable disc, and the disc is rotated at a constant velocity. The pin describes a circle on the surface of the disc concentric with the axis thereof. Both pin and disc are submerged in a vessel holding the test lubricant, so that the point of contact between the pin and the rotating disc is lubricated at all times. The pin and the disc are each made of either AISI 1020 or AISI 4140 steel. The pin has a hemispherical tip and the disc is 7.6 cm. in diameter. The force of the disc against the pin may be changed by varying the load outside of the lubricant vessel.

The lubricating fluids shown in the table were tested at a load of 8 kg. using a sliding velocity of 10 centimeters per second. The tests were run at the temperatures indicated in the table, in a base fluid of cetane (n-hexadecane). After the test was completed, the pin was microscopically examined to measure the wear scar diameter. The wear rate was determined from this measurement.

Synergism in Friction and Wear Between
Metal O,O-Diisopropylphosphorodithioates and Boroxaryl Compounds

			8 kg. 10.0 cm./sec. Cetane carrier ambient Atm.			
			AISI 1020 Steel		AISI 4140 Steel	
Example No.	Additive	Temp., °C.	Coeff. friction	Wear rate, cc./cm.	Coeff. friction	Wear rate. cc./cm.
1	Base fluid (blank)	93.3	1.2	$250,000 \times 10^{-11}$	1.0	$5,500 \times 10^{-11}$
2	0.2% 9-hydroxy-9,10 boroxarophenanthrene	93.3	0.34	15×10^{-11}	0.10	0.20×10^{-11}
3	0.38% zinc O,O-diisopropylphosphorodithioate	93.3	0.43	180.0×10^{-11}	0.28	23×10^{-11}
4	0.10% 9-hydroxy-9,10 boroxarophenanthrene/0.19% zinc O,O-diisopropylphosphorodithioate	93.3	0.06	5.8×10^{-11}	0.09	0.090×10^{-11}
5	0.48% mercuric O,O-diisopropylphosphorodithioate	93.3	0.21	7.4×10^{-11}	---	---
6	0.10% 9-hydroxy-9,10 boroxarophenanthrene/0.24% mercuric O,O-diisopropylphosphorodithioate	93.3	0.16	0.47×10^{-11}	---	---
7	0.10% 5,12-di-n-hexa-decoxy-5,12-dibora-6,13-dioxarodibenz-[a,h] anthracene	93.3	0.17	11.2×10^{-11}	0.20	1.3×10^{-11}
8	0.05% 5,12-di-n-hexadecoxy-5,12-dibora-6,13-dioxarodibenz-[a,h] anthracene/ 0.19% zinc O,O-diisopropylphosphorodithioate	93.3	0.03	2.0×10^{-11}	0.03	0.05×10^{-11}

As will be apparent from the data of the above table, the synergistic mixtures of the addition agents in accordance with the process as shown in Examples 4, 6 and 8 are markedly effective in inhibiting organic compositions, and lubricants in particular, against deterioration or in wear of metal surfaces.

Graphite Treated with Polytetrafluoroethylene

A.J. Groszek; U.S. Patent 3,518,190; June 39, 1970; assigned to The British Petroleum Company Limited, England has found that oleophilic graphite prepared by grinding graphite and polytetrafluoroethylene in an organic liquid has enhanced grease-thickening, load-carrying and high temperature stability.

Oleophilic graphite can be obtained by grinding natural or synthetic graphite in most organic liquids, but it is desirable to use one the bulk of which can be easily removed from the modified oleophilic graphite. In preparing the modified oleophilic graphite, a fluorinated hydrocarbon is added to the grinding medium. Those liquids distilling below 500°C. and having a viscosity below 600 centistokes at 100°F. (38°C.) are preferred liquids for use as the grinding medium. (Liquids having a surface tension below 72 dynes/cm., preferably from 10 to 40 dynes/cm., at 25°C. are preferred). Preferably the viscosity of the liquid is less than 30 centistokes, more preferably less than 3 centistokes and most preferably less than 1 centistoke at 38°C.

The fluorinated hydrocarbons which can be used include the polyfluorinated and perfluorinated hydrocarbons, e.g., perfluoro-alkyl-cyclohexanes such as fluoromethyl-cyclohexane and fluorocarbon polymers such as polytetrafluoroethylene (PTFE). The latter is especially useful as it can be added as a powder mixed with the graphite.

The following are examples of methods by which the grease may be prepared.

(1) The grinding liquid is filtered off from the slurry of modified oleophilic graphite produced in the grinding operation. The resulting filter cake is ground by, for example, feeding the cake through a colloid mill, and the resulting powder is stirred into the oil. The grease produced is finished by colloid milling.

(2) The grinding liquid is boiled off rapidly from the slurry of modified oleophilic graphite to avoid the formation of a graphite cake, the resulting powder is stirred into the oil and the grease is finished by colloid milling.

(3) Base oil is added to the slurry of modified oleophilic graphite and the grinding liquid distilled off.

(4) Base oil is added to the slurry of modified oleophilic graphite and the mixture circulated through a homogenizer (for example, one of the Manton-Gaulin type) so that temperatures up to or exceeding 140°C. are produced. The temperature must be high enough to drive off the grinding liquid.

(5) The grease may also be made directly by grinding the graphite in the base oil for the grease. For example,

a low surface tension, low boiling point mineral lubricating oil with a viscosity up to 600 cs. at 100°F. (38°C.) can be used. Elevated temperatures can be used during the grinding.

Methods (3), (4) and (5) are particularly preferred. In general, the modified oleophilic graphites can be incorporated into the base oil either at ambient temperatures or, if desired, at elevated temperatures, for example, up to 400°C.

These greases have remarkably high drop points. When their drop points are measured according to the IP or ASTM standard methods, they are found to be above 400°C. Such greases are described as "infusible" and are difficult to produce by conventional methods. By using carefully selected base oils, for example, synthetic oils with high oxidation and thermal stability, greases having a unique combination of properties can be produced. These greases have good oxidation and mechanical stability at temperatures up to about 130° to 140°C. when mineral oils are used as the base oils. For temperatures above this range synthetic base oils can be used.

The effect of grinding graphite in the presence of perfluoromethyl-cyclohexane and PTFE is shown in the following table. The grinding fluid used was n-heptane. The greases were formed by method (3), using a BG 160/95 base oil (an oil of VI 95 and viscosity of 160 at 210°F.). The relative surface areas were determined using the heat of adsorption of n-dotriacontane from normal heptane, the heats being expressed in calories per gram of graphite. The heats of adsorption increase with the surface area of the graphites. The greases were formed so that 15% of solids (i.e., PTFE + graphite) were added to the base oil. The amount of PTFE added to the grinding fluid was calculated as a percentage of the total weight of graphite plus PTFE.

Oleophilic graphite				Grease		
Grinding fluid	Fluorinated hydrocarbon	Amount of fluorinated hydrocarbon as a percentage of total solids	Heat of adsorption cals/gr. of n-dotriacontane from n-heptane	Amount of thickener percent	Amount of graphite, percent	¼ scale penetration worked
n-Heptane	None	Zero	1.15	15	15	335
Do	PTFE	10	1.32	15	13.5	313
Do	PTFE	25	1.43	15	11.25	313
Do	PTFE	35		15	9.7	320
Do	PTFE	50	1.76	15	7.5	339
Do	PTFE	75	0.86	15	3.25	No grease
Do	PTFE	100	Nil	15	None	No grease
Do	Perfluoro-methyl cyclohexane.	Liquid				

[1] These penetration values are ¼ scale penetrations converted to full scale readings by the general accepted correlation established for soap-thickened greases (ASTMD 1403-56T).

The grease made with a thickener consisting of 75% graphite and 25% PTFE was tested in the four ball machine for extreme pressure properties and the results are shown below. This grease was subjected to a Skefko 4A rig test and the grease ran the full 600 hours at a mean temperature of 113°C. No inhibitor was present and after the test the consistency and appearance of the grease was similar to the fresh grease. The elements were well coated and there was no sign of wear. There was only slight lacquering.

Thickener composition	Total wt. of thickener, percent	Penetration: Unworked	Penetration: Worked	Bleed, percent	Extreme pressure test: MHL	Extreme pressure test: WL	Extreme pressure test: ISL
75% / 25% PTFE	15	298	313	5.3	51	316	79

MHL is mean Hertz load, ISL is initial seizure load, WL is welding load.

These results show the excellent lubricating and temperature stable properties of the grease.

Fumarate-Hydroxyethyl-Vinyl Acetate Polymeric Additives

A process described by J. Panzer, J. Nixon and R. Kassinger; U.S. Patent 3,354,086; November 21, 1967; assigned to Esso Research and Engineering Company involves greases thickened with alkaline earth metal salts of acetic acid in combination with other fatty acids, which are improved by including a copolymer of certain unsaturated esters.

Lubricating greases containing alkaline earth metal salts, particularly calcium salts, of acetic acid and various C_6 to C_{30} fatty acids, have found widespread use in commercial applications. In general, these mixed salt lubricants have good antiwear and load-carrying properties, which properties have made them commercially successful. However, greases of this type tend to unduly harden, particularly when used at elevated temperatures, e.g., in the end caps of bearings, and also frequently plug small diameter feed lines of some centralized lubrication system where the grease is forced under pressure to the parts to be lubricated, particularly when the rate of feed is low and the feed lines are in a hot environment. It has been found that greases of this type can be inhibited against hardening and line plugging by certain copolymers of unsaturated esters.

The copolymers used in the process are those which are soluble in mineral lubricating oil and have Staudinger molecular weights of about 20,000 to 1,000,000, preferably 100,000 to 800,000. These copolymers are formed by copolymerizing (1) a mono- or diester of a fully esterified C_3 to C_4 aliphatic unsaturated mono- or dicarboxylic acid, e.g., acrylic, methacrylic, fumaric, maleic and itaconic acid, and a C_1 to C_{20}, preferably a C_3 to C_{20}, saturated, straight or branched chain aliphatic alcohol; (2) an ester of vinyl alcohol and a C_2 to C_6 saturated aliphatic monocarboxylic acid; and (3) a polar monomer such as 2-N-vinylpyrrolidone, maleic anhydride and hydroxyalkyl acrylate. The polymer will contain about 0.1 to 2.0, preferably 0.2 to 1.0 mols of ester of unsaturated acid, and about 0.01 to 0.5, preferably 0.03 to 0.3 mol of the polar monomer, per molar proportion of the vinyl alcohol ester. The following examples illustrate the process.

Example 1: A grease thickened with the mixed salts of acetic acid and tallow fatty acid in a relative molar ratio of 10:1 was prepared as follows: 7.7 parts of tallow fatty acid, 0.6 part of a polymer solution consisting of 37 weight percent polymer and 63 weight percent inert mineral lurbicating oil, and 11.4 parts of hydrated lime were added to 63.1 parts of mineral lubricating oil of 80 SUS viscosity at 210°F. inan electrically heated grease kettle. This mixture was warmed, while stirring, to 125°F. 16.2 parts of glacial acetic acid were slowly added to the kettle. The temperature of the kettle contents was then raised to about 300°F. for 15 minutes in order to dehydrate the grease. The grease was then cooled to 275°F. where 1 part of phenyl-beta-naphthylamine was added as an oxidation inhibitor. The grease was further cooled to 100°F. and homogenized by passage through a Morehouse mill having 0.003" clearance.

The tallow fatty acid (beef tallow acid) had a saponification number of about 195, an iodine number (Wijs) of about 55 and averaged a C_{16} chain. The polymer used above had a molecular weight of about 500,000 and contained 0.394 molar proportion of di-tallow fumarate, 0.424 molar proportion of di-C_8-oxo-fumarate and 0.131 molar proportion of hydroxyethyl methacrylate per molar proportion of vinyl acetate. The polymer was made by copolymerizing a mixture of the monomers in a mineral oil diluent under a nitrogen atmosphere using tertiary butyl perbenzoate as the catalyst and using about 115% excess vinyl acetate. This excess vinyl acetate was evaporated by nitrogen purging at the end of the polymerization reaction. The resulting product consisted of about 37 weight percent polymer dissolved in 63 weight percent mineral lubricating oil diluent. The hydroxyethyl methacrylate used above had the structure:

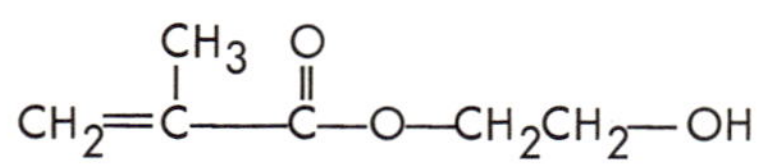

A comparison grease (Grease A) was prepared in the same manner as the grease of Example 1, except for different proportions and the fact that no polymer was used. The formulations and properties of the greases of Example 1 and the comparison Grease A are summarized in the following table.

Composition, Wt. Percent	Example I	Grease A
Glacial Acetic Acid	16.2	19.0
Hydrated Lime	11.4	13.8
Tallow Fatty Acid	7.7	10.0
Phenyl-beta-naphthylamine	1.0	1.0
Polymer Solution[1]	0.6	----------
Mineral Lubricating Oil, 80 SUS at 210° F	63.1	56.2
Properties:		
ASTM Penetration, 77° F., mm./10, 60 strokes	335	325
Trabon Life, Hrs	400	237
Hardening—Micro-cone Penetration, ASTM D-1403, 56T;		
Room temperature:		
Initial	174	186
Percent Δ pen. after 1 day	12.6	21.5
150° F.:		
Percent Δ pen, after 1 day	19.5	33.9
Percent Δ pen. after 1 week	31.0	43.5
Dispensability at 0° F., 150 p.s.i. No. 1 capillary, grams/min.:		
Initial	0.85	1.00
After 16 hrs. at 300° F	0.35	02.4
Percent reduction in flow rate	59	76

[1] 37 wt. percent polymer in 63 wt. percent oil.

The Trabon life was determined in a Trabon centralized lubrication system which included a pump circulating the grease from a reservoir under high pressure through small diameter feed tubes and back to the reservoir. The number of hours of operation before plugging of the system was determined. The hardening information was determined simply by measuring the percent change in micro-cone penetration. For example, the grease of Example 1 had an initial micro-cone penetration of 174, but after standing 24 hours its penetration had decreased to 152 or a change of 12.6%. The dispensability at 0°F. under 150 psi through a capillary opening was determined by a test procedure described in an article on page 76 et seq., in the February 1963 issue of Lubrication Engineering.

As seen by the preceding table, the grease composition of this process (Example 1) containing 0.6% of the copolymer solution, i.e., 0.22 weight percent polymer, showed less susceptability to line plugging and hardening than the comparison Grease A of a comparable ASTM penetration but which contained no polymer. This is indicated by the longer Trabon life, the lesser percent change in penetration in the Micro-cone Penetration Test, the greater lubricator cup delivery rate and the lesser percent reduction in flow rate through a No. 1 capillary. While Grease A varied somewhat from the grease of Example 1 with regard to the proportions of materials used, it can be seen from their ASTM penetrations that both greases were made up to substantially the same consistencies. Apparently, the polymer also increases the yield, i.e., the hardness of the grease, perhaps by more finely dispersing the calcium salts. Thus, with the polymer present, the grease of Example 1 had substantially the same ASTM consistency as Grease A even though it contained less of the calcium salt thickener. Since the thickener is the more expensive part of the grease, it is an economic advantage to obtain thickening to the desired consistency with the minimum amount of thickener, thus representing another advantage for using the polymer.

Example 2: As a further example of the process, Example 1 can be repeated exactly but using a polymer containing 0.131 molar proportion of 2-N-vinyl pyrrolidone as the polar monomer in place of the 0.131 molar proportion of hydroxyethyl methacrylate used in Example 1.

Extreme Pressure Grease Compositions

A process described by G.P. Caruso; U.S. Patent 3,370,007; February 20, 1968; assigned to Shell Oil Company involves greases gelled with waterproofed clays and containing extreme pressure additives.

In U.S. Patent 3,271,309, it is noted that greases gelled with oleophilic clay products may be improved with respect to their extreme pressure properties by the presence of an effective load-bearing proportion of a metal carbonate having a Moh hardness of 1 to 5 and a thermal decomposition above about 250°C. Another aspect of that process comprises the finding that an improvement can be obtained by combining molybdenum disulfide, an amorphous antimony sulfide and/or a lead naphthenate therewith, thus enabling a drastic reduction in the carbonate requirement for a given extreme pressure (EP) value.

Still another aspect of that process lies in the discovery that greases gelled with oleophilic clay products can be improved with respect to their extreme pressure properties under both wet and dry operating conditions by the combination of the abovementioned metal carbonate and molybdenum disulfide, and/or an amorphous antimony sulfide, supplemented by a lead naphthenate. This approach depends upon the finding that operation under wet conditions seriously degrades the effectiveness of the sulfide as an extreme pressure agent (even though good dry operation is obtained), but that this can be corrected and even supplemented by the presence of a lead naphthenate. However, it has also been found that the proportion of naphthenate is restricted by the deleterious effect of the naphthenate upon the consistency of the grease if it is present in more than strictly limited amounts. If greater than these limited amounts are employed, the grease structure may in fact even be destroyed.

However, it has been found that certain saturated aliphatic carboxylic acid esters have the property of interaction with either or both the metal sulfide or the metal carbonate-EP additives in such a manner that the efficacy of the EP additives is improved still further. This means, of course, that such esters may be used in either of two ways: (1) they can be added to raise the EP properties of clay thickened greases containing either or both of the foregoing two EP additives, or (2) they can be added to such greases containing smaller amounts of the EP additives while maintaining the same level of EP effectiveness. Since the cost of the esters is considerably less than either the metal sulfide or the metal carbonates, considerable savings in cost can be obtained without any sacrifice in extreme pressure properties of the grease in which they are used.

The esters which may be used are a quite narrow class of saturated aliphatic carboxylic acid esters corresponding to the formula

$$R_1\overset{\displaystyle O}{\overset{\|}{C}}OR_2$$

where R_1 is a normal alkyl radical having from 1 to 20 carbon atoms and R_2 is likewise a normal alkyl radical but having only from 1 to 8 carbon atoms, the sum of R_1 and R_2 being at least 5. The esters are therefore neither substituted nor branched. It is preferred that R_1 and R_2 contain, respectively, from 16 to 18 and 1 to 6 carbon atoms. Butyl stearate is particularly preferred.

The following examples illustrate the interaction of the carboxylic acid esters in clay thickened greases containing metal carbonate and sulfide primary EP additives.

Example 1: For the purpose of comparison, mineral oils were utilized throughout the following examples, all of the grease contained hectorite clay waterproofed with 60%, based on the weight of the clay, of an amino amide formed between tall oil fatty acids and the mixture of polyethylene polyamines described as the bottoms product obtained in the preparation of ethylene diamine. In each of the grease blends shown in the following table, the carboxylic acid ester was butyl stearate. The data show that the ester, by itself, has no EP activity in clay thickened grease. In fact, it appears by comparison of the base grease with Grease B that the ester is at least slightly detrimental to EP properties. On the other hand, by comparison of Grease A with the base grease, it is shown that the lead carbonate increased the EP properties of the base grease. However, by using the ester in conjunction with the carbonate primary EP additive, the EP properties of the base grease were more than doubled. In fact, by using slightly higher amounts of the ester, even the EP properties of the grease containing the carbonate EP agent was more than doubled. It is important, too, to note that Grease F containing the ester as well as the carbonate had quite excellent wet EP properties. The effect was also observed to practically the same magnitude when the primary EP additive was changed to calcium or sodium carbonate.

Grease	Ester, percent wt.	Metal Carbonate		Dry Timken Test[1]		Wet Timken Test[1]	
		Composition	percent, wt.	Pass	Fail	Pass	Fail
Base	None	------	None	20	25	------	------
A	None	$2Pb(CO_3)\cdot Pb(OH)_2$	2.0	25	35	------	------
B	2.5	------	None	------	20	------	------
C	0.75	$2Pb(CO_3)\cdot Pb(OH)_2$	2.0	40+	------	------	------
D	1.0	$2Pb(CO_3)\cdot Pb(OH)_2$	2.5	55	60	------	------
E	2.5	$2Pb(CO_3)\cdot Pb(OH)_2$	2.5	60	------	------	------
F	2.5	$2Pb(CO_3)\cdot Pb(OH)_2$	2.5	55+	------	20	25

[1] Timken Extreme Pressure Test, ASTM Bulletin 228, February 1958, pages 28-31.

Example 2: Samples of the basic grease of Example 1 were prepared, modified as indicated in the table below and tested for their EP properties. Again, by comparison of Grease H with the base grease, the ester exhibited essentially no EP properties without the presence of a primary EP additive. However, it raised substantially the Timken rating of the grease containing the metal carbonate primary EP agent.

Grease	Ester, percent wt.	Metal Carbonate		Dry Timken Test	
		Composition	Percent wt.	Pass	Fail
Base	None	------	None	20	25
G	None	$CaCO_3$	4.0	35	40
H	2.5	------	None	------	20
I	2.5	$CaCO_3$	2.5	40	45
J	------	Na_2CO_3	2.5	20	25
K	2.5	Na_2CO_3	2.5	40	45

The secondary EP effect of the carboxylic acid esters is also applicable to metal sulfide EP additives, such as antimony or molybdenum sulfides, as is shown in the following example.

Example 3: The same procedures of the previous examples were employed except amorphous antimony sulfide was used as the primary EP agent in place of the metal carbonates. The results of the EP tests were as follows:

Grease	Ester, percent wt.	Metal Sulfide		Dry Timken Test		Wet Timken Test	
		Composition	Percent wt.	Pass	Fail	Pass	Fail
Base	None	------	None	20	25	------	------
L	None	Sb_2S_3	0.75	50	55	------	25
M	0.75	------	None	20	25	------	------
N	2.5	------	None	20	25	------	------
O	0.75	Sb_2S_3	0.75	50	55	60+	------
P	0.75	Sb_2S_3	0.75	60	------	50	55

The above data are most interesting in that a somewhat different interaction seems to occur between the metal sulfide and the ester. Whereas the ester has little if any effect on the dry EP properties of the metal sulfide-containing grease, it nevertheless incurs truly phenomenal increases in the wet EP properties of the grease. It is also interesting to note from the above data that the amount of ester above about 0.75% had no effect on the grease containing no primary EP agents. Therefore, further confirmation is given that the carboxylic acid esters of the process are essentially ineffective to benefit the EP properties of a grease in the absence of a metal carbonate or metal sulfide primary EP agent.

METAL WORKING

Polybutene and Hydrogenated Fatty Acid Mixtures

J.S. Brown; U.S. Patent 3,298,954; January 17, 1967; assigned to Standard Oil Company describes a nonstaining metal-working lubricant which is obtained by the combination consisting essentially of (1) from about 50 to 98% of liquid polybutenes having a viscosity at 210°F. of from 30 to 1,100 SSU (Saybolt Seconds Universal); (2) from about 2 to 50% fatty materials which provide lubricity and antifriction properties to the lubricant; and (3) from 0 to 5% of an emulsifier. The polybutenes provide the requisite load-bearing properties to the lubricant. The lubricants provide the necessary adherence, boundary lubrication and freedom from surface discoloration and contamination after annealing or casting that is required by various metal working or fabricating processes. The liquid polybutenes suitable for use in the process are commercially available products and can be prepared by polymerizing butenes from a butane-butylene stream as described, for example, in U.S. Patents 2,407,873; 2,677,000; 2,677,001 and 2,677,002.

The fatty materials which provide lubricity and antifriction properties to the lubricant can be high molecular weight fatty acids and glycerides derived from animal, vegetable and marine fats and oils; esters of such fatty acids and glycerides; oxidized oils; sulfurized fatty acids and oils such as sulfurized sperm oil; halogenated materials such as chlorinated paraffin waxes; and naturally occurring and synthetic waxes such as sperm wax, montan wax and beeswax. The saturated fatty materials are preferred.

In the following metal-working lubricant formulations, the polybutenes used are identified as follows:

Polybutene A — liquid polybutene having about 300 average molecular weight and a viscosity at 210°F. of about 40 SSU.
Polybutene B — liquid polybutene having about 830 average molecular weight and a viscosity at 210°F. of about 1,100 SSU.
Polybutene C — liquid polybutene having about 460 average molecular weight and a viscosity at 210°F. of about 65 SSU.

Formula 1

Polybutene A	65%
Polybutene B	20%
Hydrogenated tallow fatty acid	15%

Formula 2

Polybutene C	96%
Hydrogenated tallow fatty acid	4%

Formula 3

Polybutene C	92%
Hydrogenated tallow fatty acid	8%

Each of the above lubricants was tested on a commercial steel cold rolling mill consisting of four 4-high mill stands 54 in. wide operating at speeds of 800 to 900 feet per minute effecting a reduction in thickness of from about 0.8 to 0.2 inch. The lubricants were applied to the hot rolled steel strip from the pickling bath as it was being wound into coils on the upcoiler. The hot rolled coils did not show any corrosion or loss of lubricant from the coils prior to cold rolling. Thereafter, the lubricated coils were passed through the cold rolling mill without further lubrication, and the rolled strips annealed at about 1000 to 1800°F. After annealing, the annealed sheets did not show any surface discoloration or surface blemishes. Rolling speeds were comparable to the roll speeds obtained with conventionally used mineral oil based lubricants.

In the production of thin tinplate, the previously annealed cold rolled steel sheets are further reduced to the desired thickness. The mechanical energy required to effect this reduction produces substantial heat in the strip. Therefore, water emulsions of the rolling lubricants are used for both lubrication and heat removal. The metal-working lubricants can be used to form the desired oil-in-water emulsions containing from 5 to 60% lubricant depending on the type of emulsion used by the specific mill. Quick breaking emulsions can be formed without the use of any emulsifier for single use service. For recirculating type mills, the addition of up to 5% emulsifier produces a stable emulsion suitable for extended service periods. Conventional emulsifiers for the preparation of oil-in-water emulsions can be used. The preferred emulsifiers are diethanolamine and diglycolamine. The following formulas illustrate lubricants that form either quick breaking or stable oil-in-water emulsions.

Formula 4

Sperm oil wax	10%
Blown sperm oil (Sperm 1000 S)	10%
Hydrogenated tallow fatty acid	15%
Triethanolamine	5%
Polybutene A	60%

Formula 5

Sperm oil wax	10%
Blown sperm oil (Sperm 1000 S)	10%
Hydrogenated tallow fatty acid	15%
Polybutene A	65%

Polybutene Stabilized with Propylene Tetramer

R.D. Guminski; U.S. Patent 3,298,951; January 17, 1967 describes a lubricant for aluminum rolling mill bearings which is substantially nonstaining of aluminum sheet and which maintains a substantially constant viscosity in use. The lubricant comprises a polybutene and a viscosity stabilizer; in some cases a light mineral oil may be present to adjust the viscosity of the composition to a desired value. Any of the butene polymers having an approximate number average molecular weight between about 500 and 5,000 may be utilized in the composition.

The mineral oil with which the polybutenes may be compounded is a light mineral oil, derived from petroleum and having a very low bromine number, so that it contains very few unsaturated groupings which are believed to lead to some staining problems. Gulf mineral seal oil and mineral colza oil have similar characteristics, typically a boiling range of from 428° to 536°F., a low viscosity of about 37.5 SSU at 100°F. and a bromine number of 0.4. Broadly, the light mineral oil has a viscosity between about 30 and 100 SSU at 100°F., usually 30 to 50 SSU at 100°F., and a bromine number not in excess of 0.8.

The viscosity stabilizer(s) present in the composition are highly reactive organic substances (compounds) of low molecular weight, which by virtue of ethylenically unsaturated bonds in their molecules can be easily polymerized. The molecular weight of the viscosity stabilizer is preferably below 250 and should not exceed about 300. Subject to this qualification of maximum molecular weight, it appears that all oil-soluble, ethylenically unsaturated carboxylic acids, esters, aldehydes, ketones and alcohols are, to a greater or lesser degree, suitable viscosity stabilizers. The especially preferred viscosity stabilizer additive is propylene tetramer in an amount of 1 to 3%, but other branched chain alkenes (either monomer or lower polymer) are suitable. Among other preferred stabilizers are: styrene, acrylic acid, mesityl oxide, isophorone, acrolein, crotonaldehyde, vinyl acetate, methacrylic acid, methyl methacrylate and methyl acrylate.

The following are examples of lubricant compositions which possess a high original viscosity, retain it after the application of shearing forces, and are free from or have only a small tendency to form brown stains on aluminum strip during annealing.

Composition 1*

Mineral colza oil	24%
Polybutene 24	75%
Styrene	1%

*Original viscosity 1,850 SSU at 100°F.; after the application of shearing forces for 15 hours, 2,180 SSU at 100°F.

Composition 2*

Mineral colza oil	48%
Polybutene 128	50%
Styrene	2%

*Original viscosity 1,300 SSU at 100°F.; after the application of shearing forces for 15 hours, 1,550 SSU at 100°F.

Composition 3*

Mineral colza oil	48%
Polybutene 128	50%
Acrylic acid	2%

*Original viscosity 1,300 SSU at 100°F.; after the application of shearing forces for 15 hours, 1,550 SSU at 100°F.

Composition 4*

Mineral colza oil	46%
Polybutene 128	50%
Styrene	2%
Lauryl alcohol	2%

*Original viscosity 1,400 SSU at 100°F.; after the application of shearing forces for 15 hours, 2,050 SSU at 100°F.

Composition 5*

Polybutene 128	52%
Propylene tetramer	2%
Stearyl alcohol	½%
Cetyl alcohol	½%
Santolube AR and Topanol BHT:	
Antioxidants	1/10%
Gulf mineral seal oil, the balance.	

*Original viscosity 1,450 SSU at 100°F.; after the application of shearing forces for 15 hours, 1,700 SSU at 100°F.

It was found that by varying the polybutene content of Composition 5 the viscosity could be increased from 200 SSU at 100°F. with 30% polybutene to 1,140 SSU at 100°F. with 50% polybutene and up to 3,200 SSU at 100°F. with 60% polybutene, the polybutene being polybutene 128 in each case. For this purpose a polybutene 128 content of 45 to 60% is preferred. It will be seen that the function of the light petroleum fraction is to reduce the viscosity of the polybutene to a desired value, at which it is stabilized by the viscosity stabilizer additive.

It will be observed that in all the above cases the viscosity of the stabilized roll bearing lubricants was actually increased after a number of hours service by virtue of the very small quantity of reactive unsaturated viscosity stabilizing additive present. The amount of styrene required to stabilize the lower molecular weight polybutene in Composition 1 was less than that required for the higher molecular weight polybutene of Composition 2.

The purpose of the addition of the long chain saturated fatty alcohols in Compositions 4 and 5 is to act as nonstaining, load-bearing additives when the mill rolls are turning slowly. Under slow speed conditions the hydrodynamic lubrication due to polybutene is only partially developed and it is therefore found desirable to add load-bearing additives to support the load under these conditions. In general, the bearing lubricant will include up to about 5% of load-bearing additive and more commonly between about 0.5 to 2% of the additive.

Example 1: When a bearing lubricating composition of this type was tested in a foil-bearing apparatus, it was found that the lubricant behaved as a non-Newtonian fluid with a coefficient of friction in which it was designed to operate. The results obtained indicated that the lubricant composition was superior to the Newtonian mineral oils of comparable viscosity which are in standard general use as roll bearing lubricants in rolling mills.

Example 2: The stabilizing effects of other unsaturated substances on the viscosity of polybutenes was tested. The test was carried out in the following manner. First, a quantity of high molecular weight polybutenes (e.g., Oronite 128) was heated to about 160°F. and mixed with a light mineral oil such as mineral colza oil, which has a viscosity of about 37.5 SSU, a boiling range of 425° to 535°F., and a low bromine number, to give a basic lubricant having a viscosity of about 1,500 SSU at 100°F.

Second, some of this basic lubricant was mixed with 2% by weight of the viscosity stabilizer to be tested, the mixing being carried out by stirring manually at ambient temperature. Third, the viscosity of a 500 gram sample of the lubricant-stabilizer mixture was measured at 100°F. using a Redwood No. 1 viscometer. This measurement of the viscosity of the freshly made and substantially nonsheared sample of the lubricant-stabilizer mixture was made with an accuracy of 5 SSU.

Fourth, the sample of the stabilized lubricant on which the viscosity determination had been carried out was placed immediately after that determination in the bowl of a Sunbeam Mixmaster food mixing machine, which was equipped for this purpose with two contrarotating 4 bladed paddles, one of which rotated the bowl. The speed of the paddles was set at 330 rpm and the sample of the stabilized lubricant was stirred continuously for 24 hours, during which time its temperature rose from its initial value of about 100°F. to 125°F. On completion of the 24 hour stirring of the sample, its viscosity was measured at 100°F. using the Redwood No. 1 viscometer. The results of this test are set out below.

Compound	Initial Viscosity SSU 100° F.	Final Viscosity SSU at 100° F.
Methyl methacrylate	1,150	1,200
Crotonaldehyde	1,160	1,205
Dipentene	1,157	1,208
Vinyl Acetate Monomer	1,100	1,120
Polybutene light polymer (Mixture of butylene tetramer and Pentamer in 1:1 ratio):	1,200	1,200
Acrylic acid	1,170	1,168
Methyl acrylate	1,100	1,108
Iso-phorone	1,150	1,175
Mesityl oxide	1,130	1,140
Kerosene (control)	1,160	850

Example 3: Using a lubricant with 50% polybutene 128 and other ingredients in the proportions given in Composition 5, excellent protection was given to white metal-steel bearings of a mill for cold rolling aluminum strip. Its leakage into the virtually nonstaining cold rolling lubricant did not give rise to any staining tendency of the strip, even when the concentration of the bearing lubricant in the rolling lubricant was allowed to rise to the unusually high level of 15%. The lubricant in the rolling lubricant system of an aluminum cold rolling mill is usually discarded before this degree of contamination occurs.

α-Bromocinnamaldehyde Antimicrobial Activity

A process described by P. Adams, E.G. Shay and A.N. Petrocci; U.S. Patent 3,515,671; June 2, 1970; assigned to Millmaster Onyx Corporation involves metal-working lubricants of antimicrobial α-halo, α,β-unsaturated carbonyl compounds of the type which may be derived by, for example, the Claisen-Schmidt condensation of a substituted or unsubstituted aromatic aldehyde with an aldehyde or a ketone, followed by halogenation and subsequent dehydrohalogenation, and having the general formula

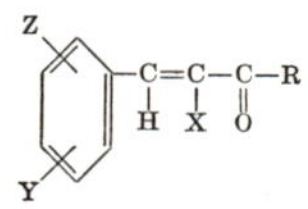

where X is a halogen, Y and Z may be hydrogen, halogen, alkyl or nitro, and R may be hydrogen or alkyl, aryl or cycloalkyl containing from 1 to 8 carbon atoms.

Methods for the preparation of these types of compounds are known. For example, benzaldehyde or its mono- or dialkyl, halo or nitro substituted derivatives may be condensed with an aldehyde such as acetaldehyde and propionaldehyde, or a ketone such as acetone, methyl ethyl ketone, acetophenone and methyl cyclohexyl ketone. The resulting substituted or unsubstituted cinnamaldehyde or its derivatives may be halogenated, in the course of which process two halogen atoms add at the double bond. On treatment with a mild alkali, one halohydrogen is split off, leaving the desired compound, namely, a substituted or unsubstituted alpha halo cinnamaldehyde. The following examples illustrate the process.

Example 1: A variety of common types of water based cutting and grinding fluids were tested as described below. Some of these are formulated commercial products. Others were formulated according to the art. To an aliquot of each there was added sufficient α-bromocinnamaldehyde such that on diluting one part of concentrate by volume with water to make 25 parts, the respective samples contained 400, 200 or lesser parts per million of the agent, as required. Sterile wide mouth, four ounce jars were charged with 100 ml. of the prepared fluids. In each case, a blank was set up containing the same fluids, but no preservative; these were examined for a background count of contaminating organisms.

A series of 24 hour broth cultures of Escherichia coli, Staphylococcus aureus, Pseudomonas aeruginosa, Bacillus species, Proteus species and Aerobacter aerogenes were pooled, diluted with sterile broth and inoculated into the 100 ml. volumes of prepared fluids to provide 1 to 10 x 10^6 bacteria cells/ml. of prepared fluids.

At weekly intervals, up to five weeks after the original inoculation, the samples were examined to determine the number of viable organisms present. At the five week point, each jar was reinoculated with the culture as originally, and the testing was continued for a total period of ten weeks. The counts are given in the following table, in which SLA represents a sulfurized lard oil composition, M.S. a mahogany soap-mineral oil composition, TEAN a triethanolamine naphthenate composition and TEAP a triethanolamine phosphate ester composition. The concentration of α-bromocinnamaldehyde is given in parts per million, and the bacterial count is to be multiplied by 10^3.

Oil Comp.	Agent p.p.m.	Weeks 0	1	2	3	4	5	6	7	8	9	10
SLA	0	2,760	280	1,600	2,700	3,700	3,200	5,800	4,000	1,500	1,600	8,400
SLA	400	2,760	0	0	0	0	0	0	0	0	0	0
M.S.	0	3,290	90,000	70,000	25,000	4,100	1,390	95,000	72,000	11,600	6,500	11,200
M.S.	200	3,290	<10	<10	0	0	0	0	0	0	0	0
TEAN	0	3,290	9,400	<1,000	7	440	1,040	9,300	1,080	1,390	540	1,700
TEAN	200	3,290	<10	<10	0	0	0	0	0	0	0	0
TEAP	0	4,140	610	540	6,100	7,400	6,300	118,000	5,100	800	1,520	1,880
TEAP	400	4,140	0	0	0	0	0	0	0	0	0	0
P.S.	0	2,760	2,500	3,500	1,800	900	650	15,000	7,600	11,000	7,400	3,800
P.S.	200	0	0	0	0	0	0	0	0	0	0	0

The mahogany soap-mineral oil composition was further tested at higher dilutions, the results being tabulated below.

Oil Comp.	Agent, p.p.m.	Weeks 0	1	2	3	4	5	6	7	8	9	10
M.S.	0	2,760	7,100	2,300	1,300	1,000	630	13,000	8,100	4,100	9,300	4,400
M.S.	200	2,760	0	0	0	0	0	0	0	0	0	0
M.S.	100	2,760	0	0	0	0	0	0	0	0	0	0

The α-bromocinnamaldehyde is thus found to be effective at low concentration in the fluids. For the sake of uniformity in testing, the 1:25 dilution was made standard. Variations are not likely to cause significant differences in view of the fixed preservative content.

Example 2: Dilutions of cutting fluids prepared as in Example 1, but with the addition to each of 5 grams of iron filings, were maintained at a temperature of 120°F. for a period of two weeks, following which they were inoculated as before. They were then examined by determining the viable organism count, which was found to be essentially similar to those of Example 1.

Alkyl Methacrylate-Methacrylic Acid Copolymers

H. Gisser and M. Petronio; U.S. Patent 3,554,913; January 12, 1971; assigned to the U.S. Secretary of the Army describe friction-reducing compositions which consist of the copolymers of n-alkyl methacrylates with methacrylic acid where the alkyl group may vary from 14 to 22 carbon atoms. A copolymer of this type enables variation of the long chain alkyl group, its concentration, and the concentration of the polar groups. The preferred copolymers consist of n-hexadecyl methacrylate and methacrylic acid and n-docosyl methacrylate and methacrylic acid.

The copolymers were prepared by heating the monomer mixture in 10% solution in benzene with 1% benzoyl peroxide in sealed glass tubes at 80°C. for 16 hours in a nitrogen atmosphere. The copolymers were precipitated from the benzene solution by adding ethanol and then recrystallized from benzene by precipitation with ethanol.

Coefficient of friction (μ) was determined on a modified Bowden-Leben machine in which the tangential force was detected by strain gauges mounted on a strain ring, the signals being amplified and recorded. All friction measurements were made with a 1/4 inch diameter 440 C stainless steel ball rider on an approximately 6 mm. thick plane substrate. Steel, copper and glass substrates were employed for the measurements. The steel substrate was 1020 steel plate prepared by rinsing in boiling benzene, polishing successively with 400, 600, 2/0, 3/0 and 4/0 silicon carbide paper, and rinsing in boiling benzene. It was then placed in a desiccator over anhydrous calcium sulfate for 24 hours before use. The copper substrate (99.98% Cu) was prepared in the same manner as the steel. The glass specimens were cut from 6 mm. thick polished plate, silvering quality soda lime glass, abraded under water successively with 400 and 600 silicon carbide paper, followed by rinsing with hot distilled water. These specimens were also stored over anhydrous calcium sulfate for 24 hours before use.

All measurements were made at a 100 gram load and 0.04 cm./second sliding speed. The length of traverse was 1.3 to 1.9 cm. for each measurement. Three drops of polymer solution were placed in a continuous line along the path to be traversed by the rider, the first drop being placed at a location so that it wet the rider. The area covered by the fluid was about 0.9 sq. cm. Friction measurements were started 5 to 10 seconds after applying the copolymer solutions. The standard deviation of the coefficient of friction measurement was 0.005.

The carboxyl groups varied as a function of the mol percent methacrylic acid. Accordingly, polymers of n-hexadecyl methacrylate were prepared by copolymerizing with 20, 40, 60 and 90 mol percent of methacrylic acid. In the case of n-docosyl methacrylate, polymers were prepared by copolymerizing with 30, 50 and 90 mol percent of methacrylic acid. The results of the friction tests for n-hexadecyl methacrylate and methacrylic acid are given below in Table 1. Results for the n-docosyl methacrylate polymer are presented in Table 2. In both cases a concentration of 1.0 mg. copolymer per ml. of xylene was used; however, benzene or various petroleum hydrocarbon fluids could also be employed.

TABLE 1:

Mol Fraction	μ on Steel	μ on Copper
0.0	0.210	0.262
0.2	0.198	0.251
0.4	0.168	0.238
0.6	0.157	0.238
0.9	0.221	0.280

TABLE 2:

Mol Fraction	μ on Steel
0.0	0.210
0.3	0.185
0.5	0.162
0.9	0.176

As shown by the data, a substrate coated with copolymer consisting of about 50 to 60 mol percent methacrylic acid will have a minimum coefficient of friction. Since there is an increase in the coefficient of friction when either the methacrylic acid or the straight chain alkyl groups are increased, it is evident that both the polar carboxyl groups and the alkyl chain cooperate in decreasing friction on a surface. It should be noted that the concentration of the polymer in xylene, benzene or other solvents may vary from 0.01 to 10.0 mg./ml. without any significant effect on the coefficient of friction. The molecular weights of the n-hexadecyl methacrylate copolymers are given below.

Mol Fraction Methacrylic Acid	Molecular Weight* $\times 10^{-5}$	Relative Polymer Chain Length**
0	0.50	0.021
0.2	0.50	0.024
0.4	1.0	0.059
0.6	1.5	0.11

*Determined by membrane osmometry.
**Relative to poly(methyl methacrylate).

A process described by H. Gisser and M. Petronio; U.S. Patent 3,496,103; February 17, 1970; assigned to the U.S. Secretary of the Army describes poly(n-alkyl methacrylate) where the alkyl group contains 22 carbon atoms which reduces the friction of solid surfaces when applied in solution or as a dry film.

OTHER LUBRICANTS

Alkyl Substituted Biphenyls or Terphenyls as Nonspreading Lubricants

H. Mertwoy and H. Gisser; U.S. Patent 3,426,076; February 4, 1969; assigned to the U.S. Secretary of the Army describe nonspreading lubricants and fluids which are oxidatively stable at low and high temperatures and comprise a mono- or di-sec-butylbiphenyl or terphenyl or mono- or di-sec-butylmethoxybiphenyl or methoxyterphenyl. In the table below spreading data are presented for the nonspreading oxidatively stable fluids, the data being obtained at room temperature under anhydrous conditions.

Compound	Percent spreading after—	
	7 days	30 days
4-sec-butylbiphenyl	(1)	------------
4,4'-di-sec-butylbiphenyl	0	0.3
4-sec-butyl-4'-methoxybiphenyl	0.4	0.9
3,4-di-sec-butyl-4'-methoxybiphenyl	0	0.8
4-sec-butyl-m-terphenyl	0	0
4,4''-di-sec-butyl-m-terphenyl	0.5	1.2
4-sec-butyl-o-terphenyl	1.2	1.5
4,4''-di-sec-butyl-o-terphenyl	0	0.5
4-sec-butyl-2''-methoxy-p-terphenyl	5.0	20.0
Bis(p-isopropylbenzyl)3-methyl adipate [2]	5.0	115.0

[1] Evaporated.
[2] Compound is a known spreading fluid and is presented for comparison purposes.

Despite the high boiling point of 4-sec-butylphenyl (107°C. at 0.7 mm.), it completely evaporated within about four days leaving no residue on the disc on which it was being tested. This compound has been found useful in closed systems. While a drop of 4-sec-butyl-4'-methoxybiphenyl only spreads 0.9% after 30 days, its volume had noticeably decreased indicating that this compound (BP 146°C. at 0.7 mm.) was also evaporating but at a much slower rate.

Spreading of an oil was determined in the following manner. One inch diameter discs of WD 52-100 steel, hardened to Rockwell C-62, were polished successively with 0, 2/0, 3/0 and 4/0 emery paper, washed under warm tap water while being scrubbed with cotton gauze, then washed with distilled water followed by redistilled methanol. The discs were then dried by clean dry air. To determine cleanliness, a drop of redistilled methanol was placed on the surface of each disc. Complete and uniform evaporation indicated surfaces acceptable for these tests. A drop of the oil to be tested, 1 to 2 mm. in diameter, was placed on each disc by means of platinum wire. The degree of spreading was obtained by measuring the change in diameter after 7 and 30 days, respectively, using a traveling microscope.

The nonspreading fluid may be prepared as following using 4-sec-butyl-4'-methoxybiphenyl and 3,4-di-sec-butyl-4'-methoxybiphenyl as examples. p-Phenylanisole (0.25 mol, 46 g.) was dissolved in 700 ml. of hot heptane contained in a 1,000 cc four necked flask, fitted with a stirrer, reflux condenser, thermometer and dropping funnel. Anhydrous aluminum chloride (14 g.) was then added. The solution was heated, and while maintaining a temperature at 75° to 85°C., 0.27 mol (25 g.) 2-chlorobutane was added dropwise over a period of 2.5 hours. After addition was complete, stirring was continued for an additional 5.5 hours, the temperature being maintained at 75° to 85°C. The reaction mixture was poured over an equal volume of cracked ice and stirred vigorously to decompose the aluminum chloride complexes. The organic layer was washed free of acid with distilled water. The unreacted p-phenylanisole was removed by distillation using a steam condenser. The remaining mixture was distilled in a molecular pot still to separate the product from solids and tarry products. The liquid was then fractionated using a Vigreux column to yield both 4-sec-butyl-4'-methoxybiphenyl (BP 137°C. at 0.65 mm.) and 3,4-di-sec-butyl-4'-methoxybiphenyl (BP 162°C. at 0.75 mm.). Both compounds were then fractionated for further purification using a Vigreux column until the distillation range was less than 2°C.

In the table below some oxidation rates of the nonspreading compound are presented. As can be readily observed from the data, there is very little oxygen uptake even after 100 hours. Oxygen consumption at 100°C. was measured by using standard Warburg manometry.

Compound	(Moles O_2/mole compound/hr.) $\times 10^4$	
	Initial rate during 5–35 hr.	Final rate during 65-100 hr.
4-sec-butylbiphenyl	0.03	0.03
4,4'-di-sec-butylbiphenyl	0.13	0.17
4-sec-butyl-4'-methoxybiphenyl	0.00	0.00
3,4-di-sec-4'-methoxybiphenyl	0.12	0.20
4-sec-butyl-m-terphenyl	0.00	0.00
4,4''-di-sec-butyl-m-terphenyl	0.32	0.21
4-sec-butyl-o-terphenyl	0.00	0.00
4,4''-di-sec-butyl-o-terphenyl	0.30	0.23
4-sec-butyl-2''-methoxy-p-terphenyl	0.00	0.00

Mold Lubricant for Casting Oxidizable Metals

R.L. Holshouser; U.S. Patent 3,034,186; May 15, 1962; assigned to The Dow Chemical Company describes a mold lubricant which provides an inexpensive and effective method of reducing the oxidation of a readily oxidizable metal, and particularly of preventing burning of magnesium, as it emerges from the mold in a continuous casting operation.

The process consists of dispersing boric acid in a suitable oily or oily base material to make an oxidation preventive lubricant which is then applied to the inner surface of a continuous mold used in the continuous casting of readily oxidizable metals, e.g., magnesium and aluminum and their alloys. Among the oily or oily base materials suitable for practicing the process are vegetable, animal and mineral oils which possess a sufficient high oxidizing temperature to resist charring, a low moisture and impurity content, safe flash and fire points, and a pour point and viscosity which impart fluidity without excessive accumulation of oil at the edge of the top surface of the molten metal and without excessive drainage at the temperatures employed in the casting operation. Palm and peanut oils, lard oil and petroleum derived fractions having a Saybolt Universal viscosity at 100°F. at about 200 seconds are examples of suitable oils.

Oil suspended graphite particles may be advantageously added to the mixed glyceride oil above. Preferably the graphite is in flake form and the suspending oil contains a thickening agent such as a soap. The oil suspended graphite which may be used usually consists of a mineral oil having a Saybolt Universal viscosity at 210°F. of between 600 and 700 seconds as determined according to the ASTM test D-88-53, thickened with a metallic soap, e.g., calcium stearate, and containing 10 to 15% of fine graphite in suspension. The oil suspended graphite usually has a work penetration of between 265 and 295 mm. as determined by ASTM penetration test D-217-52T and meeting the specification of the National Lubrication Grease Institute described as No. 1.

Example: This examples illustrates a preferred formula and method. A lubricant, consisting by weight of 2/3 lard oil falling within the values set out before and containing the glycerides set out below, 1/3 calcium stearate thickened oil suspended graphite as described before containing 12.5% electric furnace graphite and 2% anhydrous boric acid by weight, was prepared. The lard oil used contained glycerides of the following fatty acids:

	Weight Percent
Myristic acid	1
Palmitic acid	25
Stearic acid	12
Oleic acid	48
Linoleic acid	12
Linolenic acid	1
Arachnidonic acid	1

The lard oil and oil suspended graphite described above were mixed together in a 2:1 ratio by weight at room temperature. To 50 pounds of the resulting mixture thus prepared, 1 pound of technical grade boric acid powder was added and mixed into the lard oil and oil suspended graphite mixture.

A mold for continuous casting of magnesium ingot, similar to that described in U.S. Patent 2,503,819 but without the stilling plate or vibrator shown there, was prepared for operation. Magnesium metal was melted in a kettle equipped with a heating means and pumped therefrom to the continuous mold. Prior to its receiving molten magnesium into the mold from the heated kettle providing a continuous source of the metal, a thin coating of the lubricant prepared above was applied by a brush to the interior of the mold. Molten metal was then received into the mold and the casting was started. As the casting operation proceeded, additional applications of the boric acid treated lubricant were made by brushing the interior of the mold above the level of the molten metal. After coming in contact with the interior of the mold, the viscosity of the lubricant was lowered and the lubricant therefore flowed gradually downward to the point of contact of the molten magnesium and the interior of the mold wall. It was then carried along with the magnesium as the magnesium moved downward. The magnesium gradually solidified and emerged below the mold. No sulfur dioxide or other oxidation preventative was employed. No tendency to flame or burn was detected in the emerging magnesium due to the presence of the boric acid treated lubricant. The fire hazard and metal loss were therefore clearly lessened. The exudate appearing along the surface of the magnesium thus cast was definitely less oxidized as shown by a marked reduction in darkening due to the exudate when compared to magnesium ingots which were cast in a similar manner but in which no boric acid was employed in the lubricant. Irregularities due to lapping and folding were clearly less prominent in the ingots produced.

Dibasic Lead Phosphite for Solid Film Lubricant

G.P. Murphy, Jr.; U.S. Patent 3,314,885; April 18, 1967; assigned to the U.S. Secretary of the Army describes an improved resin bonded solid film lubricant which is provided by admixing a quantity of dibasic lead phosphite with an epoxy-phenolic resin system, antimony trioxide, molybdenum disulfide, magnesium bentonite and diethylene dioxide. The following examples illustrate compositions of the process.

Example 1:

	Percent by Weight
Epoxy-phenolic resin system (35% solids) (solids: 12.4)	35.4
Antimony trioxide, powdered, ACS grade	6.2
Molybdenum disulfide, natural	17.4
Dibasic lead phosphite	4.1
Modified magnesium bentonite	0.5
1,4-Diethylene dioxide	36.4
	100.0

Alternative lubricants may be prepared consisting of the following compositions.

Example 2:

	Percent by Weight
Epoxy-phenolic resin system (35% solids) (solids: 10.0)	28.5
Antimony trioxide, powdered, ACS grade	5.2
Molybdenum disulfide, natural	14.5
Dibasic lead phosphite	7.9
Modified magnesium bentonite	1.5
1,4-Diethylene dioxide	42.4
	100.0

Example 3:

	Percent by Weight
Epoxy-phenolic resin system (35% solids) (solids: 10.0)	28.5
Antimony trioxide, powdered, ACS grade	5.7
Molybdenum disulfide, natural	16.0
Dibasic lead phosphite	5.9
Modified magnesium bentonite	1.5
1,4-Diethylene dioxide	42.4
	100.0

In the above identified examples, the epoxy-phenolic resin system which has been selected by preference is that known in the trade as Epon Resin Formulation YP-100. The ingredients of this resin system are composed of 35% solids and 65% solvent taken from the following compositions:

Solids	Percent by Weight
Epon resin 1007	73.1
General Electric resin 75108	24.4
Silicone resin SR-82	1.0
Phosphoric acid	1.5

Petroleum

Solvents	Percent by Weight
Ethyl alcohol	25.0
Methyl ethyl ketone	15.0
Toluene	50.0
Pine oil	10.0

In this resin system Epon resin 1007 is commercially available and is a solid high molecular weight polyglycidyl ether of 2,2-bis(4-hydroxyphenyl) propane.

The ingredients set forth in the formulations were prepared by adding them in the following order to a one gallon friction lid can containing 20 one-half inch diameter steel balls: (1) epoxy-phenolic resin system; (2) antimony trioxide; (3) molybdenum disulfide; (4) dibasic lead phosphite; (5) modified magnesium bentonite; and (6) diethylene dioxide. There was substantially no time lag between the addition of each ingredient. The can was sealed, shaken by hand for approximately one minute and then rolled for four hours at room temperature. The formulation was prepared at room temperature and no heating was required.

It is always considered best practice to steel grit blast or sand blast steel surfaces prior to zinc phosphatizing. All test specimens were steel grit blasted prior to zinc phosphatizing. Inasmuch as the formulation contains a relatively high percentage of solvent, it was necessary to dip each specimen three times in order to obtain a coating thickness of 0.0004 to 0.0007 inch. The dipping procedure was accomplished by suspending the specimen in the lubricant for a period of 5 seconds for each successive dip, removing it after each dip and allowing it to dry for ten minutes after each of the first two dips before proceeding with the next dip. After the final dip, the specimen was allowed to dry for one-half hour followed by oven curing to bake the resin at 400°F. for one hour.

Coatings provided in accordance with the above described formulations were tested and found to provide excellent wear life as well as corrosion protection when applied to zinc phosphatized steel as well as anodized and sealed aluminum. For this purpose, test specimens coated with these lubricants consisted of American Iron and Steel Institute (AISI) C1020 steel panels 1/8" x 2" x 3" for the salt spray cabinet exposure tests, and standard Falex test specimens for the wear tests. The Falex pins were made from AISI 3135 steel hardened to Rockwell B87-90. The V-blocks were made from AISI 1137 steel with a hardness of Rockwell C20.

In order to demonstrate the effectiveness of employing dibasic lead phosphite in the lubrication composition, specimens were tested using compositions made in accordance with the formulations described here and comparison made with a composition prepared in accordance with Example 2 of U.S. Patent 3,223,626. The results of this comparison are as follows:

	Falex Wear Life (in minutes)	20% Salt Spray (hours to failure)
Example 1	500	120
Example 2	160	168
Example 3	375	168
Example 2 of U.S. Patent 3,223,626 (no dibasic lead phosphite)	365	24

It can be readily seen from this comparison that the formulations containing dibasic lead phosphite give better corrosion protection without sacrifice of wear life. The increased corrosion resistance is due to the ability of dibasic lead phosphite to accept and neutralize acid formed by the hydrolysis of the molybdenum disulfide. At the same time, the dibasic lead phosphite is responsible for a slight increase in wear life.

The dibasic lead phosphite used in this process is that whose formula is: $2PbO \cdot PbHPO_3 \cdot 0.5H_2O$. It is a white crystalline solid with a molecular weight of 743, a specific gravity of 6.94 and a lead content of 90.8%.

Monomolecular Lubricant

A process described by D.H. Ulman; U.S. Patent 3,305,482; February 21, 1967; assigned to Honeywell Inc. is related to lubrication and particularly to monomolecular lubrication which comprises the application of an extremely thin (one molecule thickness) layer of lubricant to the surface to be lubricated. An application of monomolecular lubrication is a gas or air journal bearing of the hydrodynamic type. Under starting conditions, the rotating and nonrotating portions of these journal bearings are in contact with each other; therefore, a high frictional coefficient is undesirable because of the high starting torque requirements and the wear associated with the relative movement bewteen these surfaces. Because gaseous support is being utilized in the hydrodynamic bearing, the liquid type lubrication techniques cannot be used.

This process involves mixing abrasive particles and a polar lubricant with a sufficient quantity of a nonpolar solvent to form a paste. The term paste means, for purposes of this process, a composition comprising a polar lubricant, abrasive particles and a nonpolar solvent where the amount of polar lubricant in the composition is substantially greater than the amount of polar compound in the surface film which is cleaned (removed) from the surface to be lubricated. The viscosity of the composition is not critical and will vary from very high to very low depending on various constituents. This paste (composition) is rubbed against the surface of a solid so as to simultaneously chemically clean the surface and allow the lubricant to be adsorbed thereto.

The relative movement between the paste and the surface causes the abrasive particles to remove the surface film contamination on the surface to be lubricated and exposes the high energy level substrate surface. However, it should be noted that the high energy level substrate surface is not exposed to the environment since the paste completely covers the cleaned surface. Therefore no contamination of the substrate surface can occur. The lubricant is thus positioned adjacent to the substrate surface and readily absorbs thereto.

All of the excess paste is then removed from the surface except for a monomolecular film of lubricant by suitable cleaning method such as scrubbing the surface with a detergent solution. Adhesion of the monomolecular film of lubricant to the substrate surface is much higher than the adhesion between the outer surface of the monomolecular film and any adhering molecules. Consequently by sufficiently cleaning the surface, the low adhesion between the monomolecular film and the other molecules is overcome and the other molecules are removed. However, the strong adhesion between the monomolecular film of lubricant and the substrate is not disturbed so that a single monomolecular film is retained upon the surface to be lubricated. The following examples are illustrative of the process.

Example 1: The specimen to be lubricated is a cylindrical shaft of a hydrodynamic journal bearing which has an aluminum oxide surface. Since the abrasive particles function to remove the surface film contamination from the surface to be lubricated, the abrasive is chosen with reference to the hardness of the surface to be lubricated. The abrasive which is utilized for cleaning an aluminum oxide surface is cerium oxide. The size of the abrasive particles is also important in the cleaning process. The monomolecluar films are approximately 25 Angstroms thick so that the surface which is lubricated must be extremely smooth. Excellent results have been obtained with abrasive particles having a diameter of 0.1 micron (25 Angstroms would equal 0.0025 micron).

A suitable polar lubricant must be chosen for the aluminum oxide surface to be lubricated. A polar lubricant, as utilized in this process, is defined as a polar straight chain compound having at least 10 carbon atoms which is capable of close packing (that is, no side branches). One end of the straight chain compound must be strongly polar so as to be strongly adsorbed to the surface to be lubricated and the other end must have a low energy end grouping so as to provide an oleophobic surface. Molecular chain length must be essentially constant for optimum planar alignment of the oleophobic end groups of the polar molecules.

Acids, amines or alcohols satisfying these requirements may be utilized. Examples of such an acid are stearic acid, perfluoro octanonic acid, and arachidic acid. Examples of such an amine are octadecyl amine and hexadecyl amine. Examples of such an alcohol are eicosyl alcohol and octadecyl alcohol. Excellent results were obtained in utilizing stearic acid as the polar lubricant for the aluminum oxide surface. Stearic acid is a solid material at room temperature and a suitable nonpolar solvent must be utilized to combine the abrasive particles (cerium oxide) and the polar lubricant (stearic acid) into a paste. Five parts abrasive particles and 95 parts polar lubricant have proven satisfactory, as has 50 parts abrasive particles and 50 parts polar lubricant.

The paste comprising abrasive particles, polar lubricant and nonpolar solvent is positioned upon a piece of microcloth (polishing cloth) which is in turn positioned upon a resilient material such as a sponge. The sponge is positioned over a smooth rigid surface such as a normal work table. The paste is then rubbed against the surface of the cylindrical shaft for a period of 1 to 2 minutes by moving the shaft relative to the microcloth. The resilient material allows the pressure to be applied around a large segment (approximately 90°) of the aluminum oxide surface of the cylindrical shaft. The shaft is then rotated 90° and again the paste is rubbed against the surface for a period of 1 or 2 minutes. This continues until the paste has been rubbed against the entire surface.

Rubbing the paste against the aluminum oxide surface causes the cerium oxide abrasive to abrade the surface and remove the surface film contamination thereby chemically cleaning or ultracleaning the aluminum oxide surface. As soon as the high energy substrate surface is exposed, the stearic acid compound is adsorbed. It should be noted that this cleaning of the surface to be lubricated is being accomplished under the paste, that is, there is no exposure of the ultraclean surface to the atmosphere. Since the lubricant is contained in the paste, the desired lubricant is contiguous the surface and is adsorbed. As the lubricant adsorbs to the surface, the long straight chain molecules pack closely relative to one another so as to form an oleophobic surface on the aluminum oxide shaft.

After the cleaning process, the aluminum oxide surface is tested to determine that a monomolecular film of stearic acid remains and that the film is uniform over the whole surface. The nonwetting characteristic of the oleophobic film provided by the stearic acid compound, that is the methyl group end, is utilized in this test. A low viscosity organic solvent is utilized which has a surface tension greater than 25 dyne centimeters. Benzene has proven to be a satisfactory test solvent. It is recognized that the viscosity and surface tension of the test solvent may vary over a range of values in accordance with the temperature. The aluminum oxide shaft is immersed within the test solvent and slowly removed. If the surface is dry, that is the test solvent does not wet the surface, it indicates a monomolecular film of lubricant has been adsorbed upon the surface. It is important to note that any molecules which attach to the methyl group end of the stearic acid compound and which have not been removed by the cleaning process will be indicated by wetting the surface. Thus the test indicates whether or not a monomolecular film of lubricant has been adsorbed upon the aluminum oxide surface.

Example 2: The surface to be lubricated is tungsten carbide. As stated in Example 1, a particular abrasive must be chosen which will abrade the tungsten carbide material and remove the surface film so as to expose the high energy level substrate surface to the lubricant. Because tungsten carbide is a harder surface than aluminum oxide, alumina particles may be utilized. Again the abrasive size is 0.1 micron. The cerium oxide abrasive particles may be utilized for tungsten carbide surface; however a longer period of rubbing is required to obtain a chemically clean surface. Stearic acid is used as the polar lubricant. The stearic acid is mixed with the alumina in a 10:1 ratio. The substantially nonpolar solvent to be utilized in this case was kerosene. Only as much kerosene as necessary to form a pastelike consistency is required.

This paste is rubbed against the tungsten carbide surface in the manner set forth in Example 1 so as to clean the surface and allow the stearic acid to adsorb. The excess paste is removed as previously set forth. The surface is tested with a test solvent as described in Example 1.

Alkyl Substituted Pyrazine Fluids

A process described by R.E. Dolle, Jr.; U.S. Patent 3,322,671; May 30, 1967; assigned to the U.S. Secretary of the Air Force is based upon the use as additives of a secondary aromatic amine compound in combination with an aromatic tin-containing compound in an aliphatic hydrocarbon fluid. While various amines have been used as storage stability inhibitors to protect against oxidation at relatively low temperatures, it has been only when such amines were combined with other substances that they provided any oxidation resistance at higher temperatures. The substantial improvements in high temperature oxidation resistance have been largely confined, however, to aromatic type fluids.

In the case of this process on the other hand, the addition of the combination of the amine compound and the aromatic tin-containing material has not only improved the oxidation resistance of the mineral oils and related functional lubricating fluids but also has done so to a degree that is far greater than could have been anticipated from the previously known effects of the amine compounds or of the organometallic substances used singly and not in combination with each other. As a result of such synergistic improvements provided by this process, various aliphatic hydrocarbon fluids which previously lacked adequate oxidative stability at elevated temperatures may be satisfactorily utilized in severe oxidative environments at temperatures of 400°F. and beyond as instrument oil, gear box oil and grease base fluids.

The aliphatic hydrocarbon base fluids which are applicable include the straight chain and branch chain aliphatic materials. Specific base fluids within this class include mixed alkyl substituted pyrazines and alkyl substituted silane materials. Within this class of materials are 2-n-heptyl-6-(5-tridecyl)pyrazine; 2-n-heptyl-5-(5-tridecyl)pyrazine; 2-(2-octyl)-6-(2-decyl)pyrazine; 2-n-heptyl-6-n-nonylpyrazine; 2-phenoxy-3-(5-nonyl)pyrazine; 2,6,10,14-tetramethylpentadecane and octadecyl tri-n-decyl silane. The lubricating base fluids alone are not stable under severe oxidation conditions at the high temperatures in the order of 350°F. and above, except when they are mixed with the amine and tin combinations. Specific aryl tin compounds which have yielded the improvements sought by the process include bis-(p-phenoxyphenyl)diphenyltin; 5-ethyl-5,10-dihydro-10,10-diphenylphenazastannine; tetraphenyltin and perfluorotetraphenyltin.

To demonstrate the effectiveness of the additives, various functional fluid formulations were subjected to microoxidation tests involving the passage of up to 20 liters per hour of dried air through a 20 milliliter test sample for a 24 hour period at a constant temperature. The test temperatures and air flow rates in the tests were adjusted for a particular base fluid so that under the same conditions the fluid without the additives would be severely oxidized. The criteria for determining the extent of oxidation were the percent of kinematic viscosity change at 100°F. and neutralization number change (mg. KOH/g.), based upon the viscosity and neutralization number of the fluid formulation before the test.

Aromatic Tin Compound	Weight Percent	Kinematic Viscosity Change at 100° F. (percent)	Neutralization Number Change (mg. KOH/g.)
Bis(p-phenoxyphenyl)diphenyltin	1.0	20.0	0.3
Do	2.0	13.3	0.4
Do	0.5	25.4	0.3
5-ethyl-5,10-dihydro-10,10-diphenylphenazastannine	1.0	31.7	0.4
Perfluorotetraphenyltin	1.0	31.3	0.6
Do	0.25	43.2	0.6
Tetraphenyltin	0.1	89.2	1.4
Do	1.0	15.3	0.6
Do	2.0	19.8	0.7

In one set of comparative tests, 2-n-heptyl-6-(5-tridecyl)pyrazine was employed as a base fluid to which 1.0 weight percent of phenyl alpha-naphthylamine (sometimes referred to as PANA) was added along with varying weight percents of various aryl tins as illustrated in the previous table. When these formulations were subjected to the microoxidation tests described before at temperatures of 425°F. the listed values of kinematic viscosity change and neutralization number change were achieved.

For purposes of comparison, the pyrazine base fluid involved in the tests was run without any secondary amine or aryl tin additives, and at 425°F. the sample charred badly and only a solid residue remained. Where the same test on the unmodified fluid was run at 325°F., the kinematic viscosity change at 100°F. was 63.9% and the neutralization number change was 3.4. Where the pyrazine base fluid was modified by the addition of 1 weight percent of the PANA without any aryl tin additive, the oxidation test at 425°F. showed a kinematic viscosity change of 140% and a neutralization number change of 2.0, which are both objectionably high as compared with this process. At the same time, where the base fluid was modified with the addition of 1.0 weight percent of the bis(p-phenoxyphenyl)diphenyltin, 0.25 weight percent of the perfluorotetraphenyltin, 0.1 weight percent of tetraphenyltin and 1.0 weight percent of tetraphenyltin, without any PANA, all the samples charred badly and only solid residues remained. In other tests involving the same base fluid modified with 1.0 weight percent of the PANA and comparable amounts of non-tin-containing materials, of the type generally related to the aryl tin compounds but having silicon, lead or germanium moieties, the kinematic viscosity changes were from four to eight times greater than the better formulations that were made to embody this process.

The same tests and comparisons made in connection with the combination of the secondary amine and aryl tin additives to an alkyl substituted pyrazine base fluid such as 2-(2-octyl)-6-(2-decyl)pyrazine; 2-n-heptyl-6-n-nonylpyrazine; 2-phenoxy-3-(5-nonyl)pyrazine and 2-n-heptyl-5-(5-tridecyl)pyrazine showed comparable degrees of improvement over the base fluids without any additives or with one or the other of the additives singly. Also, similar tests run on synthetic high molecular weight aliphatic hydrocarbons and 2,6,10,14-tetramethylpentadecane and octadecyl tri-n-decyl silane to which the antioxidant system of this process had been added gave overwhelming oxidative superiority over the same base fluids without any additives or with one or the other of the additives.

Comparative experimental data for the high molecular weight hydrocarbon, pentadecane and silane formulated with varying weight percents of various aryl tins combined with either PANA or p,p'-dioctyldiphenylamine (DODPA) are given in the following three tables.

Microtest Conditions

400°F.
20 liters/hour air rate
24 hours duration

Additive(s) in High Molecular Weight Hydrocarbon	Kinematic Viscosity Change at 100° F. (percent)	Neutralization Number Change (mg.KOH/g.)
1.0 wt. percent DODPA +1.0 wt. percent bis (p-phenoxyphenyl) diphenyltin	15.7	0.0
1.0 wt percent DODPA +1.0 wt. percent 5-ethyl-5,10-dihydro-10,10-diphenylphenazastannine	15.8	0.0
1.0 wt percent DODPA +0.5 wt. percent 5-ethyl-5,10-dihydro-10,10-diphenylphenazastannine	15.2	0.0
1.0 wt. percent PANA +1.0 wt percent 5-ethyl-5,10-dihydro-10-10-diphenylphenazastannine	15.4	0.0
1.0 wt. percent PANA alone—note the the objectionable high viscosity	1,380	1.4
1.0 wt. percent bis(p-phenoxyphenyl) diphenyltin	959	0.4
1.0 wt. percent 5-ethyl 5,10-dihydro-10,10-diphenylphenazastannine	134	0.8
1.0 wt. percent PANA +1.0 wt percent 5-ethyl-5,10-dihydro-10,10-diphenylphenazasiline	1,180	0.7
Base fluid with no additive	3,020	0.7

Microtest Conditions

400°F.
20 liters/hour air rate
24 hours duration

Additive(s) in 2,6,10,14-Tetramethylpentadecane	Kinematic Viscosity Change at 100° F. (percent)	Neutralization Number Change (mg.KOH/g.)
1.0 wt. percent PANA +1.0 wt percent bis(p-phenoxyphenyl) diphenyltin	0.0	0.0
1.0 wt. percent PANA +1.0 wt. percent 5-ethyl-5,10-dihydro-10,10-diphenylphenazastannine	0.0	0.4
1.0 wt. percent PANA +1.0 wt. percent triphenylallytin	0	0.1
1.0 wt. percent PANA +0.5 wt. percent tetracyclohexytin	2.0	0.3
1.0 wt. percent DODPA	(2)	(2)
1.0 wt. percent PANA submitted for contrast since it does not work because of the resulting high viscosity which is also true of the next two formulations	26,800	37.1
1.0 wt. percent bis(p-phenoxyphenyl) diphenyltin	(1)	(1)
1.0 wt. percent PANA +1.0 wt. percent 5-ethyl-5,10-dihydro-10,10-diphenylphenazasiline	1,910	28.4
Base fluid	(2)	(2)

[1] Barely fluid at room temp.
[2] Not fluid at room temperature.

Microtest Conditions

425°F.
10 liters/hour air rate
24 hours duration

Additive(s) in Octadecyl Tri-n-Decyl Silane	Kinematic Viscosity Change at 100° F. (percent)	Neutralization Number Change (mg. KOH/g.)
1.0 wt. percent PANA + 1.0 wt. percent bis(p-phenoxyphenyl)diphenyltin	64.9	1.3
1.0 wt. percent PANA + 1.0 wt. percent 5-ethyl-5,10-dihydro-10,10-diphenylphenazastannine	54.2	1.1
1.0 wt. percent DODPA + 1.0 wt. percent 5-ethyl-5,10-dihydro-10,10-diphenylphenazastannine	10.9	0.1
1.0 wt. percent PANA for contrast it does not work satisfactorily, nor do the next three (3) formulations which also are submitted for contrast	610	5.6
1.0 wt. percent bis(p-phenoxyphenyl) diphenyltin	3,610	0.2
1.0 wt. percent tetraphenylgermanium	1,980	2.7
1.0 wt. percent PANA + 1.0 wt. percent 5-ethyl-5, 10-dihydro-10,10-diphenylphenazasiline	945	0.4
1.0 wt. percent PANA + 1.0 wt. percent tetraphenylgermanium	1,136	7.4
Base fluid	2,520	1.7

While these tests demonstrate that a synergistic effect is produced strongly inhibiting the oxidation of the aliphatic hydrocarbon structures which does not occur when the secondary amine or the tin-containing compounds are used singly (i.e., not in combination with each other), it is also demonstrated that the uncombined additives do not in and of themselves produce an effective antioxidant system for aliphatic hydrocarbons at high temperatures.

COMPANY INDEX

The company names listed below are given exactly as they appear in the patents, despite name changes, mergers and acquisitions which have, at times, resulted in the revision of a company name.

INVENTOR INDEX

U.S. PATENT NUMBER INDEX

NOTICE

Nothing contained in this Review shall be construed to constitute a permission or recommendation to practice any invention covered by any patent without a license from the patent owners. Further, neither the author nor the publisher assumes any liability with respect to the use of, or for damages resulting from the use of, any information, apparatus, method or process described in this Review.

Complete copies of the patents described in this book may be obtained at a cost of $0.50 per patent prepaid. Address order to the Commissioner of Patents, U.S. Patent Office, Washington, D.C. 20231.

What and Where to Buy

POLYOLEFIN RESINS GUIDEBOOK AND DIRECTORY 1972

Blue Book No. 3

This Guidebook and Directory to olefin polymers and their producers is the third volume in our comprehensive Blue Book Series. In keeping with our policy, this book attempts to provide a complete listing of commercially available, standardized resins offered by U.S. producers.

The book describes the readily available higher-molecular-weight polymers of ethylene and propylene. Some elastomeric copolymers are also listed. Included are product specifications and applications.

Polyethylene is the most widely used plastic material today. Although its monomer, ethylene, has an extremely simple structure, namely $H_2C{=}CH_2$, the polymer, also structurally simple, can have infinite variations in its side chains which may vary in length and branching from several thousand CH_2 units to as much as a million units.

Highly branched polyethylene as seen under the electron microscope, has more open spaces between its carbon atoms than the mostly linear polymer, and therefore has a low density, e.g. between 0.910 and 0.925, while the higher density and more linear polymer varies from 0.940 to 0.967. Likewise, the higher the molecular weight, the tougher the end use plastic, but it is also harder to process and higher temperatures are required.

It is obvious from a perusal of this book that many factors, including molecular weight distribution, must be considered and that the guidelines accompanying each product must be considered carefully by the manufacturer.

An excellent balance of properties and good processing capabilities show that polypropylene is the material of choice for injection molding.

Polypropylene products are also very well suited for electroplating and electroless plating.

The book purports to be a clearly delineated guide to aid in the most difficult decisions a plastics manufacturer must make in the following processes:

Rotational Molding
Blow Molding
Compression Molding
Injection Molding
Films, including EVA (ethylene-vinyl-acetate copolymer) Films
Filaments and Fibers
Extrusion of Pipes and Sheets
Coatings
Wire and Cable Coverings and such variegated techniques as Thermoforming or Vacuum Forming

This is a guide to U.S. manufacturers with standard product lines and gives precise, complete, up-to-date and pertinent information.

All the information in this volume is taken directly from the manufacturers' hard-to-get data sheets and technical bulletins. In each case enough information is provided to enable the researcher to judge from the data presented whether or not a given polyolefin type or grade will do the job for the intended applications, which may vary between rotationally molded automotive and industrial parts, high density polyethylene gasoline tanks and coextruded EVA films or grocery bags and trays.

The data appearing in this book were selected by the publisher from each manufacturer's literature at no cost to, nor influence from, the manufacturers of the materials.

Because it lists the intended uses and physical properties, as well as the manufacturers and suppliers of polyolefin resins, this book can serve as a constant source of reference and as a real time-saver to the plastic articles manufacturer and fabricator; as well as a general source of reference for the chemical and polymer industry.

291 pages.

$36

An Inorganic in Short Supply

HYDROFLUORIC ACID MANUFACTURE 1972

by S. Weiss

Chemical Process Review No. 56

Few products of the chemical process industries have undergone the spectacular rise in demand shown by hydrofluoric acid in the past few years.

U.S. demand climbed from 129,000 tons in 1962 to 337,000 tons in 1970, with expectations of reaching 440,000 tons in 1974. A growth rate of 7.8% per year from 1959 to 1969 is expected to continue at 7.1% through 1974. Major users are the aluminum and fluorocarbon industries, which account jointly for 86% of HF consumption.

The current expansion in HF capacity appears to stress the traditional fluorsparsulfuric acid route in foreign plants built or planned. This is particularly true in Mexico, site of the world's largest fluorspar reserves, where an additional 187,000 tons of capacity will be on stream by 1974.

Domestic producers, squeezed by the rising costs and shrinking supplies of high grade fluorspar, are becoming increasingly interested in recovering the acid from waste products of phosphate fertilizer plants.

In view of the rising prices for HF, coupled with rising demands, there appear to be attractive opportunities in the manufacture of this commodity chemical.

This book reviews 166 U.S. patents pertaining to HF manufacture, including a chapter on fluosilicic acid. Partial Table of Contents showing process examples:

254 pages

$36

Actual Conditions and Techniques

DETERGENTS AND POLLUTION 1972

Problems and Technological Solutions

by H. R. Jones

Pollution Control Review No. 7

Detergents with present annual sales of over 2½ million tons in the U.S. alone, rank quantitatively among the top products of the chemical industry.

Thanks to a complicated series of scientific disputes over alleged deleterious effects on environment and human health, there is probably more confusion about how to properly satisfy this market, than there is about any other class of industrial chemical products.

This book reviews the pertinent literature and describes actual manufacturing processes germane to the soap and detergent industry, as the proper solutions begin to emerge through the ingenuity of the inventive mind.

In addition, 188 Processes based largely on the U.S. patent literature are described in detail. Because of the actuality of the problem most patents are of very recent date, although some older, basic patents are described also to completely round out the technological picture. A partial table of contents follows. Where patents relate to a given topic, the number of applicable and described patents is given in parentheses.

268 pages

$36

Separation and Purification

INDUSTRIAL MEMBRANES 1972

Design and Applications

by J. McDermott

Membrane technology is assuming an increasing role in industrial separations. Substantial amounts of government funds are being spent for prototype installations employing tube membrane units and hollow fiber separators to obtain fresh water from the sea.

The Environmental Protection Agency is sponsoring the investigation of reverse osmosis for pollution problems, such as the removal of ammonia and phosphates from municipal waste streams.

Membranes are also being used in the recovery of proteins, desalination of cheese whey, concentration and fractionation of biologicals, and water sterilization. The purification of milk and many beverages requires highly efficient filter media for rapid and reliable processing. Gas and selective solvent separations are highly dependent on the careful design of permeation membranes.

This book presents a broad spectrum of the technology as it has evolved through the patent literature since 1960. 118 processes are described. The numbers in ( ) indicate the number of processes for each topic. Partial list of contents:

To further illustrate the wealth of information presented, a breakdown of III, 7 on the Preparation of Ion Exchange Membranes is given here:

Solid Ion Exchange Structures
Polyelectrolyte Membrane Structures
Antipolarization Membrane
Laminar Membrane
Coprecipitation of Water-Soluble Electrolytes
Carbodiimide Treatment of Cellulose Acetate
Treatment of Cellulose Acetate with Polyol and Sulfolane Solutions
Casting from Butyrolactone Solutions, Nylon and Sulfonic Acid
Electrolyte Surface Treatment
Prestressed, Reinforced Membrane
Polyvinyl Pyrrolidone
Elastomeric Material
Filled Silicone Rubber
Graft Copolymers by High Energy Radiation
Perhalogenated Fluorocarbons
Polyhalide Resins
Styrene-Butadiene with Phenolsulfonic Acid
Styrene-Butadiene with Phenylenediamine
Chloromethylated and Aminated Sty-Bu
Vinyl Chloride Polymers
Styrene Monomer on Polyethylene
Paper with Ion Exchange Resin

246 pages

$36

New Antipollution Techniques

WASTE TREATMENT WITH POLYELECTROLYTES 1972

by S. Gutcho

Pollution Control Review No. 8

Polyelectrolytes are polymers with high molecular weights, often in the range of 1 to 10 million. They are water soluble or water dispersible and are classified into three types—anionic, cationic and nonionic.

Polyelectrolytes are now being used in the treatment of sewage, the clarification of industrial waste water, and the purification of contaminated rivers and lakes. Polyelectrolytes can be used to remove phosphate ions and detergents from water, they tend to produce sewage sludges with low ash content and to reduce the cost of water pollution control. The recycling of waste water to a purity acceptable and useful to population centers and industry can be accomplished through the judicious use of polyelectrolytes as flocculants.

In this volume 144 processes are considered as they evolved in the U.S. patent literature largely since 1965. The numbers in ( ) indicate a plurality of processes for each topic. Partial table of contents:

237 pages

$36

Expanding "Fields"

SYNTHETIC TURF AND SPORTING SURFACES 1972

by M. S. Casper

Two factors have combined to make synthetic sporting surfaces the growth phenomenon displayed over the past seven years:

Increased labor costs have made natural turf surfaces almost prohibitively expensive. Increased leisure time with increased interest in spectator and participation sports make it desirable to provide sport surfaces which can be used despite weather conditions and seasonal variations.

Large scale stadium installations indicate only a portion of the uses for artificial sport surfaces. Already in use are artificial ski surfaces, toboggan slides, ice skating rinks, tennis courts, race tracks, and practice golfing surfaces.

New markets offering a large potential are highway medians, gas stations, motels and restaurants.

Playgrounds, lawns for private homes, golf courses, aircraft landing strips and cemeteries are examples of other uses with high potential. This review of 84 U.S. patents describes the present state of the art. A partial Table of Contents is reproduced here. Numbers in ( ) indicate numbers of processes for each heading.

260 pages

$36

An Economic Feasibility

LIQUID FUELS FROM COAL 1972

by Dr. G. K. Goldman

Chemical Process Review No. 57

Studied for the past fifty years, the conversion of coal to highest quality gasoline is now a technical possibility.

The economic difficulty exists in raising the hydrogen content from 5 wt.% to about 14% for a typical gasoline.

This book expounds and describes 101 processes evolved in the U.S. patent literature. Numbers in ( ) indicate the number of patents allocated to each topic.

230 pages

$36

A Top Overseas Market

ELECTRONICS INDUSTRY OF JAPAN 1972

The electronics industry has been Japan's fastest growing industrial sector and one of the country's most important export earners. Over the past thirteen years the growth has averaged 24 percent per year to reach a value of production of 3,400 billion yen in 1970.

"Electronics Industry of Japan" is aimed at giving you a concise but detailed picture of the present state of this important industry, describing 854 Japanese electronics firms. The book is divided into three sections.

The **first section** gives details of 145 electronics companies listed on eight Japanese stock exchanges—Tokyo, Osaka, Nagoya, Fukuoka, Hiroshima, Kyoto, Niigata and Sapporo. The following information is given on each company:

Full name
Address
Date of establishment
Capital
Total assets, sales and profits for 1968, 1969 and 1970
Principal executives
Number of employees
Major stockholders
Bankers
Technical agreements
Products

The **second section** gives more limited information on a further 714 unlisted companies. Details given include:

Full name and address
President
Capital
Number of employees
Main products

The **third section** comprises a statistical analysis of the present state of the industry and its historical development. Figures on production, trade, research and development, and employees are included.

The book is completed by an alphabetical list of registered trade names.

To anyone concerned with electronics in any part of the world, a knowledge of the important Japanese industry is essential. This significant new guide will give you all the pertinent information in one volume, providing both a convenient source of reference and a valuable marketing aid.

This guide was prepared in Japan by one of our associates, in order to give you on-the-spot coverage.

147 pages

$24

FOOD

SAUSAGE PROCESSING 1972
by Dr. E. Karmas
Food Processing Review No. 24

A patent-based review of sausage production technology, with numbers of specific processes indicated in ( ).

Cured Color (18)
Additives (3)
Forestalling Rancidity (4)
Fermented Flavors (3)
Texturizing Agents (13)
Emulsion Stabilizers (5)
Defatting (6)
Demeating Bones (4)
Composition Control (2)
Ingredients Identification (2)
Stuffing (10)
Linking & Tying (13)
Forming & Shaping (3)
Smoking & Pumping (4)
Hot Air Cooking (6)
Cooking in Liquids (3)
With Heat Exchange
Electrical Cooking (14)
Casing Release & Removal (5)
Fresh Pork Sausage (3)
Dry Sausage (8)
Various & Novelty Products (4)

136 Processes. 218 pages. **$36**

ENZYMES IN FOOD PROCESSING AND PRODUCTS 1972
by H. Wieland
Food Processing Review No. 23

Commercial availability of enzymes has increased considerably. Enzymes applicable to food processing are now plentiful, and the alert food processor is urged not to miss this opportunity to improve his products in many ways. 101 Processes. The numbers in ( ) indicate the number of processes allocated to each topic.

FRUIT & VEGETABLE PROCESSING (13)
STARCH & SUGAR CONVERSION (9)
BAKED GOODS APPLICATIONS (12)
CHEESE MAKING (11)
MEAT TENDERIZATION (18)
SPECIAL APPLICATIONS (13)
FLAVORS THROUGH ENZYMES (13)
DEOXYGENATING AND DESUGARING (6)
ENZYME STABILIZATION (6)

269 pages. **$36**

SEAFOOD PROCESSING 1971
by M. Gillies
Food Processing Review No. 22

Describes 84 processes based on U.S. patents issued since 1960. Numbers in ( ) denotes numbers of processes in each chapter.

1. PRESERVATION (13)
   On Fishing Vessels
   Chemical Methods
   Edible Coatings
   Various Preservatives
2. CANNING PROCEDURES (11)
   Tuna and Similar Fish
   Sardines
   Forestalling Struvite
3. PROTEIN CONCENTRATES (14)
   Mechanical Means
   Chemical Means
   Biological Means
   Stickwater Proteins
4. MOLLUSKS & SHELLFISH (18)
   Squid
   Bivalves
   Crustaceans
5. CONSUMER PRODUCTS (17)
6. ANIMAL FOODS (11)

206 pages. **$36**

FRUIT PROCESSING 1971
by M. Gutterson
Food Processing Review No. 21

All 140 processes (mostly developed since 1960) were selected with the purpose of providing fruits and fruit products retaining the characteristics of freshly picked fruit, yet capable of being shipped the world over and being highly acceptable by organoleptic tests. Preventing the growth of microorganisms with a minimum of chemicals and processing equipment was another goal.

1. GENERAL TECHNIQUES (34 processes)
   Heat Treatments, Increasing Cellular Permeability, Inhibiting Discoloration, Ripening, Dehydration
2. TREATMENT OF POMES (23)
3. CITRUS FRUITS (18)
4. BERRIES (20)
5. DRUPES (16)
6. DRIED FRUITS (10)
7. OTHER FRUITS (19)
   Flavor Improvement, Delaying Senescence, Use of Enzymes and Freezing Techniques. 223 pages. **$36**

POULTRY PROCESSING 1971
by G. H. Weiss
Food Processing Review No. 20

Poultry is the most efficient and effective means for converting grain to protein. This book discusses in detail the different methods devised to assure excellent flavor, texture and tenderness concomitant with easy preservation, maximum storage time, easy handling, and consumer acceptance. 55 processes in 8 chapters:

1. **Preservation (11 processes)**
2. **Chilling and Freezing (6)**
3. **Enhancing Palatability (10)**
4. **Stuffed Products (3)**
5. **Molded Rolls and Loaves (19)**
6. **Batter-Coated Products (2)**
7. **Cooking Procedures (2)**
8. **Poultry Concentrates (2)**

168 pages. **$24**

VEGETABLE PROCESSING 1971
by M. Gutterson
Food Processing Review No. 19

Shipping vegetables from one continent to another has become a normal means of supply. But such transporting is possible only by adequate processing of the perishable vegetable goods. Many of the 184 process descriptions in this book are concerned with just such treatments. Processes for improving the stability of vegetables in regard to time, temperature and moisture are numerous. So are those where the emphasis is in making vegetables more digestible and more acceptable to children and adults:

1. **General (27 processes)**
2. **Potatoes (59)**
3. **Other Roots (12)**
4. **Bulbs (13)**
5. **Leaves & Stems (8)**
6. **Tomatoes & Others (30)**
7. **Corn (6)**
8. **Legumes (23)**
9. **Olives & Mushrooms (6)**

335 pages. **$36**

FOOD

FLAVOR TECHNOLOGY 1971
by Dr. N. D. Pintauro

Scientific and trade journals contain only limited information on practical and applied flavor research and technology. Industry and commercial operators wish to keep such information confidential. This book reviews such technology from U.S. patents since 1960. There are 99 processes in 9 chapters:

1. **Spice Technology (11)**
2. **Peppermint & Citrus (11)**
3. **Fruit Essences (9)**
4. **Dairy Flavors (6)**
5. **Bread Flavors (8)**
6. **Vanilla (9)**
7. **Meat Flavors (17)**
8. **Meat Seasonings (10)**
9. **Fixation (18)**

228 pages. **$35**

MILK, CREAM AND BUTTER TECHNOLOGY 1971
by G. Wilcox
Food Processing Review No. 18

In these days of heightened consumer awareness, the alert dairy processor cannot afford to bypass the latest developments in his field. Much emphasis now is on modified milk products which contain numerous proteins of high nutritional value. The even distribution of essential amino acids bestows on modified milk a great enticement for use with otherwise deficient diets. Methods for low sodium milk products, hypoallergenic dietary prepns. and infant milk products are given special attention. 181 Processes are described: 1. Pasteurization and Sterilization (15 processes). 2. Removal of Radioactive Contaminants (8). 3. Buttermilk and Allied Products (10). 4. Modified Milk Products (21). 5. Dehydration of Skim Milk (28). 6. Dehydration of Whole Milk, Whey and Milk Blends (47). 7. Concentrated Milk (19). 8. Cream (22). 9. Butter (11). 313 pages. **$35**

EGGS, CHEESE AND YOGURT PROCESSING 1971
by G. Wilcox
Food Processing Review No. 17

One route to discovering the latest technology is via this Food Processing Review which serves to bring you timely, useful information. Brought to you in this one easy-to-use, comprehensive volume is commercial research and development done in the field from 1960 to 1970, gathered from the U.S. patent literature.

The Table of Contents shows the processes discussed in this book. The numbers in parentheses indicate the number of processes covered for each particular process, equipment or product.

Section I—Eggs: Whole Eggs (18); Egg Yolks (7); Egg Whites (27); Egg Products (10). Section II—Cheese and Yogurt: Cottage Cheese (26); Cheddar Type and Process Cheeses (45); Cream Cheese and Bakers' Cheese (8); Mozzarella, Provolone and Parmesan Cheeses (8); Miscellaneous Cheese Processes (17); Manufacture of Yogurt (5).

280 pages. **$35**

RICE AND BULGUR QUICK-COOKING PROCESSES 1970
by R. Daniels
Food Processing Review No. 16

This salient report in our Food Processing Review series summarizes with detailed process information the pertinent U.S. patent literature relating to quick-cooking processes for both rice and bulgur. The information provides needed know-how concerning processing of raw rice and wheat to obtain the more desirable refined forms.

63 specific processes covered. The numbers in ( ) indicate their distribution.

Rice Milling—Extraction—Polishing (13)
Quick-Cooking Rice (17)
Special Rice Processes (6)
Brown and Parboiled Rice (6)
Specialty Products with Rice (9)
Quick-Cooking Wheat Bulgur Products (12)

267 pages. **$35**

FRUIT JUICE TECHNOLOGY 1970
by M. Gutterson
Food Processing Review No. 15

This publication deals with the technology of the noncarbonated fruit juice industry from 1960 through 1970 as covered in the U.S. patent literature. Modern technology has studied and overcome many processing problems, resulting in a vast output of new and improved processing methods. It is oftentimes difficult to keep up with the latest technology. This book is designed to offer you such help.

In the abbreviated Table of Contents shown below, you can see the large amount of valuable material included. The numbers in ( ) indicate the number of processes discussed.

1. Manufacturing Techniques (23), 2. Concentration of Fruit Juices (37), 3. Stabilization Processes (19), 4. Dehydration (18), 5. Freeze Drying (24), 6 Flavors from Fruit Juices (10), 7. Miscellaneous Processes (9).

206 pages. **$35**

MEAT PRODUCT MANUFACTURE 1970
by Dr. E. Karmas
Food Processing Review No. 14

This Review concerns latest technology in preparing packaged meats in ready-to-cook and ready-to-eat forms.

The Table of Contents below indicates the many areas covered in this survey.

General Processing: Curing Methods and Ingredients, Increased Water Binding and Yield, Improved Curing Formulations, Integral Meats, Smoking, Thermal Processing and Sterilization, Miscellaneous Processing Methods.

Products: Bacon Production, Patty Type Products, Dehydrated Convenience and Snack Products, Modified and Novel Products.

273 pages. **$35**

FOOD

MODERN BREAKFAST CEREAL PROCESSES 1970
by R. Daniels
Food Processing Review No. 13

Describes in detail production processes and equipment for the manufacture of modern breakfast cereals. These include both ready-to-eat and quick-cooking products.

Offers detailed practical information for the manufacture and production of these cereal products based on the U.S. patent literature. 61 processes included. Abbreviated Table of Contents follows.

Dough Cooking and Extrusion Processes
Treatment Prior To Puffing
Puffing Processes
Processes For Whole Cereal Grains
Cereal Shaping Processes
Sugarcoating Process
Fruit Incorporation and Nutritional Enrichment
Quick Cooking Cereal Products

217 pages. **$35**

STARCHES AND CORN SYRUPS 1970
by Dr. A. Lachmann

This report covers the field of starch production from many standpoints.

Wet milling is the primary method of starch production, therefore much of the material is concerned with this route to starch. Dry milling processes are also covered.

In addition, coverage of the current technological progress in hydrolyzing starches into dextrins, corn syrups and dextrose and starch fractionation into amylose is covered.

Contains 139 processes covering: The Manufacture of Starch, Treatment of Starch, Modified Starch, Pregelatinized Starch, Acid Hydrolysis of Starch to Sweeteners, Starch Hydrolyzing Enzymes, Enzymatic Starch Hydrolysates, Starch Hydrolysates Produced by Acid and Enzyme Treatments, Starch Fractionation. 275 pages. **$35**

FRESH MEAT PROCESSING 1970
by Dr. E. Karmas
Food Processing Review No. 12

This Food Processing Review, deals with 106 detailed processes covering essential developments in the fresh meat processing industry since 1960. The book provides a well-organized tour through the field; the processes included are well researched and presented as an easy-to-use guide to what is being done in this vital field today.

The material has been divided into two parts; processes for enhancing palatability, and preservation processes. The numbers in ( ) after each heading indicate the number of processes for each entry.

A. Palatability: Tenderness (33), Flavor and Tenderness (8), Flavoring (12), Color (13), Integral Texture (6). B. Preservation: Moisture Retention (9), Antimicrobial Treatment (10), Ionizing Radiation (7), Other Methods of Preservation (8). 236 pages. **$35**

SOLUBLE TEA PRODUCTION PROCESSES 1970
by Dr. N Pintauro
Food Processing Review No. 11

This book describes production processes for producing soluble tea and offers a wealth of detailed practical information based primarily on the U.S. patent literature. Describes 73 specific processes in this field with substantial background information. The Table of Contents is listed below. The numbers in ( ) indicate the number of processes in that category.

Withering and Rolling (4)
Fermentation, Firing and Sorting (8)
Extraction (13)
Recovery of Aroma (10)
Tannin-Caffeine Precipitate (Cream) (15)
Filtration and Concentration (8)
Dehydration Process (6)
Agglomeration and Aromatization (9)

Illustrations. 183 pages. **$35**

BAKED GOODS PRODUCTION PROCESSES 1969
by M. Gutterson
Food Processing Review No. 9

This book describes 201 recent processes for the production of baked goods. Based on the patent literature, it offers an up-to-date comprehensive publication of manufacturing processes.

There is a substantial amount of information in this book relating to the use of various chemicals and related additives.

Contents: Bread, Yeast Leavened Products, Chemically Leavened Products, Leavening Agents, Air Leavened Products, Non-Leavened Products, Refrigerated Doughs, Emulsifiers and Dough Improvers, Miscellaneous, Indexes. Illustrations. 353 pages. **$35**

SOLUBLE COFFEE MANUFACTURING PROCESSES 1969
by Dr. N. Pintauro
Food Processing Review No. 8

This book describes significant manufacturing processes for producing soluble coffee, and offers a wealth of detailed practical information based primarily on the U.S. patent literature. Describes 114 specific processes in this field with substantial background information.

Introduction: Roasting, Extraction, Filtration and Concentration, Recovery of Aromatic Volatiles, Spray Drying and Other Dehydration Processes, Freeze Drying Processes, Aromatization of Soluble Coffee Powder, Agglomeration Techniques for Soluble Coffee, Decaffeinated Soluble Coffee, Packaging of Soluble Coffee. Illustrations, Indexes, 254 pages. **$35**

FOOD

ALCOHOLIC MALT BEVERAGES 1969
by M. Gutcho
Food Processing Review No. 7

The traditional brewing process is a batch operation, costly and time consuming. There would be economic advantages to improved continuous processes which would require less capital investment for plant and equipment, give savings in labor, better use of raw materials, shorter processing time, and a more uniform product.

Detailed descriptive process information is found in this review, based on 157 U.S. Patents in the brewing field, issued since 1960. The 157 processes are organized in 7 chapters which tend to follow the steps in the brewing process.

Contents: Malting, Wort, Hops, Fermentation, Freeze Concentration and Reconstitution of Beer, Chillproofing, Preservation against Microbiological Spoilage, Foam, Indexes. Illustrations. 333 pages. **$35**

CONFECTIONARY PRODUCTS MANUFACTURING PROCESSES 1969
by M. Gutterson
Food Processing Review No. 6

This book is of technological significance in that it details over 200 processes for producing confections, based on the U.S. patent literature since 1960.

Based solely on new technology, this book offers substantial manufacturing information relating to this field. The wide scope of detailed data can be seen by the chapter headings indicated below:

Candy
Chocolate Products
Whipped Products
Icings
Gels
Coatings and Glazes
Gums and Stabilizers
Egg Products
Marshmallows and Meringues
Puddings
Frozen Confections
Chewing Gum
Other Confections
Indexes

Illustrations. 321 pages. **$35**

EDIBLE OILS AND FATS 1969
by Dr. N. E. Bednarcyk
Food Processing Review No. 5

This book describes in detail 225 recent process developments.

Shortenings; Fluid, Plastic, Miscellaneous: Margarine and Spreads; Margarine Oils, Highly Nutritional Oil Blends, Antispattering Agents, Fluid and Whipped Margarines, Flavor, Color, and Texture Modifications, Low Calorie Spreads: Salad Oils, Mayonnaise and Emulsified Dressings; Crystallization Inhibitors, Emulsified Dressings, Flavored Salad Oils, Low Calorie Dressings: Frying and Cooking Oils; Equipment, Breakdown Inhibitors, Antispattering Additives, Other Additives: Hard Butters; Preparation by Fractional Crystallization, Preparation by Ester Exchange, Miscellaneous: Oil Processing; Antioxidants and Stabilizers; Emulsifiers and Emulsions; Mixed Ester Emulsifiers, Dried Emulsion, Miscellaneous: Peanut Butter and Spreads; Chocolate Products; Indexes. Illustrations. 404 pages. **$35**

SNACKS AND FRIED PRODUCTS 1969
by Dr. A. Lachmann
Food Processing Review No. 4

The sales of snack foods in the U.S. may reach the two billion dollar mark in 1969. Many companies are actively working on new snack foods or on improved processes. The patent literature on french fried potatoes, potato chips, corn chips and other crisps is continually growing and it is the purpose of this book to present this literature in easy readable form.

French fried potatoes and their methods of production are described in the second chapter. The next chapter deals with potato chips, still the most popular product of the snack food industry. The U.S. market for potato chips is estimated to be approximately 600 million dollars in 1969. In Chapter Four the processes for corn chips are covered; in Chapter Five, apple crisps. Chapter Six describes processes for expanded chips and some specialty items; and the last chapter deals with batter mixes. Many illustrations. 181 pages. **$35**

PROTEIN FOOD SUPPLEMENTS 1969
by R. Noyes
Food Processing Review No. 3

The 126 Processes in this book are organized in 8 chapters by raw material source including the important newer processes for producing protein by fermentation of hydrocarbons. Another chapter on textured foods describes in detail a number of processes for producing these products that simulate meat. Indexes by company, inventors and patent number help in providing easily obtainable information.

This book is based upon the patent literature and serves a double purpose in that it supplies detailed technical information and can be used as a guide to the U.S. Patent literature on processes to obtain protein materials.

Contents: Hydrocarbon Fermentation, Fish-Based Protein, Soybeans, Cottonseed, Other Oilseeds and Legumes, Wheat and Gluten, Milk-Based Protein, Textured Foods, Miscellaneous, Indexes. Many illustrations. 412 pages. **$35**

DEHYDRATION PROCESSES FOR CONVENIENCE FOODS 1969
by R. Noyes
Food Processing Review No. 2

Describes 236 up-to-date dehydration processes for producing specific foods. Most detailed body of information ever published.

The detailed, descriptive process information in this book is based on 236 U.S. patents in the food dehydration field—issued between January 1960 and May 1968. This book serves a double purpose in that it supplies detailed technical information, and can be used as a guide to the U.S. patent literature on dehydration of foods. By indicating only information that is significant, and eliminating much of the legal jargon in the patents; this book then becomes an advanced commercially oriented review of food dehydration processes.

Dry Milk Products, Cheese and Yoghurt, Eggs, Fruit and Vegetable Juices, Fruits, Potatoes, Vegetables, Coffee, Tea, Miscellaneous. Many illustrations. 367 pages. **$35**

CHEMICAL PROCESSING

SULFURIC ACID MANUFACTURE AND EFFLUENT CONTROL 1971
by M. Sittig
Chemical Process Review No. 55

102 processes of manufacture and 39 pollution control measures are outlined in this encyclopedic survey:

SO_2 from Sulfur (11 processes)
SO_2 from Waste Gases (11)
SO_2 from Hydrogen Sulfide (2)
SO_2 from Sulfide Ores (8)
SO_x and H_2SO_4 from Sulfates (12)
H_2SO_4 from SO_2 HC1 (2)
The Chamber Process (4)
Conversion of SO_2 to SO_3 (9)
SO_3 to H_2SO_4 and Oleum (5)
Integrated Contact Processes (14)
Unconventional Processes (2)
Concentration of H_2SO_4 (2)
Dilution of H_2SO_4 (1)
Purification of H_2SO_4 (2)
Recovery of Spent H_2SO_4 (17)
Removal and Recovery of SO_x from Tail Gases (30)
Recovery of Acid Mists (9)
Future Trends

423 pages. **$48**

TRIMELLITIC ANHYDRIDE AND PYROMELLITIC DIANHYDRIDE 1971
by P. Stecher
Chemical Process Review No. 53

Trimellitic anhydride (TMA) being an anhydride and a carboxylic acid, can undergo many useful reactions; it can form esters and polyimides. Demand is rising sharply. Pyromellitic dianhydride (PMDA) yields polyimides and cured epoxy resins having very high temperature stability. This book gives 61 manufacturing processes and all the technology involved in making TMA and PMDA:

1. TM-Acid Synthesis (17)
2. TM-Acid Purification (3)
3. TMA Preparation (8)
4. TMA Purification (2)
5. TM-Double Anhydride (2)
6. PM-Acid Synthesis (6)
7. PMDA Preparation (5)
8. PMDA Purification (4)
9. Non-Hazardous Oxidations (1)
10. Derivatives of TMA & PMDA (13)

233 pages. **$35**

PHTHALOCYANINE TECHNOLOGY 1970
by Y. L. Meltzer
Chemical Process Review No. 42

Advances in phthalocyanine technology have been truly explosive during the past few years. New phthalocyanine products, processes and applications have poured forth from industrial, governmental and academic laboratories at a rapid pace. These advances in technology have made themselves felt in the market place and in government programs, and have contributed to corporate sales and profits. At the same time, however, competition has become more intense in the phthalocyanine field making it imperative to keep up with the latest technological advances.

Examines recent developments in phthalocyanine technology as reflected in U.S. patents and other literature. The first 23 chapters discuss up-to-date manufacturing processes for phthalocyanine pigments and dyes. Chapters 24 through 31 discuss unusual new applications for phthalocyanines. 390 pages. **$35**

SYNTHETIC PERFUMERY MATERIALS 1970
by M. Gutcho
Chemical Process Review No. 45

This Review shows you how to produce synthetic perfumery materials. It contains a valuable odor index.

The 152 U.S. patents included in this book are distributed among the 11 areas as shown below:

From Terpenic Materials (28)
Alcohols (11)
Esters (18)
Ethers (19)
Aldehydes (10)
Ketones (18)
Lactones, Pyrones, Substituted Phenols and Quinones (12)
Other Structures (7)
Naphthalene and Indene Derivatives (17)
Compounds with Scent of Ambergris or Irone (9)
Product Application (13)

273 pages. **$35**

ION EXCHANGE RESINS 1970
by C. Placek
Chemical Process Review No. 44

This report on ion exchange resins provides detailed information on 126 U.S. patents issued since 1960 concerning the composition and manufacture of ion exchange materials. This Review, by its organization, also provides a guide to these ion exchange resins by grouping them according to physical form, behavior characteristics, etc.

1. Anion Exchange Resins
2. Cation Exchange Resins
3. Resins For Removing Metals
4. Resins Having Mixed Properties
5. Specific Use Resins
6. Unconventional Materials
7. Process Emphasis
8. Properties of Ion Exchange
9. Ion Exchange Membranes
10. Emphasis on Shapes

329 pages. **$35**

RADIATION CHEMICAL PROCESSING 1969
by R. Whiting
Chemical Process Review No. 41

A number of radiation induced chemical processes are already operating commercially. The radiation processing of chemicals has been growing at an annual rate of about 25% per year. Currently, \$100 to 150 million worth of irradiated products are produced in the United States per year, however, it has been forecast that by 1980, the value of products receiving radiation treatment will be close to \$1,000 million per year.

This book surveys the radiation processing field and is based on the U.S. patent literature since 1960. Over 250 separate processes are described in detail in the chemical, polymer, rubber, petroleum, textile and other fields.

Contents: Polyolefins, Other Polymers, Elastomers, Hydrocarbons, Organic Chemicals, Inorganic and Organo-Metallic Compounds, Other Processes, Indexes. 377 pages. **$35**

CHEMICAL PROCESSING

PHOTOCHEMICAL PROCESSES 1969
by B. Albertson
Chemical Process Review No. 36

Describes 210 photochemical production processes in detail.

Introduction, Photohalogenation, Photonitrosation, Organic Photochemical Reactions, Inorganic Photochemical Reactions, Photopolymerization, Indexes. Illustrations. 185 pages. **$35**

CHLORINE AND CAUSTIC SODA MANUFACTURE RECENT DEVELOPMENTS 1969
by Dr. R. Powell
Chemical Process Review No. 33

Brine Electrolysis
Diaphragm and Mercury Cells
Recovery of Mercury
NaOH Production
Titanium Anodes
Sea Water Electrolysis
Cl_2 Production
Deacon Process Modifications
Numerous Illustrations,
48 processes, 265 pages. **$35**

ALKALI METAL PHOSPHATES 1969
by Dr. M. W. Ranney
Chemical Process Review No. 34

The 97 processes described are based on U.S. patents issued since 1960 and offer a comprehensive treatment of up-to-date technical information.

Contents: Orthophosphates, Metaphosphates, Pyrophosphates, Tripolyphosphates, Phosphites/Hypophosphites.

Numerous illustrations. 344 pages. **$35**

AMINES, NITRILES AND ISOCYANATES PROCESSES AND PRODUCTS 1969
by M. Sittig
Chemical Process Review No. 31

Material covered includes: Manufacture of Amines, Manufacture of Mono-Nitriles, Acrylonitrile Derivatives, Isocyanate Manufacture, Future Trends. 62 illustrations. 201 pages. **$35**

NITRIC ACID TECHNOLOGY RECENT DEVELOPMENTS 1969
by Dr. R. Powell
Chemical Process Review No. 30

Ammonia Oxidation Process, Wisconsin Thermal Process, Nitrogen Fixation by Shock Waves, Nitrogen Fixation in a Nuclear Reactor, Absorption of Nitrogen Oxides in Water, Concentration of Dilute Nitric Acid Solutions, Direct Production of Concentrated Nitric Acid, Purification of Nitric Acid, Stabilizers for Nitric Acid. Numerous illustrations. 245 pages. **$35**

CITRIC ACID PRODUCTION PROCESSES 1969
by R. Noyes
Chemical Process Review No. 37

Detailed descriptions of production processes for citric acid, based on the patent literature.

The Table of Contents is indicated below:

Processing, Iron Impurities Other Microorganisms, Recovery and Purification, Other Processes. Indexes. 157 pages. **$24**

Noyes Data Corporation provides up-to-date industrial processing publications for the Chemical, Food, Textile, Pollution Control and Allied Industries. Technology is changing very rapidly, and our method of publishing means that you are getting the very latest technical and economic information.

AROMATICS MANUFACTURE AND DERIVATIVES 1968
by M. Sittig
Chemical Process Review No. 17

Contents: Introduction, Production of Aromatics, Separation of Aromatics, Purification of Aromatics, Reactions giving Hydrocarbon Products, Other Reactions, Phenol Production, Styrene Manufacture and Derivatives, Future Trends. 73 illustrations. 232 pages. **$35**

CATALYSTS AND CATALYTIC PROCESSES 1967
by M. Sittig
Chemical Process Review No. 7

Contents: Hydrocarbon Conversion Processes, Hydrocarbon Polymerization Processes, Hydrocarbon Oxidation Processes, Future Trends. 109 illustrations. 303 pages. **$35**

SUGAR ESTERS 1968
by Research Corporation

Already approved for use in foods in a number of countries, use of sugar esters in food in the United States awaits FDA clearance.

Sugar esters are an important new raw material for the food industry.

Contains papers presented at California Symposium 1967. 134 pages. **$15**

INDUSTRIAL GASES MANUFACTURE AND APPLICATIONS 1967
by M. Sittig
Chemical Process Review No. 4

This book discusses conventional cryogenic air separation and purification techniques in considerable detail.

This book also discusses newer techniques such as adsorption using molecular sieves, and permeation using various membrane materials. 313 pages. 103 illustrations. **$35**

COATINGS

POLYURETHANE COATINGS 1972
by K. Johnson

Reviews 157 U.S. patents issued mostly since 1960 related to paint vehicles, wet look, highly glossy fabric coatings, microporous products, etc.

1. COATING VEHICLES
   Isocyanates (9 processes)
   Carboxyl, Polyols (16)
   Modified Resins (13)
   Aqueous Systems (9)
   Catalysts & Crosslinking (5)
   Miscellaneous (17)
2. MICROPOROUS MATERIALS
   Solvent Processes (22)
   Pore-Forming Agents (9)
   Suede Substitutes (3)
   Others (3)
3. COATED FABRICS & PAPERS
   Water Repellency & Ability to Dryclean (8)
   Coated Fabrics (13)
   Papers (3)
4. MAGNETIC TAPES & OTHER SUBSTRATES
   Magnetic Tapes (6)
   Wire Coatings (3)
   Other Applications (18)

338 pages. **$36**

POWDER COATINGS AND FLUIDIZED BED TECHNIQUES 1971
by Dr. M. W. Ranney

Describes 166 processes. Due to the actuality of the subject all have been developed very recently. Numbers in ( ) indicate number of processes per chapter.

1. FLUIDIZED BED—SPRAY-POURING TECHNIQUES (27)
   Fluidized Bed Designs
   Spray-Powder, etc.
2. ELECTROSTATIC PROCESSES (8)
3. EPOXIES (19)
   Curing Agents
   Modified Epoxies
   Powdering Techniques
4. POLYOLEFINS (10)
   Primers & Surface Treatment
   Use of Copolymers
5. VINYLS (6)
6. OTHER RESINS (10)
7. PIPE COATINGS (28)
8. ELECTRICAL COMPONENTS (19)
9. OTHER APPLICATIONS (24)
10. INORGANIC & PARTICLE COATINGS (15)

249 pages. **$36**

PAPER COATINGS BASED ON POLYMERS 1971
by K. Johnson

178 Processes for coating paper stock elaborated during the last 10 years. Pigment binder and barrier coatings are discussed in detail e.g. acrylics give high gloss and good ink holdout, while silicone and solvent-based coatings allow considerable latitude in formulations.

1. POLYETHYLENE ETHYLENE COPOLYMERS—HOT MELTS (34 processes)
2. POLYVINYLIDENE CHLORIDE BARRIER COATINGS (10)
3. WATER-SOLUBLE COATINGS (30)
4. STYRENE-BUTADIENE ETHYLENE-PROPYLENE LATICES (27)
5. VINYL ACETATE LATICES (13)
6. ACRYLIC LATICES (23)
7. SILICONE AND SOLVENT-BASED COATINGS (24)
8. SPECIALTY COATINGS (17)
   Photographic Paper Coatings, Opaque Coatings, Metallized Coatings, Chemical Watermark Paper, Coatings for Erasable Paper, Mulch Sheets.

313 pages. **$36**

WIRE COATINGS 1971
by D. J. De Renzo

The good insulating properties of many thermoplastic and thermosetting polymers make them suitable for coating wires to be used as conductors in electrical apparatus. Other properties include abrasion resistance, impact strength, flexibility, solvent resistance, and high temperature stability.

The U.S. patent literature of the past ten years provides an excellent description of the many types of processes for organic wire coatings. Numbers in ( ) indicate the number of processes described in each chapter. 180 processes in all.

Acrylics (12), Epoxy Resins (9), Fluorinated Resins (4), Polycarbonates (5), Polyesters (31), Polyimides and Polyamides (26), Polyolefins (41), Polyspiranes (5), Polyurethanes (8), PVA Resins (15), Silicones (7), Vinyl Chloride Polymers and Copolymers (13), Other Resins (4).

232 pages. **$35**

PAINT ADDITIVES 1970
by H. Preuss

This publication surveys the field of paint additives offered for sale by manufacturers in the United States. It gathers together for you in one useful volume a series of articles written by Mr. Preuss for METAL FINISHING from 1965 through 1970. It is designed to help lead the paint formulator through the maze of modern additives; placing at his fingertips needed information about their chemistry, properties, specifications, uses and applications.

Additives form an integral part of a coating. Some of the additives discussed in this book are: antiskinning, antifoaming, antifouling, antifreezing, dispersing, destaticizing, antilivering, antisagging, curing, antisettling or suspension and moisture resistant agents; plasticizers; antioxidants; fire retardants; corrosion inhibitors; odorants and deodorants; and many others. 249 pages. **$18**

METAL COATING OF PLASTICS 1970
by Dr. F. Lowenheim

Describes 125 processes for applying metallic coatings to plastic articles on a production basis with reasonable reliability, and in such fashion that the metal is acceptably adherent to the substrate, and that the resulting products are useful for decorative or functional purposes. 254 pages. **$35**

ELECTRODEPOSITION AND RADIATION CURING OF COATINGS 1970
by Dr. M. W. Ranney

The advantages of electrodeposition: Pinhole-free coating, eliminating of fire hazards and air pollution problems, automated operation and fast throughput make this an attractive method. Radiation curing is quick, eliminates ovens and uses solvent-free vehicles. 96 Patent-based process on 170 pages. **$35**

POLYMERS AND ADHESIVES

ADHESIVES GUIDEBOOK AND DIRECTORY 1972
Blue Book No. 2

The second volume in our comprehensive Blue Book Series. In keeping with our policy, this book attempts to provide a reasonably complete listing of commercially available, standardized products offered by American industry. The book describes the "on the line" adhesive products of 121 U.S. manufacturers arranged according to company names.

All the information was selected from the manufacturers' hard-to-get data sheets at no cost to, nor influence from, the manufacturers of the adhesives.

A primary purpose of the book is to present the significant, first line information about adhesives all in one place—saving you many hours of work trying to obtain specific information and facts. 407 pages. **$36**

RIGID FOAM LAMINATES 1972
by M. G. Halpern

This book concerns itself with the manufacture of rigid foam laminates, in which the foam cores are bonded to a great variety of outer coatings. 134 processes from U.S. patents.

1. Chemical Processes for the Preparation of Plastic Foams (7)
2. Nonplastic Expanded Materials (5)
3. Plastic Skin—Plastic Foam Laminates (15)
4. Polyurethane Foam Compositions (10)
5. Nonplastic Skin and Foamed Plastic Core Laminates (9)
6. Paper & Foam Compositions (6)
7. Containers & Packaging Materials (9)
8. Refrigerated Containers (9)
9. Acoustical Panels (5)
10. Roofing Applications (12)
11. Structural Panels (14)
12. Internal Reinforcements (12)
13. Adhesive Techniques (8)
14. Embossing & Decorating (4)
15. Miscellaneous (9)

271 pages. **$36**

POLYMER ADDITIVES GUIDEBOOK AND DIRECTORY 1972
Blue Book No. 1

Perhaps the most comprehensive listing of commercially available, protective additives ever offered to the plastics industry. Gives products of 86 U.S. manufacturers arranged according to company name. Each product is carefully indexed by chemical, generic, trivial, and trade name or registered trademark in the "one alphabet" index at the end of the book.

Abounds with antioxidants and stabilizers plus countless other protective aids. By listing the intended uses and physical properties, as well as the manufacturers and suppliers, this book intends to furnish a real service to the advancing polymer technology.

The data appearing in this book were selected by the publisher from manufacturers' literature at no cost to, nor influence from the manufacturers of the materials. 472 pages. **$36**

SEALING AND POTTING COMPOUNDS 1972
by J. A. Szilard

Sealing and potting compounds are used to protect against ingress or egress of liquids or gases. In most cases the desired protection is against the pentration of moisture. Describes 166 sealant manufacturing processes based on polymer technology including silicones.

Products intended for:
General Use
Soil Treatment
Highways and Runways
Building Construction
Aircraft Construction
Pipe Joint Sealing
Automotive Use
Shoemaking
Swimming Pool & Aquarium Sealing
Shafts & Stuffing Boxes
Carton & Container Sealing
Electrical & Electronic Instruments

Also discusses special sealants for extremely high and low temperatures, lubricating sealants, encapsulants for printed and high frequency circuits, and many more. 288 pages. **$36**

REINFORCED COMPOSITES FROM POLYESTER RESINS 1972
by Dr. M. W. Ranney

1. POLYESTER INTERMEDIATES
   Hydroxy Intermediates (9)
   Carboxylic Acids-Anhydrides (14)
   Diels-Alder Adducts (8)
   Others (10)
2. ADDITIVES
   Catalysts (9)
   Accelerators-Promoters (12)
   Inhibitors (12)
   Thickeners (16)
   Color Controllers (10)
   Miscellaneous (3)
3. FLAME RETARDANTS
   Acid-Anhydrides (9)
   Hydroxy Intermediates (6)
   Other Reactants (16)
4. FORMULATIONS
   Acrylics (3)
   Nitrogen Compounds (4)
   Other Thermoplastics (6)
   Reactive Solvents (4)
   Low Profile Shrinkage (3)
5. PROCESSING
   Glass Fiber Treating (4)
   Other Processes (8)

166 Patent-based Processes on 324 pages. **$36**

PLASTIC PRINTING PLATES MANUFACTURE AND TECHNOLOGY 1971
by M. G. Halpern

Shows easy adaptation to traditional letterpress practice and automated equipment for plastic 3-dimensional relief plates.

1. PHOTOPOLYMERIZED PRINTING PLATES (2)
2. CELLULOSE POLYMERS (5)
3. OTHER POLYMERS (6)
4. ANCHOR LAYERS (6)
5. PHOTOINITIATORS (10)
6. MODIFICATIONS (22)
   Afterexposure Treatments
   Calendering Aids
   Speed & Contrast Aids
   Increasing Sensitivity
7. PHOTOCHEMICAL CROSSLINKING (9)
8. POLYAMIDE PLATES (12)
   Sensitizers
   Photomechanical Processes
9. PHOTOREPRODUCTION LAYERS (20)
10. INTAGLIO, IMAGES AND ETCHABLE PLASTIC PLATES (7)
11. MOLDED PLATES (4)

294 pages. **$36**

POLYMERS AND ADHESIVES

VINYL AND ACRYLIC ADHESIVES INCLUDING PRESSURE SENSITIVES 1971
by K. Johnson

Includes 123 processes.

1. ACRYLICS (51)
   - Pressure Sensitives
   - Laminates
   - Tire Cord Adhesives
   - Anaerobics
2. POLYVINYL ACETATE (14)
   - Hot Melts
   - Wood Bonding
3. POLYVINYL ALCOHOL (10)
   - Paper and Corrugated Board
   - Water-Soluble Pressure Sensitives
   - Cement Compositions
4. ETHYLENE COPOLYMERS (13)
   - Hot Melts
   - Atactic Polypropylene
5. OTHER VINYL POLYMERS (25)
   - Iron-On Adhesives
   - Fabric Bonding
   - Pressure Sensitive Phenol-Aldehyde Resins
6. PRESSURE SENSITIVE TAPES (10)
   - Release Coatings

287 pages. **$36**

FLEXIBLE FOAM LAMINATES 1971
by M. McDonald

Reviews the U.S. patent literature on the technology of flexible foam laminates from 1960 through early 1971. Altogether 101 processes in 6 chapters. In 1970 about 75 million pounds of polyurethane foam in the U.S. alone were bonded, mostly to fabrics, to form flexible laminates.

1. THERMAL METHODS (16 processes)
2. ADHESIVE METHODS (15)
3. FOAM-IN-PLACE METHODS (19)
4. FABRIC TO FOAM (17)
5. FLOOR COVERINGS (14)
6. MISCELLANEOUS (20)
   Packaging Materials, Foam on Cardboard, Sealing Strips, Sound Insulation, Polyester Foam Laminates, Polypropylene Foam to Metal, Deforming Foam Surfaces, Stretchable Foam Laminates.

265 pages. **$36**

POLYMERS IN LITHOGRAPHY 1971
by D. J. De Renzo

About 50% of all printing today is done by lithography, also named the planographic method. This has led to an extensive use of polymers for making the base plates and the sensitized plate coatings. Polymers are used also in etching solutions, lacquers, deletion fluids, inks, etc. 145 processes from the U.S. patent literature since 1965.

1. DIAZO TYPE PRESENSITIZERS (46 processes)
2. NONDIAZO PHOTOSENSITIVE LAYERS (17)
3. BASE PLATES AND COATINGS (32)
4. OTHER POLYMERS FOR PLATES (9)
5. LACQUERS, ETCHANTS, ETC. (18)
6. THERMOGRAPHIC PROCESSES (13)
7. ELECTROPHOTOGRAPHIC PROCESSES (9)
8. THE DRIOGRAPHIC PROCESS.

216 pages. **$36**

HOT-MELT ADHESIVES 1971
by M. McDonald

Reviews the U.S. patent literature from 1950 through early 1971. Hot-melts set fast enough to accommodate high-speed machinery in shoemaking, paper converting, bonding textiles (replacing hand and machine sewing), metal container sealing, etc. 63 processes:

1. **Hot-Melts for Shoemaking** (14)
2. **Bonding Paper & Paperboard** (23)
3. **Bonding Metals** (7)
4. **Bonding Plastics** (6)
5. **Bonding 2 or more Materials** (7)
6. **Textiles and Coated Substrates** (6)

238 pages. **$35**

FLUOROCARBON RESINS 1971
by Dr. M. W. Ranney
Chemical Process Review No. 51

1. Polytetrafluoroethylene (42). 2. Vinylidene Fluoride Elastomers (38). 3. Vinyl Fluoride. 4. Trifluorochloroethylene (15). 5. Fluorodienes (16). 6. Fluoroethers (13). 7. Fluorinated Nitroso Polymers (6). 8. Others (23). 9. General Processing Techniques (14). 226 pages. **$35**

POLYSULFIDE MANUFACTURE 1970
by C. Placek
Chemical Process Review No. 50

This report covers 73 processes dealing with polymers possessing the disulfide (-SS-) group. Basic Processes (14); Modified Polysulfide Polymers (13); Curing (12); Process Control (4); Physical Form (3); Single-Package Compositions (6). 141 pages. **$35**

ETHYLENE-PROPYLENE-DIENE RUBBERS 1970
by Dr. M. W. Ranney
Chemical Process Review No. 49

Summarizes patent literature relating to: Polyene Monomers—Polymer Synthesis; Catalysts and Activators; MW Regulators; Process Technology; Recovery Techniques; Modified Terpolymers; Adhesives; Miscellaneous Vulcanizates; Cross-Linking Agents. 272 pages. **$35**

POLYCARBONATES—RECENT DEVELOPMENTS 1970
by K. Johnson
Chemical Process Review No. 47

Part I—Aromatic Polycarbonates: Synthesis and Polymerization (17); Halogen-Containing Polycarbonates (9); Processing (20); Modified Polycarbonates (23); Applications (5). Part II—Aliphatic Polycarbonates: Cycloaliphatics (8); Linear Aliphatics (6). 298 pages. **$35**

URETHANE FOAMS TECHNOLOGY AND APPLICATIONS 1971
by Y. Meltzer

Urethane foam production is growing at about twice the annual growth rate of the overall plastics industry. Rigid foams are used in refrigerators and freezers, and in sophisticated and efficient types of food processing and preserving equipment. By far the largest consumer is the building and construction industry; while flexible foams are leading the demand in the furniture, aviation and automotive industries. The book contains descriptions of 148 manufacturing processes of which 42 deal with application technology:

Raw Materials (39)
Special Additives (30)
Product Types (29)
Processing (8)
Applications (42)

448 pages. **$36**

EPOXY AND URETHANE ADHESIVES 1971
by Dr. M. W. Ranney

An ever increasing demand for epoxy adhesives makes them the leader of the industry. Urethane polymers and intermediates are also augmenting the adhesives market. Isocyanate monomers are applied to textile fibers, metals, and elastomers as primers for adhesion. Polymethylene polyphenyl isocyanates are used in formulations for bonding glass, metal, elastomers and wood to a variety of substrates. Polyurethanes with polyesters, polyamides, and elastomers meet many high performance requirements. 111 processes in 7 chapters:

Metal to Metal (25)
Fiber to Rubber (22)
Glass & Ceramics to Metal (18)
Polymer to Polymer (12)
Plastic to Metal (10)
Paper and Wood (13)
General Purpose Adhesives (11)

280 pages. **$36**

POLYIMIDE MANUFACTURE 1971
by Dr. M. W. Ranney
Chemical Process Review No. 54

The U.S. patent literature of the past ten years provides an excellent description of the many types of polyimides and their syntheses. This book is an attempt to collate and summarize those processes pertaining to the manufacture of polymers containing an imide grouping. Emphasis has been placed on practical, technically useful information. About 90 distinct processes of manufacture are described in eight chapters.

1. Polyamide Acids (20)
2. Polyimide-Esters from TMA (7)
3. Polyimide-Amides (11)
4. Modified Polyimides and Cross-Linking (11)
5. Specialty Intermediates (15)
6. Silicone-Fluorocarbon-Polysulfone Modifications (11)
7. Cellular Polyimides (5)
8. General Processing Techniques (9)

243 pages. **$35**

ABS RESIN MANUFACTURE 1970
by C. Placek
Chemical Process Review No. 46

ABS (acrylonitrile-butadiene-styrene) resins make up one of the most rapidly growing segments of the polymer industry. **Straight ABS Materials; ABS Modified with Acrylic Derivatives; ABS from Alpha-Methylstyrene; Miscellaneous Modifiers; Modification of Properties; Process Variations.** 233 pages. **$35**

CELLULAR PLASTICS RECENT DEVELOPMENTS 1970
by K. Johnson

Describes 189 processes:
- Polyolefins (15)
- Polyvinyl Chloride (15)
- Polystyrene (22)
- Rubber (15)
- Polyurethanes (71)
- Polyesters and Epoxides (15)
- Urea-Formaldehyde and Phenolic Resins (9)
- Other Cellular Products (27)

280 pages. **$35**

AGRICULTURAL CHEMICALS MANUFACTURE 1971
by M. Sittig
Chemical Process Review No. 52

Agricultural chemicals, properly used, are essential for supplying the food requirements of the world's evergrowing population. Current attacks on the toxicity of today's pesticides notwithstanding, our agriculture saves about 5 dollars worth of produce for every dollar spent on the war against harmful pests. 172 manufacturing process and product descriptions are given:

1. **Environmental Control in Manufacture**
2. **Manufacture of Intermediates** (4)
3. **Insecticides** (80)
4. **Herbicides** (48)
5. **Fungicides** (25)
6. **Nematocides** (5)
7. **Plant Growth Regulators** (9)
8. **Fertilizer Additives** (1)
9. **Future Trends**

264 pages. **$35**

FLAME RETARDANT POLYMERS 1970
by M. Ranney

Summarizes 144 processes for fire retardant additives and reactive intermediates for major plastic materials. Polyethylene and Polypropylene (15); Polystyrene (19); Polyurethanes (50); Polyesters (13); Other Polymer Systems (26); General Utility Additives (21). 263 pages. **$35**

COMPATIBILITY AND SOLUBILITY 1968
by I. Mellan

Normally, it requires laborious testing to determine compatibility of polymers, resins, elastomers, plasticizers, and solvents. Predictions made without testing or literature searching, are usually unreliable.

This book helps you evaluate proper materials by the use of 224 tables. 304 pages. **$20**

AGRICULTURAL

HORMONAL AND ATTRACTANT PESTICIDE TECHNOLOGY 1971
by Y. L. Meltzer

The need for non-chemical pesticides which are highly selective, nonpolluting and nontoxic, is urgent. Encouraging results are being obtained with insect hormones and hormonelike substances that interfere with the life cycle of noxious insects. 23 chapters based on the world's patent literature and technical articles: 1. Scope of the Problem. 2. Insect Hormones as Pesticides. 3. Insect Development and Hormones. 4. Juvenile Hormone. 5 Ecdysone. 6. Brain Hormone. Attractants for 7. Insects. 8. Bees. 9. Boll Weevils. 10. Cabbage Looper Moths. 11. Cockroaches. 12. Bombyx mori. 13. Flies. 14. Gypsy Moths. 15. Pink Bollworms. 16. Termites. 17. Yellow Jackets. 18. 10,12-Hexadecadiene Derivs. 19. Aliphatic Hydroxy Attractants. 20. Polyenols. 21. Review Articles. 22. Regulations. 23. Future Trends. 281 pages. **$35**

AQUATIC HERBICIDES AND ALGAECIDES 1971
by J. H. Meyer

There is a growing need for control of aquatic weeds and algae. This vegetation destroys natural waterways and water supplies. The most effective control is by chemicals. This book gathers the latest technology for producing or using aquatic herbicides and algaecides, based on U.S. patent literature. Numbers in ( ) indicate the number of processes described, 108 in all.

- Metal Compounds (22)
- Chlorinated Hydrocarbons (7)
- Other Halogenated Compounds (16)
- The Halogens Themselves (4)
- Sulfur-Containing Materials (12)
- Quaternary Ammonium Compounds (8)
- Amides and Imides (10)
- Amines (6)
- Acids and Their Derivatives (8)
- Miscellaneous Organic Compounds (9)
- Carriers (6)

177 pages. **$35**

FERTILIZER DEVELOPMENTS AND TRENDS 1968
by A. V. Slack

R&D Trends, Ammonia, Ammonium Nitrate, Sulfate, Urea, Slow Release Nitrogen, Other Nitrogen Fertilizers, Phosphoric Acid, Ammonium Phosphate, Nitric & Superphosphates, Thermal and Other Phosphate Processes, Potassium Fertilizers, Fluid Fertilizers, Bulk Blending, Minor Nutrients. 98 illustrations. 406 pages. **$35**

NEW FERTILIZER MATERIALS 1968
by C. I. E. C.

Ureaform, Crotonylidene & Isobutylidene Diurea, Triple Superphosphate, Ammonium Phosphates, Nitrophosphates, Nitrate of Potash, Potassium Phosphates and Metaphosphates, Magnesium, Sulfur Fertilizers, Oxamide, Urea Nitrate and Phosphate, Hydrides of Phosphorus, Red Phosphorus. Applications. 430 pages. **$35**

AMMONIUM PHOSPHATES 1969
by Dr. M. W. Ranney
Chemical Process Review No. 35

This book describes recent processes for production of ammonium phosphates.

Introduction; Ammonium Orthophosphates, Diammonium Orthophosphates, Ammonium Polyphosphates, Metal Ammonium Phosphates, Ammonium Phosphate—Ammonium Nitrate Mixtures. Many illustrations. 278 pages **$35**

CONTROLLED RELEASE FERTILIZERS 1968
by Dr. R. Powell
Chemical Process Review No. 15

This book offers you complete technical data on numerous processes and products in this field. The two major approaches are (a) compounds of low solubility, and (b) coated granules.

Introduction; Compounds of Low Solubility, Coated Granules, Prevention of Nitrogen Losses, Rapid-Release Fertilizer. 279 pages. **$35**

TEXTILES

FLOCKED MATERIALS 1972 Technology and Applications
by E. L. Barden

Demand for flocked materials and articles covered with fibers or similar substances is increasing. Fibers and flakes are used to flock walls and ceilings, because the flocking is decorative, soundproofs, waterproofs, acts as a thermal insulator and is highly durable.

Plush toys, velvet greeting cards, good quality carpeting, even battery plates and synthetic turfs can be produced by flocking. 111 processes:

Electrostatic Methods (23)
Mechanical Flocking (4)
Types of Flock (3)
Flock Treatments (8)
Adhesives (13)
Decorative Effects (12)
Wearing Apparel (16)
Home Furnishings (11)
Automobiles (7)
Others (14)

294 pages. **$36**

POLYESTER FIBER MANUFACTURE 1971
by M. Sittig

In 1970 polyester became the number one U.S. fiber, surpassing the consumption figure for nylon by another 100 million pounds. Predictions are that continuous processing from polymerization to finished yarn will be common in 1975. A total of 116 processes is given.

1. INTRODUCTION
2. ECONOMICS
3. VARIOUS POLYESTER COMPOSITIONS (6)
4. RAW MATERIALS AND PURIFICATION (9)
5. DIMETHYL TEREPHTHALATE (8)
6. INTEGRATED POLYESTER PRODUCTION (5)
7. BIS (HYDROXYETHYL) TEREPHTHALATE (23)
8. PREPOLYMER PRODUCTION (8)
9. POLYCONDENSATION (28)
10. POLYMER AFTER-TREATMENT (2)
11. SCRAP RECOVERY (3)
12. FIBER PRODUCTION (24)
13. FUTURE TRENDS

214 pages. **$36**

SYNTHETIC PAPER FROM SYNTHETIC FIBERS 1971
by K. Johnson

This book provides a summary of the U.S. patent literature through 1970, relating to 74 processes for producing synthetic papers on conventional papermaking machinery:

1. **Cellulosics** (37 processes)
   - Cellulose Derivatives (15)
   - Cross-Linked Cellulose (4)
   - Blends with Other Fibers (10)
   - General (8)
2. **Polyamides** (8)
   - General (5)
   - Nylon + Cellulose (3)
3. **Polyacrylonitrile** (13)
   - Bonding (7)
   - General (4)
   - Blends (2)
4. **Synthetic Fibers** (16)
   - General Processing (10)
   - Specialty Papers (2)
   - Polytetrafluoroethylene (3)
   - Polyester (1)

236 pages. **$35**

NONWOVEN FABRIC TECHNOLOGY 1971
by M. McDonald

Nonwovens are structures produced by bonding or interlocking of fibers, accomplished by mechanical, chemical, thermal, or solvent means. Low cost is the primary advantage over woven or knitted products, resulting in a wide variety of disposable items from flush-away diapers to industrial uniforms. U.S. hospitals and other medical organizations bought over $100 millions worth of disposable hospital gowns and other nonwovens in 1970 of what is thought to be a $800 million-market. This book emphasizes those processes that turn out materials for making garments, draperies, upholstery, sheets, etc. 121 processes are described:

1. Resin Bonding (22 processes). 2. Spunbonding (23). 3. Needle Punching (28). 4. Fluid Pressure (13). 5. Heat Bonding (10). 6. Web Formation (7). 7. Miscellaneous (18). 240 pages. **$35**

MULTICOMPONENT FIBERS 1971
by C. Placek

Discusses processes for producing multicomponent fibers, that is, those fibers that consist of two or more polymeric compounds spun together. The spinning of multicomponent fibers is one method used to produce bulked or crimped fibers. The difference in physical properties of the two filaments results in a bulked yarn.

This report covers 106 patents issued since 1960. The processes range over a wide spectrum of fiber technology—composition, properties, physical forms and spinning techniques. Numbers in ( ) indicate the number of processes described in each chapter.

Variations of the Same Polymer (24), Chemically Unrelated Components (10), Spontaneous Crimp (2), Permanent Crimp (12), Specific Properties of Fibers (8), Production of Sheath-Core Structures (10), Side-by-Side Components (7), Spinning Technology (33).

225 pages. **$35**

SPANDEX MANUFACTURE 1970
by M. McDonald
Chemical Process Review No. 48

This book covers methods of making spandex fibers, that is, the conversion of polyurethanes into fibers, as described in the U.S. patent literature. The U.S. patent literature has the most complete and comprehensive process information available, and as such, this publication will give you key processing information for this fast-growing fiber. The Table of Contents below gives the scope of this volume. The numbers in ( ) represent the number of processes included.

Wet Spinning Processes (13), Solvent Spinning Processes (19), Melt Spinning Processes (3), Chemical Composition and Raw Materials (19), Improving Resistance to Ultraviolet Light and Oxidation (9), Improving Dyeability of Spandex (9), Miscellaneous (8).

191 pages. **$35**

SOIL RESISTANT TEXTILES 1970
by Dr. M. W. Ranney
Textile Processing Review No. 5

Ideal soil release finishes must be capable of releasing stains readily and preventing redeposition of soil during laundering. Treatments should render manmade fibers and durable press reactants less attractive to oily stains and should be more easily wetted.

This report summarizes the developments in soil retardant and soil release finishes in both the carpet industry and in textile manufacture. It includes the newest technology associated with the use of acrylates and fluorochemical treatments. The numbers in ( ) following each treating agent indicate the number of processes covered for that particular compound.

Introduction. Metal Oxides and Salts For Carpet Treatment (15), Acrylic and Vinyl Polymers (10), Silicones (6), Fluorochemical Compounds (72), General Treatments (14). 216 pages. **$35**

WATERPROOFING TEXTILES 1970
by Dr. M. W. Ranney
Textile Processing Review No. 4

This Textile Processing Review summarizes the technology of water resistant treatments for textiles and fabrics as described in the U.S. patent literature since the early 1950's. 246 waterproofing processes are included—64 relate to use of fluorochemicals.

The numbers in ( ) after each entry in the Table of Contents, where the treatment processes are organized by the agent used, indicates the number of production processes for each agent.

Production Processes for Waterproofing Textiles using the following Agents: Metal Salts and Wax-Containing Formulations (44), Silicones and Alkyl Polysiloxanes (53), Organofunctional Silicones and Fluorosilanes (20), Acrylics (8), Nitrogen Containing Compounds (30), Fluorochemical Compounds (64), Elastomer, Vinyl, Polyolefin Vapor Permeable Fabrics (27), Miscellaneous Treatments (9). 353 pages. **$35**

TEXTILES

CREASEPROOFING TEXTILES 1970
by Dr. M. W. Ranney
Textile Processing Review No. 2

Summarizes detailed process information relating to textile creaseproofing agents used to obtain wash and wear, or permanent press fabrics. Over 300,000 words, describes 343 processes in this field. Shows you chemical agents used, and processes by which they are applied.

Dimethylolethylene Urea and Related Compounds, Aldehyde-Urea Condensates, Uron Resins, Aminoplasts—Catalyst Performance, Melamine Derivatives, Triazones, Carbamates, Other Nitrogen-Containing Compounds, Phosphorus-Amino Compounds—Aziridines, Aminoplast-Thermoplastic Resin Compositions, General Processing Techniques and Formulations, Aldehydes, Acetals, Epoxies, Epihalohydrins, Sulfones, Sulfonium Salts, Cross-Linking Agents, Miscellaneous, Polymeric Coatings—Rubber, Vinyl, Silicones, Radiation Curing, Wool, Nylon, and Others. Indexes. 460 pages. $35

FLAME RETARDANT TEXTILES 1970
by Dr. M. W. Ranney
Textile Processing Review No. 3

Describes 177 commercial processes to produce flame retardant textiles and fabrics.

Most activity is based on chemical modification of cellulose through hydroxyl groups. Use of phosphoric acid, urea-phosphates, and other phosphorylating agents all confer flame retardant properties to cellulose. A significant portion of this book is devoted to the latest in application of phosphorus containing materials.

Numbers in ( ) indicate the number of processes described.

Ammonium Salts, Borates (12); Antimony, Titanium Metal Oxides (25); Amine-Phosphorus Products (21); Aziridines, APO, APS (21); Methylol-Phosphorus Polymers, THPC (27); Phosphonitrilic Chlorides (9); Triallyl Phosphates and Phosphonates (26); Silicones Isocyanates, Miscellaneous (10); Nylon, Acrylics (18). 373 pages. $35

ELECTRONIC AND ELECTRICAL MATERIALS

RESISTOR MATERIALS 1971
by P. Conrad
Electronics Materials Review No. 12

A helpful guide to new resistor materials.

Metals and Other Elements and Alloys (48)
Metal-Metal Oxide Mixtures (9)
Inorganic Oxide Compositions (32)
Other Inorganic Compounds (21)
Organic Compositions (12)

217 pages. $35

BATTERY MATERIALS 1970
by P. Conrad
Electronics Materials Review No. 10

Describes 162 processes useful in the manufacture of modern batteries:

Aqueous Battery Systems (89)
"Dry" Battery Systems (29)
Inorganic Electrolyte Systems (6)
Organic Electrolyte Systems (6)
Solid Electrolyte Systems (18)
Molten Electrolyte Systems (11)
Radioactive Batteries (3)

171 pages. $35

PHOTOCONDUCTIVE MATERIALS 1970
by M. Sittig
Electronic Materials Review No. 8

Photoconductors have a number of important applications, such as television camera tubes, solar cells, photoelectric cells, solid state light amplifiers, electrophotographic copying machines. 86 processes. 288 pages. $35

ELECTROLUMINESCENT MATERIALS 1970
by M. Sittig
Electronics Materials Review No. 6

Illustrates and describes 113 manufacturing processes for making cathode ray tube phosphors (black & white & color), fluorescent lamp materials, and luminous diodes. Discusses future trends, e.g. diode-based lamps and flat TV screens. 306 pages. $35

SUPERCONDUCTING MATERIALS 1970
by P. Conrad
Electronics Materials Review No. 11

Gives 99 detailed manufacturing and fabrication processes. Superconductive materials can improve long distance transmission of electric power, provide more compact memories for computers, improve magnets for physics research and thermonuclear power reactors. 135 pages. $35

MAGNETIC MATERIALS 1970
by M. Sittig
Electronics Materials Review No. 9

Surveys the empirical state of the art as revealed in recent U.S. patents.

1. **Simple Oxides (2)**
2. **Ferrites and Other Complex Oxides (66)**
3. **Metallic Magnetic Materials (37)**

Index of Companies, Authors and Patents. 286 pages. $35

THERMOELECTRIC MATERIALS 1970
by M. Sittig
Electronics Materials Review No. 7

Deals with thermocouples which convert thermal energy into electrical energy. 112 processes for isolating and making thermoelectric substances with descriptions of power-type thermocouples. Precise physical and electrical data. 235 pages. $35

PRODUCING FILMS OF ELECTRONIC MATERIALS 1970
by M. Sittig
Electronics Materials Review No. 5

Reviews commercially developed techniques for film production useful in the manufacture of integrated circuits and discrete components. Practically every method with practical applications is discussed, e.g. Electron Deposition, Ion Beam Deposition, Carrier Transport in Vapor Phase, etc. 113 processes. 295 pages. $35

PHARMACEUTICAL AND COSMETIC

X-RAY CONTRAST AGENTS 1971
by M. Gutcho

Deals with the preparation of new x-ray contrast agents which are economical to produce and offer the desirable property of sharper radiographs without increased toxicity. Formulations are given for use in cholecystography, urology, myelography, lymphography, and bronchography. A special chapter is devoted to process improvements in the preparation of barium sulfate and injectable iodine contrast media. A total of 55 preparative processes is described. An abbreviated Table of Contents is shown, where the number in ( ) indicates the number of agents in each area:

1. Polyiodobenzoic Acid Derivatives (16). 2. Polyiodophthalic Acid Derivatives (5). 3. Iodophenyl Derivatives (12). 4. Radiopaque Formulations (9). 5. Miscellaneous Contrast Media (4). 6. Process Improvements (9). 7. Other Medical Uses For Radiopaque Materials. 130 pages. $20

ANTIOBESITY DRUG MANUFACTURE 1970
by Dr. B. Idson
Chemical Process Review No. 43

This book discusses 162 processes in the field of anorectic drug production technology. The numbers in ( ) in the Table of Contents show the number of preparations discussed for each heading.

N-Alkyl Amines (2), Substituted Aminopropanes—Arylaminopropanes (17), Substituted Aminopropanes—1-Heterocyclic-2-Aminopropanes (2), Substituted Aminopropanes—Bicyclic Aminopropane (1), Substituted Aminopropanes—1-Heterocyclic Oxoaminopropanes (1), Substituted Aminoalkenes (2), Aralkyl Amines (22), Aralkyl Hydrocarbon (1), Ring Substituted Heterocyclic Compounds (20), Nitrogen Substituted Heterocyclic Compounds (43), Bicyclic Compounds (2), Azacyclic Compounds (5), Miscellaneous Functional Substitutions (24), Alkaloids (1), Antibiotic (1), Resins (3), Steroids (6), Amphetamine Salts (2), Compositions (8), Appendix. 193 pages. $35

ARTIFICIAL KIDNEY SYSTEMS 1970
by M. Gutcho

The processes and apparatus reported here are concerned with the wide range of components which in one way or another form the parts of a complete hemodialysis system. Some deal with the dialyzer; others with the mechanics of the dialysate solution.

Contains significant, detailed technical data, based on the patent literature relating to manufacturing, assembly, and operation of artificial kidney systems. Numerous illustrations are included. The accompanying Table of Contents, indicates the coverage offered and by the numbers in ( ), the processes covered.

Dialyzers in Artificial Kidney Systems (17); The Design of Dialyzer Parts (9); Dialysate Modifications (12); Lung and Kidney Machines (6); Dialyzers (6); Miscellaneous Blood Purification Processes (2). 320 pages. $35

COSMETIC FILMS 1970
by M. Gutcho

The 62 processes included in this book are concerned with methods of forming cosmetic films and other related preparations.

The abbreviated Table of Contents indicates the four major films and the distribution of the patents within each area — the numbers in ( ).

Powders (21)
Creams and Lotions (12)
Nail Preparations (19)
Lipsticks (14)

143 pages. $20

TETRACYCLINE MANUFACTURING PROCESSES 1969
(2 Volumes)

CTC, Oxytetracycline, TC, DMTC and DMCTC, 2N-Derivatives, Position 4 Derivatives, 6-Methylene Derivatives, 6-Deoxy Derivatives, Anhydrotetracyclines, 7-and/or 9-Derivatives, 11a-Halo Derivatives, 5a, 11a-Dehydrotetracyclines, 12a-Derivatives, Epimers, Mechanism Study intermediates. Indexes. 931 pages. **(2 Volumes) $45**

SUSTAINED RELEASE PHARMACEUTICALS 1969
by A. Williams

This book is based on the U.S. patent literature, and presents substantial technical information for production of these products.

TABLETS-PILLS
- Cellulosic Coatings
- Lipid Coating
- Gels-Gums
- Specific Medicaments
- Tablet Design
- Miscellaneous

CAPSULES
- General Coatings
- Particle Size Control
- Capsule Design & Manufacture

INJECTABLES
- Antibiotics
- Miscellaneous

MISCELLANEOUS RELEASE PREPARATIONS
- Suppositories
- Powders
- General

Indexes. 273 pages. $35

NUCLEOTIDES AND NUCLEOSIDES 1970
by S. Gutcho

Organic Synthesis of Nucleotides in General, and of Specific Nucleotides, Fermentation Procedures for Nucleotides in General and for Specific Nucleotides, Enzymatic Digestion of Nucleic Acids, Nucleotides Coenzymes, Cyclic Nucleotides, Dinucleoside Phosphates, Purification Techniques, General Procedures, Nucleotides as Flavor Enhancers. 200 pages. $35

DENTIFRICES 1970
by T. Jefopoulos

A guide to information available from U.S. Patent literature to therapeutic and cosmetic agents in dentifrices.

The numbers in ( ) indicate the number of processes covered.

Cleaning Agents (17)
Polishing Agents (14)
Prophylactic Compositions (27)
Fluorides (30)
Dentifrices for Dentists (6)
Other Dentifrices (30)
Improved Processing (2)

191 pages. $35

PHARMACEUTICAL AND COSMETIC

HAIR PREPARATIONS 1969
by A. Williams

Provides a detailed technological summary of recent developments based on 138 U.S. patents, since 1960, covering all aspects of hair preparations for the head, beard, eyelashes, and eyebrows.

Introduction, Dyeing, Bleaching, Waving, Setting, Shampoos-Rinses, Grooming-Tonics, Shaving Assistants, Other, Indexes. 208 pages. $35

FIBRINOLYTIC ENZYME MANUFACTURING 1969
by T. Rubel
Chemical Process Review No. 38

Methods for production and purification of fibrinolytic agents, their precursors and activators. Emphasis is on urokinase and streptokinase which activate plasminogen to form plasmin.

Plasminogen and Fibrinolysin, Urokinase, Streptokinase and Streptodornase, Other Fibrinolytic Enzymes. Indexes. 139 pages. $24

VITAMIN B_{12} MANUFACTURE 1969
by R. Noyes
Chemical Process Review No. 40

Vitamin B_{12} active substances are important therapeutic products for treatment of pernicious anemia. Also used for treatment of various other human ailments, and as a veterinary growth factor. This book offers various methods of producing vitamin B_{12} active substances. 327 pages. $35

VITAMIN E MANUFACTURE 1969
by T. Rubel
Chemical Process Review No. 39

This review relates the known methods for the preparation of tocopherols from natural products or by synthetic means, and conversion of non-alpha tocopherols.

Introduction; Tocopherols From Deodorizer Sludge, Conversions to Alpha Tocopherol, Synthesis of Tocopherols, Miscellaneous Related Processes, Indexes. 114 pages. $24

U. S. MARKETS

POLLUTION CONTROL COMPANIES U.S.A. 1972

Provides a marketing guide to the U.S. pollution control industry.

The first section is an alphabetical listing of ca. 1,500 companies or company units (divisions, subsidiaries, etc.) which manufacture or supply products useful in the areas of air, water, noise, and radiation pollution control and waste management.

The second section lists more than 500 companies and company units which provide such professional services as consulting, design engineering, and analyses of air or water pollutants.

The address and telephone number of each company is listed together with a brief description of the company's pollution control products or services, although these may constitute only a small portion of the company's business. 239 pages. $24

FOOD AND BEVERAGE PROCESSING INDUSTRIES 1971

The usefulness of this book derives from its organization. It is divided into 2 sections. The first section is an alphabetical listing of approximately 3,500 U.S. food firms giving the correct name, address and zip code. Also concise, pertinent information (where available) such as:

(a) Annual Sales
(b) Number of Employees
(c) Name of Chief Executive
(d) Product Types

The second section is arranged numerically according to zip code with the companies once again listed alphabetically within their proper zip code numbers, thereby providing you with an easy-to-use geographical index to the U.S. food industry.

It is a great help in marketing efforts by providing the means for forward geographical planning.

169 pages. $20

CHEMICAL GUIDE TO THE UNITED STATES 1971
Sixth Edition

Describes over 400 of the largest chemical firms in the U.S.: Those who actually carry out chemical syntheses in their plants. Companies and factual data about them are listed in alphabetic order, followed by an index which gives companies, subsidiaries and divisions, again by strict alphabetic arrangement.

Whenever available the following information is given in detail:

Name and Address with Zip Code
Ownership
Annual Sales
Number of Employees
Principal Executives and Titles
Plant Locations
Products
Subsidiaries and Affiliates
Internal Structure

Also gives information on closely held firms, joint ventures, and others that do not publish annual reports. 191 pages. $20

PHARMACEUTICAL AND COSMETIC FIRMS U.S.A. 1970

Describes the 700 leading pharmaceutical and cosmetic firms in the United States. Includes (1) ethical, (2) proprietary, (3) veterinary, (4) private formula, (5) cosmetics, and (6) toiletries firms.

Contains (where available): name, address, and telephone; ownership, annual sales figure; number of employees; names of executives; subsidiaries and affiliates; plant locations; products.

Has two indexes—a subsidiary and division index and a zip code index. The zip code index is an extremely important sales tool; lists companies in numerical order by zip code, providing an easy-to-use, invaluable, geographical index.

This guide will help you; concentrate on the big buyers, prepare market reports, increase sales effectiveness, research potential acquisitions and divestitures, and serve as an employment and personnel guide. 212 pages. $20

OVERSEAS MARKETS

FOOD GUIDE TO EUROPE 1972

Describes 1,600 of the leading food processing companies located in 18 countries of Western Europe. Where available and pertinent, the following information is given:

Name
Address
Telephone Number
Telex Number
Ownership
Sales
Number of Employees
Plant Locations
Principal Executives
Products
Domestic Subsidiaries
Foreign Subsidiaries

Each chapter includes a short resume of the food industry in the country concerned, giving the salient features of the industry and production and trade statistics. Prepared in our London office. 325 pages. $24

EUROPEAN CHEMICAL DISTRIBUTORS 1971

A considerable proportion of the sales of the world's major chemical companies are made through distributors, particularly in foreign countries. The book is divided into a main section, arranged by European distributors, and an index of worldwide companies the 1,300 distributors are representing.

Includes:

Austria	Luxembourg
Belgium	Netherlands
Denmark	Norway
Finland	Portugal
France	Spain
Germany	Sweden
Greece	Switzerland
Iceland	Turkey
Ireland	United Kingdom
Italy	

The chemical companies represented by the distributors are cross-referenced in the index. 264 pages. $24

WORLD PHARMACEUTICAL FIRMS 1972

Lists alphabetically by country the names and full addresses of over 6,000 manufacturers of ethical drugs, pharmaceutical specialties and basic pharmaceutical materials in all the major producing areas of the world.

For easy reference the 52 countries are arranged in geographical areas:

North America
South America
Western Europe
Eastern Europe
Africa
Australasia
Asia
North Africa and Middle East

This book will enable companies dealing with the pharmaceutical sector to mail sales literature, help plan and evaluate area representation, assist in the preparation of market reports and provide useful references.

Prepared in our London office. 122 pages. $24

EUROPEAN PHARMACEUTICAL MARKET REPORT 1971
Second Edition

Gives statistical information from 1960 to the end of 1969 on the 5,000 million dollar drug industry of the EEC (European Economic Community) and of the EFTA (European Free Trade Association). Type of information included:

General and Historical
Growth Potentials
Production and Sales
Domestic and Foreign Trade
Distribution—Pricing
Market Structure
Structure of the Industry
Names and Addresses of Trade Associations and of over 1,000 Manufacturers
Research and Development
Government Legislation
Health Service Expenditures

This book, a greatly expanded version of the earlier successful study, will bring you up to date with the European pharmaceutical scene and will give you a coherent picture of this complex market. 158 pages. $36

EUROPEAN PAINT MANUFACTURERS 1971

Developing technology in the field of polymer-based and solventless coatings has resulted in European acquisitions and mergers. Describes about 1,000 of the most important paint manufacturers in 17 countries of Western Europe. Entries are listed alphabetically by country and the following data are presented as fully as they were obtainable:

Name
Address
Telephone Numbers
Telex Numbers
Number of Employees
Principal Executives
Products
Plant Locations
Domestic Subsidiaries
Foreign Subsidiaries
Sales Volume

Prepared in our London Office. 155 pages. $24

TEXTILE INDUSTRY OF JAPAN 1971

Relates significant developments of nearly 900 companies:

Name and Address
Principal Officers
Employees
Capital and Sales
Total Assets
Bankers & Stockholders

Statistically analyzes the whole Japanese textile industry:

Production
Trade
Raw Materials
Consumption
Employment Pattern
Wages & Productivity
Important Trademarks

Increasing pressures from expanding developing countries and restricted imports by the U.S. are producing many changes evaluated here. 205 pages. $35

OVERSEAS MARKETS

CHEMICAL GUIDE TO EUROPE 1971
Fifth Edition

Prepared in our London Office to give on-the-spot coverage. Describes ca. 1,100 companies in the 19 countries of Western Europe which together constitute a market almost as large as that of the U.S. Includes all major European Chemical Companies. Gives this information (where pertinent and available):

Name and Address
Telephone and Telex Numbers
Ownership
Plant Locations and Products
Internal Structure
Local Subsidiaries and Affiliates
Foreign Subsidiaries and Affiliates
Principal Executives
Annual Sales
Number of Employees

Also companies which are predominantly non-chemical, but have important chemical interests, are included. 288 pages. **$20**

KEY EUROPEAN INDUSTRIALS 1970

This important new directory describes 1,000 leading manufacturers in 19 countries of Western Europe. Prepared by our London office, it will provide the hard-pressed executive with a unique, on-the-spot guide to the activities of Europe's key industrial companies and groups. Find out who owns whom.

Full name and address
Telephone and telex numbers
Share capital, sales, profit
Number of employees
Principal executives
Range of products and activities
Domestic and foreign subsidiaries

Volume I — (EFTA) — Austria, Denmark, Finland, Iceland, Ireland, Norway, Portugal, Sweden, Switzerland, United Kingdom. 180 pages.

Volume II — (EEC) — Belgium, France, Germany, Greece, Italy, Luxembourg, Netherlands, Spain, Turkey. 182 pages.

2 Volumes—$35

ELECTRONICS MANUFACTURERS OF WESTERN EUROPE AND U.S.A. 1970

This book provides the reader with a comprehensive and up-to-date guide to names and addresses of electronics manufacturers in the following 16 countries:

Austria	Netherlands
Belgium	Norway
Denmark	Portugal
Finland	Spain
France	Sweden
Germany	Switzerland
Ireland	United Kingdom
Italy	United States

It will enable companies dealing with the electronics sector to: mail sales literature and promotional material as effectively as possible; help plan and evaluate area representation more accurately; assist in the preparation of market reports; and provide a useful day-to-day work of reference. 131 pages. **$19**

PETROCHEMICAL INDUSTRY OF JAPAN 1970

Japanese petrochemical industry output by 1969 reached a record level of $2,500 million. Japanese petrochemicals now constitute a major force in the world chemical market.

The bulk of the book is a detailed guide to the 205 manufacturers in the industry. Companies are listed alphabetically, giving their address, capital, sales, number of employees, president and, where relevant, ownership. For every company, the existing and planned capacity of each product is listed, together with its plant location and expected completion date.

To those associated with developments in the petrochemical sector, this important new guide to one of the world's leading producers will be both a convenient source of reference and an invaluable marketing aid. **147 pages. $35**

EUROPEAN FOOD MARKET RESEARCH SOURCES 1970

Enables you to pinpoint publications most likely to be of assistance.

References are classified by Government statistics and reports, other statistics and reports, trade associations, food trade journals; other newspapers and periodicals, directories, advertising statistics, bank reviews. 111 pages. **$19**

ELECTRONICS GUIDE TO EUROPE 1969

Contains company profiles of over 600 leading electronics manufacturers in 14 countries. Arranged for easy comparison: Full name and address, ownership, principal executives, product range, domestic and overseas subsidiaries, plant location, latest sales figures, number of employees. 150 pages. **$20**

KNITWEAR AND HOSIERY GUIDE TO EUROPE 1970

Entries are listed alphabetically under country and are accompanied, wherever possible, by the following key data: full name and address; telephone and telex numbers; principal executives; product range; domestic and foreign subsidiaries and affiliates; plant location; latest sales figures and number of employees. 201 pages. **$20**

EUROPEAN KNITWEAR AND HOSIERY MARKET REPORT 1970

17 countries included. Trend of activity established. Size structure of domestic market determined. Information on yarn consumption, knitwear and hosiery production, foreign trade, number of companies and employees engaged in manufacture. Comprehensive selection of statistical material. 166 pages. **$35**

ENVIRONMENT AND POLLUTION

ENVIRONMENTAL CONTROL IN THE INORGANIC CHEMICAL INDUSTRY 1972
by H. R. Jones
Pollution Control Review No. 6

Inorganic pollution control is still a field that must receive greater attention and action. Be greater attention. Because of the nature of the industry obsolete units remain sufficiently profitable to continue in use.

Provides helpful directions for adequate pollution control. Commercial processes are shown, as well as detailed technology from the U.S. patent literature:

1. Wastewater Characteristics
2. Water Pollution Problems in the Manufacture of 33 Specific Chemicals
3. Actual Pollution Control Processes for Specific Wastewaters
4. Air Pollution Problems in the Manufacture of 15 Specific Chemicals
5. Actual Pollution Control Processes for Atmospheric Emissions
6. Awareness and Future Trends.

249 pages. **$36**

ENVIRONMENTAL CONTROL IN THE ORGANIC AND PETROCHEMICAL INDUSTRIES 1971
by H. R. Jones
Pollution Control Review No. 3

1. WASTE SOURCES
2. WATER USE
3. WATER POLLUTION
4. RECEIVING WATERS
5. REUSE OF WATER
6. PHYSIOLOGICAL EFFECTS
7. POLLUTION PARAMETERS
8. MONITORING
9. CHEMICAL CLASSIFICATION
10. SPECIFIC PRODUCTS
11. WASTE WATERS
12. INDUSTRIAL-MUNICIPAL TREATMENT PLANTS
13. INTERNAL IMPROVEMENTS
14. PHYSICAL TREATMENTS
15. CHEMICAL TREATMENTS
16. BIOLOGICAL TREATMENTS
17. OTHER DISPOSAL METHODS
18. WATER ECONOMICS
19. AIR POLLUTION
20. AIR POLLUTANTS REMOVAL
21. IMMEDIATE RECOMMENDATIONS
22. REFERENCES-INDEXES

257 pages. **$36**

HYDROGEN SULFIDE REMOVAL PROCESSES 1972
by P. G. Stecher
Pollution Control Review No. 5

Furnishes reliable and efficient methods of H_2S removal from gases, air, and liquids. 80 Patent-based processes.

1. ABSORPTION FROM GASES WITH INORGANICS
2. WITH ORGANICS
- Alkanolamines
- Other Amines
- Esters and Ethers
- Other Compounds
- Recycling Absorbents

3. REMOVAL FROM GASES BY OTHER METHODS
- Oxidation
- Adsorption
- Hydrotreating
- Use of Electrolysis
- H_2S Recycle in White Liquor Regeneration
- Ion Exchange Resins
- Molecular Sieves
- Adsorption plus Oxidation

4. REMOVAL FROM LIQUIDS
- From Fluid Hydrocarbons
- From Aqueous Solutions

288 pages. **$36**

SULFUR DIOXIDE REMOVAL FROM WASTE GASES 1971
by A. V. Slack
Pollution Control Review No. 4

Reviews the problems of smelter operators, power plants, refineries, sulfur acid plants, and Claus process sulfur plants.

1. THE PROBLEM
   Alternatives of Control
2.-3. THROWAWAY PROCESSES
   Dry Systems
   Wet Systems
4. ECONOMIC FACTORS IN RECOVERY
5. RECOVERY PROCESSES
6. ALKALIS AS ABSORBENTS
7. ALKALINE EARTH ABSORBENTS
8. METAL OXIDE SORPTION
9. ADSORPTION PROCESSES
10. CATALYTIC OXIDATION & REDUCTION
11. OTHER RECOVERY METHODS
   Organic Sorbents
   Fuel Gasification
   Gas Cleaning

200 pages. **$36**

PAPER RECYCLING AND THE USE OF CHEMICALS 1971
by M. McDonald

Paper accounts for almost half of our solid waste (ca. 40 million tons per year accumulate in the U.S. alone). But paper can be reused. Treatment is mechanical and chemical, using many substances for removing printer's ink, plastic coatings, wax, adhesives, etc. The secondary fibers industry, on the threshold of an era of expanded growth, is consuming a steadily increasing amount of solvents, bleaches, and other chemicals discussed in this book. 63 processes.

Removing Ink (23)
Removing Coatings and Impregnants (19)
Dispersing Coatings (11)
Repulpable Adhesives and Coatings (5)
Miscellaneous (10)

304 pages. **$36**

CATALYTIC CONVERSION OF AUTOMOBILE EXHAUST 1971
by J. McDermott
Pollution Control Review No. 2

The need for controlling the exhaust gas emissions from gasoline, diesel, and jet engines is becoming increasingly urgent. The emission of carbon monoxide, nitrogen oxides, lead compounds, reactive olefins, and even of carcinogens such as benzopyrene, has become the concern of almost every legislature.

This book summarizes the U.S. patent literature relating to combustion catalysts, such as various metal oxides, platinum, and palladium. In general, the catalytic unit, mounted after the exhaust manifold, completes the oxidation of unburned hydrocarbons and converts the carbon monoxide to carbon dioxide. 94 processes and devices are described.

Catalytic Converter Design (31)
Catalyst Bed Design (17)
Catalysts (32)
Catalytic Units (14)

208 pages. **$36**

ENVIRONMENT AND POLLUTION

CORROSION RESISTANT MATERIALS HANDBOOK 1971
Second Edition
by I. Mellan

Corrosion, always an urgent and persistent problem, bothers and baffles us even more today, because of the quantity and complexity of chemicals in our polluted biosphere. This book will help you cut losses by enabling you to choose the proper commercially available corrosion resistant material. The index lists thousands of corrosive substances and refers you to specific recommendations in the 147 tables:

Synthetic Resins (90 tables)
Elastomers (17 tables)
Cements (11 tables)
Glass & Ceramics (4 tables)
Wood (3 tables)
Metals & Alloys (22 tables)

487 pages. **$25**

MERCURY POLLUTION CONTROL 1971
by H. R. Jones
Pollution Control Review No. 1

In 1970 mercury pollution hit the headlines. Mercury ions were expected to react with other inorganic ions in the water, form precipitates, and sink harmlessly to the bottom. But when organic molecules from sewage or dead organisms are present, mercury reacts with these molecules to form toxic methyl mercury compounds which are excreted very slowly by fish and man. This book explains in detail what measures can be taken to prevent further pollution and to remove existing contamination. 13 Chapters: 1. The Fish Episode. 2. Production of Mercury. 3. Its Properties. 4. Uses of Mercury. 5. Its Toxicology. 6. Detection and Determination. 7. Air Pollution by Mercury. 8. Cleanup of Spilled Mercury. 9. Removal from Gases. 10. From Liquids. 11. From other Materials. 12. Legislation. 13. Trends and Problems. 251 pages. **$35**

HAZARDOUS CHEMICALS HANDLING AND DISPOSAL 1970
The Institute of Advanced Sanitation Research International

This publication is the record of a symposium of hazardous chemicals handling and disposal.

- Hygiene Control—Handling Certain Hazardous Chemicals
- Hazardous Chemicals Handling in "The Pharmaceutical Chemical Industry"
- Effects of Hazardous Chemicals on Biochemical Oxygen Demand Tests of Stream Water Samples
- Pesticide Handling in an Industrial Plant
- Thermal Method for the Disposal of Hazardous Wastes
- Land Disposal of Hazardous Chemicals
- Hazardous Chemicals Disposal in a Large Chemical Complex
- Research, Development and Application of New Biological Methods for Toxic Wastes Degradation and Disposal
- Design and Tests of a Portable Cask for Explosive Chemicals

130 pages. **$15**

DESALINIZATION BY DISTILLATION 1971 RECENT DEVELOPMENTS
by J. McDermott

The prospect of revitalizing arid areas, together with the problem of providing increased quantities of potable water for the developed areas has produced considerable research in desalinization processes. In the past two years alone, over one hundred patents relating to distillation processes have been issued in the U.S.

Flash Evaporators: Multistage Multieffect Evaporators, Direct Contact Condensation, Use of Combustion Gases. Flash Evaporation and Power Generation. Vapor Compression: General Processes, Contact Condensation. Control of Scale Formation: Process Techniques, Chemical and Miscellaneous Techniques. Membranes and Refrigerant Circuits: Semiporous Membrane, Silicone Rubber, Distilland Heating, Heat Exchange Circuits. Miscellaneous: Latent Heat Distillation, Partially Submerged Unit, Water Purification Units. 194 pages. **$35**

DESALINIZATION BY FREEZE CONCENTRATION 1971
by J. McDermott

This volume is the second of a series of three, dealing with the major desalinization processes. The information contained will provide needed know-how concerning the renewed interest in freeze concentration processes.

The Table of Contents indicates the coverage of processes and equipment offered by this volume. The numbers in ( ) indicate the processes in a particular area.

Direct Contact Volatile Refrigerants
- General Processes (24)
- Crystal Washing Techniques (11)
- Direct Contact Low Volatility Heat Transfer Mediums
  - Kerosene and Other Hydrocarbons (9)
  - Other Coolants (2)

Vacuum Freezing
- Zarchin Process (9)
- Other Processes (6)

Hydrate Processes (11)

Surface Freezing
- General Processes (7)
- Other Processes (7)

207 pages. **$35**

OTHERS

ANTISTATIC AGENTS 1972 Technology and Applications
by K. Johnson

Antistatic agents described in 160 U.S. patents since 1965 are covered in this book with specific sections devoted to two large volume applications—Plastics and Textiles.

Antistatic agents are used in plastics such as phonograph records, bottles, and film wraps to reduce pickup of dust and dirt.

In the textile industry, charged fibers interfere with the spinning process, and the attraction of dust particles produces marks and soiling in weaving.

1. Polyolefins (27)
2. Records and Films (17)
3. Fibers & Fabrics Treatment (29)
4. Further Fibers & Fabrics (27)
5. Techniques of Treatment (22)
6. Softener & Lubricant Formulations (20)
7. Other Applications and Syntheses (18)

307 pages. **$36**

HAZARDOUS CHEMICALS HANDLING AND DISPOSAL 1970
The Institute of Advanced Sanitation Research International

This is the record of the Second Symposium.

- Pesticide Container Decontamination
- Fate and Effects of Pesticides
- Separation of Organic and Inorganic Chemicals in a Waste Stream
- Vector Problems in Waste Disposal
- Plants Poisonous to Livestock
- Land Application and Anaerobic Lagoon Disposal of Waste
- Reverse Osmosis for Reclamation and Reuse of Chemical and Metal Waste Solutions
- Thermal Methods for Destruction of Chemical Waste
- Specialty Gases and Hazardous Wastes
- Education of the Public
- Hazardous Chemicals from Natural Sources
- Maximum Allowable Concentrations and Water Quality
- Biochemical Oxygen Demand and its Meaning

163 pages. **$20**

WASTE WATER CLEANUP EQUIPMENT 1971
by C. Pazar
Pollution Control Handbook No. 2

This book describes commercial equipment used for water pollution control. It is based directly on information supplied by manufacturers. Appropriate, detailed, technical information and illustrations were selected by the author (at no cost to, nor influence from, the manufacturers of this equipment). An abbreviated Table of Contents is shown below. The numbers in ( ) indicate the number of companies whose equipment is described.

Sewage Pretreatment (8), Clarification—Flocculation—Sludge Collection (25), Aerobic Treatment (28), Sludge Processing and Handling (28), Treatment of Plant Effluents (14), Treatment for Specific Contaminants (40), Sewage Systems and Hydraulic Components (20), Packaged and Prefabricated Treatment Plants (11), Anaerobic Processing (4), Tertiary Treatment (22). 551 pages. **$35**

DESALINIZATION BY REVERSE OSMOSIS 1970
by J. McDermott

Summarizes the patent literature through June 1970 relating to reverse osmosis processes with particular emphasis on membrane technology and equipment design. The abbreviated Table of Contents shows material covered. The numbers in ( ) indicate number of patents.

MEMBRANE PREPARATION
- Cellulosic (6)
- Discrete Particles (4)
- Miscellaneous Membrane Systems (5)
- Membrane Supports (8)

EQUIPMENT DESIGN
- Tubular Units (7)
- Spirally Wound Membrane Modules (7)
- Miscellaneous (5)

MODIFIED REVERSE OSMOSIS TECHNIQUES
- Vapor Permeation (2)
- Other Energy Sources (6)
- Solution Modifications (3)

SPECIAL UNITS
- Submerged Units (4)
- Portable Units (3)
- General (3)

209 pages **$35.**

AIR AND GAS CLEANUP EQUIPMENT 1970
by C. Pazar
Pollution Control Handbook No. 1

Supplies detailed technical data, diagrams and specifications for air pollution control equipment.

An abbreviated Table of Contents is shown. The number of companies whose equipment is described is the number in ( ).

DRY DUST COLLECTORS
- Electrostatic Precipitators (9)
- Inertial Separators (25)
- Filtration Through Fabric (35)
- Components, Auxiliaries (20)

WET SCRUBBERS
- Packed Towers and Beds (13)
- Cyclones (12)
- Venturi Scrubbers (20)
- Impingement—Inertial—Capillary (18)
- Gas Coolers, Separators (6)

INCINERATION AND GAS PROCESSING
- Incineration of Gases (10)
- Sludge, Liquid Incineration (6)
- Solid Refuse (15)
- Incineration Components (7)
- Oxidation (10)
- Carbon Adsorption, Gases (4)

637 pages. $35

SOUNDPROOF BUILDING MATERIALS 1970
by Dr. M. W. Ranney

Table of Contents shows scope of this volume.

Fibrous Glass and Mineral Wool
- Fibrous Glass
- Surface Coating Techniques for Fibrous Glass
- Mineral Wool
- Flake Glass Panels

Fiberboard
- Perforation Techniques
- Panel Design
- Surface Coatings
- Miscellaneous

Gypsum—Ceramic—Perlite
- Gypsum
- Ceramic
- Perlite, Vermiculite and Silicate
- Miscellaneous

Plastic Products
- Polystyrene
- Polyurethane
- Miscellaneous

General Processes
- Flexible Partitions
- General Construction

217 pages. $35

OPTICAL BRIGHTENERS 1972 Technology and Applications
by T. Rubel

Describes synthesis and use of fluorescent dyes that have the property of absorbing ultraviolet radiant energy and emitting energy in the visible range.

To the unaided human eye the fluorescence is not noticeable. Its presence, however, has the effect of brightening colors and making things whiter than white.

Based on 136 U.S. patents issued since Jan. 1965:

1. Optical Brighteners Added to Natural Materials (6 Processes)
2. To Synthetic Polymers (52)
3. To Natural & Synthetic Substances (42)
4. To Detergents (8)
5. To Fabric Softeners (5)
6. Brighteners for Papers and Photographic Materials (19)
7. Other Applications (4)
   Company Index
   Inventor Index
   U.S. Patent Number Index.

281 pages. **$36**